# Handbook of Green Chemistry

**Volume 5
Reactions in Water**

*Edited by
Chao-Jun Li*

# Related Titles

Tanaka, K.

## Solvent-free Organic Synthesis

Second, Completely Revised and Updated Edition
2009
ISBN: 978-3-527-32264-0

Lefler, J.

## Principles and Applications of Supercritical Fluid Chromatography

2009
ISBN: 978-0-470-25884-2

Wasserscheid, P., Welton, T. (eds.)

## Ionic Liquids in Synthesis

Second, Completely Revised and Enlarged Edition
2008
ISBN: 978-3-527-31239-9

Sheldon, R. A., Arends, I., Hanefeld, U.

## Green Chemistry and Catalysis

2007
ISBN: 978-3-527-30715-9

Li, C.-J., Chan, T.-H.

## Comprehensive Organic Reactions in Aqueous Media

Second Edition
2007
ISBN: 978-0-471-76129-7

Lindstrom, U. M. (ed.)

## Organic Reactions in Water

Principles, Strategies and Applications
2008
ISBN: 978-1-4501-3890-1

# Handbook of Green Chemistry

Volume 5
Reactions in Water

*Edited by*
*Chao-Jun Li*

WILEY-VCH Verlag GmbH & Co. KGaA

**The Editor**

**Prof. Dr. Paul T. Anastas**
Yale University
Center for Green Chemistry & Green Engineering
225 Prospect Street
New Haven, CT 06520
USA

**Volume Editor**

**Prof. Dr. Chao-Jun Li**
McGill University
Department of Chemistry
801 Sherbrooke Street West
Montreal, QC H3A 2K6
Canada

**Handbook of Green Chemistry – Green Solvents**
Vol. 4:  Supercritical Solvents
ISBN:    978-3-527-32590-0
Vol. 5:  Reactions in Water
ISBN:    978-3-527-32591-7
Vol. 6:  Ionic Liquids
ISBN:    978-3-527-32592-4

Set II (3 volumes):
ISBN: 978-3-527-31574-1

**Handbook of Green Chemistry**
Set (12 volumes):
ISBN:    978-3-527-31404-1

All books published by Wiley-VCH are carefully produced. Nevertheless, authors, editors, and publisher do not warrant the information contained in these books, including this book, to be free of errors. Readers are advised to keep in mind that statements, data, illustrations, procedural details or other items may inadvertently be inaccurate.

**Library of Congress Card No.:**
applied for

**British Library Cataloguing-in-Publication Data**
A catalogue record for this book is available from the British Library.

**Bibliographic information published by the Deutsche Nationalbibliothek**
The Deutsche Nationalbibliothek lists this publication in the Deutsche Nationalbibliografie; detailed bibliographic data are available on the Internet at http://dnb.d-nb.de.

© 2010 WILEY-VCH Verlag GmbH & Co. KGaA, Boschstr. 12, 69469 Weinheim, Germany

All rights reserved (including those of translation into other languages). No part of this book may be reproduced in any form – by photoprinting, microfilm, or any other means – nor transmitted or translated into a machine language without written permission from the publishers. Registered names, trademarks, etc. used in this book, even when not specifically marked as such, are not to be considered unprotected by law.

**Typesetting**   Thomson Digital, Noida
**Printing and Binding**   betz-druck GmbH, Darmstadt
**Cover Design**   Adam-Design, Weinheim

Printed in the Federal Republic of Germany
Printed on acid-free paper

**ISBN:** 978-3-527-31591-7

# Contents

**About the Editors**  *XIII*
**List of Contributors**  *XV*

| | | |
|---|---|---|
| **1** | **The Principles of and Reasons for Using Water as a Solvent for Green Chemistry**  *1* | |
| | *Ronald Breslow* | |
| 1.1 | Introduction  *1* | |
| 1.2 | Binding of Two Species Together Driven by the Hydrophobic Effect in Water  *2* | |
| 1.3 | Aromatic Chlorination  *3* | |
| 1.4 | Acylation of Cyclodextrins by a Bound Ester  *4* | |
| 1.5 | Mimics of Metalloenzymes Using the Hydrophobic Effect in Water  *5* | |
| 1.6 | Mimics of the Enzyme Ribonuclease  *7* | |
| 1.7 | Mimics of Enzymes that Use Pyridoxamine Phosphate and Pyridoxal Phosphate as Coenzymes  *9* | |
| 1.8 | Artificial Enzymes Carrying Mimics of Thiamine Pyrophosphate  *15* | |
| 1.9 | Enolizations and Aldol Condensations  *17* | |
| 1.10 | Hydrophobic Acceleration of Diels–Alder Reactions  *18* | |
| 1.11 | Selectivities in Water Induced by the Hydrophobic Effect – Carbonyl Reductions  *18* | |
| 1.12 | Selectivities in Water Induced by the Hydrophobic Effect – Oxidations  *19* | |
| 1.13 | Using Hydrophobic Effects in Water to Determine the Geometries of Transition States for Some Important Reactions  *21* | |
| 1.14 | Conclusion  *25* | |
| | References  *25* | |
| | | |
| **2** | **Green Acid Catalysis in Water**  *31* | |
| | *Chikako Ogawa and Shū Kobayashi* | |
| 2.1 | Introduction  *31* | |
| 2.2 | Lewis Acids in Water  *31* | |
| 2.2.1 | Introduction. Lewis Acids in Aqueous Media: Possible?  *31* |

*Handbook of Green Chemistry, Volume 5: Reactions in Water.* Edited by Chao-Jun Li
Copyright © 2010 WILEY-VCH Verlag GmbH & Co. KGaA, Weinheim
ISBN: 978-3-527-31591-7

| 2.2.2 | Lewis Acid Catalysis in Water as "Sole Solvent" 32 |
| --- | --- |
| 2.2.2.1 | LASC: Lewis Acid–Surfactant Combined Catalyst 33 |
| 2.2.2.2 | Polymer-supported Scandium Triflate 34 |
| 2.2.2.3 | Silica Gel-supported Scandium with Ionic Liquid 35 |
| 2.3 | Chiral Lewis Acid-catalyzed Asymmetric Reactions in Water 36 |
| 2.3.1 | Mannich-type Reactions in Water 36 |
| 2.3.2 | Michael Reaction in Water 37 |
| 2.3.3 | Epoxide Ring-opening Reaction in Water 39 |
| 2.3.4 | Hydroxymethylation in Water 40 |
| 2.4 | Brønsted Acid Catalysis in Pure Water 45 |
| 2.4.1 | Surfactant-type Brønsted Acid Catalysts 45 |
| 2.4.2 | Polymer-supported Brønsted Acid Catalysts 50 |
| 2.5 | Conclusion and Perspective 53 |
| | References 53 |
| | |
| **3** | **Green Bases in Water** 57 |
| | *José M. Fraile, Clara I. Herrerías, and José A. Mayoral* |
| 3.1 | Introduction 57 |
| 3.2 | Types of Bases and Their Use from a Green Point of View 58 |
| 3.3 | Liquid–Liquid Processes 59 |
| 3.4 | Solid–Liquid Processes 65 |
| | References 70 |
| | |
| **4** | **Green Oxidation in Water** 75 |
| | *Roger A. Sheldon* |
| 4.1 | Introduction 75 |
| 4.2 | Water-soluble Ligands 76 |
| 4.3 | Oxidations Catalyzed by Metalloporphyrins and Metallophthalocyanines 76 |
| 4.4 | Epoxidation and Dihydroxylation of Olefins in Aqueous Media 78 |
| 4.5 | Alcohol Oxidations in Aqueous Media 85 |
| 4.6 | Aldehyde and Ketone Oxidations in Water 94 |
| 4.7 | Sulfoxidations in Water 96 |
| 4.8 | Conclusion 97 |
| | References 98 |
| | |
| **5** | **Green Reduction in Water** 105 |
| | *Xiaofeng Wu and Jianliang Xiao* |
| 5.1 | Introduction 105 |
| 5.2 | Water-soluble Ligands 106 |
| 5.2.1 | Water-soluble Achiral Ligands 107 |
| 5.2.2 | Water-soluble Chiral Ligands 108 |
| 5.3 | Hydrogenation in Water 108 |
| 5.3.1 | Achiral Hydrogenation 109 |
| 5.3.1.1 | Hydrogenation of Olefins 109 |

| | | |
|---|---|---|
| 5.3.1.2 | Hydrogenation of Carbonyl Compounds | 113 |
| 5.3.1.3 | Hydrogenation of Aromatic Rings | 114 |
| 5.3.1.4 | Hydrogenation of Other Organic Groups | 115 |
| 5.3.1.5 | Hydrogenation of $CO_2$ | 116 |
| 5.3.2 | Asymmetric Hydrogenation | 118 |
| 5.3.2.1 | Asymmetric Hydrogenation of Olefins | 118 |
| 5.3.2.2 | Asymmetric Hydrogenation of Carbonyl and Related Compounds | 121 |
| 5.4 | Transfer Hydrogenation in Water | 123 |
| 5.4.1 | Achiral Transfer Hydrogenation of Carbonyl Compounds | 124 |
| 5.4.2 | Asymmetric Transfer Hydrogenation | 126 |
| 5.4.2.1 | Asymmetric Transfer Hydrogenation of Ketones | 126 |
| 5.4.2.2 | Asymmetric Transfer Hydrogenation of Imines | 132 |
| 5.4.2.3 | Asymmetric Transfer Hydrogenation with Biomimetic Catalysts | 133 |
| 5.4.3 | Water-facilitated Catalyst Separation and Recycle | 135 |
| 5.5 | Role of Water | 137 |
| 5.5.1 | Coordination to Metals | 138 |
| 5.5.2 | Acid–Base Equilibrium | 139 |
| 5.5.3 | H–D Exchange | 140 |
| 5.5.4 | Participation in Transition States | 141 |
| 5.6 | Conclusion | 142 |
| | References | 142 |
| | | |
| **6** | **Coupling Reactions in Water** | **151** |
| | *Lucie Leseurre, Jean-Pierre Genêt, and Véronique Michelet* | |
| 6.1 | Introduction | 151 |
| 6.2 | Reaction of Carbonyl Compounds and Derivatives | 151 |
| 6.2.1 | Grignard-type Reactions | 151 |
| 6.2.1.1 | Allylation Reaction | 152 |
| 6.2.1.2 | Propargylation and Allenylation Reaction | 156 |
| 6.2.1.3 | Alkylation Reaction | 157 |
| 6.2.1.4 | Arylation and Vinylation Reactions | 158 |
| 6.2.1.5 | Alkynylation Reaction | 161 |
| 6.2.2 | Pinacol Coupling | 162 |
| 6.3 | Reaction of Alkenes and Alkynes | 163 |
| 6.3.1 | Reaction of Unconjugated Alkenes and Alkynes | 163 |
| 6.3.1.1 | Hydroformylation Reaction | 163 |
| 6.3.1.2 | Hydroxycarbonylation | 164 |
| 6.3.1.3 | Metathesis, Polymerization Reactions, and Carbene Reactivity | 165 |
| 6.3.1.4 | Isomerization of Alkenes | 165 |
| 6.3.1.5 | Coupling of Alkynes | 166 |
| 6.3.1.6 | Mizoroki–Heck Reaction and Related Hydroarylation Reactions | 166 |

| | | |
|---|---|---|
| 6.3.1.7 | Cyclization and Cyclotrimerization of Polyfunctional Unsaturated Derivatives | 167 |
| 6.3.2 | Reaction of Conjugated Alkenes | 170 |
| 6.3.2.1 | Telomerization of Dienes | 170 |
| 6.3.2.2 | 1,4-Addition to α,β-Unsaturated Derivatives | 171 |
| 6.4 | Reaction of Organic Halides and Derivatives | 172 |
| 6.4.1 | Homo- and Heterocoupling Reactions | 172 |
| 6.4.2 | Suzuki–Miyaura (S–M) Reaction | 173 |
| 6.4.2.1 | Palladium-catalyzed Reactions (Aryl and Vinyl Iodides, Triflates, Bromides, and Diazoniums salts) | 173 |
| 6.4.2.2 | Nickel- and ruthenium-catalyzed Reactions | 180 |
| 6.4.3 | Stille Reaction | 180 |
| 6.4.4 | Sonogashira Reaction, Alkyne Oxidative Dimerization | 181 |
| 6.4.5 | Tsuji–Trost Reaction | 186 |
| 6.4.6 | Hartwig–Buchwald Coupling | 189 |
| 6.4.7 | Hiyama Reaction | 189 |
| 6.5 | Conclusion | 191 |
| | References | 191 |

**7     "On Water" for Green Chemistry     207**
*Li Liu and Dong Wang*

| | | |
|---|---|---|
| 7.1 | Introduction | 207 |
| 7.2 | Pericyclic Reactions | 208 |
| 7.3 | Addition of Heteronucleophiles to Unsaturated Carbonyl Compounds | 211 |
| 7.4 | Enantioselective Direct Aldol Reactions | 213 |
| 7.5 | Coupling Reactions | 215 |
| 7.5.1 | Transition Metal-catalyzed Cross-coupling Reactions | 215 |
| 7.5.2 | Dehydrogenative Coupling Reactions | 216 |
| 7.6 | Oxidation | 221 |
| 7.7 | Bromination Reactions | 223 |
| 7.8 | Miscellaneous Reactions | 224 |
| 7.8.1 | Nucleophilic Substitution | 224 |
| 7.8.2 | Functionalization of SWNTs | 225 |
| 7.9 | Theoretical Studies | 226 |
| 7.10 | Conclusion | 227 |
| | References | 227 |

**8     Pericyclic Reactions in Water. Towards Green Chemistry     229**
*Jaap E. Klijn and Jan B.F.N. Engberts*

| | | |
|---|---|---|
| 8.1 | Introduction | 229 |
| 8.1.1 | Pericyclic Reactions | 229 |
| 8.1.2 | Water, the Ultimate Green Solvent | 232 |
| 8.2 | Pericyclic Reactions in Aqueous Media | 234 |

| 8.2.1 | Introduction  234 |
| --- | --- |
| 8.2.2 | Normal and Inverse Electron-demand Diels–Alder Reactions in Water  234 |
| 8.2.3 | Intramolecular Diels–Alder Reactions  242 |
| 8.2.4 | Retro-Diels–Alder Reactions  243 |
| 8.2.5 | Forward and Retro-hetero-Diels–Alder Reactions  244 |
| 8.2.6 | Photocycloadditions  248 |
| 8.2.7 | 1,3-Dipolar Cycloadditions  249 |
| 8.2.8 | Claisen Rearrangements  251 |
| 8.2.9 | Mixed Aqueous Binary Mixtures  252 |
| 8.2.10 | "On Water" Pericyclic Reactions  254 |
| 8.2.11 | (Bio)catalysis, Cyclodextrins, Surfactant Aggregates, Molecular Cages, Microwaves, Supercritical Water  255 |
| 8.2.11.1 | (Bio)catalysis  256 |
| 8.2.11.2 | Catalysis by Cyclodextrins  259 |
| 8.2.11.3 | Catalysis by Surfactant Aggregates  261 |
| 8.2.11.4 | Microwave-assisted Aqueous Pericyclic Reactions  266 |
| 8.2.11.5 | Supercritical Water  267 |
| 8.3 | Conclusion  267 |
| | References  268 |

| **9** | **Non-conventional Energy Sources for Green Synthesis in Water (Microwave, Ultrasound, and Photo)**  273 |
| --- | --- |
| | *Vivek Polshettiwar and Rajender S. Varma* |
| 9.1 | Introduction  273 |
| 9.2 | MW-assisted Organic Transformations in Aqueous Media  274 |
| 9.2.1 | Carbon–Carbon Coupling Reactions  274 |
| 9.2.2 | Nitrogen-containing Heterocycles  277 |
| 9.2.3 | Oxygen-containing Heterocycles  280 |
| 9.2.4 | Heterocyclic Hydrazones  282 |
| 9.2.5 | Other Miscellaneous Reactions  282 |
| 9.3 | Sonochemical Organic Transformations in Aqueous Media  285 |
| 9.3.1 | Synthesis of Heterocycles  285 |
| 9.3.2 | Pinacol Coupling Reaction  286 |
| 9.4 | Photochemical Transformations in Aqueous Media  287 |
| 9.5 | Conclusion  288 |
| | References  288 |

| **10** | **Functionalization of Carbohydrates in Water**  291 |
| --- | --- |
| | *Marie-Christine Scherrmann, André Lubineau, and Yves Queneau* |
| 10.1 | Introduction  291 |
| 10.2 | C–C Bond Formation Reactions  292 |
| 10.2.1 | Knoevenagel Condensations  292 |

| 10.2.2 | Barbier-type Reactions  *297* |
|---|---|
| 10.2.3 | Baylis–Hillman Reactions  *305* |
| 10.2.4 | Electrophilic Aromatic Substitution Reactions  *306* |
| 10.2.5 | Mukaiyama Aldol Reaction  *308* |
| 10.3 | C–N Bond Formation Reactions  *309* |
| 10.3.1 | Glycosylamines and Glycamines  *309* |
| 10.3.2 | Aza Sugars  *311* |
| 10.4 | Functionalization of Hydroxy Groups  *312* |
| 10.4.1 | Esterification, Etherification, Carbamation: the Example of Sucrose  *313* |
| 10.4.2 | Oxidation Reactions  *318* |
| 10.4.3 | Bioconversions  *319* |
| 10.5 | Glyco-organic Substrates and Reactions in Aqueous Sugar Solutions  *319* |
| 10.6 | Conclusion  *324* |
|  | References  *324* |

**11  Water Under Extreme Conditions for Green Chemistry**  *331*
*Phillip E. Savage and Natalie A. Rebacz*

| 11.1 | Introduction  *331* |
|---|---|
| 11.2 | Background  *332* |
| 11.2.1 | Properties of HTW  *332* |
| 11.2.2 | Process Engineering Considerations  *334* |
| 11.2.3 | Theoretical, Computational, and Experimental Methods  *335* |
| 11.2.3.1 | Classical Theory  *336* |
| 11.2.3.2 | Molecular and Computational Modeling  *337* |
| 11.2.3.3 | Experimental Methods  *338* |
| 11.2.4 | pH Effects  *339* |
| 11.3 | Recent Progress in HTW Synthesis  *343* |
| 11.3.1 | Hydrogenation  *343* |
| 11.3.2 | C–C Bond Formation  *344* |
| 11.3.2.1 | Friedel–Crafts Alkylation  *344* |
| 11.3.2.2 | Heck Coupling  *345* |
| 11.3.2.3 | Nazarov Cyclization  *346* |
| 11.3.3 | Condensation  *346* |
| 11.3.4 | Hydrolysis  *348* |
| 11.3.5 | Rearrangements  *350* |
| 11.3.6 | Hydration/Dehydration  *351* |
| 11.3.7 | Elimination  *353* |
| 11.3.8 | Partial Oxidation to Form Carboxylic Acids  *354* |
| 11.3.9 | C–C Bond Cleavage  *355* |
| 11.3.10 | H–D Exchange  *355* |
| 11.3.11 | Amidation  *355* |
|  | References  *356* |

| | | |
|---|---|---|
| **12** | **Water as a Green Solvent for Pharmaceutical Applications** 363 | |
| | *Peter Dunn* | |
| 12.1 | Introduction – Is Water a Green Solvent? 363 | |
| 12.2 | Water-based Enzymatic Processes 363 | |
| 12.2.1 | The Pregabalin (Lyrica) Process 364 | |
| 12.2.2 | Enzymatic Routes to Statins 366 | |
| 12.2.2.1 | The Enzymatic Process to Make Rosuvastatin (Crestor) Intermediate 5 366 | |
| 12.2.2.2 | Enzymatic Routes to Atorvastatin (Lipitor) 367 | |
| 12.2.3 | The Enzymatic Process to Make LY300164 369 | |
| 12.2.4 | The Enzymatic Process to Prepare 6-Aminopenicillanic Acid 371 | |
| 12.2.5 | Enzymatic Routes to Oseltamivir Phosphate (Tamiflu) 372 | |
| 12.3 | Processes in Which the Product is Isolated by pH Adjustment to the Isoelectric Point 374 | |
| 12.3.1 | Process to Prepare the Sildenafil Citrate (Viagra) Intermediate 24 374 | |
| 12.3.2 | The Sampatrilat Process 374 | |
| 12.4 | Carbon–Carbon Bond-forming Cross-coupling Reactions in Water 375 | |
| 12.4.1 | Process to Make Compound 29 an Intermediate for a Drug Candidate to Treat Depression 376 | |
| 12.4.2 | An Aqueous Suzuki Reaction to Prepare Diflusinal 33 377 | |
| 12.5 | Pharmaceutical Processes Using Mixed Aqueous Solvents 378 | |
| 12.5.1 | The Lumiracoxib Process 379 | |
| 12.5.2 | An Environmentally Friendly Baylis–Hillman Process 379 | |
| 12.6 | Conclusion 380 | |
| | References 382 | |
| | | |
| **13** | **Water as a Green Solvent for Bulk Chemicals** 385 | |
| | *Ferenc Joó and Ágnes Kathó* | |
| 13.1 | Introduction 385 | |
| 13.2 | Hydroformylation – an Overview 386 | |
| 13.2.1 | General Aspects of Hydroformylation 386 | |
| 13.2.2 | Industrial Hydroformylation Processes in Non-aqueous Systems 388 | |
| 13.2.2.1 | Cobalt-based Hydroformylation Catalysts and Processes 388 | |
| 13.2.2.2 | Rhodium-based Hydroformylation Catalysts and Processes 389 | |
| 13.2.2.3 | Ligands Used for Catalyst Modification 391 | |
| 13.2.3 | Central Questions in Hydroformylation Processes 392 | |
| 13.3 | Water as Solvent for Hydroformylation 393 | |
| 13.3.1 | Aqueous–Organic Biphasic Catalysis 393 | |
| 13.3.2 | Aqueous–Organic Biphasic Hydroformylation 394 | |
| 13.3.2.1 | The Ruhrchemie–Rhône-Poulenc Process 395 | |
| 13.3.2.2 | Green Features of the Ruhrchemie–Rhône-Poulenc Process 398 | |

| | | |
|---|---|---|
| 13.3.2.3 | Hydroformylation of Longer Chain Alkenes in Aqueous–Organic Biphasic Systems *399* | |
| 13.3.2.4 | Developments in Reactor Design for Aqueous–Organic Biphasic Hydroformylations *400* | |
| 13.3.3 | Catalyst Recovery by Water-induced Phase Separation *401* | |
| 13.4 | Water as Solvent in the Production of 2,7-Octadien–1-ol (Kuraray Process) *403* | |
| 13.5 | Conclusion *405* | |
| | References *406* | |

**Index** *409*

# About the editors

### Series Editor

**Paul T. Anastas** joined Yale University as Professor and serves as the Director of the Center for Green Chemistry and Green Engineering there. From 2004–2006, Paul was the Director of the Green Chemistry Institute in Washington, D.C. Until June 2004 he served as Assistant Director for Environment at the White House Office of Science and Technology Policy where his responsibilities included a wide range of environmental science issues including furthering international public-private cooperation in areas of Science for Sustainability such as Green Chemistry. In 1991, he established the industry-government-university partnership Green Chemistry Program, which was expanded to include basic research, and the Presidential Green Chemistry Challenge Awards. He has published and edited several books in the field of Green Chemistry and developed the 12 Principles of Green Chemistry.

### Volume Editor

**Chao-Jun Li** (FRSC, UK) received his PhD at McGill University (1992) and was an NSERC Postdoctoral fellow at Stanford University (1992–1994). He was an Assistant Professor (1994), Associate Professor (1998) and Full Professor (2000–2003) at Tulane University, where he received a NSF CAREER Award (1998) in organic synthesis and the 2001 US Presidential Green Chemistry Challenge Award (Academic). In 2003, he became a Canada Research Chair (Tier I) in Organic/Green Chemistry and a Professor of Chemistry at McGill University in Canada. He serves as the Co-Chair of the Canadian Green Chemistry and Engineering Network, the Director of CFI Infrastructure for Green Chemistry and Green Chemicals, and Co-Director the FQRNT Center for Green Chemistry and

Catalysis (Quebec). He is the current Associate Editor for Americas for the journal of *Green Chemistry* (published by the Royal Society of Chemistry). He has been widely recognized as the leader in Green Chemistry for Organic Synthesis in developing innovative and fundamentally new organic reactions that defy conventional reactivities and have high synthetic efficiency.

# List of Contributors

**Ronald Breslow**
Columbia University
Department of Chemistry
3000 Broadway
New York, NY 10027
USA

**Peter Dunn**
Pfizer UK
Pfizer Green Chemistry Lead
Ramsgate Road
Sandwich
Kent CT13 9NJ
UK

**Jan B.F.N. Engberts**
University of Groningen
Stratingh Institute
Nijenborgh 4
9747 AG Groningen
The Netherlands

**José M. Fraile**
Universidad de Zaragoza-CSIC
Instituto de Ciencia de Materiales de Aragón
Facultad de Ciencias
Departamento de Química Orgánica
50009 Zaragoza
Spain

**Jean-Pierre Genêt**
ENSCP
UMR 7573
Laboratoire de Synthèse Sélective Organique et Produits Naturels
11 rue P. et M. Curie
72231 Paris Cedex 05
France

**Clara I. Herrerías**
Universidad de Zaragoza-CSIC
Instituto de Ciencia de Materiales de Aragón
Facultad de Ciencias
Departamento de Química Orgánica
50009 Zaragoza
Spain

**Ferenc Joó**
Hungarian Academy of Sciences
Research Group of Homogeneous Catalysis
and
University of Debrecen
Institute of Physical Chemistry
1 Egyetem tér
4010 Debrecen
Hungary

**Ágnes Kathó**
University of Debrecen
Institute of Physical Chemistry
1 Egyetem tér
4010 Debrecen
Hungary

**Jaap E. Klijn**
University of Groningen
Stratingh Institute
Nijenborgh 4
9747 AG Groningen
The Netherlands

**Shū Kobayashi**
The University of Tokyo
School of Science and Graduate School of Pharmaceutical Sciences
Department of Chemistry
Hongo
Bunkyo-ku
Tokyo 113-0033
Japan

**Lucie Leseurre**
ENSCP
UMR 7573
Laboratoire de Synthèse Sélective Organique et Produits Naturels
11 rue P. et M. Curie
72231 Paris Cedex 05
France

**Li Liu**
Chinese Academy of Sciences
Institute of Chemistry
Center for Chemical Biology
Beijing National Laboratory for Molecular Science (BNLMS)
Beijing 100080
China

**André Lubineau**
Université Paris-Sud 11
Institut de Chimie Moléculaire et des Matériaux d'Orsay
CNRS UMR 8182
Laboratoire de Chimie Organique Multifonctionnelle, Bâtiment 420
15 rue Georges Clémenceau
91405 Orsay
France

**José A. Mayoral**
Universidad de Zaragoza-CSIC
Instituto de Ciencia de Materiales de Aragón
Facultad de Ciencias
Departamento de Química Orgánica
50009 Zaragoza
Spain

**Véronique Michelet**
ENSCP
UMR 7573
Laboratoire de Synthèse Sélective Organique et Produits Naturels
11 rue P. et M. Curie
72231 Paris Cedex 05
France

**Chikako Ogawa**
The University of Tokyo
School of Science and Graduate School of Pharmaceutical Sciences
Department of Chemistry
Hongo
Bunkyo-ku
Tokyo 113-0033
Japan
Present address:
Eisai Research Institute
Lead Identification
4 Corporate Drive
Andover, MA 01810
USA

## List of Contributors

**Vivek Polshettiwar**
US Environmental Protection Agency
National Risk Management Research Laboratory
Sustainable Technology Division
26 W. Martin Luther King Drive
Cincinnati, OH 45268
USA

**Yves Queneau**
INSA-Lyon
Laboratoire de Chimie Organique
Bâtiment J. Verne
20 avenue A. Einstein
69621 Villeurbanne
France
and
Université Lyon 1
Institut de Chimie et Biochimie
Moléculaires et Supramoléculaires
INSA-Lyon
CNRS, UMR 5246
CPE-Lyon
Bâtiment CPE
43 boulevard du 11 novembre 1918
69622 Villeurbanne
France

**Natalie A. Rebacz**
University of Michigan
Chemical Engineering Department
2300 Hayward
Ann Arbor, MI 48109
USA

**Phillip E. Savage**
University of Michigan
Chemical Engineering Department
2300 Hayward
Ann Arbor, MI 48109
USA

**Marie-Christine Scherrmann**
Université Paris-Sud 11
Institut de Chimie Moléculaire et des Matériaux d'Orsay
CNRS UMR 8182
Laboratoire de Procédés et Substances Naturelles
Bâtiment 410
15 rue Georges Clémenceau
91405 Orsay
France

**Roger A. Sheldon**
Delft University of Technology
Department of Biotechnology
Julianalaan 136
2628 BL Delft
The Netherlands

**Rajender S. Varma**
US Environmental Protection Agency
National Risk Management Research Laboratory
Sustainable Technology Division
26 W. Martin Luther King Drive
Cincinnati, OH 45268
USA

**Dong Wang**
Chinese Academy of Sciences
Institute of Chemistry
Center for Chemical Biology
Beijing National Laboratory for Molecular Science (BNLMS)
Beijing 100080
China

**Xiaofeng Wu**
University of Liverpool
Liverpool Centre for Materials and Catalysis
Department of Chemistry
Liverpool L69 7ZD
UK

**Jianliang Xiao**
University of Liverpool
Liverpool Centre for Materials and
Catalysis
Department of Chemistry
Liverpool L69 7ZD
UK

# 1
# The Principles of and Reasons for Using Water as a Solvent for Green Chemistry
*Ronald Breslow*

## 1.1
## Introduction

Chemical reactions used to manufacture important compounds such as medicinals are essentially always carried out in solution, and this is also true of the research work that is used to invent the new compounds and to develop appropriate ways to manufacture them. In the past, continuing into the present, the solvents used are normally volatile organic compounds (VOCs), and these pose an environmental problem. Their vapors can contribute to the greenhouse effect that causes global warming, and in some cases the solvent vapors can catalyze the destruction of the ozone layer that protects the Earth and its living inhabitants from short-wavelength ultraviolet solar radiation. The vapors may also be toxic to humans, plants, or animals, or they may cause diseases.

The liquids themselves can be a problem. If they are released into the earth, rivers or the ocean, they can cause direct environmental damage, while also slowly releasing their vapors. In principle, the solvents can be completely captured and purified for reuse during manufacturing, but it is difficult to prevent some loss to the environment. Hence there is interest in using environmentally benign liquids as the solvents in chemical reactions.

One possibility is supercritical carbon dioxide, which is a liquid under pressure and which has attractive solvent properties. However, unless it is completely contained and reused, it will release gaseous carbon dioxide, a greenhouse gas. Thus interest has increasingly turned to water as the solvent for chemical reactions.

Water is the solvent in which biochemical reactions are performed in Nature, and it is environmentally benign. However, it is a good solvent only for organic chemicals that have polar groups, such as alcohols and carboxylic acids. This may not be an insuperable problem. Over 20 years ago we reported that the special selectivities seen in water solution (see below) were also seen in some water suspensions, where one soluble component reacted with one that was poorly soluble [1, 2]. We pointed out that such suspensions in water could well be generally more practical ways to use water in

*Handbook of Green Chemistry, Volume 5: Reactions in Water.* Edited by Chao-Jun Li
Copyright © 2010 WILEY-VCH Verlag GmbH & Co. KGaA, Weinheim
ISBN: 978-3-527-31591-7

manufacturing [2]. Recently, Sharpless and co-workers described a remarkable acceleration of a reaction in such a suspension, which they called reactions ON water [3, 4]. The large reported rate effect was seen in only one particular case, but even without a large acceleration the selectivities that we describe below could perhaps make suspensions in water a practical way for the environmentally benign properties of water to be generally useful even with insoluble reaction components.

One industry that has switched from VOCs to water is the paint industry. We are all familiar with the water-based paints that no longer emit strong solvent odors, and these have been widely adopted for painting automobiles, for instance. It is essentially impossible to capture all the solvent vapors that are released when a vehicle is spray painted, but when the solvent is water there is no problem.

Water is not simply an environmentally benign solvent; it has special properties that are essentially unique, related to what is called the "hydrophobic effect." This is the tendency for hydrocarbons or molecules with hydrocarbon components to avoid contact with water, and to associate instead with other hydrocarbon species in water. This is what makes aqueous soap solutions dissolve grease, and it is the driving force in biology for the associations that produce cell membranes, and that cause nucleic acids to form the famous double helix. It drives the folding of proteins into their shapes in enzymes and antibodies, and it also promotes the binding of biological substrates into enzymes and antibodies [5].

As described below, the hydrophobic effect has now been used to mimic biological chemistry and to provide remarkable selectivities in the field called biomimetic chemistry. It has even been used to permit the discovery of the geometries of the transition states for some interesting reactions, information that is otherwise inaccessible. The remainder of this chapter describes examples of the use of the unique property of water to achieve not just solubility but also selectivity, but the examples will be mainly chosen from our own work. Hence it is important to refer to a number of sources in which other authors have also described their use of water and the hydrophobic effect in chemical studies.

Some of the work of our group has been presented as chapters in the books *Structure and Reactivity in Aqueous Solution* [6], *Green Chemistry* [7], and most recently *Organic Reactions in Water* [8]. In addition, in various review articles our work has been placed in context with that of other groups [2, 5, 9–20]. The remainder of this chapter describes the various contexts in which we have seen the special properties of water as a solvent.

## 1.2
**Binding of Two Species Together Driven by the Hydrophobic Effect in Water**

Cyclodextrins are molecules composed of glucose units linked in rings, the most common being $\alpha$-cyclodextrin (six glucose units), $\beta$-cyclodextrin (seven glucose units) and $\gamma$-cyclodextrin (eight glucose units) (Scheme 1.1). The three exposed hydroxyl groups on each glucose unit make then water soluble, but they have an internal cavity that is less polar, and that will bind hydrocarbons such as aromatic rings using the hydrophobic effect in water. In later sections it is described how such cyclodextrin–substrate complexes can catalyze reactions, imitating enzymes. Here the cases where binding alone was studied are described.

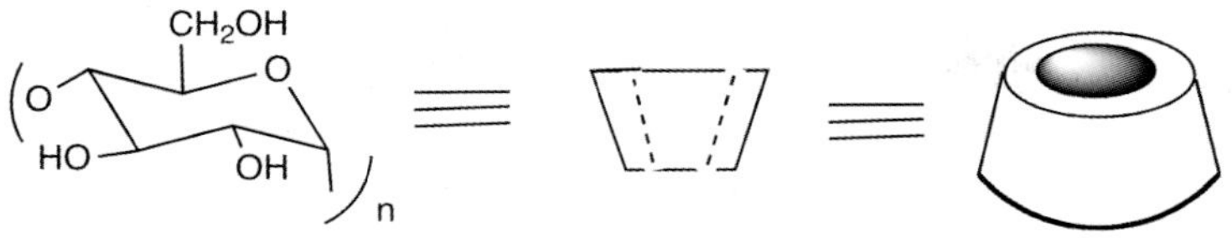

**Scheme 1.1** The three cyclodextrins used – α, β, and γ-cyclodextrin – and two ways in which they are symbolized.

alpha-cyclodextrin n = 6
beta-cyclodextrin  n = 7
gamma-cyclodextrin n = 8

In one example, we saw that some dipeptides would selectively bind into simple β-cyclodextrin in water [21], and that the large steroid lithocholic acid bound strongly [22], as did cocaine [23]. When we linked two β-cyclodextrins together, we achieved even better binding of cholesterol [24], and such cyclodextrin dimers also showed strong and selective hydrophobic binding of compounds with two phenyl groups [25], of peptides with two hydrophobic amino acid components [26], and of oligopeptides whose binding promoted the formation of a helix [27].

We also tied two β-cyclodextrins with *two* links, which made a hinge that could let the two cyclodextrins close around a substrate, and also another geometry in which they were prevented from cooperating [28]. As hoped, the dimer with the correct geometry was a very strong binder of hydrophobic substrates, since the double link had frozen out the incorrect geometries. Interestingly, in one study we saw that such strong double binding was reflected in a better enthalpy, rather than entropy [29]. Our early work with cyclodextrin dimers has been reviewed [30]. We also examined a dimer of a cyclophane, another species with an internal cavity that binds hydrophobic groups [31]. The findings were similar to those with the cyclodextrin dimers. In addition, we examined some trimers of cyclodextrins, but did not see as much cooperativity as one might expect [32].

We synthesized some cyclodextrin dimers with photocleavable links as potential carriers of anticancer photodynamic sensitizers [33, 34]. We also saw that some cyclodextrin dimers could bind to proteins and prevent their aggregation [35]. Furthermore, we saw that some of our cyclodextrin dimers and trimers could bind to amyloid protein and prevent the aggregation that causes Alzheimer's disease [36]. Some other studies with cyclodextrin dimers will be presented in Section 1.5 on mimics of metalloenzymes.

## 1.3
## Aromatic Chlorination

In our earliest work using cyclodextrins to bind substrates, we examined the chlorination of anisole by hypochlorous acid in water with and without added α-cyclodextrin [37, 38]. We saw that the anisole in solution was chlorinated in both the *ortho* and *para* positions, but in the complex with α-cyclodextrin only the *p*-chloroanisole was formed. The kinetic studies showed that the chlorination involved the prior attachment of chlorine to a hydroxyl group of the cyclodextrin, and then its transfer to the bound anisole (Scheme 1.2).

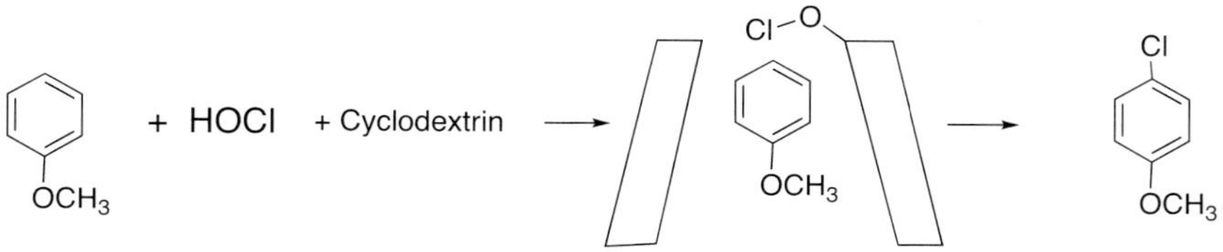

**Scheme 1.2** α-Cyclodextrin catalyzed the selective chlorination of anisole in water by an intra-complex transfer of a chlorine atom.

We also examined other substrates, whose behavior reflected this same mechanism [38]. In this case, the cyclodextrin is acting as a mimic of the enzyme chlorinase, except that interestingly the enzyme mimic was more selective than was the enzyme itself. In a later study, we established which hydroxyl group was the chlorine transfer agent, and showed that a cyclodextrin polymer could perform the selective chlorination in a flow reactor [39].

## 1.4
## Acylation of Cyclodextrins by a Bound Ester

Komiyama and Bender examined the reaction of *m*-nitrophenyl acetate with cyclodextrins, and saw that they transferred the acetyl group to a hydroxyl of the cyclodextrin, with a modest 250-fold rate enhancement over the hydrolysis rate in water under the same conditions [40]. Our modeling of this process indicated that the starting material could occupy the cyclodextrin cavity, but that the tetrahedral intermediate for acetyl transfer would have its nitrophenyl group largely pulled from the cavity. This picture was confirmed by a study of the effect of high pressure on the reaction rate, which indicated that the volume of the transition state was larger than that of the starting complex, as such a geometric change would cause [41]. Such a loss of binding would be energetically unfavorable for the reaction, accounting for the very modest rate of the acetyl transfer process. We therefore created a series of substrates that could avoid this problem.

Molecular models indicated that compound **1**, based on a ferrocene core, would be able to acylate a cyclodextrin hydroxyl while still retaining most of the binding of the ferrocene unit in the cyclodextrin cavity. We synthesized **1**, and saw that indeed it acylated β-cyclodextrin with a 51 000-fold rate acceleration [42]. However, high-pressure studies [41] indicated that although indeed the transition state for the reaction retained most of the binding into the cyclodextrin cavity, it was not yet the ideal substrate. By modifying the cyclodextrin itself – adding a floor to the cavity – and adjusting the substrate further, we achieved a rate acceleration of ca $10^6$-fold [43].

With an even better substrate geometry, we achieved a rate acceleration of $10^8$-fold, and the reaction was also enantioselective (cyclodextrin is composed of chiral glucose units), with a 20:1 preference for one substrate enantiomer over the other [44]. The optimizations and their explanations were described in a full paper [45], and theoretical calculations on the geometric factors involved were described in another publication [46].

The very high rates of the best substrates reflected a rigid geometry that favored the first step of acylation – addition of the cyclodextrin hydroxyl to the ester carbonyl to form a intermediate – but in the next step, departure of the *p*-nitrophenoxide ion to form the product acylated cyclodextrin, this rigidity was undesirable. Some flexibility was needed in the substrate to permit the rotation involved in this second step. When we incorporated such flexibility, both steps were well catalyzed even with an ordinary ester, where the second step could be rate limiting [47]. Thus, these studies on cyclodextrin acylation by bound substrates indicated the enormous rate accelerations that can be achieved using the hydrophobic effect to promote catalyst–substrate binding in a well-designed geometry.

Such work accomplishes two goals. It indicates that incorporating the factors we believe play a role in enzymatic catalysis does indeed lead to very good catalysis, approaching the rates of the best enzymes. This helps confirm our ideas about how enzymes are able to function so effectively. At the same time, these studies strengthen ordinary chemistry. They show how to make effective catalysts with good rates and selectivities, adopting the principles but not the details of enzymatic reactions.

## 1.5
## Mimics of Metalloenzymes Using the Hydrophobic Effect in Water

For hydrolytic enzymes, the formation of an acyl-enzyme intermediate is only a first step; for catalysis, the intermediate must hydrolyze to regenerate the catalyst and liberate the product carboxylate ion. In many hydrolytic enzymes, including the most effective ones, substrate binding involves both the hydrophobic effect induced by water and some binding to the metal ion itself, which is held in the enzyme by typical coordinating groups. In our first study of mimics for such enzymes, we constructed an artificial enzyme **2** comprised of an α-cyclodextrin ring for hydrophobic binding and an attached pyridinecarboxylate to bind a Ni(II) ion [48]. The nickel also bound a nucleophilic oxime group.

We found that *p*-nitrophenyl acetate was hydrolyzed in a two-step process, after it was hydrophobically bound into the cyclodextrin. First the nucleophilic oxime removed the acetyl group, in a nickel-catalyzed reaction, and then the nickel ion catalyzed the hydrolysis of this intermediate, regenerating the catalyst. The geometry permitted this process, not the direct acylation of cyclodextrin as in the systems in the previous section. However, the rate acceleration was modest, reflecting the many degrees of flexible freedom in the catalyst.

We constructed some metal ligands mirroring those in metalloenzymes such as carbonic anhydrase, and studied their ability to bind zinc(II), the metal ion in carbonic anhydrase and in carboxypeptidase, and other metal ions [49]. In a study of the hydrolysis of a phosphate triester, we saw evidence that a bound Zn(II) acted as a bifunctional catalyst, delivering a hydroxide ion to the phosphorus while coordinating to the phosphate oxygen atom to stabilize the phosphorane intermediate in hydrolysis [50]. We have seen such a process in many enzyme mimics that also use hydrophobic binding of substrates, as discussed below.

Carboxypeptidase uses metal ion catalysis in the hydrolysis of an amide group, a peptide bond. In a relevant study we used Co(III) to lock the amide oxygen to a metal ion [cobalt(III) is substitution inert], and saw hydrolysis of the amide with the assistance of phenol groups of the catalyst [51]. Apparently the phenol group and some others that we examined play a role in the second step of amide hydrolysis, fragmentation of the tetrahedral intermediate. We also attached α-cyclodextrin to a macrocyclic zinc ligand that held the metal so strongly that it could exist as the zinc hydroxide without losing the zinc [52]. The compound bound phosphate esters into the cyclodextrin using the hydrophobic effect in water, and then used the bifunctional zinc hydroxide mechanism to hydrolyze the substrate. In related work, we catalyzed the cyclization/cleavage of a model for conversion of RNA to its cyclic phosphate using the well-bound zinc macrocycle with an attached thiol or imidazole second catalytic group [53].

We constructed cyclodextrin dimers with a catalytic metal ion bound to the linking group. In the first example, esters that could hydrophobically bind into both cyclodextrins, stretching along the linking, were hydrolyzed by bound copper(II) hydroxide using the bifunctional nucleophilic bound hydroxide plus electrophilic metal ion mechanism, in one case achieving a 220 000-fold rate acceleration over uncatalyzed hydrolysis in water [54]. We also saw that such a cyclodextrin dimer could bind a bis-*p*-nitrophosphate anion to an La(III) ion coordinated to the linking group and then achieve catalytic cleavage of the phosphate ester with added hydrogen peroxide [55]. In a full paper describing such cyclodextrin dimer catalysts for ester hydrolysis, we saw as much as a $10^7$-fold rate acceleration [56]. Using cyclodextrin dimers with bound metal ions, we saw a $10^3$-fold acceleration of the hydrolysis of a bound benzyl ester, less reactive than some of the *p*-nitrophenyl esters used in earlier studies [57].

In another approach, we constructed a β-cyclodextrin that had both a metal ion binder and an imidazole general base catalyst attached to the cyclodextrin, and examined the hydrolysis of a *tert*-butylcatechol cyclic phosphate **3** that hydrophobically bound to the cyclodextrin [58]. The hydrolysis was accelerated ca $10^3$-fold. We describe other studies of such a hydrolysis in the next section.

## 1.6
## Mimics of the Enzyme Ribonuclease

Ribonucleic acid (RNA) is cleaved by the enzyme ribonuclease in an overall two-step process (Scheme 1.3). In the first step there is a cyclization/fragmentation in which the hydroxyl group on C-2 of the ribose attacks the phosphate group of the RNA chain and produces a cyclic phosphate **4** while breaking the chain at that point. In the second step, this cyclic phosphate is hydrolyzed to release the C-2 hydroxyl again while opening the cyclic phosphate ring. The enzyme can catalyze both of these rather different steps.

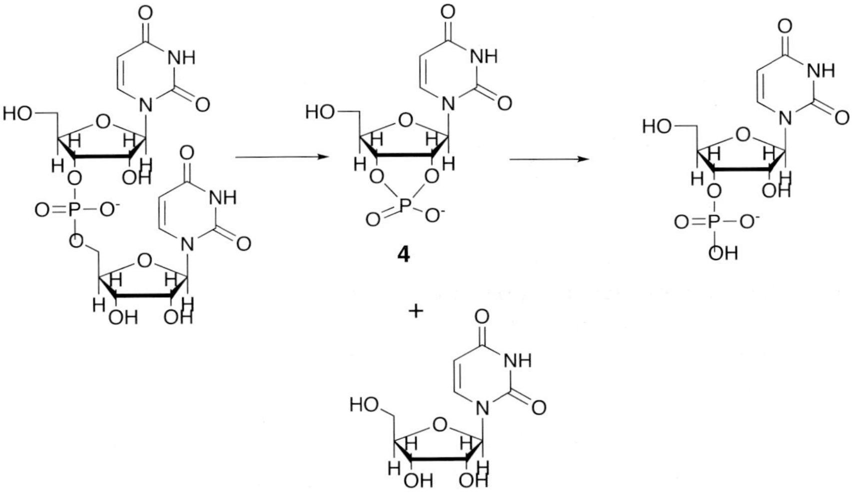

**Scheme 1.3** The enzyme ribonuclease cleaves RNA by a cyclization, then a hydrolysis of the cyclic phosphate. It is shown for uridyluridine, a dinucleotide component of RNA.

The major catalytic groups in bovine ribonuclease A are two imidazole rings of the amino acid histidine, although an ammonium ion of lysine also plays a role. At the optimum pH for the enzyme, one imidazole is protonated and serves as a general acid catalyst, whereas the other imidazole is unprotonated and acts as a general base. We decided to produce a mimic of this enzyme by using hydrophobic binding of a substrate into the cyclodextrin cavity in water, in which the cyclodextrin also had two imidazole rings replacing two hydroxyls of the cyclodextrin.

In our first study (Scheme 1.4), we attached the imidazoles on opposite sides of the cyclodextrin cavity and examined the ability of this catalyst to hydrolyze compound **3**, a cyclic phosphate as a rough mimic of the cyclic phosphate that is hydrolytically cleaved in the second step of the enzymatic process [59]. We saw that there was a pH optimum for this hydrolysis that was essentially identical with that of the enzyme itself, indicating that both the general base and the general acid versions of the imidazoles were cooperating in the hydrolysis process. The substrate was selectively cleaved to **5**, leaving the phosphate group *meta* to the *tert*-butyl group. By moving the imidazoles out slightly, we could reverse the selectivity, now leaving the phosphate group *para* to the *tert*-butyl group [60].

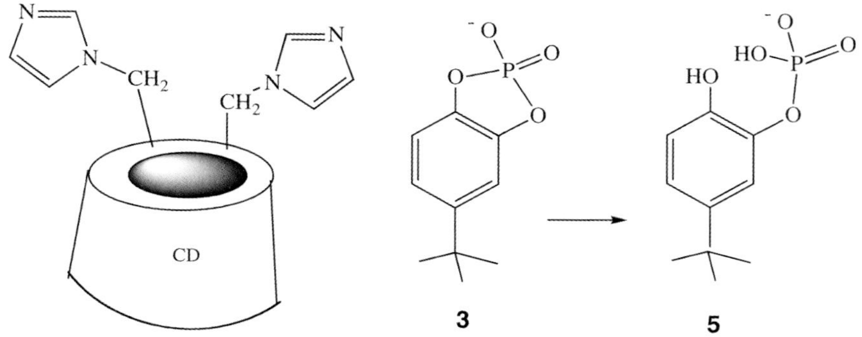

**Scheme 1.4** β-Cyclodextrin with two attached imidazole rings catalyzes the hydrolysis of a bound cyclic phosphate ester in water with specificity, and with geometric and isotopic evidence that indicates a process involving a phosphorane intermediate.

A method called proton inventory had been applied to ribonuclease [61]. By observing the rate with different ratios of $H_2O$ and $D_2O$, it is possible to deduce whether one or two protons are moving in the rate-determining step, and it was concluded that two protons were moving. This means that as the general base imidazole is removing the proton from the C-2 hydroxyl group the general acid imidazolium ion is transferring its proton to the substrate in a simultaneous bifunctional process. We applied this test to our bisimidazolecyclodextrin enzyme mimic, and saw the same result, and with almost the same data as had been seen with the enzyme [62].

We also varied the structure of the bisimidazolecyclodextrin catalyst. We were able to synthesize isomers with the two imidazoles on neighboring glucose units, which we called the A,B isomer, and also an isomer with imidazoles on the A,C units and on the A,D units [63]. If the cleavage mechanism had involved direct attack on the phosphate group while a proton was being placed on the leaving oxygen, the A,D isomer should have been the best. However, we saw that the most active isomer was A,B, with the acid and base groups on neighboring glucose units. This absolutely requires a mechanism in which the hydrolysis proceeds through an intermediate with five oxygens on phosphorus, a phosphorane, which later fragments to the final product. As we shall describe, we saw evidence for the same mechanism with a different model system.

We examined the hydrolysis of a simple dinucleotide, uridyluridine, in water solution with imidazole buffer. Since this process does not involve the hydrophobic special effects of water, it will not be described in detail and rather the relevant references are listed [64–73]. The evidence points to a phosphorane intermediate for this simple buffer-catalyzed process, and we suggested that the enzyme may well be using the same mechanism, rather than a direct cleavage. There is not general agreement on this idea for the enzyme. We have published an account of both the cyclodextrin studies and the buffer studies in ribonuclease mimics [74], and an account of the result of variation in the geometries of the bisimidazolecyclodextrins and the substrate, which made it clear how important it is to have a relatively tight fit of the substrate in the hydrophobic cavity of the cyclodextrin [75].

## 1.7
## Mimics of Enzymes that Use Pyridoxamine Phosphate and Pyridoxal Phosphate as Coenzymes

We have constructed a number of such mimics. In general, they use the hydrophobic effect in water to bind the substrates for the reactions, with the pyridoxal or pyridoxamine unit coenzyme mimics covalently attached to a cyclodextrin or a hydrophobic polymer. In a few cases we have also used the hydrophobic effect to bind reversibly a coenzyme mimic itself. Some of the resulting rate effects are truly enormous.

In our first study, we covalently linked a pyridoxamine unit to β-cyclodextrin in compound **6** and examined its ability to convert α-keto acids to amino acids in water (Scheme 1.5) [76]. This directly mimics the process used by enzymes to synthesize most amino acids. We compared the conversion of phenylpyruvic acid **7** to phenylalanine **8** and of indolepyruvic acid **9** to tryptophan **10**, both of which can exhibit hydrophobic binding into the cyclodextrin cavity, with the conversion of pyruvic acid **11** to alanine **12**, in which there was no hydrophobic binding of the small methyl group.

The two aromatic compounds had essentially the same rate as did simple pyruvic acid when pyridoxamine was used without the attached cyclodextrin, but hydrophobic binding of keto acids **7** and **9** led to about a 100-fold preference over alanine with the enzyme mimic **6**. With **6**, the tryptophan was formed with a 33% enantiomeric excess (*ee*) of the L-isomer, induced by the chirality of the cyclodextrin unit, and phenylalanine was formed with a 67% *ee* [77]. We saw similar results with compound **13**, in which the pyridoxamine was attached on one of the primary carbons of β-cyclodextrin and the other primary carbons were converted to methyl groups, forming a deeper hydrophobic pocket [78]. Even higher selectivities were seen in some of our later work using the hydrophobic effect in amino acid synthesis.

In **6**, the pyridoxamine is attached on the primary side of the cyclodextrin, but we also examined a compound **14** in which it was attached to the secondary side [77]. All the acylations of cyclodextrin described earlier were directed to the secondary side. We found that with **14** there was also a preference for the formation of phenylalanine and tryptophan, rather than alanine, reflecting the hydrophobic binding of the two aromatic ring substrates in water. However, the preference was less than with the original primary-side linked compound **6**, and the chiral inductions were also smaller. Apparently in this class of compounds, the primary side of the cyclodextrin is the better place for attachment of the pyridoxamine unit. A full paper summarized these results [79]. We also prepared a couple of transaminase mimics with *two* links between the pyridoxamine and the cyclodextrin [80]. The much better geometric control that this afforded led to very high selectivities among substrates with different geometries themselves.

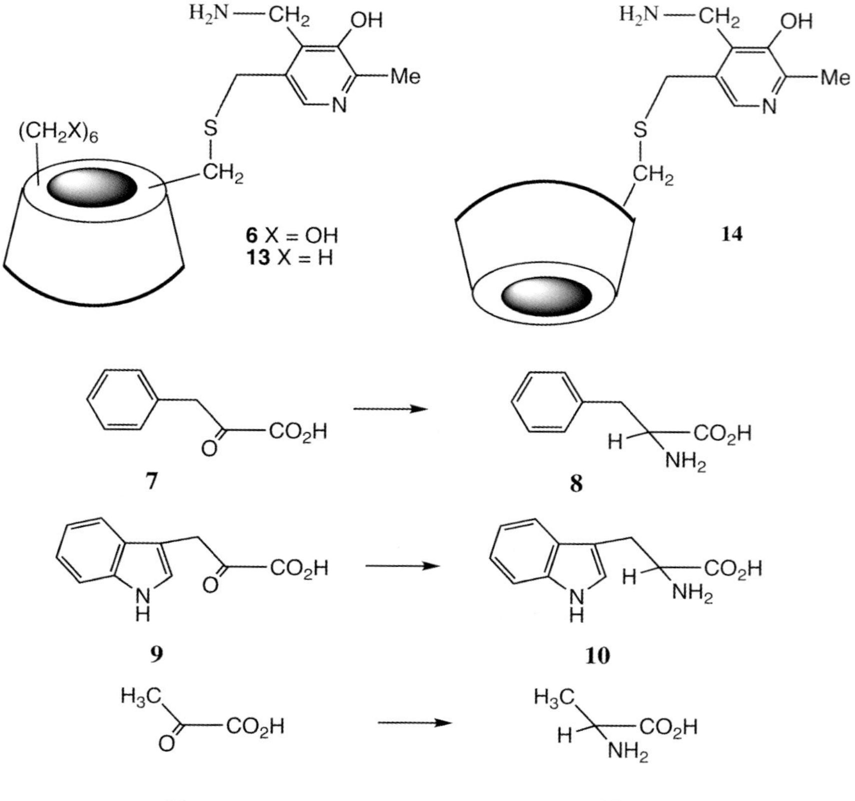

**Scheme 1.5** Three different versions of a pyridoxamine attached to a cyclodextrin convert keto acids to amino acids in water with a preference for those hydrophobic substrates that can bind into the cyclodextrin cavity, imitating in part a biological process by which the amino acids are formed.

Although the cyclodextrins are conveniently available compounds for incorporating hydrophobic binding in water into enzyme mimics, they are not unique. We also used some novel synthetic macrocycles that could carry a pyridoxamine unit and bind hydrophobic substrates into their cavity in water solution [81]. We saw that compound **15** converted phenylpyruvic acid to phenylalanine 15 times more rapidly than did simple pyridoxamine, again reflecting acceleration by hydrophobic binding of the substrate. However, the effect was not as large as with **6**, carrying a pyridoxamine attached to the primary side of β-cyclodextrin. Also, there was of course no chiral induction with the achiral synthetic macrocycle.

In a successful attempt at chiral induction, we synthesized compound **16**, which has no hydrophobic binding group but has a chain carrying a basic unit that can deliver the new proton of the amino acid with geometric control [82]. We saw that as much as a 94 : 6 ratio of D- to L-tryptophan was seen with the enantiomer of **16** that

## 1.7 Mimics of Enzymes that Use Pyridoxamine Phosphate and Pyridoxal Phosphate as Coenzymes

we used. The studies on amino acid synthesis up to this point were summarized in a full paper [83].

**15**

**16**

Subsequently, we combined these ideas in enzyme mimics that had a pyridoxamine with a chirally mounted rigid proton transfer group, that was also attached to a cyclodextrin for hydrophobic binding [84]. As much as a 96% *ee* was obtained in the best case. We also showed that we could reverse the optical preferences in some cyclodextrin–pyridoxamine enzyme mimics as we moved an attached basic group into different positions [85].

The principle biochemical function of pyridoxamine phosphate as a coenzyme is to convert keto acids to amino acids, as above, but the result of this process is also to convert the pyridoxamine unit to an aldehyde, pyridoxal phosphate. This aldehyde has biochemical functions in addition to reversing the transaminations process, so we synthesized compound **17** – with a pyridoxal unit attached to a primary carbon of β-cyclodextrin – to examine such reactions in water using hydrophobic binding [86]. In Nature, tryptophan is synthesized by an enzyme with a pyridoxal phosphate coenzyme in which the pyridoxal forms an imine with the amino acid serine, and catalyzes its dehydration to a reactive olefin that then adds to the beta position of indole. The product is then hydrolyzed to tryptophan, with regeneration of the pyridoxal phosphate (Scheme 1.6).

Metzler *et al.* had tried to perform this reaction with simple pyridoxal, and had obtained tryptophan in a 1% yield [87]. We obtained a somewhat better yield substituting β-chloroalanine for the serine, since HCl elimination was more facile than dehydration. With compound **17** carrying a hydrophobic binding group the process was three to five times better, but still tryptophan was produced in only a few percent yield. Furthermore, the chiral induction was not as large as in our previous transamination procedures for tryptophan production. The new system shows that two substrates can be coupled if one is hydrophobically bound into the cyclodextrin cavity and the other is covalently linked to the pyridoxal unit, but side reactions still predominate.

As the previous example indicates, there is a competition between reactions catalyzed by pyridoxal groups – in which the pyridoxal is recovered again at the end, so there can be catalytic turnover – and reactions with amino acids that reverse the amination process and produce α-keto acids and pyridoxamine. The latter

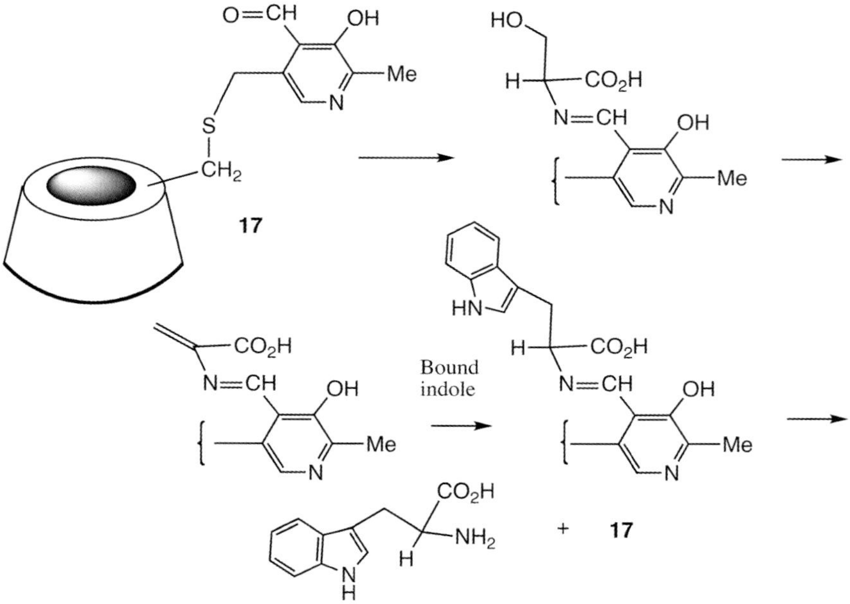

**Scheme 1.6** A cyclodextrin carrying a pyridoxal unit catalyzes the conversion of indole to tryptophan, aided by hydrophobic binding, in imitation of a biochemical process.

processes do not permit turnover catalysis by the pyridoxal. Thus in part we have strived to control this dichotomy. For example, pyridoxal units can catalyze the racemization of amino acids, but the process is in competition with transamination to form pyridoxamines. We saw that an added rigid base group could do the proton transfers needed for racemization and increase the selectivity for this process [88]. An even more extensive study with rigid bases was published later, again showing how racemization could be increased relative to transaminations [89]. In a full paper, we described how the attachments of rigid bases could be used to control both the reactions performed and the stereochemistry achieved [90].

Pyridoxal phosphate is also the coenzyme for the coupling of glycine to acetaldehyde to form threonine (Scheme 1.7). We synthesized a pyridoxal with a loop across its face, and examined this as catalyst for the glycine–acetaldehyde reaction [91]. The catalysis was successful, and an interesting reversal of optical selectivity occurred when the pH of the medium was changed.

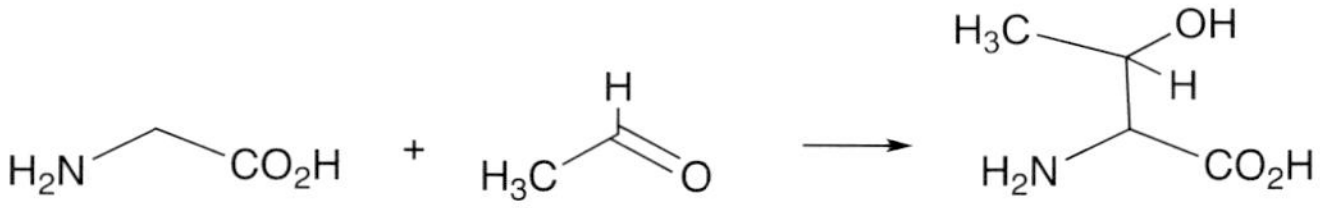

**Scheme 1.7** Glycine condenses with acetaldehyde to form threonine.

Cyclodextrins and related synthetic macrocycles have been used in many artificial enzyme systems, as described above, but Klotz's group pioneered a different

## 1.7 Mimics of Enzymes that Use Pyridoxamine Phosphate and Pyridoxal Phosphate as Coenzymes

approach. They reasoned that real enzymes are long-chain macromolecules, so they examined the use of synthetic polymers to mimic them. The particular polymers on which they concentrated are polyamines, polyethylenimines, in which the nitrogens are separated by two ethano groups and there are both linear and branched structures [92]. They and Kirby and co-workers [93] had used such a polymer to perform hydrolysis reactions, but we adopted it as the basis for mimics of synthetic enzymes, the transaminases. We have pursued artificial enzymes based on various polyamines of different sizes and produced in different ways to achieve some spectacular rate accelerations, and good chiral selectivities.

In our first approach, we used a commercial polyethylenimine with a molecular weight of about 60 000 that is fairly polydisperse and has extensive branching, with about 25% of tertiary nitrogens [94]. We attached a pyridoxamine to it covalently, and then added some lauryl chains to nitrogens to produce a hydrophobic interior. The polyamine is ca 50% protonated at pH 8, so it has both general base amino groups and general acid ammonium groups, and can use these to catalyze transaminations. With this artificial enzyme **18**, we saw the conversion of pyruvic acid **11** to alanine **12** in water with an 8300-fold rate acceleration over the same reaction with simple pyridoxamine, and a 10 000-fold acceleration relative to unbuffered pyridoxamine (the polymer needs no buffer, since the amino and ammonium groups provide the equivalent of buffer catalysis).

**18**

On further study, we saw that this polymer–pyridoxamine compound converted indolepyruvic acid **9** to tryptophan **10** in water with a rate acceleration of 240 000-fold relative to pyridoxamine itself, reflecting the hydrophobic binding of the substrate into the polymer [95]. The large effect reflected the hydrophobic character of the lauryl chains in the polymer, and with simple methyl groups instead the rate acceleration was much smaller.

We also examined other polyamines, including polyallylamine [94], and polyethylenimines of various sizes [96]. Interestingly, even fairly small polyamines were almost as effective as the large one, probably indicating that the smaller ones clustered to form an effectively large structure.

Dendrimers are an interesting class of polymers that branch from a central point. We examined a group of dendrimers **19** that branched from a pyridoxamine unit, by adding two methyl acrylate ester units to each amino group, then reacting each ester group with ethylenediamine, then repeating the process again and again. We saw that the largest ones were similar in catalytic activity to the polyethylenimines that did not carry attached hydrophobic lauryl groups [97]. In later work, we compared the two classes [98], and saw that attachment of chiral amino acids to the exterior of the dendrimers induced chiral amino acids [99]. In this latter study, we also showed, by calculation, that the pyridoxamine that was the formal center of the dendrimer could actually lie on the surface, near the surface chiral groups.

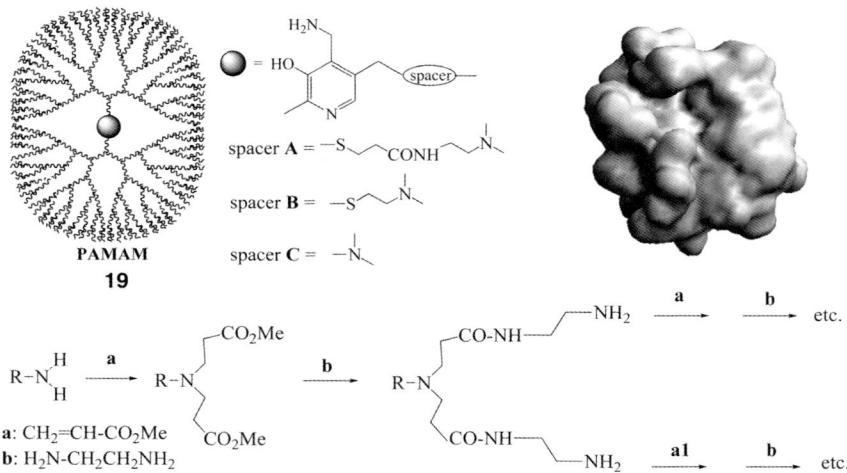

In the reactions in this section described so far, pyridoxamine units were reacting with keto acids to form amino acids, but the other products were pyridoxal units so the pyridoxamines were not regenerated. Hence we did not refer to such processes as "catalyses." In Nature, the pyridoxals are converted back to pyridoxamines by reversing the amination process with a sacrificial amino acid, but neither we nor others had been able to achieve significant catalytic turnovers no matter which amino acids we tried to sacrifice. We finally solved this problem with an unusual class of amino acids, those such as α-methylphenylglycine **20** that could not undergo simple transaminations, but that instead performed a process we called transaminative decarboxylation [100].

In this process (Scheme 1.8), the pyridoxal forms an imine with the novel amino acid, and then promotes decarboxylation to form a delocalized anion. This anion protonates next to the pyridine ring, where most of the negative charge is centered, and in water the resultant compound hydrolyzes to generate pyridoxamine and a ketone, which in the case shown is acetophenone. We saw that this process could occur in the same medium in which transaminations by pyridoxamine was being performed, and simultaneously. Thus we obtained as many as 100 turnovers by the pyridoxamine unit, true catalysis.

In the same paper, we reported another important advance: the pyridoxamine was not attached to the polymer, but was reversibly bound in water when the pyridoxamine carried a hydrophobic side chain [100]. The reversible system was somewhat faster than the covalently attached one, presumably because now both the substrate and the pyridoxamine bound into the same hydrophobic region in the polymer. This is similar to the real enzymatic case, in which pyridoxal phosphate is reversibly bound to the enzyme, not covalently attached. In a full paper, we examined

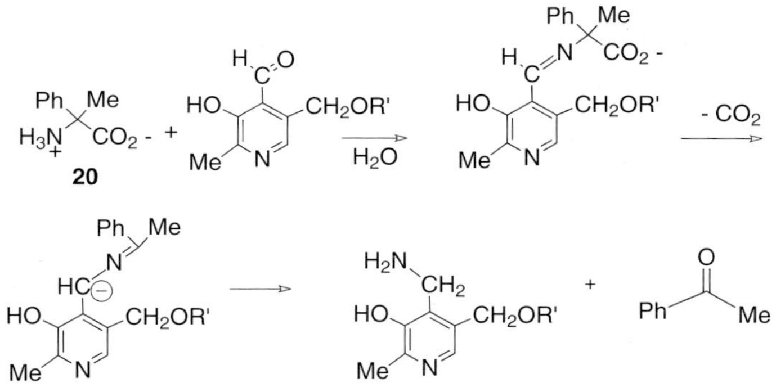

**Scheme 1.8** Decarboxylative transaminations that convert pyridoxal derivatives to pyridoxamines, permitting turnover catalysis of amino acid synthesis.

the hydrophobic and electronic factors involved in decarboxylative transaminations reactions [101]. It might be mentioned that we used a related decarboxylative transamination process in a mimic of a possible prebiotic reaction that could form natural amino acids with enantioexcesses, as part of an explanation of the origin of chirality on Earth [102].

In Nature, the transamination of keto acids forms amino acids as single enantiomers. As mentioned above [82], we had been able to perform transaminations with a chirally mounted basic group and achieve some enantioselectivity. We now turned to the production of polyamines with themselves carrying chirally attached side chains to see how well these could induce chiral selectivity in the polyamine-catalyzed reactions.

In our first study, we reported that chiral polyamines could be synthesized by the borane reduction of polypeptides, and that these induced some chiral selectivity in transamination processes [103]. In a different approach, we synthesized chiral polyamines by the polymerization of oxazolines, and saw that these too induced some chiral preferences when they catalyzed transaminations [104]. Further work towards the goal of high enantioselectivity in such catalyses is currently under way. A summary of the situation up to the near present was published [105], as was a general review of transamination mimics [20].

## 1.8
## Artificial Enzymes Carrying Mimics of Thiamine Pyrophosphate

Many years ago, we had shown that the coenzyme thiamine pyrophosphate acted by losing the C-2 proton of the thiazolium ring to form a compound that could be described as a thiazolium zwitterion **21** with a carbene resonance form **22** [106]. Such zwitterion/carbene hybrid species were also formed with imidazolium ions (**23** and **24**), and they are now important ligands in metal ion catalysts, often referred to as "stabilized carbenes" (which we had described in an early paper [107]).

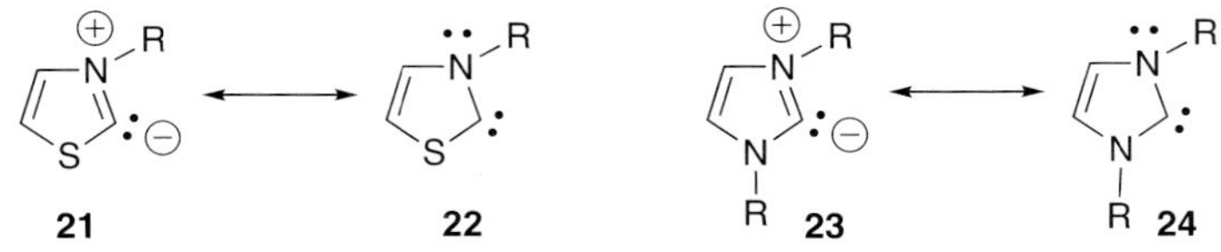

The biochemical decarboxylations that thiamine pyrophosphate performs can also be performed in model systems by simple cyanide ion, not a practical choice for biology. They are electronically related to the benzoin condensation, for which both cyanide ion and thiazolium salts can perform as catalysts. In contrast to pyridoxamine, thiamine pyrophosphate is a true catalyst that does not need regeneration. Thus we synthesized artificial enzymes linking thiazolium ions to cyclodextrins, to use again hydrophobic binding of substrates in water.

In our first study, we attached a thiazolium salt to β-cyclodextrin in compound **25**, and saw that it bound *tert*-butylbenzaldehyde to the hydrophobic cavity in water and promoted the ionization of the benzaldehyde proton, an important step in the benzoin condensation [108]. However, the cavity of β-cyclodextrin is too small to accommodate a second benzaldehyde species, so the benzoin condensation was not catalyzed.

γ-Cyclodextrin – with a ring of eight glucose units – is able to bind *two* benzaldehydes side-by-side, so we prepared a related compound with a thiazolium ion linked to γ-cyclodextrin [109]. This was a very effective catalyst. The flexible thiazolium ion link added a thiazolium ion to one benzaldehyde, and the deprotonated aldehyde group then added to the second benzaldehyde in the cavity. The thiazolium ion addition was then reversed and benzoin was liberated from the cavity (Scheme 1.9). Benzoin itself is linear and rigid, so the two phenyls can no longer occupy the same cavity. Thus this "enzyme" showed no inhibition by the product, and the transition state for the reaction did not resemble the geometry of the product. We are currently studying such enzyme mimics in which thiazolium, imidazolium, and triazolium ions cooperate with the polyamines described in the previous section [110].

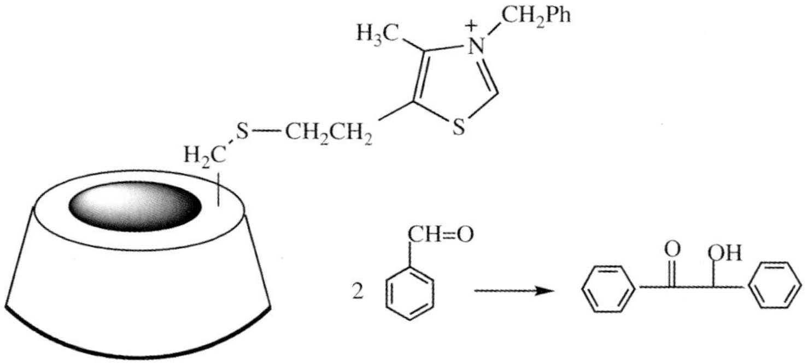

**25** beta-cyclodextrin
**26** gamma-cyclodextrin

**Scheme 1.9** A thiazolium salt coupled to β-cyclodextrin binds one benzaldehyde in the cavity, in water, and catalyzes deuterium exchange, but with the γ-cyclodextrin derivative *two* benzaldehydes are bound into the cavity and converted to benzoin.

## 1.9
## Enolizations and Aldol Condensations

In the section on mimics of the enzyme ribonuclease, we described the use of cyclodextrins carrying two imidazole groups, and the at first surprising results that the cooperative base/acid hydrolysis of a substrate bound into the cyclodextrin cavity in water by the hydrophobic effect was preferentially performed when the two imidazoles were linked to adjacent glucose residues, what we called the A,B isomer. The availability of the group of A,B and A,C, and A,D isomers let us examine the geometric preference for enolization of a bound ketone with bifunctional catalysis by the two imidazoles. Thus we studied deuterium exchange into *p-tert*-butylacetophenone **27** by the set of bisimidazolecyclodextrins, and found that the preferred catalyst in this case was the A,D isomer [111]. The pH versus rate profile showed that this was again bifunctional catalysis, with one imidazole acting as a base to remove the proton while the other, as the imidazolium ion, was hydrogen bonded to the carbonyl oxygen and delivered its proton to form the enol during deprotonation of the methyl group.

Models show that all three isomers could reach the carbonyl oxygen and the methyl hydrogen, but the direction of approach was different. The preferred isomer had a preferred direction of attack on the methyl hydrogen, pushing the electrons towards the carbonyl group (Scheme 1.10). The defined geometries of the three isomeric bisimidazolecyclodextrins could be a useful tool in discovering such details of mechanism.

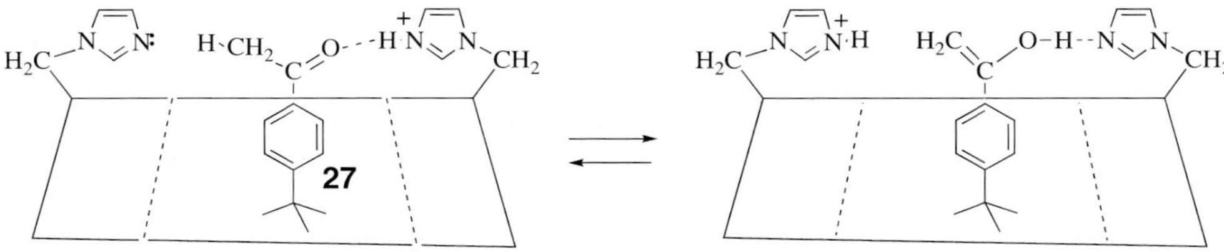

**Scheme 1.10** A bisimidazolecyclodextrin catalyzes the enolization of a hydrophobic ketone in water. The most effective catalyst had the imidazoles far apart, indicating the geometry used for proton abstraction, the first evidence for this detail.

We used the same bisimidazolecyclodextrin isomers to perform an internal aldol condensation of a keto aldehyde. Again there was a preference for the A,D isomer, and the product of the aldol condensation reflected the substrate geometry in the cyclodextrin cavity. When we used the bisimidazolecyclodextrin to catalyze an aldol condensation of a dialdehyde substrate that could hydrophobically bind into the cyclodextrin cavity, an otherwise random aldol reaction was turned into a selective one, again because of the geometry of the bound substrate in the cavity [112]. We also catalyzed the condensation of a hydrophobically bound group of various benzaldehydes with ketones that formed enamines with amino groups linked to the cyclodextrin [113]. The catalyst captured the benzaldehydes into the cavity by hydrophobic binding, then captured the ketones with the catalyst amino group, then held them together and linked them in the product.

## 1.10
### Hydrophobic Acceleration of Diels–Alder Reactions

In the benzoin condensation described earlier, we were able to bind two benzaldehydes into a γ-cyclodextrin ring, and then couple them with a thiazolium ion attached to the cyclodextrin. It had earlier occurred to us that the hydrophobic binding of two substrates in water into a cyclodextrin cavity could accelerate their reaction without the use of other catalytic groups, so we examined a group of Diels–Alder reactions [114]. Indeed, we found that the Diels–Alder reaction of cyclopentadiene with butanone or acrylonitrile in water was accelerated by added β-cyclodextrin, where the two substrates could both fit.

However, we needed to determine their reaction rates in water without the cyclodextrin, as a comparison, and we discovered a remarkable fact: these Diels–Alder reactions were much faster in water than in other solvents. With β-cyclodextrin they were faster still, but with α-cyclodextrin they were slowed relative to water. Both substrates cannot fit into the smaller cavity, so the cyclodextrin is hidden from the dienophiles by binding into α-cyclodextrin.

We concluded that in the reactions using water alone, the reaction between diene and dienophiles was being promoted by the hydrophobic effect, and this was confirmed by the changes in rate produced with added substances that were known to increase, or decrease, the hydrophobic effect. We then examined other Diels–Alder reactions in water, including those of 9-hydroxymethylanthracene [115]. Again, we saw large accelerations with this solvent, and in other cases we also saw a significant preference for the production of the *endo* product.

In that paper [115], we also reported that we saw such increased preference for *endo* addition even when the Diels–Alder components were present as suspensions, not solutions, in water. We pointed out that this indicated that the special hydrophobic effects of water can be seen even with rather insoluble reactants, and that this use of suspensions might be a more general and practical way to apply the water effect. We also made this point in our review describing the use of the hydrophobic effect in organic chemistry [2].

We showed that the preference for *endo* addition was a simple result of the hydrophobic effect, by using prohydrophobic and antihydrophobic additives [1]. We saw that the effects were much smaller when "water-like" solvents were used instead of water [116], and explored the range of antihydrophobic effects further [117]. As is described below, we were eventually able to use the effects of antihydrophobic additives in water to determine the geometries of a number of important transition states, including those of Diels–Alder reactions.

## 1.11
### Selectivities in Water Induced by the Hydrophobic Effect – Carbonyl Reductions

Since the bringing together of a diene and a dienophiles in water was promoted by the hydrophobic effect, it seemed likely that this effect would also promote the reaction of a hydrophobic reagent with a hydrophobic substrate. As one example, we examined

## 1.12 Selectivities in Water Induced by the Hydrophobic Effect – Oxidations

the competition in reduction of a methyl ketone with a phenyl or naphthyl ketone in water by borohydride anion, and by phenylborohydride anion and pentafluorophenylborohydride anion (Scheme 1.11) [118]. The data indicated that the intrinsically less-reactive aryl ketones were selectively reduced by the arylborohydrides in water, but not in methanol. Thus in water the association of the aryl groups of reagent and substrate was hydrophobically promoted.

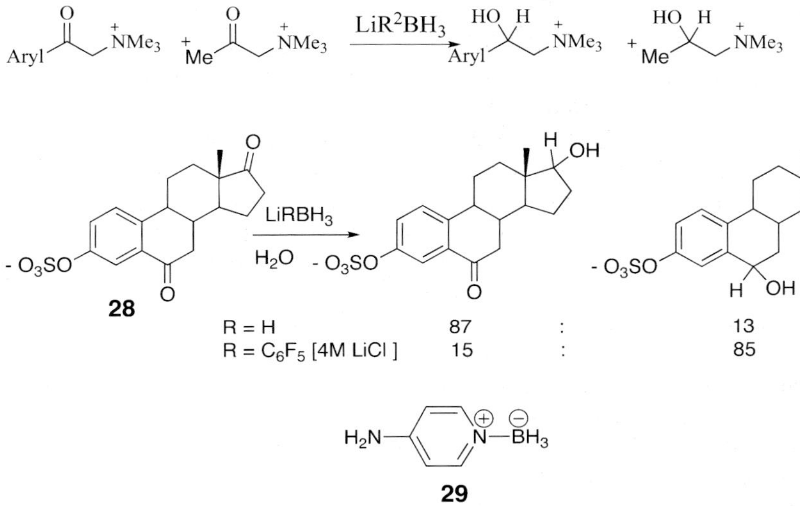

**Scheme 1.11** In water, hydrophobic borohydrides and amine-boranes selectively reduce those carbonyls accessible in hydrophobic complexes.

A striking example was seen in the reversal of selectivity in reducing a steroid diketone **28** in water [119]. Simple borohydride anion selectively reduced the intrinsically more reactive cyclopentanone carbonyl, but pentafluorophenylborohydride anion preferred to reduce the other one in a transition state with the reagent perfluorophenyl ring hydrophobically packed on the steroid phenyl ring.

The synthesis of arylborohydrides is not trivial, and it turns out that there is a simpler way to link a borohydride to an aryl ring – coordinate borane with a pyridine nitrogen to form an amine-borane with a positive nitrogen and a negative boron. We showed that this kind of species (**29**) could also perform selective reductions in water by promoting the hydrophobic association of the aryl rings of substrates and reagents [120].

## 1.12
### Selectivities in Water Induced by the Hydrophobic Effect – Oxidations

The enzymes called cytochrome P450s oxidize many substrates biochemically by binding them next to a heme group, and converting the heme to an iron–oxo species that transfers the oxygen atom from iron to substrate. One such classic biochemical reaction is the epoxidation of olefinic groups in substrates such as squalene. We set out to mimic such enzymes by attaching two or more binding groups to metalloporphyrins and salen [121]. In this first work, we used 8-hydroxyquinoline binding groups to link to substrates with two pyridine or bipyridyl groups through bridging Cu(II) ions. We saw that indeed we could achieve substrate selectivity for those substrates that could use such copper bridging to the catalysts. In this work, the solvent was acetonitrile.

We then performed similar studies in water solution using attached cyclodextrins to doubly bind the hydrophobic substrates to metalloporphyrins, and again saw substrate selectivity [122]. After this, we showed that such metalloporphyrins with attached cyclodextrin groups (e.g. **30**) could selectively hydroxylate a saturated steroid **31** on a well-defined position, attaching a hydroxyl to the C-6 methylene group [123]. When these systems were further examined, we saw that the epoxidation of olefinic substrates occurred with as many as 650 catalytic turnovers. However, the hydroxylation of the saturated carbon in the steroid substrate was slower, and the catalyst was oxidatively destroyed after only three to five catalytic turnovers [124].

The porphyrin ring was being oxidized, and work by others had suggested that this could be suppressed if the phenyl rings used to link the cyclodextrin to the porphyrin were replaced with perfluorophenyl groups. Therefore, we synthesized catalyst **32**, and used it to catalyze the hydroxylation of substrate **31** in water solution [125]. We saw that now the steroid was selectively hydroxylated at C-6 with 187 catalytic turnovers. Furthermore, the perfluoro derivative **32** was particularly easy to synthesize, attaching the cyclodextrin groups by selective displacement of *para*-fluorines on pentafluorophenyl groups.

We then saw that we could move the selective hydroxylation position in the steroid to C-9 by using *three* reversible hydrophobic binding groups to attach a steroid to the catalyst in water [126]. The product is a useful precursor to important steroids such as corticoids. Full descriptions of the extension of this work to other substrates and to other catalyst geometries were also published [127, 128]. We also examined a catalyst carrying synthetic cyclic hydrophobic binding groups instead of cyclodextrins, and saw that these catalysts were chemically slightly more stable, but less selective than the cyclodextrin analogs [129]. A full paper summarized all the results from such mimics of cytochrome P450 [130].

In these oxidations, we used iodosobenzene to supply oxygen atoms to the catalyst, but in Nature the oxidant is molecular oxygen itself. Also, in Nature the iron atom of cytochrome P450 has a thiolate ligand from cysteine, in addition to the four nitrogens of the porphyrin. Hence we added a thiolate from thiophenol, either by covalent attachment to the catalyst or by hydrophobic binding using the two cyclodextrins on the catalyst that were not involved in substrate binding [131]. We now found that the cheap and convenient oxidant hydrogen peroxide could be used instead of iodosobenzene for the steroid oxidations.

We also returned to our previous use of metal ion bridging as a way to coordinate substrates to the metalloporphyrins. We saw that a new catalyst with Cu(II) bridges between substrate and catalyst gave higher turnovers and selectivities than we had obtained with the cyclodextrin-based catalysts [132].

In this work, we used the hydrophobic effect in water to bind substrates to catalysts, but we also examined using the hydrophobic effect to bind a reagent to a substrate, as we had done with carbonyl reductions. Houk had shown that the transition-state geometry for epoxidations of styrene-type olefins would not permit hydrophobic packing of a reagent such as perbenzoic acid onto the phenyl of the substrate [133], and we confirmed that this did not occur [134]. However, he had calculated that epoxidation of styrene by phenyldioxirane would permit such phenyl–phenyl stacking in water. The same geometry permits epoxidation with hydrophobic stacking by aryl-bearing oxaziridinium ions, and by competition experiments we showed that this was indeed the case [134]. We achieved good selectivities for the epoxidations of aryl-substituted olefins in water, but not in methanol, indicating that hydrophobic effects were involved.

In recently published work, we have examined the correlation of enantioselectivity in epoxidations by dioxiranes with the calculated transition-state energies in water. The preferred geometries again involve hydrophobic packing of reagent to substrate [135].

## 1.13
## Using Hydrophobic Effects in Water to Determine the Geometries of Transition States for Some Important Reactions

Throughout this chapter, we have invoked ways in which the hydrophobic effect produces rate accelerations by lowering the energy of the transition state relative to that of the reactants. For example, in the previous section we saw that the predicted difference in transition-state geometry for epoxidation by peracids and by dioxiranes could be confirmed by our experiments, and explained why the dioxirane oxidations showed packing of a hydrophobic reagent to a hydrophobic substrate in water, whereas that did not occur with the peracids.

Hydrophobic effects can be modulated by added solutes in water. Simple salts such as lithium chloride increase the hydrophobic effect by contracting the water volume. However, antihydrophobic salts such as guanidinium chloride *decrease* the hydrophobic effect in water. We showed that this reflected an increased solvation of the hydrophobic surfaces by the guanidinium ion [136], rather than some effect on water structure, as had commonly been proposed. As we had described for the Diels–Alder reaction, the antihydrophobic effect is even greater if both the cation and the anion of the salts are able to solvate hydrophobic surfaces [117].

We realized that we could determine whether the transition states for reactions in water had some hydrophobic packing of non-polar components in the reactants by comparing the effects of the antihydrophobic additives on the starting materials – increasing their water solubility [137] – and on the transition states – determined from the effects of the additives on the rates of the reactions. As Figure 1.1 shows, an antihydrophobic additive will lower the energy of the reactants and will lower the energy of the transition state by the same amount if it has the same amount of solvent-exposed hydrophobic surface. However, the rate will be slowed if the transition state has hidden some hydrophobic surface. From a comparison of the solubility effect on reactants and the rate effect, we could say *how much* of the reactant hydrophobic surface was hidden in the transition state.

This simple idea furnished the first way to determine the geometries of transition states by experiment, rather than by theory. In order to put it into practice, we abandoned salts, as bringing too many effects from ionic strength, and investigated antihydro-

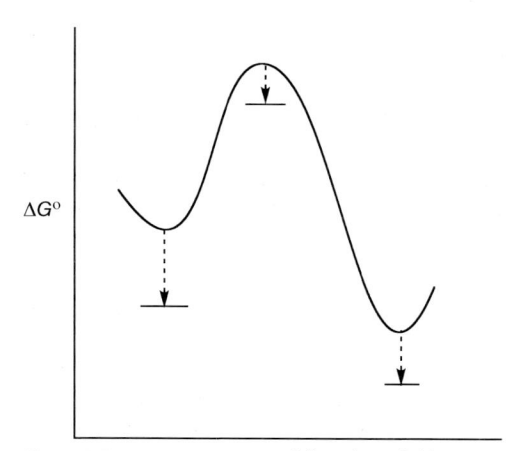

**Figure 1.1** Reactions in water follow the solid line energy curve, and with an added antihydrophobic material the energies are lowered in proportion to the amount of accessible hydrophobic surface in the starting material, the transition state, and the product. From solubility and rate data we can deduce how much hydrophobic surface is exposed in the transition state, the first evidence about the geometry of this elusive species.

## 1.13 Using Hydrophobic Effects in Water to Determine the Geometries of Transition States

phobic solvents. Our first studies were done by examining solubilities and rates in water, and then with a few percent of added ethanol. We checked the method by determining the geometries of transition states where the results were predictable. For example, we examined the effects in the Diels–Alder dimerization of cyclopentadiene and in the Diels–Alder addition of a maleimide to an anthracene [138]. The results were completely consistent with simple ideas and with a quantum mechanical calculation of the transition-state geometry for the dimerization of cyclopentadiene.

We then applied our method to the cyanide-catalyzed benzoin condensation of benzaldehyde, and saw that it predicted a transition state with only partial overlap of the phenyls in the transition state (Scheme 1.12) [139]. This was consistent with a transition state in which the nucleophilic mandelonitrile anion approached the electrophilic benzaldehyde from the rear of the carbonyl group. This new result was completely consistent with modern chemical ideas. In the same paper, we also described the reactions of nucleophiles such as thiophenoxide ion and N-methylaniline with the p-carboxylate of benzyl chloride (the carboxylate is there to achieve water solubility, as our method requires.) We saw that the two phenyl groups overlapped in this displacement by N-methylaniline, but not with the thiophenoxide ion. We pursued such displacements further in subsequent papers.

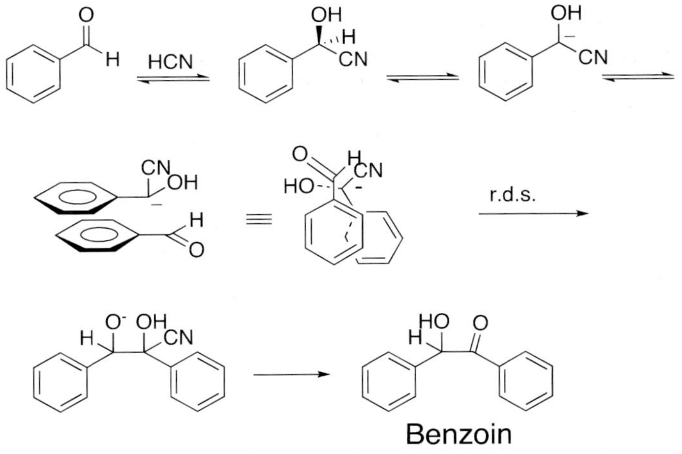

**Scheme 1.12** From this technique, we deduce that the hydrophobic surfaces of benzaldehydes are partially hidden from solvent water in the transition state for the benzoin reaction.

Our most striking result was that the displacement on p-carboxybenzyl chloride by phenoxide ion showed no evidence of hydrophobic overlap in the transition state, with no significant decrease in the rate with added ethanol [140]. This indicated that the phenoxide ion is attacking with its n-electrons rather than with the electrons that are part of the π-system, so the two phenyl rings are far apart, whereas with the aniline nucleophile there is no such choice and the phenyls overlap in the transition state (Scheme 1.13). We also confirmed the lack of slowing with nucleophilic thiophenoxide ion, and proposed that this nucleophile used a single electron transfer

(SET) mechanism, for which we offered some evidence. Both ideas were further pursued, and the results to date were described in a full paper [141].

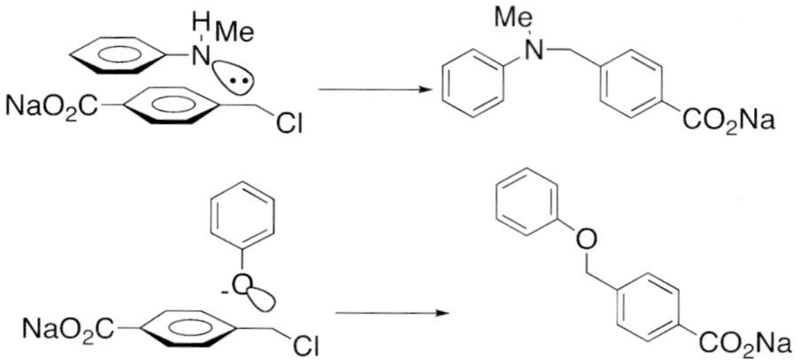

**Scheme 1.13** From this technique, we deduce that the transition state for displacement by an aniline on a benzylic chloride in water involves substantial phenyl–phenyl overlap. However, when phenoxide ion is the nucleophile the n-electrons are used, and there is no overlap of the phenyl groups.

We saw further evidence for the SET mechanism with thiophenoxide ions [142]. It was a dominant mechanism with iodide ion as the leaving group, as expected from the ease of one-electron cleavage of the carbon–iodine bond, and the SET mechanism showed remarkably low rate effects from steric hindrance of the nucleophile, as expected of the transition state for electron transfer if it could occur over longer distances than would be needed for an $S_N2$ reaction. In a subsequent full paper, we showed that quantum mechanics predicted the geometry of phenoxide ion attack, using the n-electrons as we had concluded [143]. However, we did raise the concern that with such ionic reactions it was critical to determine how important the decreased polarity of water with added ethanol could be, since that could also slow the rates. We found a way to deal with this concern in later work.

A powerful tool came from using dimethyl sulfoxide (DMSO) instead of ethanol as an additive to water [144]. We saw that DMSO was even better than ethanol, on a molar basis, in solvating hydrophobic surfaces and thus increasing the solubilities of hydrophobic materials in water. However, DMSO is rather polar, and had a much smaller effect than ethanol on the dielectric constant of the water-based solvent. Using this, we were able to show that the slowing of displacement reactions indeed reflected the modulation of hydrophobic effects, not simply changes in solvent polarity.

In the same paper, we introduced a critical piece of evidence, the solvent effects on two different reactions of the same substrates. 2,6-Dimethylphenoxide ion can be alkylated by *p*-carboxybenzyl chloride in two different spots, on the oxygen and on the *para*-position. Others had studied this and saw, as we did, that only in water was there alkylation of the benzene ring, and that simple phenoxide ion showed only oxygen alkylation. They had proposed that the alkylation of the benzene ring reflected hindrance of the oxygen by the methyl groups at positions 2 and 6. Our results showed that something completely different was involved.

The alkylation of oxygen showed no slowing by ethanol or DMSO cosolvents, consistent with the idea that again there is no phenyl–phenyl overlap in the transition state for phenoxide oxygen alkylation. However, the alkylation of the phenyl ring showed significant slowing with the added cosolvents, indicating that such phenyl alkylations have transition states with phenyl overlaps, as indeed they must. The methyl groups do not slow oxygen alkylation; they add an extra hydrophobic group to promote the hydrophobic packing needed for carbon alkylation.

In a full paper, we described further results in detail [145]. The most striking new result was that moving the two methyl groups to the 3- and 5-positions also caused C-alkylation that was not seen without the methyl groups. They were now next to the carbon being alkylated, away from the phenoxide ion, and this completely disproved the idea that the methyls were causing C-alkylation by shielding the oxygens. Some alkylation now occurred at C-2, not just at C-4 of the phenoxide ion, and even a single methyl group in the *para*-position of phenoxide ion was enough, by adding hydrophobic surface, to promote alkylation at C-2 while without that methyl group no C-alkylation occurred. Thus here the hydrophobic effect can take over a simple O-alkylation of phenoxide ion and divert a significant fraction to the phenyl ring. The use of our method to determine the geometries of transition states was described in a review [146].

## 1.14
## Conclusion

As this chapter describes, there are many reasons to use water as a solvent for chemical reactions. In manufacturing, it is both a "green" solvent and an inexpensive one. The finding that some of the effects seen in water solution can be seen also in water suspensions suggests that even insoluble compounds might be better used in water. However, the major real reason to pursue water as a solvent is that the hydrophobic effect leads to such remarkable new chemistry not otherwise achievable. We saw at least some evidence for this in a water suspension, not just a solution. Process chemists are urged to examine whether they can achieve similar special chemical results in reactions of interest while at the same time solving the environmental problems that so many other solvents produce.

## References

1. Breslow, R. and Maitra, U. (1984) *Tetrahedron Lett.*, **25**, 1239–1240.
2. Breslow, R. (1991) *Acc. Chem. Res.*, **24**, 159–164.
3. Narayan, S., Finn, M., Fokin, V., Kolb, H. and Sharpless, K. (2005) *Angew. Chem. Int. Ed.*, **44**, 3275.
4. Narayan, S., Fokin, V. and Sharpless, K. (2007) *Organic Reactions in Water*, in (ed. M. Lindstrom), Blackwell, Oxford, pp. 350–365.
5. Breslow, R. (2006) *J. Phys. Org. Chem.*, **19**, 813–822.
6. Breslow, R. (1994) *Structure and Reactivity in Aqueous Solution* (eds C.J. Cramer and D.G. Truhlar), American Chemical Society, Washington, DC, pp. 291–302.
7. Breslow, R. (1998) (eds P.T. Anastas and T.C. Williamson), *Green Chemistry*, Oxford University Press, New York, Chapter 13.

8 Breslow, R. (2007) (ed. M. Lindstrom), *Organic Reactions in Water*, Blackwell, Oxford, pp. 1–28.
9 Breslow, R. (1992) *Supramol. Chem.*, 411–428.
10 Breslow, R. (1994) *J. Mol. Catal.*, **91**, 161–174.
11 Breslow, R. (1995) *Acc. Chem. Res.*, **28**, 146–153.
12 Breslow, R. (1995) *Supramol. Chem.*, **66**, 41–47.
13 Breslow, R. (1996) *NATO ASI Ser., Ser. E.*, **320**, 113–135.
14 Breslow, R. (1997) Proceedings of the Robert A Welch Foundation 40th Conference on Chemical Research, 1–11.
15 Breslow, R. and Dong, S.D. (1998) *Chem. Rev.*, **98**, 1997–2011.
16 Breslow, R. (1999) *Templated Organic Synthesis* (eds P. Stang and F. Diederich), Wiley-VCH Verlag GmbH, Weinheim, Chapter 6.
17 Breslow, R., Belvedere, S., Gershell, L. and Leung, D.K. (2000) *Pure Appl. Chem.*, **72**, 333–342.
18 Breslow, R. (ed.) (2005) *Artificial Enzymes*, Wiley-VCH Verlag GmbH, Weinheim.
19 Breslow, R. (2005) (ed. R. Breslow), *Artificial Enzymes*, Wiley-VCH Verlag GmbH, Weinheim, pp. 1–35.
20 Liu, L. and Breslow, R. (2005) (ed. R. Breslow), *Artificial Enzymes*, Wiley-VCH Verlag GmbH, Weinheim, pp. 37–62.
21 Maletic, M., Wennemers, H., McDonald, D.Q., Breslow, R. and Still, W.C. (1996) *Angew. Chem. Int. Ed. Engl.*, **35**, 1490–1492.
22 Yang, Z. and Breslow, R. (1997) *Tetrahedron Lett.*, **38**, 6171–6172.
23 Nesnas, N., Lou, J. and Breslow, R. (2000) *Bioorg. Med. Chem. Lett.*, **10**, 1931–1933.
24 Breslow, R. and Zhang, B. (1996) *J. Am. Chem. Soc.*, **118**, 8495–8496.
25 Breslow, R., Greenspoon, N., Guo, T. and Zarzycki, R. (1989) *J. Am. Chem. Soc.*, **111**, 8296–8297.
26 Breslow, R., Yang, Z., Ching, R., Trojandt, G. and Odobel, F. (1998) *J. Am. Chem. Soc.*, **120**, 3536–3537.
27 Wilson, D., Perlson, L. and Breslow, R. (2003) *Bioorg. Med. Chem.*, **11**, 2649–2653.
28 Breslow, R. and Chung, S. (1990) *J. Am. Chem. Soc.*, **112**, 9659–9660.
29 Zhang, B. and Breslow, R. (1993) *J. Am. Chem. Soc.*, **115**, 9353–9354.
30 Breslow, R., Halfon, S. and Zhang, B. (1995) *Tetrahedron*, **51**, 377–388.
31 Breslow, R., Duggan, P.J., Wiedenfeld, D. and Waddell, S.T. (1995) *Tetrahedron Lett.*, **36**, 2707–2710.
32 Leung, D.K., Atkins, J.H. and Breslow, R. (2001) *Tetrahedron Lett.*, **42**, 6255–6258.
33 Ruebner, A., Yang, Z., Leung, D.K. and Breslow, R. (1999) *Proc. Natl. Acad. Sci. USA*, **96**, 14692–14693.
34 Baugh, S.P.D., Yang, Z., Leung, D.K., Wilson, D.M. and Breslow, R. (2001) *J. Am. Chem. Soc.*, **123**, 12488–12494.
35 Leung, D.K., Yang, Z. and Breslow, R. (2000) *Proc. Natl. Acad. Sci. USA*, **97**, 5050–5053.
36 Moore, S., Askew, J.A., Gibson, G.L., Aborgrein, A., El-Agnaf, O., Allsop, D., Leung, D.K. and Breslow, R. (2002) *Neurobiol. Aging*, **23** (S105), 397–404.
37 Breslow, R. and Campbell, P. (1969) *J. Am. Chem. Soc.*, **91**, 3085.
38 Breslow, R. and Campbell, P. (1971) *Bioorg. Chem.*, **1**, 140–156.
39 Breslow, R., Kohn, H. and Siegel, B. (1976) *Tetrahedron Lett.*, 1645–1646.
40 Komiyama, M. and Bender, M. (1978) *J. Am. Chem. Soc.*, **100**, 4576–4579.
41 Le Noble, W.J., Srivastava, S., Breslow, R. and Trainor, G. (1983) *J. Am. Chem. Soc.*, **105**, 2745–2748.
42 Czarniecki, M.F. and Breslow, R. (1978) *J. Am. Chem. Soc.*, **100**, 7771–7772.
43 Breslow, R., Czarniecki, M.F., Emert, J. and Hamaguchi, H. (1980) *J. Am. Chem. Soc.*, **102**, 762–770.
44 Trainor, G. and Breslow, R. (1981) *J. Am. Chem. Soc.*, **103**, 154–158.
45 Breslow, R., Trainor, G. and Ueno, A. (1983) *J. Am. Chem. Soc.*, **105**, 2739–2744.
46 Thiem, H.-J., Brandl, M. and Breslow, R. (1988) *J. Am. Chem. Soc.*, **110**, 8612–8616.

47 Breslow, R. and Chung, S. (1990) *Tetrahedron Lett.*, **31**, 631–634.
48 Breslow, R. and Overman, L.E. (1970) *J. Am. Chem. Soc.*, **92**, 1075–1077.
49 Tang, C.C., Davalian, D., Huang, P. and Breslow, R. (1978) *J. Am. Chem. Soc.*, **100**, 3918–3922.
50 Gellman, S., Petter, R. and Breslow, R. (1986) *J. Am. Chem. Soc.*, **108**, 2388–2394.
51 Schepartz, A. and Breslow, R. (1987) *J. Am. Chem. Soc.*, **109**, 1814–1826.
52 Breslow, R. and Singh, S. (1988) *Bioorg. Chem.*, **16**, 408–417.
53 Breslow, R., Berger, D. and Huang, D.-L. (1990) *J. Am. Chem. Soc.*, **112**, 3686–3687.
54 Breslow, R. and Zhang, B. (1992) *J. Am. Chem. Soc.*, **114**, 5882–5883.
55 Breslow, R. and Zhang, B. (1994) *J. Am. Chem. Soc.*, **116**, 7893–7894.
56 Zhang, B. and Breslow, R. (1997) *J. Am. Chem. Soc.*, **119**, 1676–1681.
57 Yan, J. and Breslow, R. (2000) *Tetrahedron Lett.*, **41**, 2059–2062.
58 Dong, S.D. and Breslow, R. (1998) *Tetrahedron Lett.*, **39**, 9343–9346.
59 Breslow, R., Doherty, J., Guillot, G. and Lipsey, C. (1978) *J. Am. Chem. Soc.*, **100**, 3227–3229.
60 Breslow, R., Bovy, P. and Hersh, C.L. (1980) *J. Am. Chem. Soc.*, **102**, 2115–2117.
61 Matta, M. and Vo, D. (1986) *J. Am. Chem. Soc.*, **108**, 5316–5321.
62 Anslyn, E. and Breslow, R. (1989) *J. Am. Chem. Soc.*, **111**, 8931–8932.
63 Anslyn, E. and Breslow, R. (1989) *J. Am. Chem. Soc.*, **111**, 5972–5973.
64 Breslow, R. and Labelle, M. (1986) *J. Am. Chem. Soc.*, **108**, 2655–2659.
65 Breslow, R., Huang, D..-L. and Anslyn, E. (1989) *Proc. Natl. Acad. Sci. USA*, **86**, 1746–1750.
66 Anslyn, E. and Breslow, R. (1989) *J. Am. Chem. Soc.*, **111**, 4473–4482.
67 Breslow, R. and Huang, D..-L. (1990) *J. Am. Chem. Soc.*, **112**, 9621–9623.
68 Breslow, R. and Huang, D..-L. (1991) *Proc. Natl. Acad. Sci. USA*, **88**, 4080–4083.
69 Breslow, R. (1991) *Acc. Chem. Res.*, **24**, 317–324.
70 Breslow, R. and Xu, R. (1993) *Proc. Natl. Acad. Sci. USA*, **90**, 1201–1207.
71 Breslow, R. (1993) *Proc. Natl. Acad. Sci. USA*, **90**, 1208–1211.
72 Breslow, R. and Xu, R. (1993) *J. Am. Chem. Soc.*, **115**, 10705–10713.
73 Breslow, R., Dong, S.D., Webb, Y. and Xu, R. (1996) *J. Am. Chem. Soc.*, **118**, 6588–6600.
74 Breslow, R., Anslyn, E. and Huang, D.-L. (1991) *Tetrahedron*, **47**, 2365–2376.
75 Breslow, R. and Schmuck, C. (1996) *J. Am. Chem. Soc.*, **118**, 6601–6605.
76 Breslow, R., Hammond, M. and Lauer, M. (1980) *J. Am. Chem. Soc.*, **102**, 421–422.
77 Breslow, R. and Czarnik, A.W. (1983) *J. Am. Chem. Soc.*, **105**, 1390–1391.
78 Czarnik, A.W. and Breslow, R. (1984) *Carbohydr. Res.*, **128**, 133–139.
79 Zimmerman, S.C., Czarnik, A.W. and Breslow, R. (1983) *J. Am. Chem. Soc.*, **105**, 1694–1695.
80 Breslow, R., Canary, J.W., Varney, M., Waddell, S.T. and Yang, D. (1990) *J. Am. Chem. Soc.*, **112**, 5212–5219.
81 Winkler, J., Coutouli-Argyropoulou, E., Leppkes, R. and Breslow, R. (1983) *J. Am. Chem. Soc.*, **105**, 7198–7199.
82 Zimmerman, S.C. and Breslow, R. (1984) *J. Am. Chem. Soc.*, **106**, 1490–1491.
83 Breslow, R., Czarnik, A.W., Lauer, M., Leppkes, R., Winkler, J. and Zimmerman, S. (1986) *J. Am. Chem. Soc.*, **108**, 1969–1979.
84 Breslow, R., Chmielewski, J., Foley, D., Johnson, B., Kumabe, N., Varney, M. and Mehra, R. (1988) *Tetrahedron*, **44**, 5515–5524.
85 Fasella, E., Dong, S.D. and Breslow, R. (1999) *Bioorg. Med. Chem.*, **7**, 709–714.
86 Weiner, W., Winkler, J., Zimmerman, S.C., Czarnik, A.W. and Breslow, R. (1985) *J. Am. Chem. Soc.*, **107**, 4093–4094.
87 Metzler, D., Ikawa, M. and Snell, E. (1954) *J. Am. Chem. Soc.*, **76**, 648–650.
88 Chmielewski, J. and Breslow, R. (1987) *Heterocycles*, **25**, 533–540.
89 Liu, L. and Breslow, R. (2001) *Tetrahedron Lett.*, **42**, 2775–2777.

90 Liu, L., Rozenman, M. and Breslow, R. (2002) *Bioorg. Med. Chem.*, **10**, 3973–3979.
91 Koh, J.T., Delaude, L. and Breslow, R. (1994) *J. Am. Chem. Soc.*, **116**, 11234–11240.
92 Klotz, I. and Suh, J. (2005) *Artificial Enzymes* (ed. R. Breslow), Wiley-VCH Verlag GmbH, Weinheim, pp. 63–88.
93 Hollfelder, F., Kirby, A. and Tawfik, D. (2001) *J. Org. Chem.*, **66**, 5866–5871.
94 Liu, L. and Breslow, R. (2002) *J. Am. Chem. Soc.*, **124**, 4978–4979.
95 Liu, L., Rozenman, M. and Breslow, R. (2002) *J. Am. Chem. Soc.*, **124**, 12660–12661.
96 Zhou, W., Liu, L. and Breslow, R. (2003) *Helv. Chim. Acta*, **86**, 3560–3567.
97 Liu, L. and Breslow, R. (2003) *J. Am. Chem. Soc.*, **125**, 12110–12111.
98 Liu, L. and Breslow, R. (2004) *Bioorg. Med. Chem.*, **12**, 3277–3287.
99 Breslow, R., Wei, S. and Kenesky, C. (2007) *Tetrahedron*, **63**, 6317–6321.
100 Liu, L., Zhou, W., Chruma, J.J. and Breslow, R. (2004) *J. Am. Chem. Soc.*, **126**, 8136–8137.
101 Chruma, J.J., Liu, L., Zhou, W. and Breslow, R. (2005) *Bioorg. Med. Chem.*, **13**, 5873–5883.
102 Levine, M., Kenesky, C., Zheng, S., Quinn, J. and Breslow, R. (2008) *Tetrahedron Lett.*, **49**, 5746–5750.
103 Zhou, W., Yerkes, N., Chruma, J., Liu, L. and Breslow, R. (2005) *Bioorg. Med. Chem. Lett.*, **15**, 1351–1355.
104 Bandyopadhyay, S., Zhou, W. and Breslow, R. (2007) *Org. Lett.*, **9**, 1009–1012.
105 Breslow, R., Bandyopadhyay, S., Levine, M. and Zhou, W. (2006) *ChemBioChem*, **7**, 1491–1496.
106 Breslow, R. (1957) *J. Am. Chem. Soc.*, **79**, 1762.
107 Breslow, R. (1958) *J. Am. Chem. Soc.*, **80**, 3719–3726.
108 Hilvert, D. and Breslow, R. (1984) *Bioorg. Chem.*, **12**, 206–220.
109 Breslow, R. and Kool, E. (1988) *Tetrahedron Lett.*, **29**, 1635–1638.
110 Zhag, H., Foss, F. and Breslow, R. (2008) *J. Am. Chem. Soc.*, **130**, 12590–12591.
111 Breslow, R. and Graff, A. (1993) *J. Am. Chem. Soc.*, **115**, 10988–10989.
112 Breslow, R., Desper, J. and Huang, Y. (1996) *Tetrahedron Lett.*, **37**, 2541–2544.
113 Yuan, D..-Q., Dong, S.D. and Breslow, R. (1998) *Tetrahedron Lett.*, **39**, 7673–7676.
114 Rideout, D. and Breslow, R. (1980) *J. Am. Chem. Soc.*, **102**, 7816–7817.
115 Breslow, R., Maitra, U. and Rideout, D. (1983) *Tetrahedron Lett.*, **24**, 1901–1904.
116 Breslow, R. and Guo, T. (1988) *J. Am. Chem. Soc.*, **110**, 5613–5617.
117 Breslow, R. and Rizzo, C.J. (1991) *J. Am. Chem. Soc.*, **113**, 4340–4341.
118 Biscoe, M. and Breslow, R. (2003) *J. Ame. Chem. Soc.*, **125**, 12718–12719.
119 Biscoe, M., Uyeda, C. and Breslow, R. (2004) *Org. Lett.*, **6**, 4331–4334.
120 Uyeda, C., Biscoe, M.R., LePlae, P. and Breslow, R. (2005) *Tetrahedron Lett.*, **47**, 127–130.
121 Breslow, R., Brown, A.B., McCullough, R.D. and White, P.W. (1989) *J. Am. Chem. Soc.*, **111**, 4517–4518.
122 Breslow, R., Zhang, X., Xu, R., Maletic, M. and Merger, R. (1996) *J. Am. Chem. Soc.*, **118**, 11678–11679.
123 Breslow, R., Zhang, X. and Huang, Y. (1997) *J. Am. Chem. Soc.*, **119**, 4535–4536.
124 Breslow, R., Huang, Y., Zhang, X. and Yang, J. (1997) *Proc. Natl. Acad. Sci. USA*, **94**, 11156–11158.
125 Breslow, R., Gabriele, B. and Yang, J. (1998) *Tetrahedron Lett.*, **39**, 2887–2890.
126 Yang, J. and Breslow, R. (2000) *Angew. Chem. Int. Ed.*, **39**, 2692–2694.
127 Breslow, R., Yang, J. and Yan, J. (2002) *Tetrahedron*, **58**, 653–659.
128 Breslow, R., Yan, J. and Belvedere, S. (2002) *Tetrahedron Lett.*, **43**, 363–365.
129 Breslow, R. and Fang, Z. (2002) *Tetrahedron Lett.*, **43**, 5197–5200.
130 Yang, J., Gabriele, B., Belvedere, S., Huang, Y. and Breslow, R. (2002) *J. Org. Chem.*, **67**, 5057–5067.
131 Fang, Z. and Breslow, R. (2005) *Bioorg. Med. Chem. Lett.*, **15**, 5463–5466.

132 Fang, Z. and Breslow, R. (2006) *Org. Lett.*, **8**, 251–254.
133 Washington, L. and Houk, K. (2000) *J. Am. Chem. Soc.*, **122**, 2948–2949.
134 Biscoe, M.R. and Breslow, R. (2005) *J. Am.Chem. Soc.*, **127**, 10812–10813.
135 Schneebeli, S., Hall, M., Breslow, R. and Friesner, R. (2009) *J. Am. Chem. Soc.*, **131**, 12590–12591.
136 Breslow, R. and Guo, T. (1990) *Proc. Natl. Acad. Sci. USA*, **87**, 167–169.
137 Breslow, R. and Halfon, S. (1992) *Proc. Natl. Acad. Sci. USA*, **89**, 6916–6918.
138 Breslow, R. and Zhu, Z. (1995) *J. Am. Chem. Soc.*, **117**, 9923–9924.
139 Breslow, R. and Connors, R.V. (1995) *J. Am. Chem. Soc.*, **117**, 6601–6602.
140 Breslow, R. and Connors, R. (1996) *J. Am. Chem. Soc.*, **118**, 6323–6324.
141 Breslow, R., Connors, R. and Zhu, Z. (1996) *Pure Appl. Chem.*, **68**, 1527–1533.
142 Mayer, M.U. and Breslow, R. (1998) *J. Am. Chem. Soc.*, **120**, 9098–9099.
143 Breslow, R., Groves, K. and Mayer, M.U. (1998) *Pure Appl. Chem.*, **70**, 1933–1938.
144 Breslow, R., Groves, K. and Mayer, M.U. (1999) *Org. Lett.*, **1**, 117–120.
145 Breslow, R., Groves, K. and Mayer, M.U. (2002) *J. Am. Chem. Soc.*, **124**, 3622–3635.
146 Breslow, R. (2004) *Acc. Chem. Res.*, **3**, 471–478.

# 2
# Green Acid Catalysis in Water
*Chikako Ogawa and Shū Kobayashi*

## 2.1
## Introduction

In the past centuries, organic chemistry has influenced many fields as a tool to create new products that are useful, variable, and closely bound to human life. It can be said that human life has changed as organic chemistry has been developed. Despite the great impact on human society and the dramatic progress of organic chemistry, environmentally benign "green chemistry" has not yet become completely established. Products and wastes exist together and chemists have responsibilities to take care of them. The best way to achieve green chemistry is to design reaction systems that do not require excessive amounts of chemical reagents, solvents, and catalysts.

In this chapter, "green acid catalysis in water" is described. Organic reactions are usually carried out in organic solvents, which means that huge amounts of organic wastes are produced. In contrast to common organic solvents, water is a safe, cheap, and environmentally friendly solvent. Therefore, the replacement of organic solvents with water would be a first step to achieving "green chemistry."

## 2.2
## Lewis Acids in Water

### 2.2.1
### Introduction. Lewis Acids in Aqueous Media: Possible?

The use of Lewis acid catalysts is one of the most powerful strategies in organic reactions. In general, conventional Lewis acids such as $AlCl_3$, $TiCl_4$, $BF_3 \cdot (OEt)_2$, and $SnCl_4$ require strictly anhydrous conditions, since they immediately react with water rather than substrates and cause serious decomposition of the catalysts or even substrates and products. In addition, the recovery and reuse of such Lewis acids are difficult, and these disadvantages have restricted the use of Lewis acids in organic synthesis. To address these issues and aim towards true green chemistry, 'water-compatible' Lewis acids have been developed. The first discovery goes back to 1991 [1] and it was found that rare earth triflates [Sc(OTf)$_3$, Yb(OTf)$_3$, etc.] could be used as Lewis acid catalysts in water or water-containing solvents (water-compatible Lewis acids) [2]. For example, the Mukaiyama aldol reaction of benzaldehyde with silyl enol ether **1** was catalyzed by Yb(OTf)$_3$ in water–THF (1 : 4) to give the corresponding aldol adduct in high yield (Scheme 2.1).[1, 3] Interestingly, when this reaction was carried out in dry THF (without water), the yield of the aldol adduct was very low (ca. 10%). Thus, this catalyst was not only compatible with water but also was activated by water, probably due to dissociation of the counter anions from the Lewis acidic metal. Furthermore, in this example, the catalyst can be easily recovered and reused.

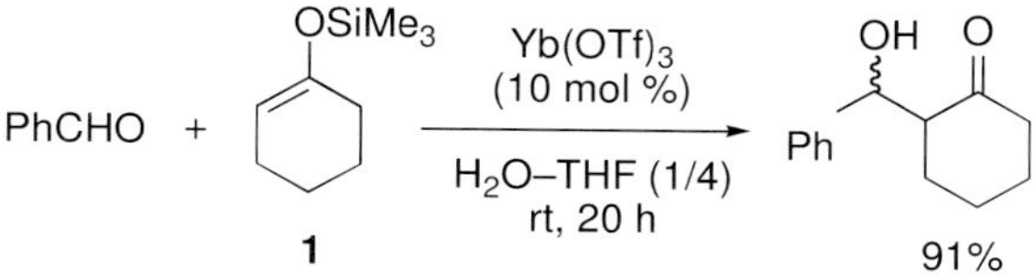

PhCHO + [silyl enol ether **1**, OSiMe$_3$] → [aldol adduct, Ph, OH O] 91%
Yb(OTf)$_3$ (10 mol %)
H$_2$O–THF (1/4)
rt, 20 h

**Scheme 2.1**

Metal salts other than those derived from rare earth elements were also found to be water-compatible Lewis acids. In order to find other Lewis acids which can be used in aqueous solvents and to find criteria for water-compatible Lewis acids, Group 1–15 metal chlorides, perchlorates, and triflates were screened in the aldol reaction of benzaldehyde with silyl enol ether **2** in water–THF (1 : 9) (Scheme 2.2)[4]. This screening revealed that not only Sc(III), Y(III), and Ln(III) but also Fe(II), Cu(II), Zn(II), Cd(II), and Pb(II) worked as Lewis acids in this medium to afford the desired aldol adduct in high yields.

PhCHO + [OSiMe$_3$, Ph, **2**] → [Ph, OH O, Ph]
MX$_n$ (20 mol %)
H$_2$O–THF (1/9)
rt, 12 h

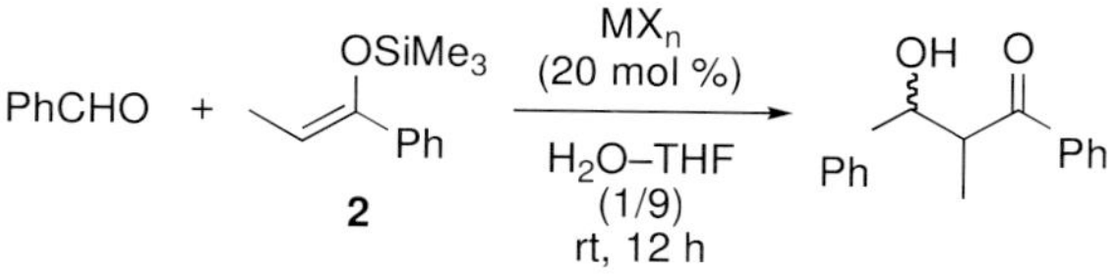

**Scheme 2.2**

## 2.2.2
### Lewis Acid Catalysis in Water as "Sole Solvent"

To achieve truly environmentally benign chemistry, it is necessary to use water *as sole solvent*. This idea faces a serious issue: how can organic substrates be dissolved in water? In general, most organic compounds are not soluble in water, and therefore the device of reaction media is required. The strategies for this issue focus on how to create a hydrophobic area in water.

## 2.2 Lewis Acids in Water

### 2.2.2.1 LASC: Lewis Acid–Surfactant Combined Catalyst

Lewis acid–surfactant combined catalyst (LASC) is an idea to create a hydrophobic area in water. LASCs designed from water-compatible Lewis acids are expected to act as both surfactants and Lewis acids in water. One example of LASCs which can be readily prepared from scandium chloride and sodium tris(dodecyl sulfate) in water, that is, scandium tris(dodecyl sulfate) [$Sc(DS)_3$], is shown in Scheme 2.3.

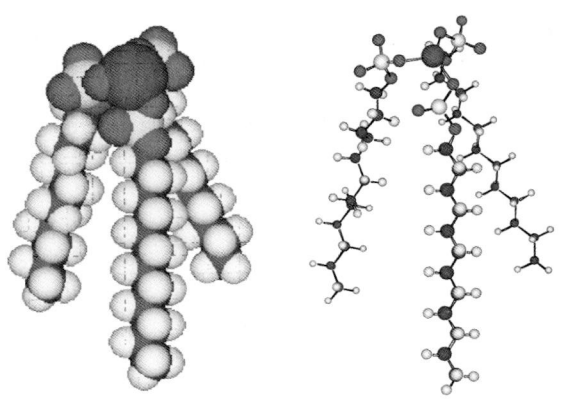

**Scheme 2.3**

$Sc(DS)_3$ functioned very efficiently in the aldol reaction of benzaldehyde with the silyl enol ether derived from propiophenone in water, whereas the reaction proceeded sluggishly when $Sc(OTf)_3$ was used as a catalyst (Scheme 2.4) [5].

PhCHO + Ph-C(OSiMe$_3$)=CH(Me) → catalyst (0.1 equiv.), H$_2$O, rt, 4 h → Ph-CH(OH)-CH(Me)-C(O)-Ph

$Sc(DS)_3$; 92% Yield
$Sc(OTf)_3$; 3% Yield

**Scheme 2.4**

A kinetic study on the initial rate of this reaction revealed that the reaction in water was about 100 times faster than that in dichloromethane. On the other hand, the reaction proceeded much more slowly, giving a low yield, under neat conditions (Table 2.1).

A key to the success in this system is assumed to be the formation of stable emulsions. The size and shape of emulsion droplets were examined by transmission electron microscopy (TEM), and it was found that only about 0.08 mol% of $Sc(DS)_3$ was sufficient to form monolayers (Scheme 2.5).

Based on these observations, the advantageous effect of water is attributed to the following factors: (1) concentration of substrates and catalysts, (2) defense of hydrolysis of silyl enol ethers in hydrophobic environments formed by water and LASC, and (3) high catalytic turnover caused by hydrolysis of scandium aldolates.

**Table 2.1** Solvent effect on aldol reaction in water.

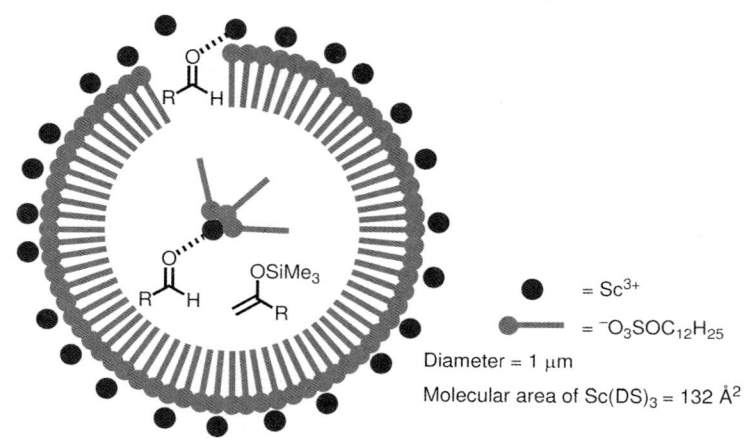

| Solvent | Yield (%) | Solvent | Yield (%) |
| --- | --- | --- | --- |
| $H_2O$ | 92 | THF | Trace |
| MeOH | 4 | $Et_2O$ | Trace |
| DMF | 14 | Toluene | Trace |
| DMSO | 9 | Hexane | 4 |
| MeCN | 3 | (Neat) | 31 |
| $CH_2Cl_2$ | 3 | | |

Diameter = 1 μm
Molecular area of $Sc(DS)_3$ = 132 Å$^2$

● = $Sc^{3+}$
●— = $^-O_3SOC_{12}H_{25}$

**Scheme 2.5**

In addition, several LACS-catalyzed reactions, such as Mannich-type reactions [6], α-aminophosphate synthesis [7], Michael reactions [8], and Friedel–Crafts-type reactions [9] have been reported ($FeCl_3$ was also used in water in the presence of SDS for a diastereoselective aldol reaction [10]).

#### 2.2.2.2 Polymer-supported Scandium Triflate

Polystyrene-based polymers are expected to be highly hydrophobic materials. Since the discovery of scandium triflate as a water-compatible Lewis acid, several supported scandium catalysts that function efficiently in water have been developed. Polymer-supported scandium-based Lewis acid **3** worked well in several carbon–carbon-forming reactions such as allylations to ketones, Diels–Alder reactions, Mukaiyama aldol reactions, and Strecker-type reactions [11]. It was found that the spacer could help to form a hydrophobic area in water. Moreover, as expected, **3** was easily recovered and reused {a new type of PS-Sc(OTf)$_3$ simply prepared from PS-SO$_3$H and Sc(OTf)$_3$ was also found to be an effective catalyst in several useful reactions in water [12]}.

**3**

### 2.2.2.3 Silica Gel-supported Scandium with Ionic Liquid

A novel heterogeneous scandium catalyst system, silica gel-supported scandium with ionic liquid (silica-Sc-IL) has been developed (Scheme 2.6) [13]. The catalyst **4a** coated with an ionic liquid, [DBIm]SbF$_6$, worked efficiently in the Mukaiyama aldol reaction in water (Scheme 2.7). The reaction proceeded much faster in water than in organic solvents (entries 2, 4, and 5). In the absence of IL (entry 2) or under neat conditions (entry 3), the catalyst did not work well. These experiments clearly suggested that silica-Sc-IL and an ionic liquid formed hydrophobic reaction environments in water (Scheme 2.8). It should be noted that water-labile reagents such as **5** can work well in water under these conditions, and this is the first demonstration of the combination of silica gel-supported metal catalysts with ionic liquids to create efficient hydrophobic environments for organic reactions in water.

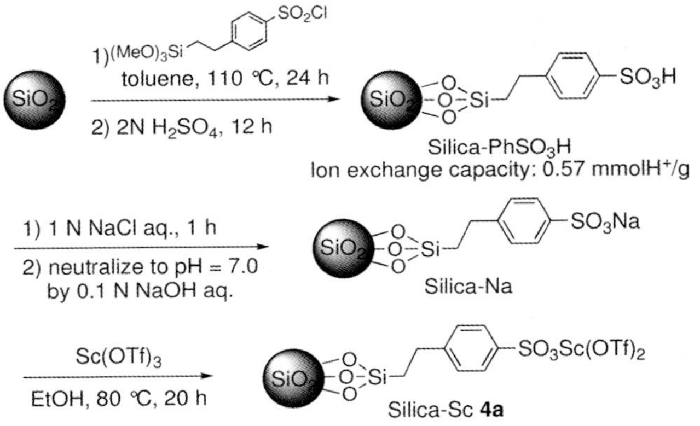

**Scheme 2.6**

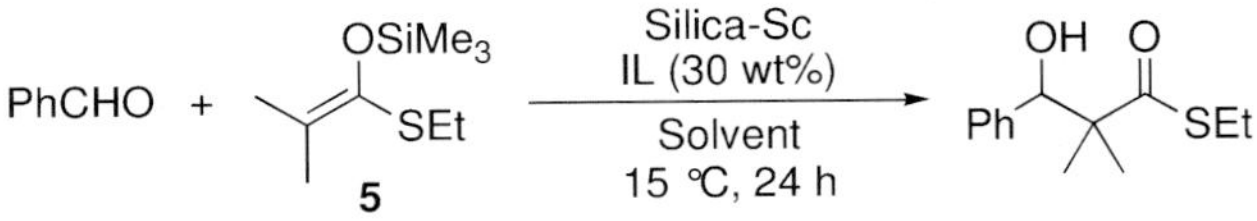

| Entry | Catalyst (Silica-Sc) | IL | Solvent | Yield (%) |
|---|---|---|---|---|
| 1 | none | [DBIm]SbF$_6$ | H$_2$O | 0 |
| 2 | 4a (0.27 mmol/g, 5.4 mol%) | none | H$_2$O | 31 |
| 3 | 4a (0.27 mmol/g, 5.4 mol%) | [DBIm]SbF$_6$ | none | 25 |
| 4 | 4a (0.27 mmol/g, 5.4 mol%) | [DBIm]SbF$_6$ | Hexane | 26 |
| 5 | 4a (0.27 mmol/g, 5.4 mol%) | [DBIm]SbF$_6$ | Et$_2$O | 22 |
| 6 | 4a (0.27 mmol/g, 5.4 mol%) | [DBIm]SbF$_6$ | H$_2$O | 97 |

**Scheme 2.7**

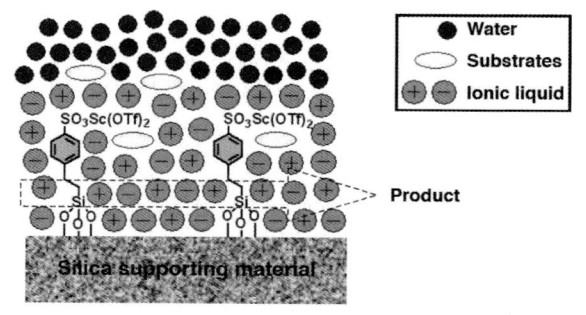

**Scheme 2.8**

Furthermore, in the Michael reaction of indoles with enones, the reaction proceeds smoothly in water to afford the desired Michael adducts in high yields. It is remarkable that in the reaction of indole and methyl vinyl ketone, the turnover number reached 4650 and the turnover number frequency reached to 15 and 387.5 h$^{-1}$ (Scheme 2.9).

**4b**; Silica-Sc(OSO$_2$CF$_3$)$_2$ (0.08 mmol/g, 1.6 mol%), 4 h, 96%
**4c**; Silica-Sc(OSO$_2$C$_8$F$_{17}$)$_2$ (0.001 mmol/g, 0.02 mol%), 12 h, 93%

**Scheme 2.9**

## 2.3
## Chiral Lewis Acid-catalyzed Asymmetric Reactions in Water

Three devices to create a hydrophobic area in water have been described above. As a next step, the development of asymmetric reactions in water as sole solvent is of great interest. In general, the formation of a chiral Lewis acid complex is much more difficult in water than in organic media, since a chiral ligand competes with water in the coordination to Lewis acid. Although there are successful reports of chiral Lewis acid-catalyzed asymmetric reactions in *aqueous media*, it is still very challenging to use *water as sole solvent* [14, 15].

### 2.3.1
### Mannich-type Reactions in Water

Asymmetric Mannich reactions provide useful routes for the synthesis of optically active β-amino ketones and esters, which are versatile chiral building blocks for the preparation of many nitrogen-containing biologically important compounds [16]. Recently, diastereo- and enantioselective Mannich-type reactions of α-hydrazono ester **6** with silicon enolates in aqueous media have been achieved with a ZnF$_2$–chiral diamine **7** complex (Scheme 2.10) [17]. This complex enabled reactions in water without any organic co-solvents or additives to proceed smoothly, affording the corresponding products in high yields and high stereoselectivities (Scheme 2.10, entries 1–3) [18]. Although the reaction of α-monosubstituted ketone-derived silyl enol ether with **6** proceeded very sluggishly (Scheme 2.10, entry 4), the addition of cetyltrimethylammonium bromide (CTAB) accelerated the reaction (Scheme 2.10, entry 8). It is also noted that, in contrast to most asymmetric Mannich-type reactions, either *syn-* or *anti-*adducts were stereospecifically obtained from silicon (*E*)- or (*Z*)-enolates in the present reaction (Scheme 2.10, entries 9 and 10). Finally, the amount of $ZnF_2$ and **7** could be reduced to 10 and 5 mol%, respectively, maintaining the same level of result (Scheme 2.10, entries 8 and 11).

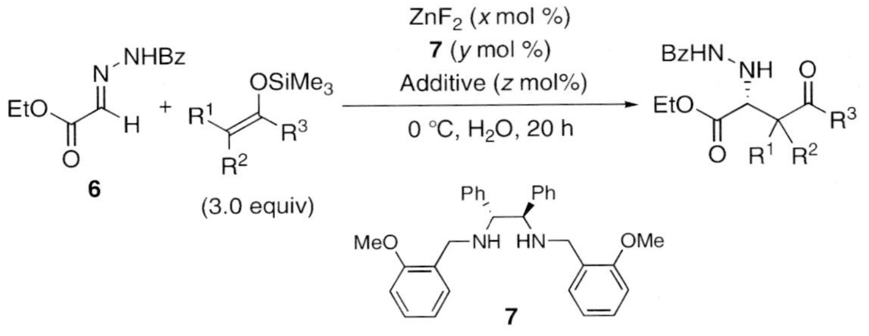

| Entry | $R^1$ | $R^2$ | $R^3$ | x | y | z | Yield (%) | syn/anti | ee (%) (syn) |
|---|---|---|---|---|---|---|---|---|---|
| 1 | H | H | 4-Me-$C_6H_5$ | 100 | 10 | - | 91 | - | 95 |
| 2 | H | H | 4-MeO-$C_6H_5$ | 100 | 10 | - | 93 | - | 91 |
| 3 | H | H | 4-Cl-$C_6H_5$ | 100 | 10 | - | 94 | - | 95 |
| 4[a] | Me | H | $C_6H_5$ | 100 | 10 | - | 6 | 91/9 | 94 |
| 5[a] | Me | H | $C_6H_5$ | 100 | 10 | TfOH (1) | 5 | 91/9 | 90 |
| 6[a] | Me | H | $C_6H_5$ | 100 | 10 | SDS (5) | 9 | 91/9 | 78 |
| 7[a] | Me | H | $C_6H_5$ | 100 | 10 | Triton X-405 (5) | 10 | 91/9 | 93 |
| 8[a] | Me | H | $C_6H_5$ | 100 | 10 | CTAB (2) | 93 | 94/6 | 96 |
| 9[b,d] | Me | H | Et | 100 | 10 | CTAB (2) | 57 | 86/14 (98.5/1.5)[e] | 97 (>99.5)[e,f] |
| 10[c,d] | H | Me | Et | 100 | 10 | CTAB (2) | 94 | 12/88 | 94[f] |
| 11[a,g] | Me | H | $C_6H_5$ | 10 | 5 | CTAB (2) | 87 | 93/7 | 96 |

[a] E/Z = <1/>99. [b] E/Z = 76/24. [c] E/Z = 2/98. [d] Time = 72 h. [e] After recrystallization. [f] ee of major diastereomer. [g] Time = 40 h.

**Scheme 2.10**

### 2.3.2
### Michael Reaction in Water

AgOTf–$PPh_3$ complex-catalyzed Michael additions of β-keto esters to nitroalkenes in water have been reported. The reaction proceeded efficiently only in water, and interestingly the reaction did not proceed well in organic solvents (Scheme 2.11).

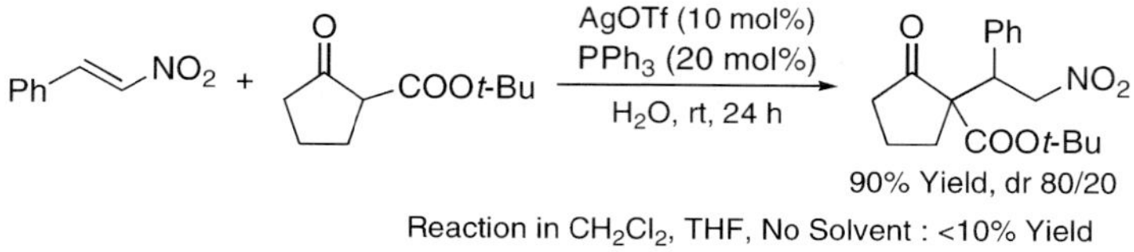

**Scheme 2.11**

Ph–CH=CH–NO₂ + [cyclopentanone-COOt-Bu] → [product with Ph, NO₂, COOt-Bu substituents]

Conditions: AgOTf (10 mol%), PPh₃ (20 mol%), H₂O, rt, 24 h
90% Yield, dr 80/20
Reaction in CH₂Cl₂, THF, No Solvent : <10% Yield

Based on these results, a plausible mechanism is shown in Scheme 2.12. In the formation of metal enolate **B**, TfOH is generated and the reaction mixture becomes heterogeneous, where metal enolate **B** stays in organic phase, while TfOH is excluded to the water phase because of the difference in hydrophobicity between them. On the other hand, in the case of a normal organic solvent system, the reaction mixture becomes homogeneous, leading the reverse reaction from **B** to **A to be** fast. As a result, metal enolate **B** does not make contact with TfOH, and the reverse reaction from **B** to **A** is suppressed. Metal enolate **B** and nitrostyrene would thus combine in high concentration and the Michael addition step (**B** to **C** in Scheme 2.12) may proceed smoothly. Moreover, this reaction system could be applied to catalytic asymmetric synthesis in water (Scheme 2.13) [19].

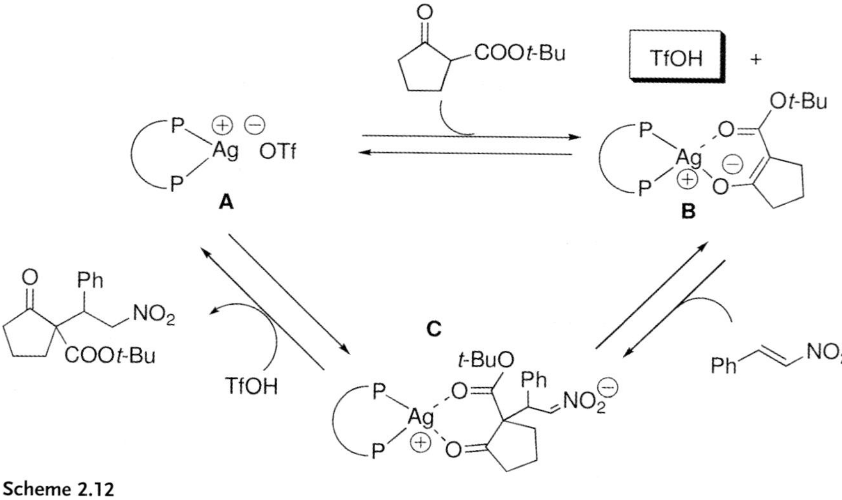

**Scheme 2.12**

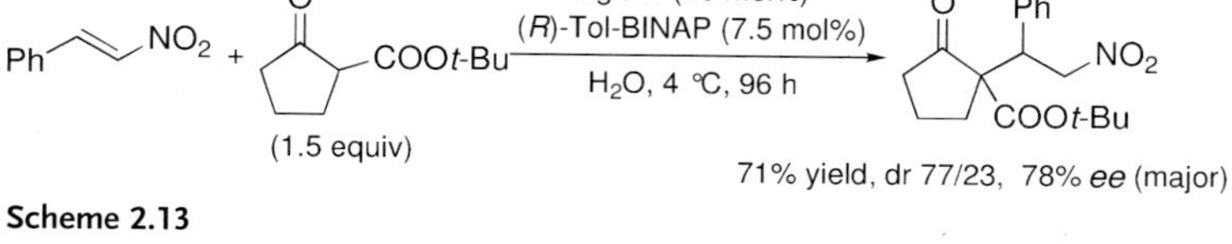

**Scheme 2.13**

Ph–CH=CH–NO₂ + [cyclopentanone-COOt-Bu] (1.5 equiv) → [product]

Conditions: AgOTf (10 mol%), (R)-Tol-BINAP (7.5 mol%), H₂O, 4 °C, 96 h
71% yield, dr 77/23, 78% ee (major)

### 2.3.3
### Epoxide Ring-opening Reaction in Water

Chiral β-amino alcohol units are found in many biologically active compounds and chiral auxiliaries/ligands used in asymmetric reactions [20]. Catalytic enantioselective synthesis of these chiral building blocks relies mainly on asymmetric ring opening of *meso*-epoxides. Indeed, several examples using a chiral catalyst (typically a chiral Lewis acid) have been reported [21]; however, all these reactions proceeded in organic solvents. Probably, epoxides are readily decomposed under acidic conditions in water. Sc(DS)$_3$ functions both as a Lewis acid and as a surfactant, as discussed above. To explore this catalyst further, chiral Sc(DS)$_3$ catalyst has been investigated. The complex Sc(OTf)$_3$·8 was found to be effective in asymmetric hydroxymethylation using aqueous formaldehyde solution in DME–H$_2$O co-solvent conditions [22] {Sc(OTf)$_3$·8 complex has been found to be effective in several reactions using organic solvents after the report in ref. [22]; see [21d, 23]}. Therefore, there was a possibility that Sc(DS)$_3$ could form a chiral complex with 8 in water. First, asymmetric ring opening of *cis*-stilbene oxide with aniline in water was investigated. The reaction proceeded smoothly in high yield with high enantioselectivity using 1 mol% of Sc(DS)$_3$ and 1.2 mol% of 8 in water (Scheme 2.14). It is noted that the ring-opening reaction proceeded smoothly in water, and that no diol formation was observed. This is the first example to date of an asymmetric epoxide ring opening in water as sole solvent [24, 25] {similarly, catalytic asymmetric ring-opening reactions of *meso*-epoxides with aromatic amines also proceeded in the presence of a catalytic amount of bismuth triflate [Bi(OTf)$_3$], chiral bipyridine ligand 8, and sodium dodecylbenzene sulfonate (SDBS) in pure water to give the corresponding α-amino alcohols in good yields with high enantioselectivities}.

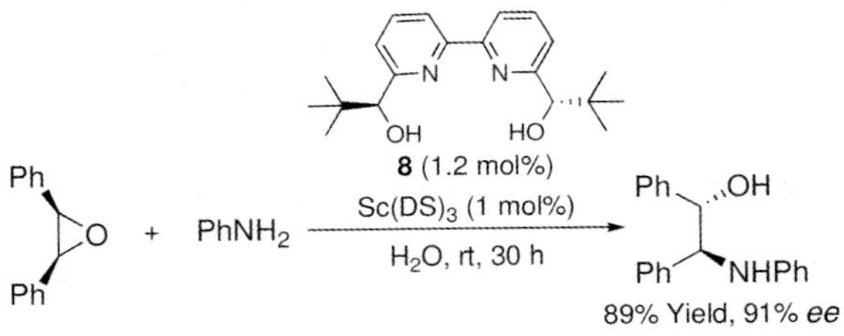

**Scheme 2.14**

Moreover, catalytic asymmetric ring-opening reactions of *meso*-epoxides with indoles, alcohols, and thiols proceeded smoothly in the presence of catalytic amounts of Sc(DS)$_3$ and chiral bipyridine ligand 8 in water to afford β-amino alcohols in high yields with high enantioselectivities (Schemes 2.15 and 2.16) [26, 27]. These results suggest that an excellent asymmetric environment can be created in water.

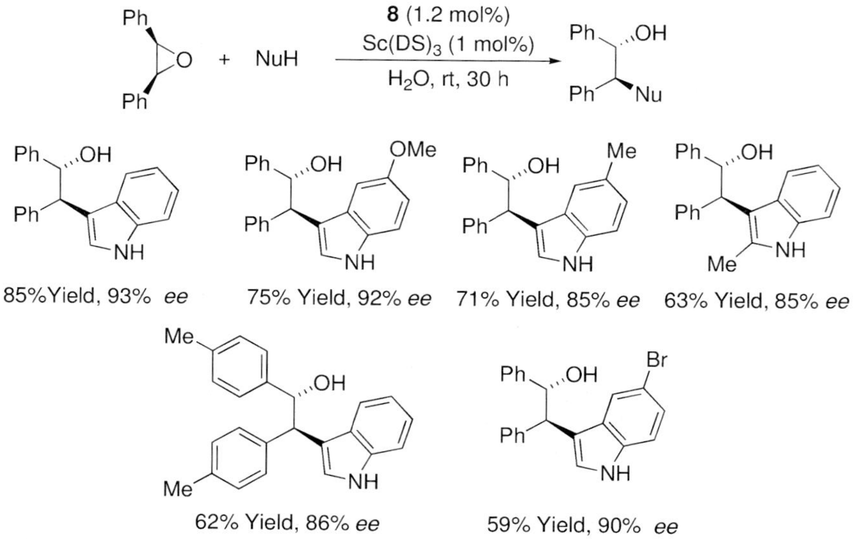

## Scheme 2.15

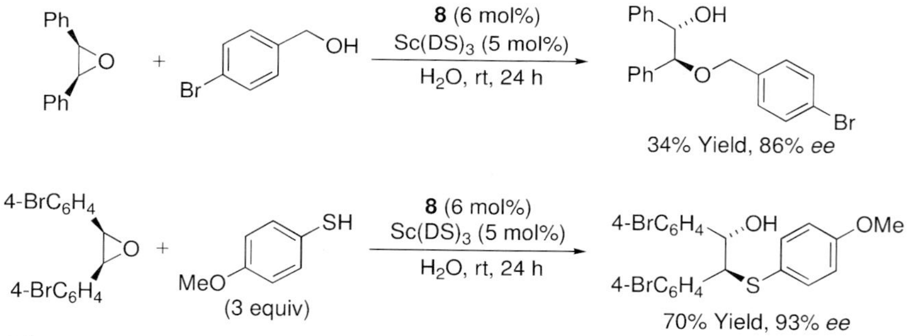

Ph₂C(O)CHPh (epoxide) + NuH → Ph(OH)CH-C(Ph)(Nu)
Conditions: 8 (1.2 mol%), Sc(DS)₃ (1 mol%), H₂O, rt, 30 h

Products:
- Indol-3-yl adduct: 85% Yield, 93% ee
- 5-OMe-indol-3-yl: 75% Yield, 92% ee
- 5-Me-indol-3-yl: 71% Yield, 85% ee
- N-Me-indol-3-yl: 63% Yield, 85% ee
- 2-(4-methylphenyl)-indol-3-yl with p-tolyl: 62% Yield, 86% ee
- 5-Br-indol-3-yl: 59% Yield, 90% ee

## Scheme 2.16

Ph₂C(O)CHPh + 4-Br-C₆H₄-CH₂OH → Ph(OH)CH-C(Ph)(OCH₂-C₆H₄-Br)
Conditions: 8 (6 mol%), Sc(DS)₃ (5 mol%), H₂O, rt, 24 h
34% Yield, 86% ee

(4-BrC₆H₄)₂C(O)C(4-BrC₆H₄) + 4-MeO-C₆H₄-SH (3 equiv) → (4-BrC₆H₄)(OH)CH-C(4-BrC₆H₄)(S-C₆H₄-OMe)
Conditions: 8 (6 mol%), Sc(DS)₃ (5 mol%), H₂O, rt, 24 h
70% Yield, 93% ee

### 2.3.4
### Hydroxymethylation in Water

Several asymmetric organic reactions have been achieved in water without any organic co-solvents. These reactions proceeded smoothly by creating a hydrophobic area in water to stabilize and concentrate organic substrates or by suppressing the undesired pathway in the reaction mechanism by water. One of the key factors for these successes is *hydrophobicity* of substrates. Therefore, asymmetric reactions in water with *hydrophilic substrates* are very challenging.

Aqueous formaldehyde solution (formalin), is one of the most important $C_1$ electrophiles and also a representative of hydrophilic substrates. Asymmetric hydroxymethylation using aqueous formaldehyde solution has been investigated in water–organic co-solvent systems [22, 28, 29]. Since hydrophobicity of a substrate is very important, as described above, it is assumed that hydrophilic substrates are difficult to handle in water.

Hydroxymethylation of silicon enolate **9** with 1 equiv. of 36% aqueous formaldehyde solution (aq. HCHO) was performed in the presence of 2 mol% of Sc(DS)$_3$ at 1.0 M at 20 °C for 1 h. Although the reaction proceeded, the yield of the desired hydroxymethylated ketone **10** was very low. Increasing the amount of aq. HCHO from 1 to 3 equiv. improved the yield to 25%; however, 5 and 10 equiv. of aq. HCHO did not improve the yield further (Scheme 2.17, Figure 2.1a). It was also found that catalyst loading, time, and concentration affected the yield of **10** (Scheme 2.17, Figure 2.1b and c). These results indicate that the population of HCHO in the hydrophobic area may increase in the presence of Sc(DS)$_3$ due to Lewis acid–Lewis base interaction between Sc(DS)$_3$ and HCHO. Under optimized conditions [5 mol% of Sc(DS)$_3$, 1 M, 6 h], **10** was obtained in 87% yield. Furthermore, the hydroxymethylation of various silyl enol ethers proceeded smoothly (Scheme 2.18). Consequently,

**Scheme 2.17**

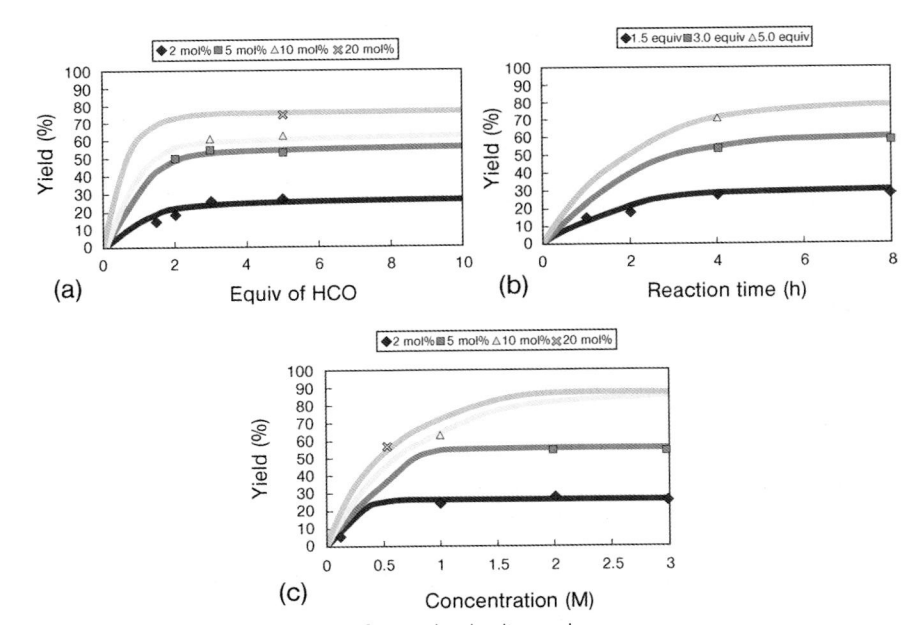

**Figure 2.1** (a) Hydroxymethylation of **9** (catalyst loading and HCHO equivalents; reaction concentration was 1.0 M, reaction time was 1 h). (b) Hydroxymethylation of **9** (reaction time and HCHO equivalents; reaction concentration was 1.0 M).
(c) Hydroxymethylation of **9** (catalyst loading and concentration; reaction time was 8 h).

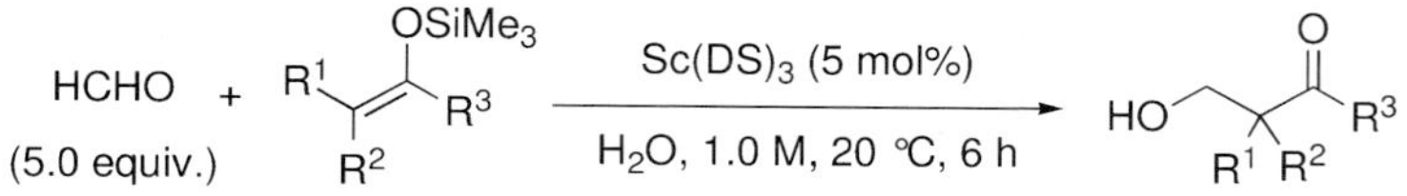

$$\text{HCHO} + \underset{R^2}{\overset{R^1}{\diagdown}}\!\!=\!\!\underset{R^3}{\overset{OSiMe_3}{\diagup}} \xrightarrow[\text{H}_2\text{O, 1.0 M, 20 °C, 6 h}]{\text{Sc(DS)}_3\ (5\ \text{mol\%})} \text{HO}\!\!-\!\!\underset{R^1\ R^2}{\overset{O}{\diagdown}}\!\!-\!\!R^3$$

(5.0 equiv.)

| Entry | Silicon enolate | Yield (%)[a] | Entry | Silicon enolate | Yield (%)[a] |
|-------|----------------|--------------|-------|----------------|--------------|
| 1 | 9 | 87 | 5 | 12 | 75 |
| 2 | 10 | 88 | 6 | 13 | 68 |
| 3 | 2 | 89 | 7 | 14 | 69[b,c] |
| 4 | 11 | 94 | | | |

[a] Isolated yield.
[b] Portionwise addition of silicon enolate (0 h and 3 h, 0.5 equiv each).
[c] Yield was determined after benzoylation.

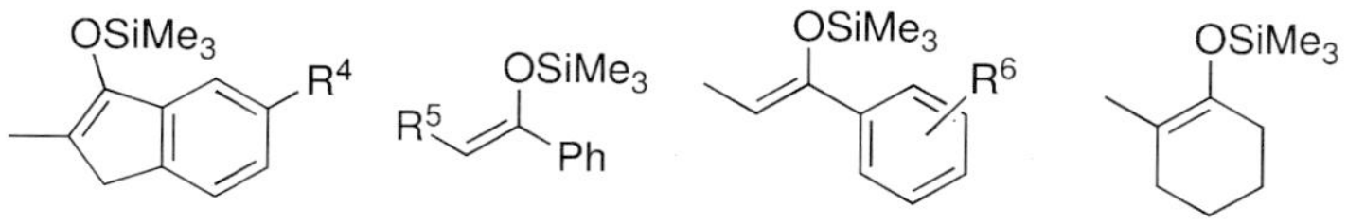

9  R⁴ = Me
10 R⁴ = H

2  R⁵ = Me
11 R⁵ = Et

12 R⁶ = p-Cl
13 R⁶ = p-OMe

14

**Scheme 2.18**

these experiments suggest that the LASC reaction system can be applied to both hydrophilic and hydrophobic substrates.

As discussed, asymmetric reactions in water and using hydrophilic substrates in water are assumed to be difficult to realize. Therefore, the development of chiral Lewis acid-catalyzed hydroxymethylation using aq. HCHO in water would impact greatly on the following research on green acid catalysis in water.

Encouraged by promising results in Scheme 2.16, an asymmetric version of hydroxymethylation using aq. HCHO was investigated. It was found that the addition of Triton X-705 suppressed the competitive hydrolysis of silicon enolates. After optimization of the reaction conditions, catalytic asymmetric hydroxymethylation reactions were successfully carried out in the presence of 10 mol% of Sc(DS)$_3$ and 12 mol% of chiral bipyridine **8a** in the presence of Triton X-705 to afford the desired products in high yields with high selectivities (Scheme 2.19). Two chiral ligands, **8a** and **8b**, were employed; **8b** gave better selectivities in some cases (entries 2, 5, 11, 14, and 17). Using 5 mol% of the catalyst combined with 6 mol% of ligand **8a**, the same levels of yield and selectivity were retained (entry 8). It is noteworthy that thioketene silyl acetals such as **20** and **21** are tolerant under these conditions even though it is known that thioketene silyl acetals (thioester-derived silicon enolates) are much less stable than silyl enol ethers (ketone-derived silicon enolates) in water. Changing the trimethylsilyl to a dimethylsilyl group on the silicon atom of silyl enolates improved

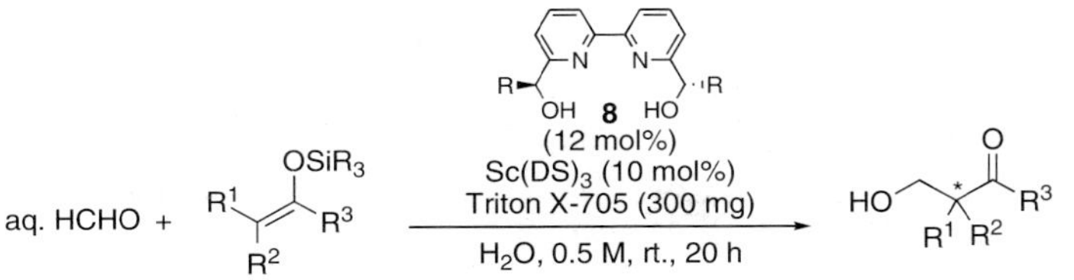


| Entry | Enolate | Yield (%)[a] | *ee* (%)[b] |
|---|---|---|---|
| 1 | **9** | 81 | 91 (*R*) |
| 2[c] | **10** | 79 | 81 |
| 3 | **10** | 69 | 83 |
| 4 | **2** | 72 | 80 (*S*) |
| 5[c] | **2** | 66 | 86 (*S*) |
| 6[d] | **22** | 70 | 89 (*S*) |
| 7 | **22** | 58 | 90 |
| 8[e] | **22** | 58 | 89 |
| 9 | **15** | 73 | 90 |
| 10 | **12** | 71 | 75 |
| 11[c] | **12** | 64 | 79 |
| 12[d] | **16** | 65 | 85 |
| 13 | **13** | 65 | 79 |
| 14[c] | **13** | 54 | 74 |
| 15[d] | **17** | 53 | 79 |
| 16 | **18** | 65 | 90 |
| 17[c] | **18** | 62 | 90 |
| 18 | **14** | 64[f] | 74 (*S*) |
| 19[d] | **19** | 56[f] | 79 (*S*) |
| 20 | **20** | 53 | 91 (*S*) |
| 21 | **21** | 48 | 90 |

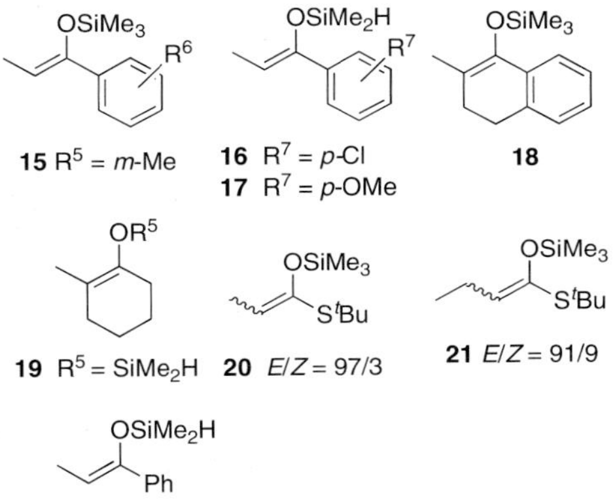

**15** $R^5$ = *m*-Me **16** $R^7$ = *p*-Cl **17** $R^7$ = *p*-OMe **18**

**19** $R^5 = SiMe_2H$ **20** *E*/*Z* = 97/3 **21** *E*/*Z* = 91/9

**22**

**8a**: R = *t*Bu.
[a] Isolated yield.
[b] Determined by chiral HPLC analysis.
[c] **8b**: R = Diethylethyl.
[d] 2,2'-Bipyridine (12 mol%) was added. Divided addition of silicon enolate (0 h and 8 h, 0.5 equiv each).
[e] 5 mol% of catalyst was used.
[f] The yield was determined after benzoylation of alcohol.
[g] Determined by chiral HPLC of it's benzoate.

**Scheme 2.19**

the selectivities, although dimethylsilyl enolates are less stable than trimethylsilyl enolates. After tuning the reaction conditions, it was found that by adding the dimethylsilyl enolate in portions and 2,2′-bipyridine as an additive, eventually high yields and enantioselectivities were obtained (entries 6, 12, 15, and 19).

This method could be applied to the synthesis of an artificial odorant **23** [30]. A synthetic route to **23** is summarized in Scheme 2.20.

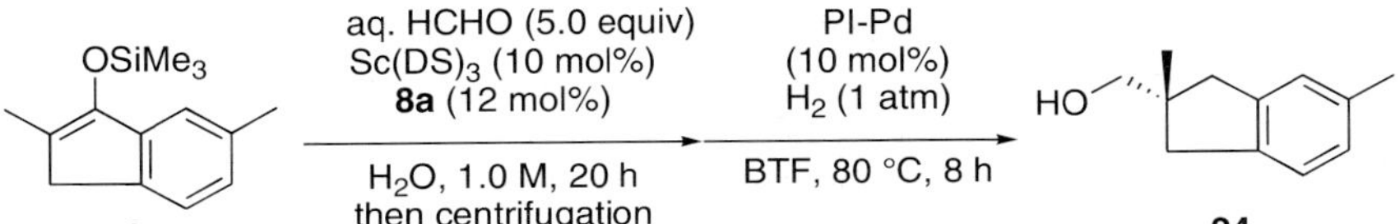


**9**

**24**
56%, 91% *ee*

**Scheme 2.20**

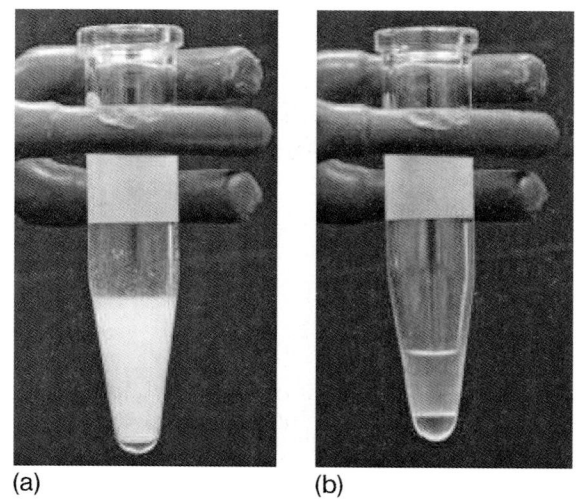

**Figure 2.2** (a) Before centrifugation of the hydroxymethylation reaction mixture. (b) Separated phases after centrifugation of the hydroxymethylation reaction mixture. Upper phase, water and formaldehyde; middle phase, Sc(DS)$_3$; bottom phase, mixture of organic compounds.

Hydroxymethylation of **23** was performed using Sc(DS)$_3$–chiral bipyridine **8a** as a catalyst. After the reaction, the reaction mixture was centrifuged (3000 rpm, 20 min) to separate the colloidal white dispersion into three phases (Figure 2.2a and b). The upper phase was a water layer, the middle phase was surfactant and the bottom phase contained organic compounds. The bottom phase was removed and the organic mixture was hydrogenated in the presence of polymer incarcerated palladium (PI-Pd) [31] in benzotrifluoride (BTF) to give compound **24** in 56% yield with 91% *ee* over two steps. It should be noted that whole process of asymmetric hydroxymethylation was performed in water alone. Furthermore, in both reactions the catalysts could be recovered and reused.

As a preliminarily study, the direct hydroxymethylation of *tert*-butyl 2-oxocyclopentanecarboxylate using aq. HCHO was investigated. It was found that in the presence of 5 mol% of Sc(DS)$_3$–**8a**, *tert*-butyl 2-oxocyclopentanecarboxylate **25** reacted with aq. HCHO in water to afford the corresponding hydroxymethylated compound. The compound was converted to its benzoate **26** in 84% yields with 49% *ee* in two steps [32].

**Scheme 2.21**

## 2.4
### Brønsted Acid Catalysis in Pure Water

In contrast to Lewis acids, Brønsted acids are easy to handle, since they are, in general, stable in water. Several types of Brønsted acid-catalyzed organic reactions in water have been investigated.

### 2.4.1
#### Surfactant-type Brønsted Acid Catalysts

Dehydration is one of the most fundamental, but still very important reactions in organic synthesis and also in biology. A representative dehydration is an esterification reaction. Esterification of carboxylic acids is a very simple and very useful reaction in synthetic organic chemistry [33]. Generally, direct esterification of carboxylic acids with alcohols is carried out in organic solvents and needs either of two methods to shift the equilibrium between reactants and products. One is removal of water (azeotropically or using dehydrating agents) generated as reactions proceed, and the other is the use of large excess amounts of one of the reactants. On the other hand, a new concept has demonstrated that the esterification could be carried out even in water without using a large excess of reactants (Scheme 2.22). The direct esterification is a dehydration step, and it is remarkable that dehydration reactions in water proceeded smoothly in the presence of a catalytic amount of a Brønsted acid such as dodecylbenzenesulfonic acid (DBSA).

$$CH_3(CH_2)_{10}CO_2H + CH_3(CH_2)_{13}OH \xrightarrow[H_2O,\ 40\ ^\circ C,\ 48\ h]{DBSA\ (10\ mol\%)} CH_3(CH_2)_{10}CO_2(CH_2)_{13}CH_3$$

$>99\%$ Yield

**Scheme 2.22**

The key to the success of this esterification is attributed to the fact that the surfactant-type catalysts and organic substrates (carboxylic acids and alcohols) in water would form droplets whose interior is hydrophobic. The surfactants would concentrate a catalytic species such as proton on the droplets' surfaces, where the reaction takes place.

When a 1 : 1 mixture of lauric acid and acetic acid was esterified with dodecanol in the presence of DBSA in water, the laurate ester was predominantly obtained in 81% yield, whereas the laurate ester was obtained in 63% yield in neat conditions (Scheme 2.23).

This unique selectivity is attributed to the hydrophobic nature of lauric acid and to the high hydrophilicity of acetic acid. This result led to the investigation of transesterification (Schemes 2.24 and 2.25), etherification (Scheme 2.26) and thioacetalization (Scheme 2.27) in water [34]. It is noted again that dehydration proceeded smoothly in water in these reactions.

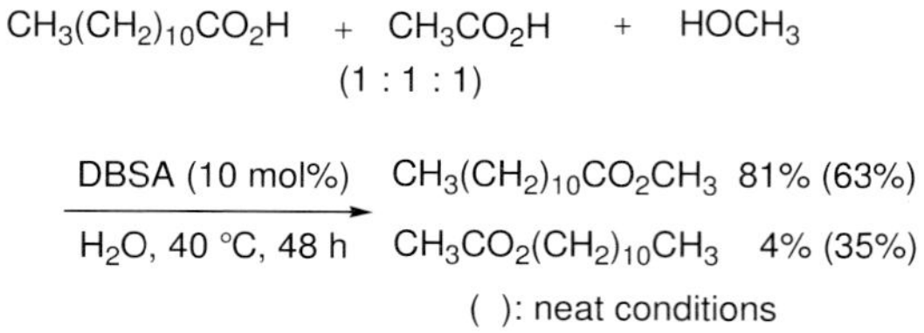

$$CH_3(CH_2)_{10}CO_2H \ + \ CH_3CO_2H \ + \ HOCH_3$$
$$(1:1:1)$$

$$\xrightarrow[\text{H}_2\text{O, 40 °C, 48 h}]{\text{DBSA (10 mol\%)}} \ \begin{array}{l} CH_3(CH_2)_{10}CO_2CH_3 \ \ 81\% \ (63\%) \\ CH_3CO_2(CH_2)_{10}CH_3 \ \ 4\% \ (35\%) \end{array}$$

( ): neat conditions

**Scheme 2.23**

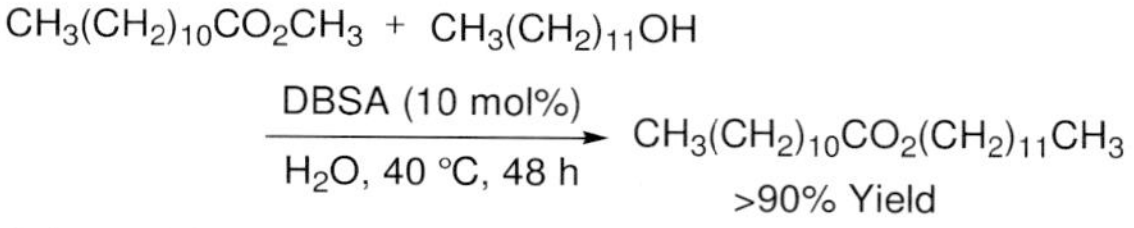

$$CH_3(CH_2)_{10}CO_2CH_3 \ + \ CH_3(CH_2)_{11}OH$$

$$\xrightarrow[\text{H}_2\text{O, 40 °C, 48 h}]{\text{DBSA (10 mol\%)}} \ CH_3(CH_2)_{10}CO_2(CH_2)_{11}CH_3$$
$$>90\% \text{ Yield}$$

**Scheme 2.24**

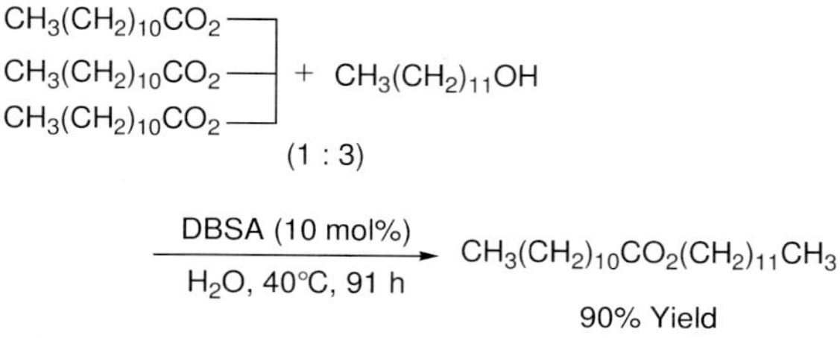

$$\left. \begin{array}{l} CH_3(CH_2)_{10}CO_2 \\ CH_3(CH_2)_{10}CO_2 \\ CH_3(CH_2)_{10}CO_2 \end{array} \right] \ + \ CH_3(CH_2)_{11}OH$$
$$(1:3)$$

$$\xrightarrow[\text{H}_2\text{O, 40°C, 91 h}]{\text{DBSA (10 mol\%)}} \ CH_3(CH_2)_{10}CO_2(CH_2)_{11}CH_3$$
$$90\% \text{ Yield}$$

**Scheme 2.25**

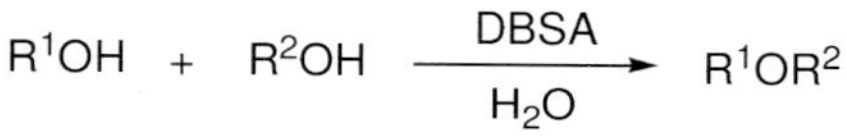

$$R^1OH \ + \ R^2OH \ \xrightarrow[\text{H}_2\text{O}]{\text{DBSA}} \ R^1OR^2$$

**Scheme 2.26**

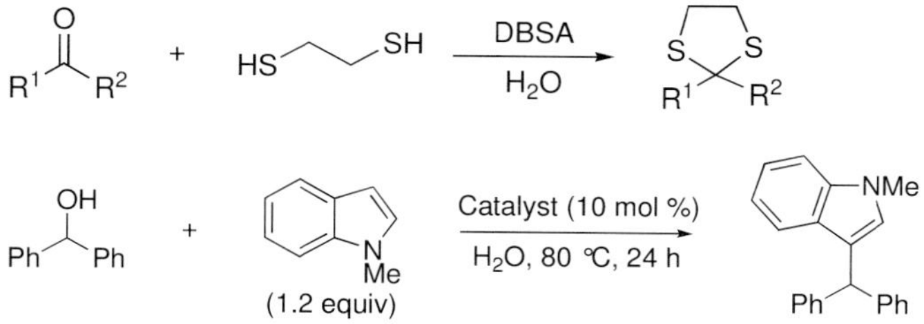

[Reaction scheme: R¹C(O)R² + HSCH₂CH₂SH → 1,3-dithiolane, DBSA/H₂O]

[Reaction scheme: Ph₂CHOH + N-methylindole (1.2 equiv) → 3-(diphenylmethyl)-N-methylindole (27), Catalyst (10 mol %), H₂O, 80 °C, 24 h]

| Entry | Catalyst | Yield (%) |
|---|---|---|
| 1 | none | 0 |
| 2 | AcOH | 0 |
| 3 | TFA | 3 |
| 4 | TfOH | 8 |
| 5 | TsOH (4-CH$_3$-C$_6$H$_4$-SO$_3$H) | 3 |
| 6 | DBSA (C$_{12}$H$_{25}$-C$_6$H$_4$-SO$_3$H) | 85 |
| 7 | C$_9$H$_{19}$COOH | 0 |

**Scheme 2.27**

C—C bond formation reactions via dehydrations in water have been investigated. Coupling reactions of alkyl halides with nucleophiles are among the most useful methods for C—C bond formation in organic synthesis. The use of alcohols as substrates instead of alkyl halides would be very valuable from the viewpoint of atom efficiency, since water is the sole by-product. However, the catalytic activation of alcohols is difficult due to the poor leaving ability of the hydroxyl group, and as a result an excess amount of a Brønsted acid or a stoichiometric amount of a Lewis acid is often required to promote the reactions [35]. Therefore, the development of nucleophilic substitutions of alcohols using catalytic amounts of Brønsted or Lewis acids is highly desirable [36].

The catalytic nucleophilic substitution of benzyl alcohols with various carbon nucleophiles in water has been studied. Several Brønsted acids were screened as catalysts in the Friedel–Crafts-type substitution reaction of benzhydrol with 1-methylindole as model substrates in water (Scheme 2.28). Commonly used Brønsted acids such as AcOH, TFA, TfOH and TsOH were not effective for this reaction (entries 2–5). On the other hand, a surfactant-type Brønsted acid such as DBSA was found to catalyze the reaction efficiently to give the product **27** in good yield (entry 6).

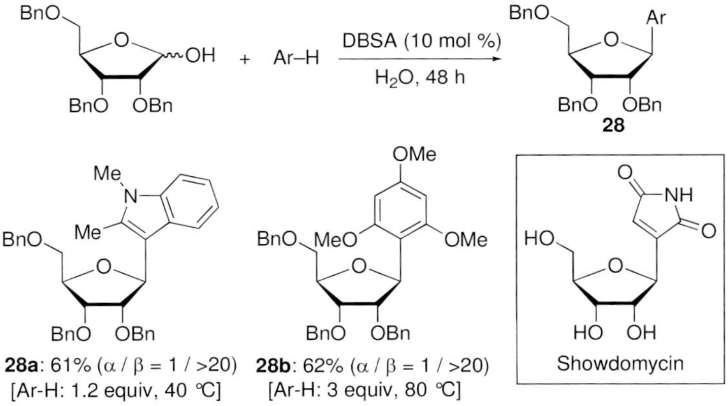

**Scheme 2.28**

As an extension of DBSA-catalyzed C—C bond formation in water, C-glycosylations of 1-hydroxy sugar in water have been investigated [37]. The reaction of 1-hydroxy-D-ribofuranose with electron-rich heteroaromatic or aromatic compounds proceeded smoothly in the presence of 10 mol% of DBSA to afford the corresponding C-nucleosides **28a** [38] and **28b** [39] in good yields with excellent β-selectivity (Scheme 2.29). The synthesis of C-nucleoside **28b** is of particular note as it is a key intermediate in the total synthesis of showdomycin [40].

Several Brønsted acids were tested as catalysts in a model Mannich-type reaction in water, and it was found that DBSA afforded the desired product in high yield (Scheme 2.30). It should be noted that *p*-toluenesulfonic acid, which has a shorter alkyl chain than DBSA, gave only a trace amount of the product. Based on this result, it was suggested that the long alkyl chain of the acid was indispensable for efficient

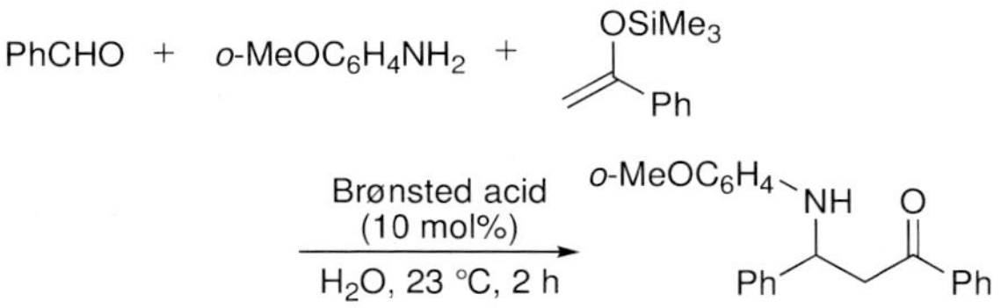

PhCHO + o-MeOC$_6$H$_4$NH$_2$ + CH$_2$=C(OSiMe$_3$)Ph

$\xrightarrow[\text{H}_2\text{O, 23 °C, 2 h}]{\text{Brønsted acid (10 mol\%)}}$ o-MeOC$_6$H$_4$-NH-CH(Ph)-CH$_2$-C(O)-Ph

| Brønsted acid | Yield (%) |
|---|---|
| DBSA | 83 |
| TsOH | Trace |
| C$_{11}$H$_{25}$COOH | 6 |

**Scheme 2.29**

catalysis. probably due to the formation of hydrophobic reaction environments, and that the strong acidity of DBSA was essential for the catalysis because a carboxylic acid having a long alkyl chain, lauric acid, was much less effective than DBSA.

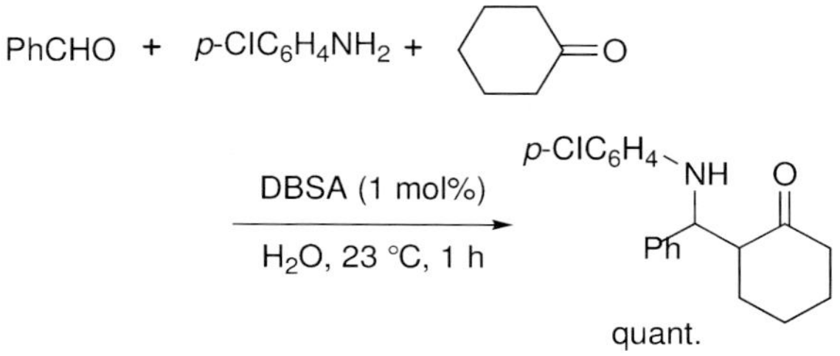

PhCHO + p-ClC$_6$H$_4$NH$_2$ + cyclohexanone

$\xrightarrow[\text{H}_2\text{O, 23 °C, 1 h}]{\text{DBSA (1 mol\%)}}$ p-ClC$_6$H$_4$-NH-CH(Ph)-(2-oxocyclohexyl)

quant.

**Scheme 2.30**

From atom economical and environmental points of view, it is desirable to develop a new, efficient system for Mannich-type reactions in which the parent carbonyl compounds are used directly [41]. Remarkably, DBSA catalyzes Mannich-type reactions in a colloidal dispersion system using ketones as nucleophilic components (Scheme 2.31) [42].

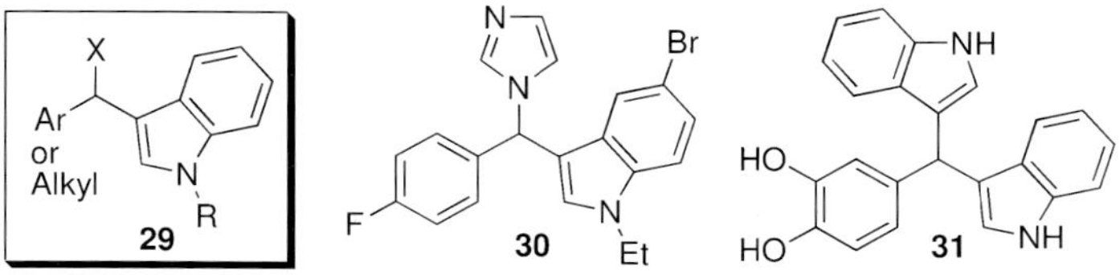

**Scheme 2.31**

3-Substituted indoles of type **29** have attracted much attention due to the broad scope of their biological activity (Scheme 2.32) [43]. For example, indole derivatives such as **29** and **30** were reported in a recent study of medicinal chemistry to act as a non-steroidal aromatase inhibitor which decreases the production of sex hormones (estrogen or testosterone) and slows the growth of tumors that need sex hormones to grow; therefore, these analogues have been investigated for anti-breast cancer therapy [44], and also **31** was reported to work as an HIV-1 integrase inhibitor [45]. Furthermore, many natural products incorporating this structural key element are known [46]. As a result of their biological and synthetic importance, a variety of methods have been reported for the preparation of 3-substituted indoles [47]. Although these methods constitute a valuable addition to the chemical literature, a truly efficient, diverse synthetic scheme [48] for type **29** indoles is still required in order to construct further indole libraries for medicinal chemistry.

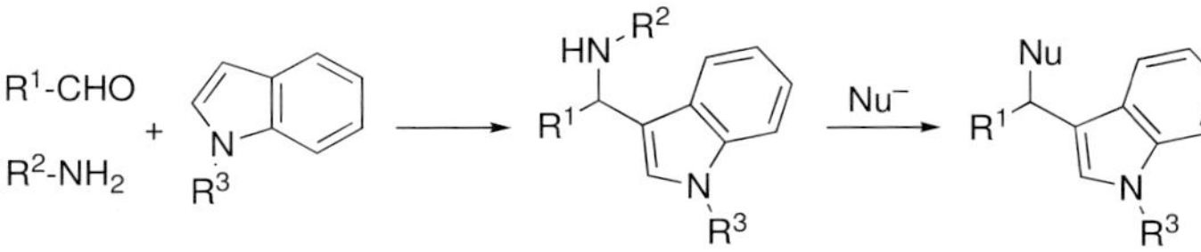

**Scheme 2.32**

To achieve these 3-substituted indoles efficiently, the synthetic route was designed as shown in Scheme 2.33. Three-component AFC reactions of aromatic aldehydes, primary amines, and indoles are unknown, since the initial product **32** is highly reactive and further addition of indoles gives undesired adduct **33** [49, 51]; a few examples of selective AFC reactions of indoles to aryl imines have been reported [50].

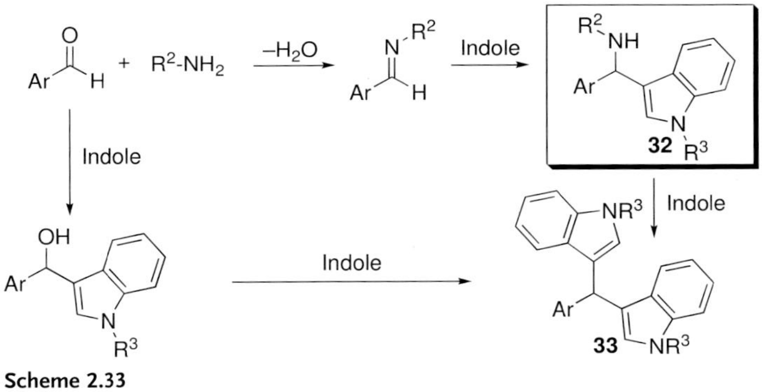

**Scheme 2.33**

To address these issues, selective three-component AFC reactions with aromatic aldehydes, primary amines, and indoles have been investigated (Scheme 2.34). Whereas AcOH and TFA did not show any catalytic ability for this reaction (entries 2 and 3), Sc(DS)$_3$ and DBSA promoted the reaction effectively; however, a certain amount of undesired adduct **35** was formed (entries 4 and 5). It was revealed that decanoic acid (C$_9$H$_{19}$COOH) efficiently promoted the reaction to afford the desired compound **34** predominantly without the formation of **35** (entry 6). Interestingly, this efficient catalysis occurred sluggishly in organic solvents such as CH$_2$Cl$_2$ and THF (entry 6, in parentheses). Furthermore, the length of the alkyl chains of the carboxylic acids was found to be crucial for this reaction. Decanoic acid (C$_9$H$_{19}$COOH) gave the best result, and it was noted that the use of carboxylic acids with shorter or longer alkyl

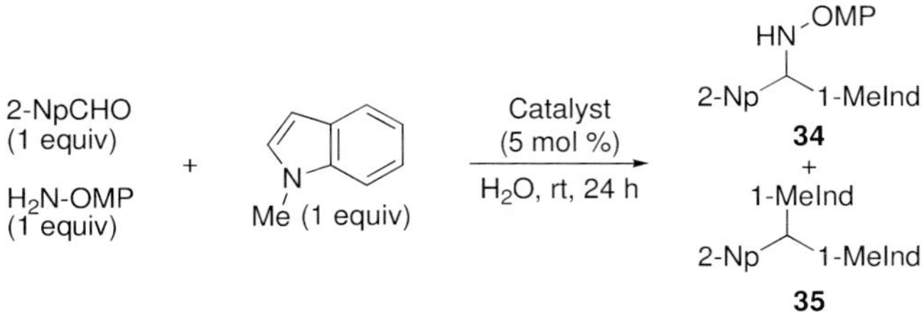

2-NpCHO (1 equiv) + [indole: Me (1 equiv)] + H₂N-OMP (1 equiv) →[Catalyst (5 mol %), H₂O, rt, 24 h] 

2-Np-CH(HN-OMP)-1-MeInd  **34**
+
1-MeInd
2-Np-CH(1-MeInd)-1-MeInd  **35**

| Entry | Catalyst | 34 / 35$^a$ | Yield of 34 (%)$^a$ |
|---|---|---|---|
| 1 | none | >20 : 1 | 5 |
| 2 | AcOH | >20 : 1 | 5 |
| 3 | TFA | >20 : 1 | 6 |
| 4 | Sc(DS)$_3$ | 3.3 : 1 | 46 |
| 5 | DBSA | 5.2 : 1 | 65 |
| 6 | C$_9$H$_{19}$COOH | >20 : 1 | 80 (3$^b$, 4$^c$) |
| 7 | C$_5$H$_{11}$COOH | >20 : 1 | 6 |
| 8 | C$_7$H$_{15}$COOH | >20 : 1 | 43 |
| 9 | C$_8$H$_{17}$COOH | >20 : 1 | 63 |
| 10 | C$_{10}$H$_{21}$COOH | >20 : 1 | 78 |
| 11 | C$_{11}$H$_{23}$COOH | >20 : 1 | 45 |
| 12 | C$_{13}$H$_{27}$COOH | >20 : 1 | 28 |
| 13 | C$_9$H$_{19}$COOH$^d$ | >20 : 1 | 91 |

$^a$ Determined by $^1$H NMR analysis.
$^b$ Reaction in CH$_2$Cl$_2$.
$^c$ Reaction in THF.
$^d$ 10 mol% of catalyst was used.

**Scheme 2.34**

chains resulted in lower yields (entries 7–12). Finally, it was discovered that increasing the catalyst loading could further improve the yield of the desired product (entry 13).

Due to the high reactivity of the C−N bond, AFC product **34** was readily converted to various nucleophiles in the presence of DBSA or Sc(OTf)$_3$ as a catalyst (Scheme 2.35). Friedel–Crafts-type substitution reactions occurred cleanly by treatment with electron-rich heteroaromatic or aromatic compounds afford to the unsymmetrical triarylmethanes **36** and **37**, respectively [50]. Substitution using thiol, allyltin, and silyl enol ether nucleophiles also proceeded smoothly, and gave the valuable compounds **38**, **39**, and **40** in good yields. Furthermore, when **34** was treated with Sc(OTf)$_3$ without any nucleophiles in toluene, triarylmethane **40** was obtained via cleavage of the C−N bond of **34** followed by Friedel–Crafts-type addition of the cleaved o-anisidine.

## 2.4.2
### Polymer-supported Brønsted Acid Catalysts

Styrene-based polymer is recognized to provide highly hydrophobic physical properties. This character can be applied to organic reactions in water. It was found that

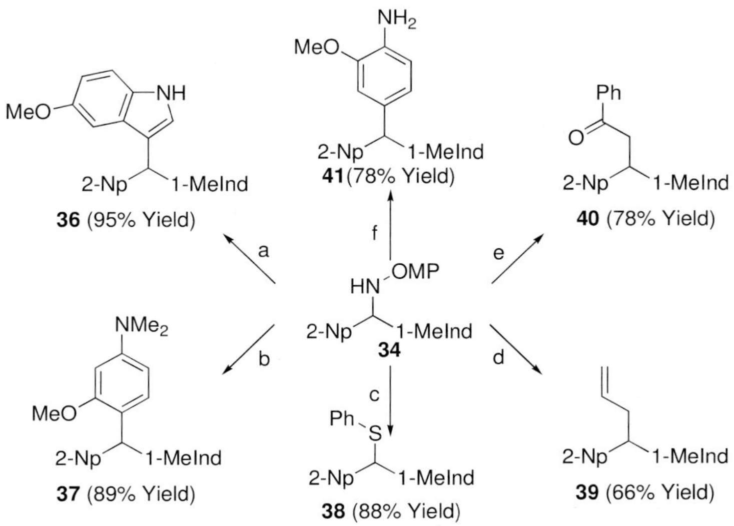


Conditions : (a) 5-MeO-Indole (1.2 equiv), DBSA (10 mol%), $H_2O$, rt, 24 h; (b) *N,N*-Dimethyl-*m*-anisidine (1.2 equiv), $Sc(OTf)_3$ (10 mol%), Toluene, 70 °C, 24 h; (c) PhSH (3 equiv), $Sc(OTf)_3$ (10 mol%), Toluene, 70 °C, 24 h; (d) $Sn(CH_2CH{=}CH_2)_4$ (1.2 equiv), $Sc(OTf)_3$ (10 mol%), Toluene, 70 °C, 24 h; (e) $Me_3SiO(Ph)C{=}CH_2$ (2 equiv), $Sc(OTf)_3$ (10 mol%), Toluene, 70 °C, 24 h; (f) $Sc(OTf)_3$ (10 mol%), Toluene, 70 °C, 24 h.

**Scheme 2.35**

dehydrative hydrophobic polystyrene-supported sulfonic acids catalyzed the esterification of lauric acid with 3-phenyl-1-propanol in water effectively, whereas commercially available Dowex 500W-X2 (H form) (**42**) did not (Scheme 2.36) [52]. This result indicates that the highly hydrophobic nature of polymer-supported catalysts plays an

$$C_{11}H_{23}COOH + HO(CH_2)_3Ph \xrightarrow[H_2O,\ 40\ °C,\ 24\ h]{catalyst} C_{11}H_{23}COO(CH_2)_3Ph$$

(1 : 1)

| Entry | Catalyst (equiv.) | Yield (%)[a] |
|---|---|---|
| 1 | DBSA (0.10) | 60 (83, 85[b])[c] |
| 2 | **42** (0.10) | 0 |
| 3 | **42** (0.10) | 2 |
| 4 | **43** (0.10) | 41 |
| 5 | **44** (0.10) | 74 (81, 84[b])[c] |
| 6 | **45** (0.10) | 72 |

[a] Isolated yield. [b] NMR yield. [c] For 120 h.

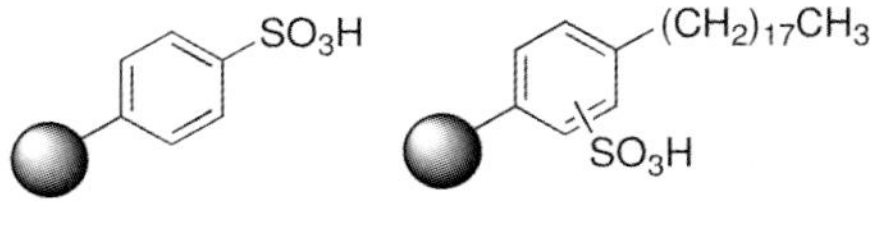


**43** (0.879 mmol/g)
**44** (0.352 mmol/g)
**45** (1.01 mmol/g)

**Scheme 2.36**

important role for activity in the dehydration reaction in water. It was observed that resin **42** swelled significantly in water due to its high sulfonic acid content. On the other hand, both **44** and **45** scarcely swelled in water but worked as efficient catalysts. Moreover, resin **44** was easily recovered by simple filtration after the esterification was completed, and the catalyst was continuously reused at least four times without loss of catalytic activity (Scheme 2.37).

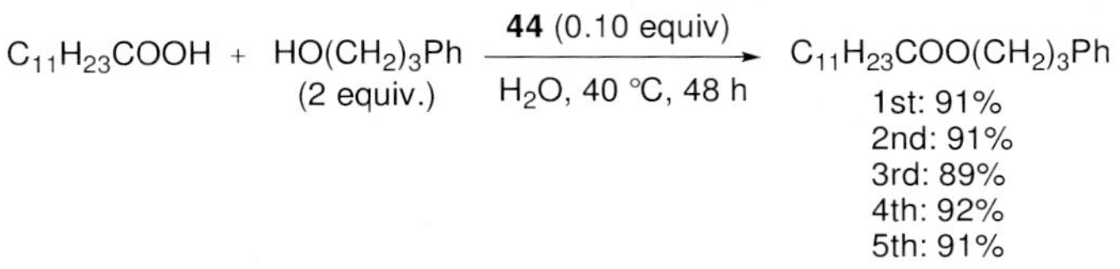

$$C_{11}H_{23}COOH + HO(CH_2)_3Ph \xrightarrow[H_2O,\ 40\ °C,\ 48\ h]{44\ (0.10\ equiv)\ (2\ equiv.)} C_{11}H_{23}COO(CH_2)_3Ph$$

1st: 91%
2nd: 91%
3rd: 89%
4th: 92%
5th: 91%

**Scheme 2.37**

Three-component Mannich-type reactions of aldehydes, amines, and silicon enolates also proceeded smoothly using PS-SO$_3$H in water (Scheme 2.38) [53]. In general, ketene silyl acetals are known to be easily hydrolyzed in the presence of water; however, such water-labile compounds could be successfully used in this reaction. Moreover, a remarkable effect of the loading levels of the polystyrene-supported sulfonic acid on yields was observed. It was suggested that the hydrophobic environments created by the catalyst might suppress hydrolysis of ketene silyl acetals.

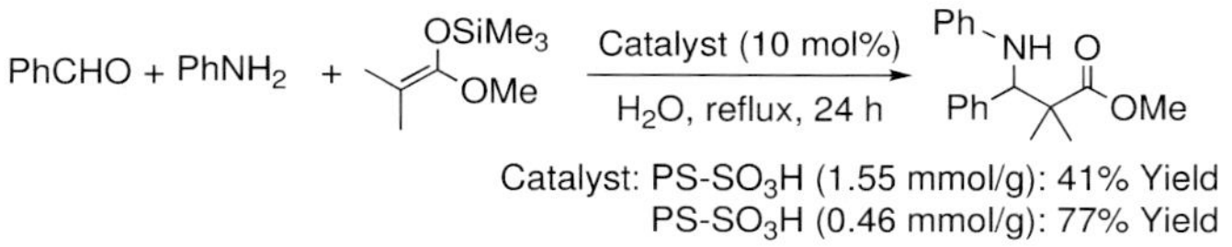

$$PhCHO + PhNH_2 + \underset{OMe}{\overset{OSiMe_3}{\diagup\!\!\!\diagdown}} \xrightarrow[H_2O,\ reflux,\ 24\ h]{Catalyst\ (10\ mol\%)} \underset{Ph}{\overset{Ph\diagdown NH\ O}{\diagup\!\!\!\diagdown}} OMe$$

Catalyst: PS-SO$_3$H (1.55 mmol/g): 41% Yield
PS-SO$_3$H (0.46 mmol/g): 77% Yield

**Scheme 2.38**

Furthermore, low-loading (e.g. 0.17 mmol g$^{-1}$) and alkylated polystyrene-supported sulfonic acids (LL-ALPSSO$_3$H) such as **46** successfully catalyzed deprotection of *tert*-butyldimethylsilyl (TBS)-protected alcohols in water without using organic co-solvents (Scheme 2.39) [54a] (a hydrophobic polymer-supported catalyst for acid-catalyzed hydrolysis of thioesters and trans-protection of thiols in water has also been reported [54b]).

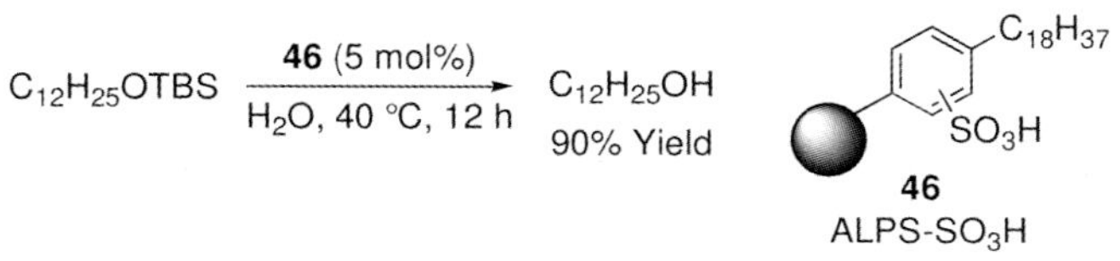

$$C_{12}H_{25}OTBS \xrightarrow[H_2O,\ 40\ °C,\ 12\ h]{46\ (5\ mol\%)} C_{12}H_{25}OH$$

90% Yield

**46** ALPS-SO$_3$H (polymer-supported C$_{18}$H$_{37}$-aryl-SO$_3$H)

**Scheme 2.39**

## 2.5
## Conclusion and Perspective

Conventionally, most organic reactions are carried out in organic solvents, since most organic materials are not soluble in other solvents. However, organic reactions in other, non-conventional solvents are now of interest from many aspects related to green sustainable chemistry. The first choice of a non-conventional solvent is water. Water is a clean, non-toxic, inexpensive, and the most environmentally friendly solvent. In addition, acid catalysis has occupied major parts of organic transformations. Therefore, acid catalysis in water as discussed here will play a key role in benign chemical processes.

As we surveyed, organic reactions in water are difficult because most organic materials are not soluble and many reactive intermediates and catalysts are not stable in water. In addressing these issues, many focused research efforts have led to rapid progress in this field, exemplified by elegant catalysis including successful asymmetric catalysis in water, which was believed to be impossible 10 years ago. Moreover, systems have been developed for recovery and reuse of catalysts utilizing biphasic reaction conditions that incorporate water.

As for the future of organic synthesis in water, several important developments are predicted. Stereoselective C—C bond-forming reactions are still an important challenge in organic synthesis, and in particular, control of the stereogenic centers of products is crucial. In this respect, the development of catalytic asymmetric reactions in water is still a major task, which represents significant challenges since most asymmetric catalysts, except for some late transition metal-based systems, are not stable in water. As such, the design of water-compatible catalysts will be the key to the future of this work. Aerobic oxidation in water is an important research target for the future. It will not be necessary to mention the importance of environmentally benign oxidation processes to the present readership, but due to the high heat capacity and stability of water, the process is really promising. The use of small water-soluble molecules, such as formaldehyde and ammonia, will be another critical topic. These molecules are inexpensive and potentially useful carbon and nitrogen building blocks; however, they are not well utilized in conventional organic synthesis in organic solvents.

Water is stunning in Nature; indeed, Nature chooses water as a "solvent." Many elegant *in vitro* reactions, mainly catalyzed by enzymes, are carried out in an aqueous environment in our bodies. Given that Nature so gracefully exploits water, why should humans not perform syntheses in water too?

## References

1 Kobayashi, S. (1991) *Chem. Lett.*, 2187.
2 (a) Kobayashi, S. (1999) Lanthanide triflate-catalyzed carbon–carbon bond-forming reactions in organic synthesis. in *Lanthanides: Chemistry and Use in Organic Synthesis* (ed. S. Kobayashi), Springer, Heidelberg, p. 63; (b) Kobayashi, S. (1999) *Eur. J. Org. Chem.*, 15; (c) Kobayashi, S. (1994) *Synlett*, 689; (d) Kobayashi, S., Sugiura, M., Kitagawa, H. and Lam, W.W.-L. (2002) *Chem. Rev.*, **102**, 2227.

3. (a) Kobayashi, S. and Hachiya, I. (1992) *Tetrahedron Lett.*, **33**, 1625; (b) Kobayashi, S. and Hachiya, I. (1994) *J. Org. Chem.*, **59**, 3590.
4. Kobayashi, S., Nagayama, S. and Busujima, T. (1998) *J. Am. Chem. Soc.*, **120**, 8287.
5. Manabe, K., Mori, Y., Wakabayashi, T., Nagayama, S. and Kobayashi, S. (2000) *J. Am. Chem. Soc.*, **122**, 7202.
6. Kobayashi, S., Busujima, T. and Nagayama, S. (1999) *Synlett*, 545.
7. Manabe, K. and Kobayashi, S. (2000) *Chem. Commun.*, 669.
8. Mori, Y., Kakumoto, K., Manabe, K. and Kobayashi, S. (2000) *Tetrahedron Lett.*, **41**, 3107.
9. Manabe, K., Aoyama, N. and Kobayashi, S. (2001) *Adv. Synth. Catal.*, **343**, 174.
10. Aoyama, N., Manabe, K. and Kobayashi, S. (2004) *Chem. Lett.*, **33**, 312.
11. Nagayama, S. and Kobayashi, S. (2000) *Angew. Chem. Int. Ed.*, **39**, 567.
12. Iimura, S., Manabe, K. and Kobayashi, S. (2004) *Tetrahedron*, **60**, 7673.
13. Gu, Y., Ogawa, C., Kobayashi, J., Mori, Y. and Kobayashi, S. (2006) *Angew. Chem. Int. Ed.*, **45**, 7217.
14. For reviews on asymmetric reactions in aqueous media, see: (a) Kobayashi, S. and Manabe, K. (2000) *Pure. Appl. Chem.*, **72**, 1373; (b) Sinou, D. (2002) *Adv. Synth. Catal.*, **344**, 221; (c) Manabe, K. and Kobayashi, S. (2002) *Chem. Eur. J.*, **8**, 4094; (d) Manabe, K. and Kobayashi, S. (2002) *Acc. Chem. Res.*, **35**, 209; (e) Kobayashi, S. and Ogawa, C. (2006) *Chem. Eur. J.*, **12**, 5954;(f) Kobayashi, S. and Ogawa, C. (2007) in *Asymmetric Synthesis* (eds M. Christmann and S. Braese), Wiley-VCH Verlag GmbH, Weinheim, p. 110;(g) Ogawa, C. and Kobayashi, S. (2007) In: *Organic Reactions in Water* (ed. U.M. Lindstroem), Blackwell Publishing Ltd., Oxford, UK, p. 60–91. (h) Kobayashi, S. (2007) *Pure. Appl. Chem.* **79**, 235; (i) Ogawa, C. and Kobayashi, S. (2008) *Process Chemistry in the Pharmaceutical Industry*, Vol. 2 (eds K. Gadamasetti and T. Braish), CRC Press, Boca Raton, FL, p. 249.
15. For reviews focused on general organic reactions in aqueous media, see: (a) Lindstroem, U.M. (2002) *Chem. Rev.*, **102**, 2751; (b) Li, C.-J. (1993) *Chem. Rev.*, **93**, 2023; (c) Li, C.-J. (2005) *Chem. Rev.*, **105**, 3095; (d) Li, C.-J. and Chen, L. (2006) *Chem. Soc. Rev.*, **35**, 68.
16. For reviews on asymmetric Mannich reactions, see: (a) Kobayashi, S. and Ishitani, H. (1999) *Chem. Rev.*, **99**, 1069; (b) Taggi, A.E., Hafez, A.M. and Lectka, T. (2003) *Acc. Chem. Res.*, **36**, 10; (c) Kobayashi, S. and Ueno, M. (2004) In: *Comprehensive Asymmetric Catalysis* (eds E.N. Jacobsen, A. Pfaltz, and H. Yamamoto), Springer, Berlin, Supplement 1, Chapter 29. 5, pp. 143–159.
17. Hamada, T., Manabe, K. and Kobayashi, S. (2004) *J. Am. Chem. Soc.*, **126**, 7768.
18. In the previous report, the additive was necessary to promote the reactions, see: (a) Kobayashi, S., Hamada, T. and Manabe, K. (2002) *J. Am. Chem. Soc.*, **124**, 5640; (b) Hamada, T., Manabe, K. and Kobayashi, S. (2006) *Chem. Eur. J.*, **12**, 1205.
19. Shirakawa, S. and Kobayashi, S. (2006) *Synlett*, 1410.
20. For reviews on the asymmetric synthesis and use of vicinal amino alcohols, see: (a) Ager, D.J., Prakash, I. and Schaad, D.R. (1996) *Chem. Rev.*, **96**, 835; (b) Bergmeier, S.C. (2000) *Tetrahedron*, **56**, 2561; (c) Yamashita, M., Yamada, K. and Tomioka, K. (2005) *Org. Lett.*, **7**, 2369; (d) Kolb, H.C. and Sharpless, K.B. (1998) In: *Transition Metals for Organic Synthesis* (eds M. Beller and C. Bolm), Wiley-VCH Verlag GmbH, Weinheim, p. 243.
21. (a) Hou, X.L., Wu, J., Dai, L.X., Xia, L.J. and Tang, M.H. (1998) *Tetrahedron: Asymmetry*, **9**, 1747; (b) Sagawa, S., Abe, H., Hase, Y. and Inaba, T. (1999) *J. Org. Chem.*, **64**, 4962; (c) Sekine, A., Ohshima, T. and Shibasaki, M. (2002) *Tetrahedron*, **58**, 75; (d) Schneider, C., Sreekanth, A.R. and Mai, E. (2004) *Angew. Chem. Int. Ed.*, **43**, 5691;

(e) Carrée, F., Gil, R. and Collin, J. (2005) *Org. Lett.*, **7**, 1023.

22 Ishikawa, S., Hamada, T., Manabe, K. and Kobayashi, S. (2004) *J. Am. Chem. Soc.*, **126**, 12236.

23 (a) Ogawa, C., Kizu, K., Shimizu, H., Takeuchi, M. and Kobayashi, S. (2006) *Chem. Asian J.*, **1**, 121; (b) Ogawa, C., Wang, N. and Kobayashi, S. (2007) *Chem. Lett.*, **36**, 34; (c) Ogawa, C., Wang, N., Boudou, M., Azoulay, S., Manabe, K. and Kobayashi, S. (2007) *Heterocycles*, **72**, 589.

24 For examples of racemic epoxide ring opening in water, see: (a) Iranpoor, N., Firouzabadi, H. and Shekarize, M. (2003) *Org. Biomol. Chem.*, **1**, 724; (b) Fan, R.H. and Hou, X.L. (2003) *J. Org. Chem.* **68**, 726; (c) Ollevier, T. and Lavie-Compain G. (2004) *Tetrahedron Lett.*, **45**, 49.

25 (a) Azoulay, S., Manabe, K. and Kobayashi, S. (2005) *Org. Lett.*, **7**, 4593; (b) Ogawa, C., Azoulay, S. and Kobayashi, S. (2005) *Heterocycles*, **66**, 201.

26 Boudou, M., Ogawa, C. and Kobayashi, S. (2006) *Adv. Synth. Catal.*, **348**, 2585.

27 Azoulay, S., Manabe, K. and Kobayashi, S. (2005) *Org. Lett.*, **7**, 4593.

28 Kobayashi, S., Ogino, T., Shimizu, H., Ishikawa, S., Hamada, T. and Manabe, K. (2005) *Org. Lett.*, **7**, 4729.

29 Ozasa, N., Wadamoto, M., Ishihara, K. and Yamamoto, H. (2003) *Synlett*, 2219.

30 (a) Vial, C., Bernardinelli, G., Schneider, P., Aizenberg, M. and Winter, B. (2005) *Helv. Chim. Acta*, **88**, 3109; (b) Winter, B. and Gallo-Flückiger S. (2005) *Helv. Chim. Acta*, **88**, 3118.

31 (a) Okamoto, K., Akiyama, R. and Kobayashi, S. (2004) *J. Org. Chem.*, **69**, 2871; (b) Akiyama, R., Sagae, T., Sugiura, M. and Kobayashi, S. (2004) *J. Organomet. Chem.*, **689**, 3806; (c) Okamoto, K., Akiyama, R., Yoshida, T. and Kobayashi, S. (2005) *J. Am. Chem. Soc.*, **127**, 2125.

32 Kokubo, M., Ogawa, C. and Kobayashi, S. (2008) *Angew. Chem. Int. Ed.*, **47**, 6909.

33 (a) Manabe, K., Sun, X.-M. and Kobayashi, S. (2001) *J. Am. Chem. Soc.*, **123**, 10101;

(b) Manabe, K., Iimura, S., Sun, X.-M. and Kobayashi, S. (2002) *J. Am. Chem. Soc.*, **124**, 11971, and references therein.

34 Kobayashi, S., Iimura, S. and Manabe, K. (2002) *Chem. Lett.*, 10.

35 For example: (a) Gullickson, G.C. and Lewis, D.E. (2003) *Aust. J. Chem.*, **56**, 385; (b) Bisaro, F., Prestat, G., Vitale, M. and Poli, G. (2002) *Synlett*, 1823; (c) Coote, S.J., Davies, S.G., Middlemiss, D. and Naylor, A. (1989) *Tetrahedron Lett.*, **30**, 3581; (d) Khalaf, A.A. and Roberts, R.M. (1972) *J. Org. Chem.*, **37**, 4227.

36 For recent example, see: (a) Motokura, K., Fujita, N., Mori, K., Mizugaki, T., Ebitani, K. and Kaneda, K. (2006) *Angew. Chem. Int. Ed.*, **45**, 2605; (b) Yasuda, M., Somyo, T. and Baba, A. (2006) *Angew. Chem. Int. Ed.*, **45**, 793; (c) Zhan, Z.-P., Yang, W.-Z., Yang, R.-F., Yu, J.-L., Li, J.-P. and Liu, H.-J. (2006) *Chem. Commun.*, 3352; (d) Sanz, R., Martínez, A., Miguel, D., Álvarez-Gutiérrez J.M. and Rodríguez, F. (2006) *Adv. Synth. Catal.*, **348**, 1841; (e) Rueping, M., Nachtsheim, B.J. and Ieawsuwan, W. (2006) *Adv. Synth. Catal.*, **348**, 1033; (f) Iovel, I., Mertins, K., Kischel, J., Zapf, A. and Beller, M. (2005) *Angew. Chem. Int. Ed.*, **44**, 3913; (g) Noji, M., Ohno, T., Fuji, K., Futaba, N., Tajima, H. and Ishii, K. (2003) *J. Org. Chem.*, **68**, 9340.

37 Shirakawa, S. and Kobayashi, S. (2007) *Org Lett.*, **9**, 311.

38 Yokoyama, M., Nomura, M., Togo, H. and Seki, H. (1996) *J. Chem. Soc., Perkin Trans.*, **1**, 2145.

39 He, W., Togo, H., Waki, Y. and Yokoyama, M. (1998) *J. Chem. Soc., Perkin Trans.*, **1**, 2425.

40 (a) Stewart, A.O. and Williams, R.M. (1985) *J. Am. Chem. Soc.*, **107**, 4289; (b) Kalvoda, L., Farkas, J. and Sorm, F. (1970) *Tetrahedron Lett.*, **11**, 2297.

41 (a) Blatt, A.H. and Gross, N. (1964) *J. Org. Chem.*, **29**, 3306; (b) Yi, L., Zou, J., Lei, H., Lin, X. and Zhang, M. (1991) *Org. Prep. Proced. Int.*, **23**, 673.

42 Manabe, K. and Kobayashi, S. (1999) *Org Lett.*, **1**, 1965.

**43** (a) Sundberg, R.J. (1996) *Indoles*, Academic Press, San Diego, CA; (b) Sundberg, R.J. (1970) *The Chemistry of Indoles*, Academic Press, New York.

**44** (a) Lézé, M.-P., Le Borgne M. Marchand, P., Loquet, D., Kogler, M., Le Baut G. Palusczak, A. and Hartmann, R.W. (2004) *J. Enzym. Inhib. Med. Chem.*, **19**, 549; (b) Le Borgne M. Marchand, P., Delevoye-Seiller B. Robert, J.-M., Le Baut G. Hartmann, R.W. and Palzer, M. (1999) *Bioorg. Med. Chem. Lett.*, **9**, 333.

**45** (a) Deng, J., Sanchez, T., Neamati, N. and Briggs, J.M. (2006) *J. Med. Chem.*, **49**, 1684; (b) Contractor, R., Samudio, I.J., Estrov, Z., Harris, D., McCubrey, J.A., Safe, S.H., Andreeff, M. and Konopleva, M. (2005) *Cancer Res.*, **65**, 2890.

**46** (a) Gul, W. and Hamann, M.T. (2005) *Life Sci.*, **78**, 442; (b) Somei, M. and Yamada, F. (2005) *Nat. Prod. Rep.*, **22**, 73; (c) Lounasmaa, M. and Tolvanen, A. (2000) *Nat. Prod. Rep.*, **17**, 175.

**47** (a) Bandini, M., Melloni, A., Tommasi, S. and Umani-Ronchi A. (2005) *Synlett*, 1199; (b) Bandini, M., Melloni, A. and Umani-Ronchi A. (2004) *Angew. Chem. Int. Ed.*, **43**, 550; (c) Comins, D.L. and Stroud, E.D. (1986) *Tetrahedron Lett.*, **27**, 1869.

**48** (a) Schreiber, S.L. (2003) *Chem. Eng. News*, **81**, 51; (b) Spring, D.R. (2003) *Org. Biomol. Chem.*, **1**, 3867.

**49** (a) Ke, B., Qin, Y., He, Q., Huang, Z. and Wang, F. (2005) *Tetrahedron Lett.*, **46**, 1751; (b) Mi, X., Luo, S., He, J. and Cheng, J.-P. (2004) *Tetrahedron Lett.*, **45**, 4567; (c) Xie., W., Bloomfield, K.M., Jin, Y., Dolney, N.Y. and Wang, P.G. (1999) *Synlett*, 498.

**50** (a) Jia, Y.-X., Xie, J.-H., Duan, H.-F., Wang, L.-X. and Zhou, Q.-L. (2006) *Org. Lett.*, **8**, 1621; (b) Esquivias, J., Gómez Arrayás R. and Carretero, J.C. (2006) *Angew. Chem. Int. Ed.*, **45**, 629; (c) Wang, Y.-Q., Song, J., Hong, R., Li, H. and Deng, L. (2006) *J. Am. Chem. Soc.*, **128**, 8156.

**51** For recent examples, see: (a) Deb, M.L. and Bhuyan, P.J. (2006) *Tetrahedron Lett.*, **47**, 1441; (b) Yadav, J.S., Reddy, B.V.S. and Sunitha, S. (2003) *Adv. Synth. Catal.*, **345**, 349.

**52** Iimura, S., Manabe, K. and Kobayashi, S. (2003) *Org. Biomol. Chem.*, **1**, 2416.

**53** Iimura, S., Nobuto, D., Manabe, K. and Kobayashi, S. (2003) *Chem. Commun.*, 164.

**54** (a) Iimura, S., Manabe, K. and Kobayashi, S. (2003) *J. Org. Chem.*, **68**, 8723; (b) Iimura, S., Manabe, K. and Kobayashi, S. (2003) *Org. Lett.*, **5**, 101.

# 3
# Green Bases in Water

*José M. Fraile, Clara I. Herrerías, and José A. Mayoral*

## 3.1
## Introduction

The use of bases to promote organic reactions has been known for centuries; in fact, the production of soap through the saponification of animal and vegetable fats was already documented in the Middle Ages. Nowadays bases are used in a wide range of industrial chemical processes, either as reagents consumed in stoichiometric amounts or as catalysts. Conventional strong bases, such as NaOH or KOH, present a number of disadvantages such as corrosion of reactors, difficulties in separation from products and the disposal of basic or acidic (if neutralized) waste water streams [1].

The concept of a "green base" is difficult to define, and even more so when working in water. A base could be considered "green" when it solves the problems outlined above, and therefore it should not be soluble in water or at least it should be easily separated from the aqueous medium or from the products. As an example, in a reaction in water in which the product precipitates at the end of the reaction, it is advisable to use a soluble base that will remain in aqueous solution, rather than an insoluble base that will be mixed with the precipitated product. In this context, the term "green process with base in water" seems more suitable than directly "green base in water." Another important point is that in water the basicity limit is that of $OH^-$, which limits the basic strength of the base used.

The relationship between base and water can be considered from different points of view. One is the amount of water, given that it can be either the reaction solvent ("in water"), or one of the solvents if the reaction is carried out in a mixture of them ("in aqueous medium"). On the other hand, the reaction system can admit a limited amount of water, with or without participation in the reaction ("in the presence of water") [2].

Another limiting point is the role of the base. It is well known that organic bases can act as organocatalysts in processes such as enantioselective aldol condensations. However, they do not play a role as a base, but as a nucleophile to form the intermediate enamine. Even supported bases (amino acids) have been used in

*Handbook of Green Chemistry, Volume 5: Reactions in Water.* Edited by Chao-Jun Li
Copyright © 2010 WILEY-VCH Verlag GmbH & Co. KGaA, Weinheim
ISBN: 978-3-527-31591-7

combination with solid acids to perform one-pot sequential reactions in the presence of water [3]. As this use cannot be considered as a true example of base use, it will not be considered here.

In this chapter, we review the small number of examples that can be considered as green efforts towards the use of bases in combination with water, irrespective of the amount of it in the reaction medium.

## 3.2
## Types of Bases and Their Use from a Green Point of View

There is general agreement in considering solid reagents and catalysts within the group of green substances, owing to the ease of separation and the possibility of reuse.

Hattori reviewed the applications of solid bases, and classified them into five groups [4]. We can simplify this classification to only three groups, but it is necessary to include a fourth group for anchored organic bases, not included in Hattori's classification:

1. Metal oxides: either single metal (e.g. MgO and $Al_2O_3$) or mixed oxides (e.g. Mg–Al oxides, usually prepared by calcination of hydrotalcites).

2. Alkaline ion-exchanged silicates: either crystalline (zeolites) or layered (clays).

3. Supported bases: this group includes alkali metals and hydroxides on supports, such as silica or alumina, supported imides and nitrides, and the classical $KF/Al_2O_3$. This group is incompatible with the use of water or highly polar solvents because of leaching problems, together with limitations in the basicity commented on above.

4. Anchored amines and analogs, such as amidines, guanidines, and proton sponges [5]. The organic bases are covalently bonded to organic (polymeric) or inorganic (silica) supports.

The bases soluble in water can be envisaged as part of a green process when this solubility allows a simple separation and, if possible, recovery and reuse. In this category we can differentiate two main groups:

1. Inorganic bases, such as $K_2CO_3$.
2. Organic bases: they may be soluble themselves (e. g. ethanolamine) or require modification to make them soluble.

Regarding their use, bases can be used either in stoichiometric amounts or as catalysts. Many organic reactions, such as substitutions or eliminations, generate acids as concomitant products and the practical application requires the use of a base to neutralize them. These reactions are intrinsically not green, as the atom economy is usually not very high. The use of one or another base can only facilitate separation, preventing the use of additional solvents or other auxiliary substances.

In the case of catalysts, bases are regenerated at the end of the catalytic cycle and ideally a separable base should be recoverable and reusable. In those cases, the role of the basic catalyst is usually the abstraction of a proton from the substrate. The presence of water, with its two protons, will compete with the substrate, inhibiting the reaction if the acidity of the substrate is not significantly higher than that of water. This fact limits enormously the use of basic catalysts in water, or even in the presence of water. This limitation is even more drastic due to the trend of strong solid bases to react with atmospheric $CO_2$, leading to surface carbonates. In such cases, the surface basicity is greatly reduced and treatments at high temperature (over 400 °C) are required to regenerate the strongly basic sites, together with the use of those solids under a $CO_2$-free atmosphere.

In this chapter, the processes using bases in water have been classified into those having the base dissolved or those including the use of a solid base. In each section, the use of stoichiometric or catalytic amounts of base is clearly outlined, and also the amount of water present in the process.

## 3.3
## Liquid–Liquid Processes

As we indicated earlier, the use of a base that is soluble in water does not determine the sustainability of the base and much less the process. Inorganic bases, such as $K_2CO_3$, $Na_2CO_3$, and $Cs_2CO_3$, or even organic bases, have been widely used in stoichiometric, or even over-stoichiometric, amounts in base-mediated reactions using water as solvent with good results. This is the case, for instance, with C–C cross-coupling reactions, such as Suzuki (Scheme 3.1) [6], Heck [7], Sonogashira [8], and Stille (Scheme 3.2) reactions [9], and the use of phase-transfer catalysts for the asymmetric synthesis of α-amino acids, their derivatives, and other compounds [10, 11]. However, the final treatment of the reaction usually involves the extraction of the products with an organic solvent and the elimination of the aqueous phase that includes the neutralized base.

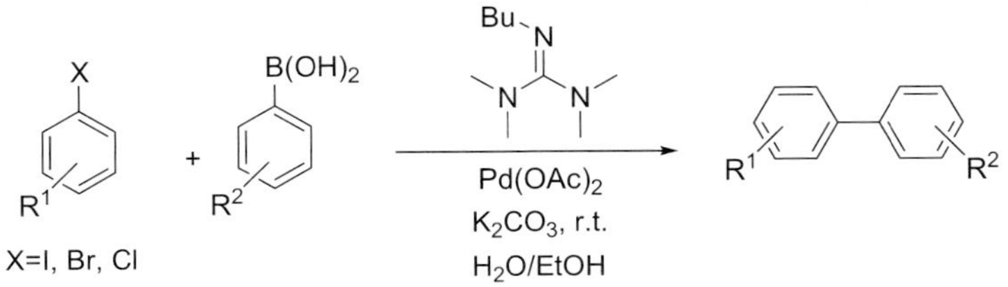

**Scheme 3.1** Suzuki coupling in aqueous medium [6h].

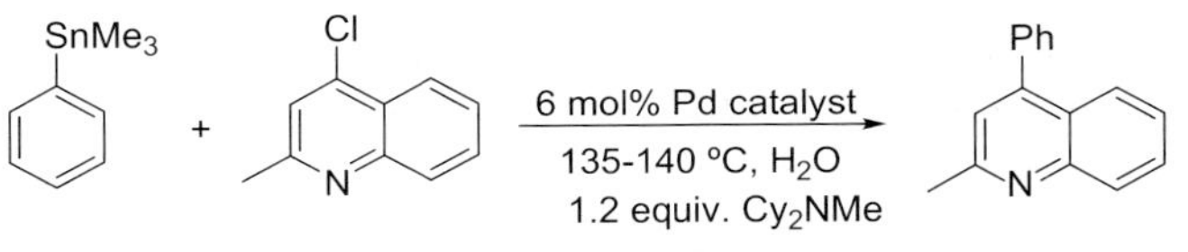

**Scheme 3.2** Stille coupling in aqueous medium [9a]

A special approach is the utilization of an ionic liquid together with water [12, 13]. This two-component solvent enables to adjust their miscibility after dissolution of reagents and products for a more efficient final separation and recycling. For example, Welton and co-workers reported the change in the miscibility of water with [bmim][$BF_4$], an ambient temperature ionic liquid, within the reaction time [13]. They carried out a palladium-catalyzed Suzuki cross-coupling reaction using $Na_2CO_3$ (2.1 equiv.) as a base (Scheme 3.3). They found that the by-products generated in the reaction ($NaHCO_3$ and $Na[XB(OH)_2]$) modify the miscibility of water with the ionic liquid and two phases were formed at the end of the reaction. The by-products were preferentially soluble in the water phase and this allowed their removal. However, the aqueous phase was again disposed of as waste and only the ionic liquid, containing the palladium catalyst, was reused.

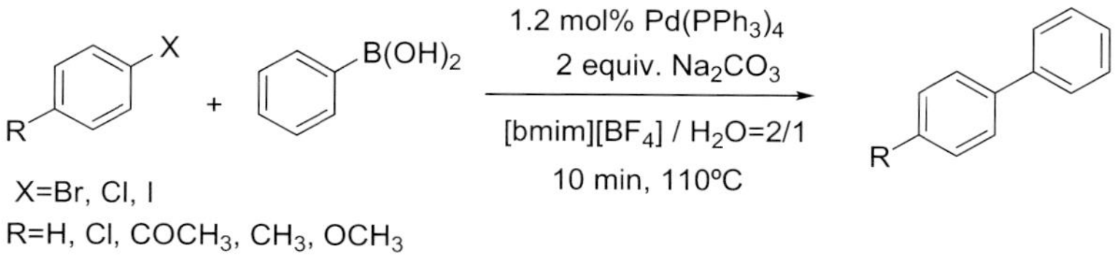

**Scheme 3.3** Suzuki coupling in aqueous–ionic liquid biphasic medium.

In all these previous cases, the base is only an auxiliary and the green character of the process relates only to the palladium or the phase-transfer catalysts, which are frequently recovered.

In other cases, the base is used as a catalyst, allowing improvement of the atom economy. The widely employed C−C bond formation reactions, namely aldol, Knoevenagel, Michael, and Mannich-type reactions, are among the most common base-catalyzed organic reactions. In this context, Hosomi and co-workers reported the use of $CaCl_2$ as a Lewis base for the aqueous aldol reaction of silyl enolates (Scheme 3.4) [14]. After testing different salts, they found that the counteranion played a crucial role in promoting the aldol reaction and that the rate-accelerating ability of the metal salt increased when the intrinsic nucleophilicity of the counteranion increased: $TfO^- < I^- \leq Br^- < Cl^-$. This study led to high yields in all the cases tested and showed the regiospecificity of the reaction when cyclic substrates were used. For an analogous reaction, Mukaiyama and co-workers utilized AcOLi (10 mol%) as Lewis base with good results [15]. However, the amount of water used in the reaction was minimal, actually, the reaction was carried out in a water-containing DMF solvent (DMF : $H_2O = 50:1$). In subsequent work [16], they were able to obtain moderate yields with lower DMF : $H_2O$ volume ratios (5 : 1 and 2 : 1).

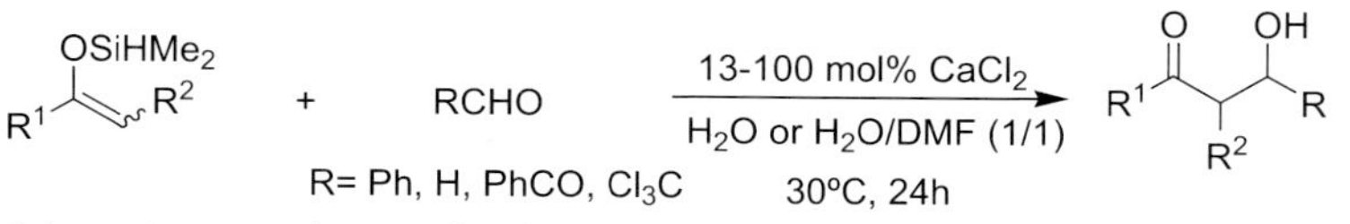

**Scheme 3.4** Lewis base-catalyzed aqueous Mukaiyama aldol reaction.

In 2002, Kobayashi et al. reported the first catalytic asymmetric Mannich-type reaction in aqueous media using $ZnF_2$, a chiral diamine, and triflic acid (TfOH) (Scheme 3.5) [17]. The reactions proceeded with high yields and good values for diastereoselectivity and enantioselectivity. The reaction solvent was a mixture of $THF–H_2O$ (9:1). They postulated a double activation in which the $Zn^{2+}$ cation acts as a Lewis acid and the fluoride anion as a Lewis base to activate the silicon enolates. After this publication, more studies on these Mannich-type reactions were published by the same group, improving on the previous results with a reduction of the load of $ZnF_2$ and avoiding the use of an organic co-solvent [18].

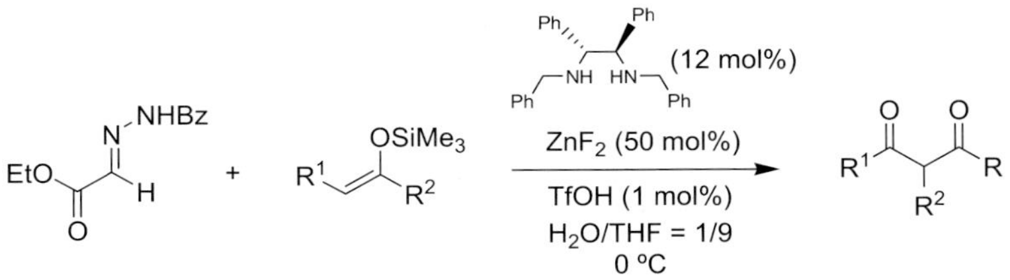

**Scheme 3.5** Catalytic asymmetric Mannich-type reaction in an aqueous medium.

On the other hand, several base systems are able to catalyze Knoevenagel condensations performed in water. The reactions of aromatic aldehydes with active methylene compounds in aqueous media have been promoted by bases, such as NaOH [19] and $(NH_4)_2HPO_4$ [20]. The reaction has been also carried out under phase-transfer conditions [21]. An interesting example describes the use of an equimolecular mixture of $I_2$, $K_2CO_3$, and KI as an effective, inexpensive, and non-toxic catalytic system. This method was applied to Knoevenagel and nitroaldol condensations, affording the corresponding products in excellent yields in very short reaction times (Scheme 3.6) [22].

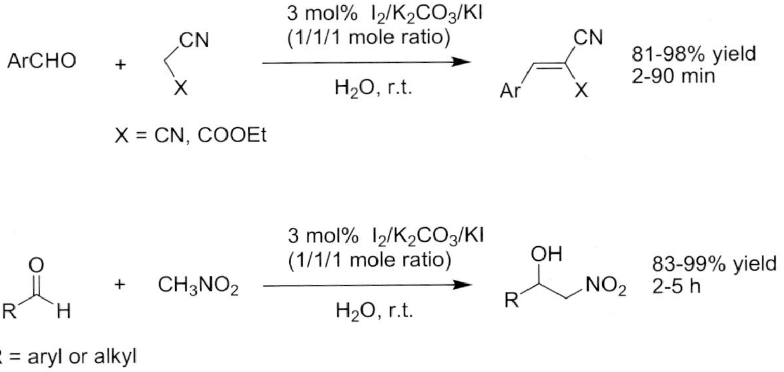

R = aryl or alkyl

**Scheme 3.6** $I_2$–$K_2CO_3$–KI as catalyst for Knoevenagel and nitroaldol reactions.

Aliphatic nitro compounds can also be obtained by base-catalyzed Michael additions in aqueous media [23]. Nitroalkanes have shown evidence of a remarkable reactivity in water, in particular in the formation of C—C bonds via their conjugate addition to electron-deficient alkenes. Bases such as $K_2CO_3$ and NaOH can conduct the reactions to obtain outstanding results and even carry out domino reactions to obtain more complicated structures [24].

Alternatively, Michael additions of β-keto esters to α,β-unsaturated carbonyl compounds can take place in water by means of catalytic non-ionic bases. Rodriguez and co-workers described the use of phosphazenes, guanidines, and more classical nitrogen-containing derivatives or triphenylphosphine as bases for the Michael reaction in water (Scheme 3.7) [25]. These compounds led to better yields in water than in THF, and in some cases there was even an almost total conversion in water whereas no reaction was detected in THF.

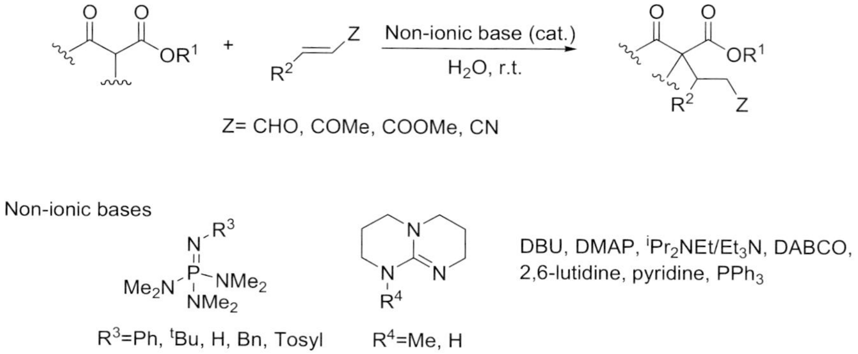
Non-ionic base (cat.)
$H_2O$, r.t.
Z= CHO, COMe, COOMe, CN
Non-ionic bases
DBU, DMAP, $^iPr_2NEt/Et_3N$, DABCO, 2,6-lutidine, pyridine, $PPh_3$
$R^3$=Ph, $^tBu$, H, Bn, Tosyl
$R^4$=Me, H

**Scheme 3.7** Michael reactions catalyzed by non-ionic bases in water.

A reaction related to Michael addition is the Baylis–Hillman reaction, which takes place between activated olefins and aldehydes in the presence of tertiary amines, generating synthetically useful allyl alcohols (Scheme 3.8). This reaction has also been carried out using an aqueous medium with different bases as catalysts [26]. The use of water as solvent has overcome problems such as low reaction yields and long reaction times.

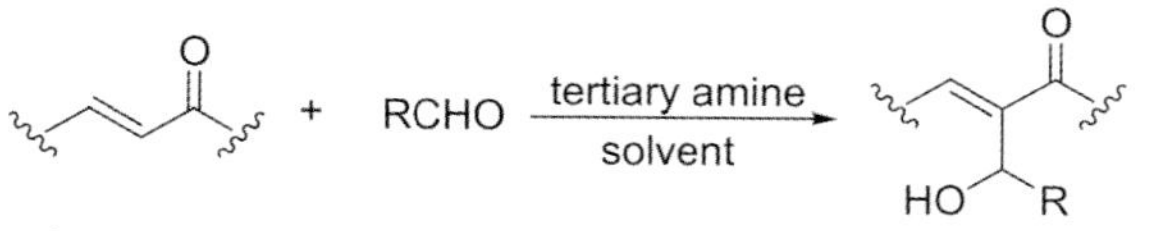
RCHO
tertiary amine
solvent

**Scheme 3.8** Baylis–Hillman reaction.

Another typical base-catalyzed organic reaction is the epoxidation of electron-deficient alkenes, usually α,β-unsaturated ketones or aldehydes, which takes place through the conjugated addition of a hydroperoxide anion (Scheme 3.9). When 30% $H_2O_2$ is used as an oxidant, concentrated aqueous NaOH solution is used as a base and the reaction takes place in a biphasic system. The use of phase transfer co-catalysts improves the reaction performance, as in the case of dodecyltrimethylammonium bromide in a biphasic heptane–water (1 : 1) system [27]. In such a case, stirring and/or sonication are crucial to obtain fast reactions. In some other cases, the use of an organic solvent can be avoided, for example, with the sodium perborate–NaOH system coupled with a phase-transfer catalyst (tetrahexylammonium hydrogensulfate) [28]. Whereas some products were directly purified by distillation, others required extraction with an organic solvent. The need for a phase-transfer catalyst can be avoided by the use of the previously mentioned biphasic ionic liquid–water system [29]. In this case, the reaction takes place in the interface and the products are separated by extraction with an organic solvent. The main advantage of this system is the easy recovery of the ionic liquid by simple decantation from the aqueous phase.

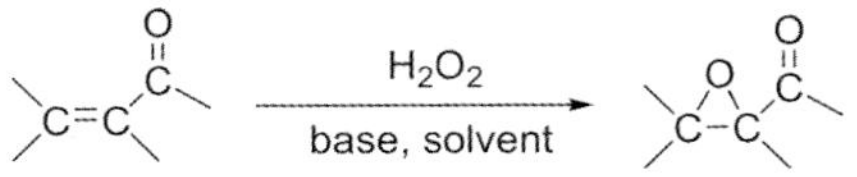

**Scheme 3.9** Epoxidation of electron-deficient alkenes under basic conditions.

The organic–water biphasic system has been adapted for enantioselective epoxidation of chalcones (Scheme 3.10) by using polyamino acids [30] (Juliá–Colonna method), usually polyleucine. In these reactions, the valuable component is not the base but the chiral moiety, which was recovered in some occasions [31].

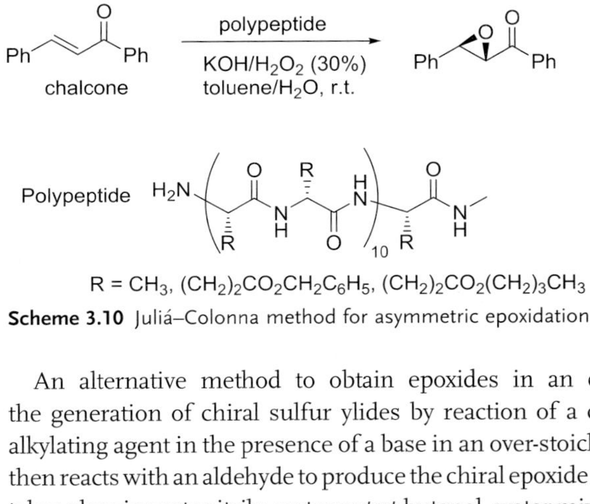

R = $CH_3$, $(CH_2)_2CO_2CH_2C_6H_5$, $(CH_2)_2CO_2(CH_2)_3CH_3$

**Scheme 3.10** Juliá–Colonna method for asymmetric epoxidation of chalcones.

An alternative method to obtain epoxides in an enantioselective manner is the generation of chiral sulfur ylides by reaction of a chiral cyclic sulfide with an alkylating agent in the presence of a base in an over-stoichiometric amount. The ylide then reacts with an aldehyde to produce the chiral epoxide (Scheme 3.11). This reaction takes place in acetonitrile–water or *tert*-butanol–water mixtures and the recovery of the valuable component, the chiral sulfide, is highly efficient in some cases [32].

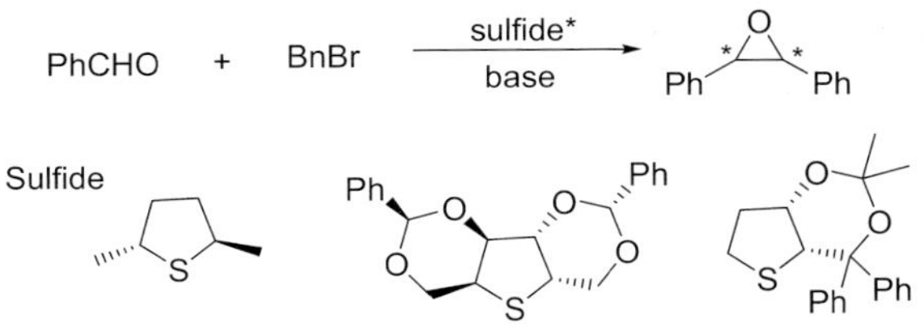

Base: NaOH, $K_2CO_3$, $Cs_2CO_3$, NaH

**Scheme 3.11** Asymmetric synthesis of epoxides using chiral sulfur ylides.

Although in the previously cited works, namely C—C bond formation reactions and epoxidations, the bases were utilized in catalytic amounts, the work-up in all cases required the separation of the products from the reaction medium, usually by extraction of the products with an organic solvent, and the subsequent disposal of the aqueous phase. Fortunately, several workers have also developed more environmentally friendly chemical processes which entail the recovery of the aqueous phase together with the catalytic base. One of these procedures was carried out by Wang et al. for the preparation of β-hydroxy ketones [33]. They reported the cross-aldol reaction of 2-acetylpyridine, acetophenone, and cyclohexanone with 4-, 3-, and 2-nitrobenzaldehydes in water as a solvent and catalyzed by $Na_2CO_3$ (Scheme 3.12). The yields obtained under these conditions were very high (87–98%), despite the low solubility of both ketones and aldehydes in water. The products were obtained by simple filtration at the end of the reaction and the aqueous phase was reused in several cycles without loss of activity. This synthetic method is very attractive from a green point of view since the generation of waste has been reduced by recycling of the catalytic $Na_2CO_3$, solving pollution problems.

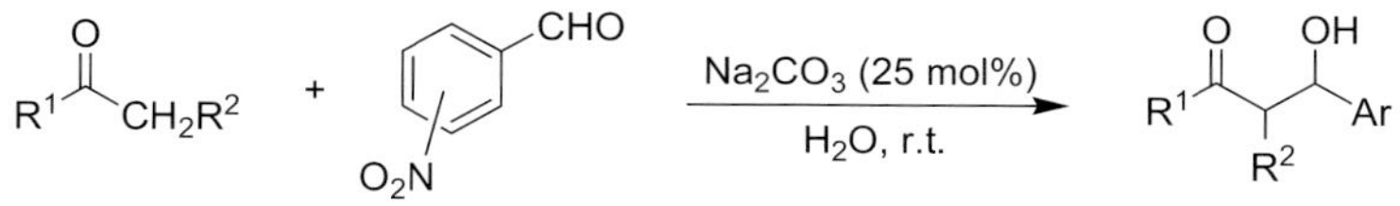

**Scheme 3.12** Reusable basic aqueous phase for cross-aldol condensations.

Malhotra and co-workers used the same approach for the synthesis of 2-chlorobenzylidenemalononitrile and its analogs in water as a solvent [34]. The reactions were carried out using 1-methylimidazole as organic basic catalyst (Scheme 3.13). The main advantage, as in the previous case, is that the product is insoluble in water and can be isolated by simple filtration. Furthermore, the aqueous mother liquors were reused in repeated runs (at least 10) with no change in either the yield or the purity of the product.

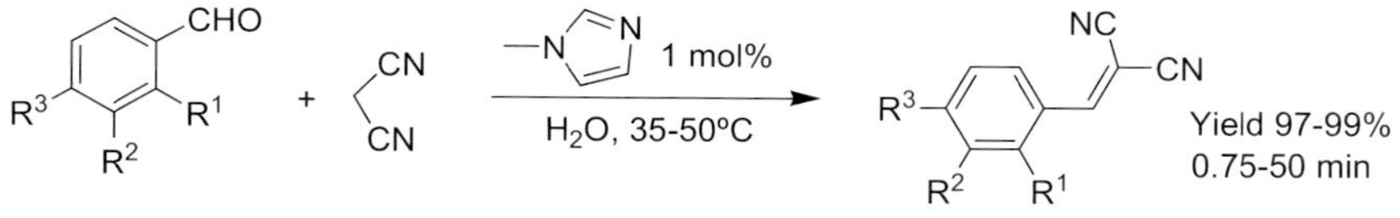

**Scheme 3.13** Reusable basic aqueous phase for Knoevenagel condensation.

Another good methodology from a green perspective was developed by Song and co-workers, who reported the use of an amino-functionalized ionic liquid as catalyst for the Knoevenagel condensation of aromatic aldehydes with malononitrile and ethyl cyanoacetate in aqueous media (Scheme 3.14) [35]. The use of an imidazolium group helps to improve the solubility of the base in water, and only 0.8 mol% of catalyst was needed. In this case, the product precipitates during the reaction and allows isolation by Büchner filtration with high purity. The described protocol is simple, efficient (85–96% yield) and eco-friendly since the aqueous phase with the ionic liquid catalyst can be recycled up to six times, maintaining the yields at 91–94% for the case of the model reaction between benzaldehyde and ethyl cyanoacetate. Recently, Gao et al. [36] applied the same strategy to promote identical Knoevenagel condensations in water. They utilized the same basic ionic liquid as catalyst but with a different counterion ([2-aemim][BF$_4$], 5 mol%), obtaining similar results.

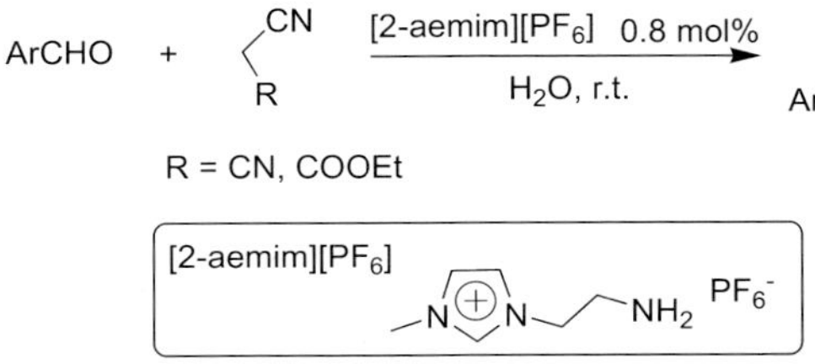

$$\text{ArCHO} + \underset{R}{\overset{CN}{\diagdown\!\!\!\diagup}} \xrightarrow[\text{H}_2\text{O, r.t.}]{\text{[2-aemim][PF}_6\text{] 0.8 mol\%}} \underset{\text{Ar}}{\diagdown}\!\!=\!\!\underset{R}{\overset{CN}{\diagup}}$$

R = CN, COOEt

[2-aemim][PF$_6$]: imidazolium cation with aminoethyl substituent, PF$_6^-$ counterion

**Scheme 3.14** Knoevenagel condensation catalyzed by a basic ionic liquid.

The greenest approach to the use of bases in water, that is, the use of water itself as a base, must not be forgotten. Under near-critical or supercritical conditions, the ionization constant of water increases with temperature, reaching a maximum near 250 °C. The increase in the dissociation of liquid water into hydronium and hydroxide ions at high temperatures has allowed acid- or base-catalyzed organic reactions to be carried out without the addition of acids or bases [37]. Near-critical or supercritical water has been found suitable to accomplish processes that require a base to catalyze the reaction, such as aldol condensations [38], aldehyde disproportionations [39], alkaline degradation of sugars [40], ester hydrolysis [41], benzil–benzilic acid rearrangement [42], and synthesis of p-isopropenylphenol from bisphenol A [43]. In some cases, the reaction mechanism has been difficult to evaluate owing to the dual behavior of water in the sub- or supercritical regions, with it remaining unclear whether it behaves as an acid or as a base. On the other hand, as an additional advantage, the solubility of organic compounds in water under these conditions is similar to that in polar organic solvents, such as acetone, at room temperature. When the reaction is finished and the medium has cooled, the separation of the products is immediate since they are usually insoluble in water at room temperature.

It should be noted that there are many other examples where bases have been used to promote aqueous liquid–liquid processes in both over-stoichiometric and catalytic amounts [44], for instance, reactions that involve acidic $C_{sp}$–H bonds. However, it is not the purpose of this chapter to provide a compilation of all of them, but to give a general overview of the use of bases from a more or less green point of view.

## 3.4
### Solid–Liquid Processes

The use of solid bases provides the opportunity for the development of greener syntheses since they can be easily separated from the crude reaction mixture and, in some cases, they can be reutilized [45, 46]. Furthermore, the employment of solid bases also avoids the utilization of harmful and unrecoverable reactants such as NaOH or KOH. However, typical solid Lewis bases do not function in aqueous media because the surface base sites are severely poisoned by water, and consequently the development of water-tolerant solid bases constitutes an important challenge in the field of heterogeneous catalysis. Only a few examples of their use in aqueous conditions have been reported in the literature. One of them was published by Parlow and co-workers, who carried out Suzuki couplings using a polymer-supported carbonate in a mixed solvent, DMF–H$_2$O (20 : 1) (Scheme 3.15) [47]. The polymer-supported tetramethylammonium carbonate acted as an inorganic carbonate base and, therefore, it activated the boronic acid for the coupling and neutralized the boric acid by-product. Additionally, any unreacted boronic acid was sequestered by the polymer-supported base. When the reaction was finished, the excess of boronic acid and the borane-containing by-products were eliminated via simple filtration, eliminating the need for work-up. The anthracene-tagged palladium catalyst was also removed from the crude reaction mixture by a Diels–Alder cycloaddition with polymer-bound maleimide. The desired product was obtained in high yield and purity.

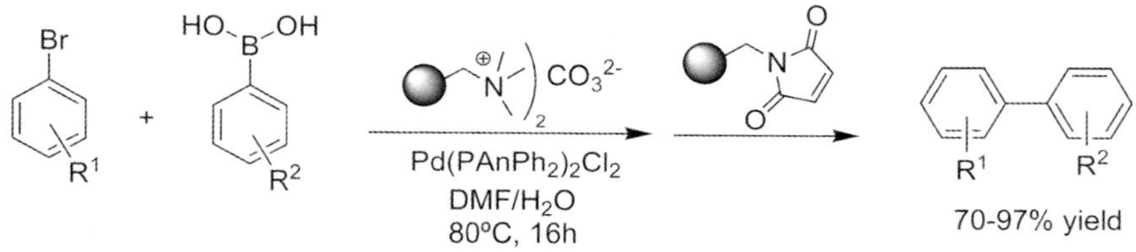

**Scheme 3.15** Suzuki coupling using polymer-supported carbonate as a base.

Another example was reported by Bigi *et al.* They used montmorillonite KSF as a water-stable and reusable catalyst for the Knoevenagel synthesis of coumarin-3-carboxylic acids [48]. The results suggested that the clay behaved as a ditopic catalyst containing both acid and basic sites. It was proposed that the basic sites dispersed over the sheets of oxygen atoms activated the Knoevenagel condensation, whereas the acid sites promoted α-pyrone ring formation by intramolecular esterification (Scheme 3.16). The catalyst was recovered by Büchner filtration and reused five times without significant loss of activity.

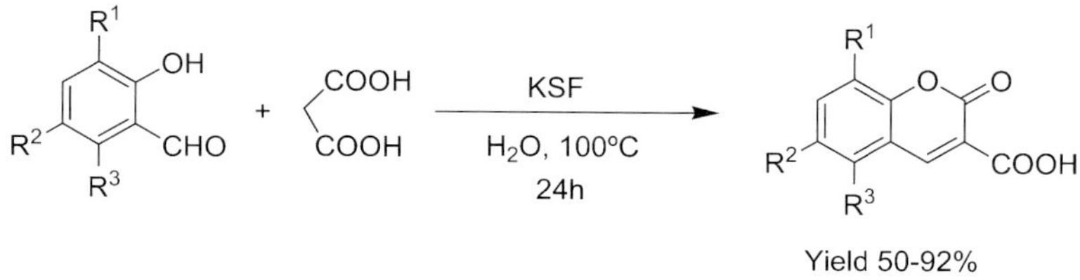

**Scheme 3.16** Synthesis of coumarin-3-carboxylic acids.

Other solid bases have also been utilized as catalysts in Knoevenagel condensations using water as solvent. For example, γ-alumina has proved to be a good catalyst for the synthesis of substituted 2-amino-2-chromenes via a cascade reaction with three components (Scheme 3.17) [49]. This solid possesses a large surface area and highly porous exterior that makes γ-alumina a very common solid to develop surface organic chemistry. When the reaction is finished, the solid can be recovered by filtration and reused four times with good yield and selectivity. Strongly basic anion-exchange resins constitute another type of solid base that can promote the condensation of malononitrile with aromatic aldehydes in water [50]. The resin was prepared from the original chloride-exchanged resin by washing with 10% aqueous NaOH until chloride was absent. After the reaction, the catalyst needed no regeneration and could be reused five times without loss of activity.

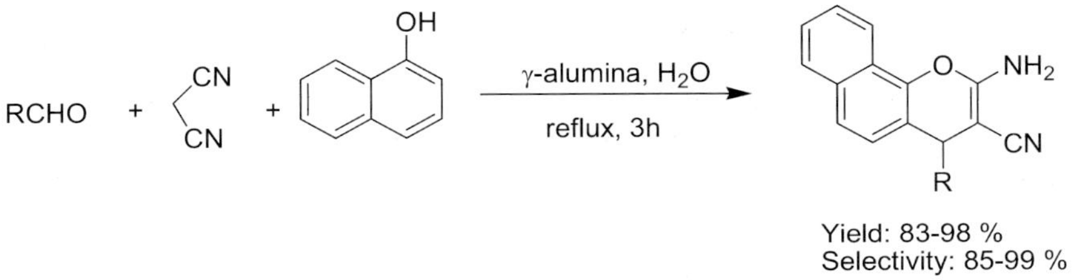

Yield: 83-98 %
Selectivity: 85-99 %

**Scheme 3.17** Synthesis of substituted 2-amino-2-chromenes by cascade reactions.

Kaneda and co-workers reported the use of a reconstructed hydrotalcite as a heterogeneous base for the Knoevenagel and Michael reactions of nitriles with carbonyl compounds in the presence of decarbonated water [51]. They also described the first example of the use of a solid base to promote aldol reactions in the presence of water with good results. They attributed the high activity of the basic catalyst to the surface base sites created during the organization of the layered structure. The base strength of the reconstructed hydrotalcites in the presence of water was estimated within a $pK_a$ range of 10.72–11.2. The procedures were operated under an argon atmosphere to avoid adsorption of $CO_2$ on the surface. The yields of the reactions were very high in most cases (Scheme 3.18) and, in several of them, the solid catalyst was separated by filtration from the reaction mixture and reused.

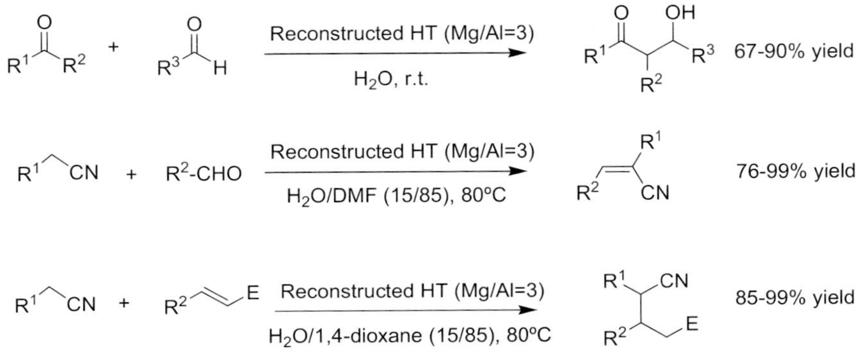

**Scheme 3.18** Aldol, Knoevenagel, and Michael reactions catalyzed by hydrotalcites.

In addition to the use of basic solids, it is also possible to functionalize with a basic moiety other type of solids. For instance, the grafting of 1,6-diaminohexane to a polyacrylamide (Scheme 3.19) allowed an efficient heterogeneous catalyst to be obtained for the Knoevenagel condensation of aromatic aldehydes with ethyl cyanoacetate in water as solvent [52]. The low cost and the easy preparation of the polymeric catalyst were additional advantages to the possibility of recovery. The catalyst was reused in three consecutive occasions without a decrease in its activity.

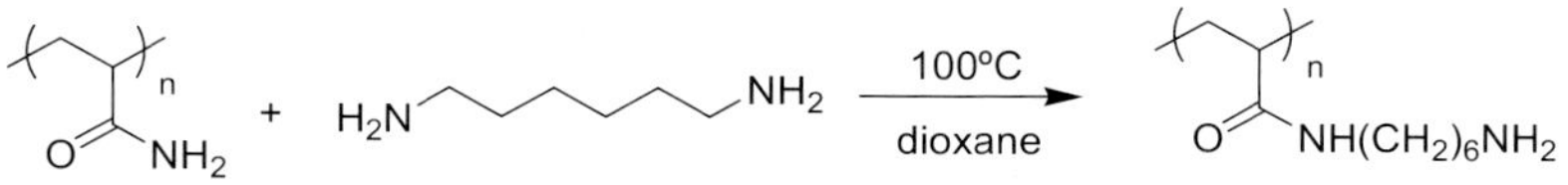

**Scheme 3.19** Grafting of an amine moiety to polyacrylamide.

Using the same methodology, several aminopropyl(trimethoxy)silanes were immobilized on powder or pellets of silica gel [53]. Silica can be a good support due to its chemical and mechanical stability in addition to a large surface area. The supported amine catalysts were tested in the aqueous Knoevenagel condensations of several aldehydes with ethyl cyanoacetate, malononitrile, or α-arylsulfinylacetonitriles and in the aqueous Michael addition of 1,3-dicarbonyl compounds to α,β-unsaturated carbonyl compounds with good results (Scheme 3.20). In all cases the catalyst could be recovered and efficiently reused several times.

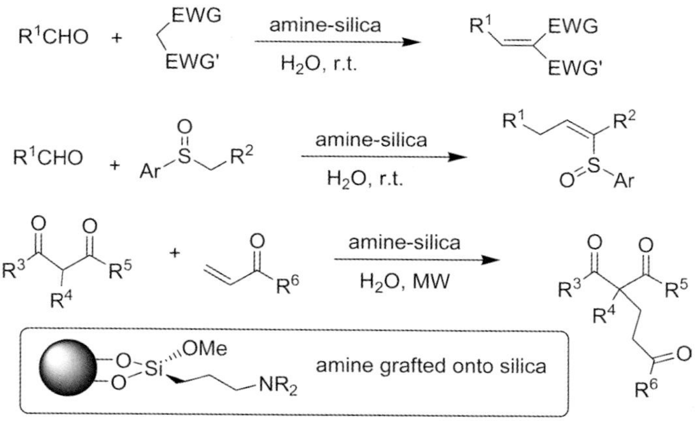

**Scheme 3.20** Michael-type reactions catalyzed by silica-grafted amine.

In addition to Knoevenagel reactions, solid bases can also be used in other types of reactions. Kaneda and co-workers reported the use of a hydroxyapatite [$Ca_{10}(PO_4)_6(OH)_2$] treated with $La(OTf)_3$ as a recyclable catalyst to promote Michael additions in water [54]. The solid proved to be an efficient catalyst for this reaction (64–99% yield). Physicochemical characterization of the catalyst established that a monomeric $La^{3+}$–phosphate complex was created on the hydroxyapatite surface, generating a heterogeneous catalyst that combined potential vacant coordination sites with robust stability. The authors proposed a mechanism that included the participation of La–OH species (Brønsted base sites), which are able to abstract an α-proton from the donor to form a 1,3-diketonate–La species. Then, an acceptor coordinates to vacant sites of the La species at the carbonyl oxygen to form a ternary La complex, followed by alkylation and hydrolysis to yield the Michael product and regenerate the initial La–OH species.

Recently, several groups have used mixed oxides to promote some aldol reactions in water. Chheda and Dumesic proved that it is possible to convert biomass-derived carbohydrates to produce liquid alkanes as an alternative to petroleum resources [55]. The procedure involves a series of reaction steps including an aldol reaction in water. Initially, the reaction was carried out by mixed Mg–Al oxide prepared by hydrotalcite-like syntheses, but the catalyst failed to recycle. Subsequently, they tested other base catalysts for good recycling ability, including MgO, CaO, La–ZrO$_2$, Y–ZrO$_2$, MgO–ZrO$_2$–I, MgO–ZrO$_2$, MgO–TiO$_2$, and N-methyl-3-aminopropylated silica. The results indicated that MgO–ZrO$_2$, obtained by sol–gel techniques, showed the best recycling ability together with high activity. On the other hand, Figueras and co-workers used a highly basic magnesium–lanthanum mixed oxide as catalyst for the aldol-type reaction of aldehydes and imines with ethyl diazoacetate in water as solvent (Scheme 3.21) [56]. This mixed oxide, prepared by coprecipitation, was an effective heterogeneous catalyst, in contrast to other solid base catalysts that were not very efficient in aqueous phase. Mg–La mixed oxide was recovered quantitatively by centrifugation and reused for four cycles with consistent activity.

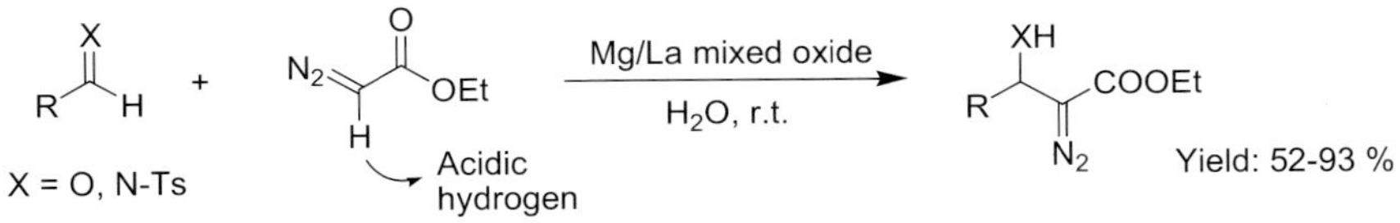

**Scheme 3.21** Aldol-type reaction of ethyl diazoacetate with aldehydes or tosylamides.

With respect to the epoxidation reactions of electron-deficient alkenes, efforts to recover the base have also been linked to the use of solid bases [57]. In the first examples, catalysts were mixed oxides prepared by calcination of hydrotalcites, and epoxidation reactions were carried out in methanol as a solvent, with water present as co-solvent of the oxidant (30% v/v H$_2$O$_2$, methanol : water ratio 75 : 25) [58]. More recently biphasic heptane–water systems and even reactions without any organic solvent (5% H$_2$O$_2$ as an oxidant) have also been described with the same type of catalyst [59]. Good recovery of these solids was demonstrated in several cases. Other strongly basic solids, such as supported guanidine-type bases [60] and natural phosphates modified by impregnation with NaNO$_3$ and calcination at 900 °C [61], are also highly active for the epoxidation of electron-deficient alkenes.

In addition, bases can also be used for the epoxidation of electron-rich alkenes in combination with a nitrile through the formation of a peroxymidic acid intermediate (Payne mechanism). The epoxidation of styrene (Scheme 3.22) using the H$_2$O$_2$–hydrotalcite system has been described in a ternary mixture of solvents (acetone–acetonitrile–water) with excellent results [62].

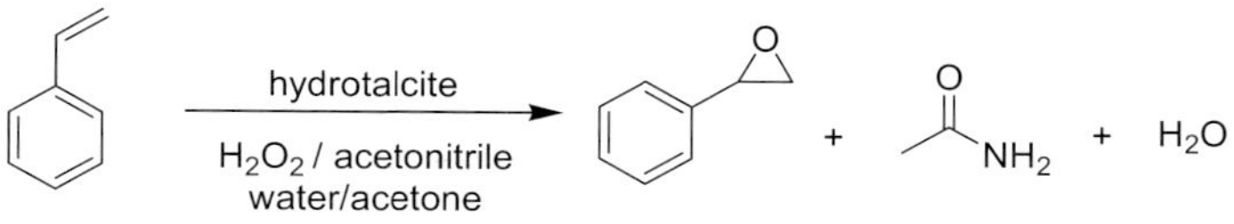

**Scheme 3.22** Epoxidation of styrene using the $H_2O_2$–acetonitrile method.

Several solid bases have been used in the reaction of isomerization of glucose to fructose in water [63]. This reaction was performed in order to evaluate the basicity of solid base catalysts such as cation-exchanged zeolites and anion-modified hydrotalcites, in water as solvent. The corresponding basic strength was determined from the plot of the logarithm of the observed reaction rates against pH in sodium hydroxide solutions. Using this plot, the corresponding values for pH were calculated for NaA (11.1) and Na- (10.8) and Cs- (11.0) exchanged X zeolites, hydrotalcite in a mixed carbonate–hydroxide (12.1) and only the carbonated (11.3) form, and meixnerite (12.4). However, the results were influenced by the reaction conditions, with an increase in pH when, for example, the catalyst weight was increased. Additionally, a leaching effect was observed for the A and X zeolites although it was not detected for hydrotalcite.

To sum up, in spite of the efforts made in recent years, the development of green processes with bases in water or aqueous media is still an area of active research and further studies are needed to increase the number of successful uses for this green approach.

# References

1 For a review on the problems associated with the use of bases in industrial condensation reactions, see: Kelly, G.J., King, F. and Kett, M. (2002) *Green Chem.*, **4**, 392–399.
2 The differences between "in water" and "in the presence of water" are discussed for the case of organocatalysis in: (a) Brogan, A.P., Dickerson, T.J. and Janda, K.D. (2006) *Angew. Chem. Int. Ed.*, **45**, 8100–8102; (b) Hayashi, Y. (2006) *Angew. Chem. Int. Ed.*, **45**, 8103–8104.
3 Akagawa, K., Sakamoto, S. and Kudo, K. (2007) *Tetrahedron Lett.*, **48**, 985–987.
4 (a) Hattori, H. (1995) *Chem. Rev.*, **95**, 537–558; (b) Hattori, H. (2004) *J. Jpn. Petrol. Inst.*, **47**, 67–81.
5 (a) Gelbard, G. and Vielfaure-Joly, F. (2001) *React. Funct. Polym.*, **48**, 65–74; (b) Benaglia, M., Puglisi, A. and Cozzi, F. (2003) *Chem. Rev.*, **103**, 3401–3429;
(c) Corma, A., Iborra, S., Rodríguez, I. and Sánchez, F. (2002) *J. Catal.*, **211**, 208–215.
6 For some recent examples, see: (a) Leadbeater, N.E. (2005) *Chem. Commun.*, 2881–2902; (b) Arvela, Riina K., Leadbeater, N.E., Mack, T.L. and Kormos, C.M. (2006) *Tetrahedron Lett.*, **47**, 217–220; (c) Crozet, M.D., Castera-Ducros, C. and Vanelle, P. (2006) *Tetrahedron Lett.*, **47**, 7061–7065; (d) Wu, W.-Y., Chen, S.-N. and Tsai, F.-Y. (2006) *Tetrahedron Lett.*, **47**, 9267–9270; (e) Chanthanvong, F. and Leadbeater, N.E. (2006) *Tetrahedron Lett.*, **47**, 1909–1912; (f) Li, S., Lin, Y., Cao, J. and Zhang, S. (2007) *J. Org. Chem.*, **72**, 4067–4072; (g) Sayah, R., Glegola, K., Framery, E. and Dufaud, V. (2007) *Adv. Synth. Catal.*, **349**, 373–381.
7 For some recent examples, see: (a) Botella, L. and Nájera, C. (2005) *J. Org. Chem.*, **70**,

4360–4369; (b) Senra, J.D., Malta, L.F.B., de Souza, A.L.F., Medeiros, M.E., Aguiar, L.C.S. and Antunes, O.A.C. (2007) *Tetrahedron Lett.*, **48**, 8153–8156.

8 For some recent examples, see: (a) Appukkuttan, P., Dehaem, W. and Van der Eycken, E. (2005) *Eur. J. Org. Chem.*, 4713–4716; (b) Brea, R.J., López-Deber, M.P., Castedo, L. and Granja, J.R. (2006) *J. Org. Chem.*, **71**, 7870–7873; (c) Li, Y., Zhou, P., Dai, Z., Hu, Z., Sun, P. and Bao, J. (2006) *New J. Chem.*, **30**, 832–837; (d) Fleckenstein, C.A., Plenio, H., *Chem. Eur. J.*, **13**, (2007) 2701–2716

9 For some recent examples, see: (a) García-Martínez, J.C., Lezutekong, R. and Crooks, R.M. (2005) *J. Am. Chem. Soc.*, **127**, 5097–5103; (b) Coelho, A.V., de Souza, A.L., de Lima, P.G., Wardell, J.L. and Antunes, O.A.C. (2007) *Tetrahedron Lett.*, **48**, 7671–7674.

10 For some recent reviews, see: (a) Nelson, A. (1999) *Angew. Chem. Int. Ed.*, **38**, 1583–1585; (b) Maruoka, K. (2003) *Chem. Rev.*, **103**, 3013–3028; (c) O'Donnell, M.J. (2004) *Acc. Chem. Res.*, **37**, 506–517; (d) Cativiela, C. and Díaz-de-Villegas, M.D. (2007) *Tetrahedron: Asymmetry*, **18**, 569–623; (e) Hashimoto, T. and Maruoka, K. (2007) *Chem. Rev.*, **107**, 5656–5682.

11 (a) Ooi, T., Taniguchi, M., Kameda, M. and Maruoka, K. (2002) *Angew. Chem. Int. Ed.*, **41**, 4542–4544; (b) Ooi, T., Kameda, M., Taniguchi, M. and Maruoka, K. (2004) *J. Am. Chem. Soc.*, **126**, 9685–9694.

12 (a) Miao, W. and Chan, T.H. (2003) *Org. Lett.*, **5**, 5003–5005; (b) Zou, G., Wang, Z., Zhu, J., Tang, J. and He, M.Y. (2003) *J. Mol. Catal. A*, **206**, 193–198; (c) Xin, B., Zhang, Y., Liu, L. and Wang, Y. (2005) *Synlett*, 3083–3089.

13 Mathews, C.J., Smith, P.J. and Welton, T. (2000) *Chem. Commun.*, 1249–1250.

14 Miura, K., Nakagawa, T. and Hosomi, A. (2002) *J. Am. Chem. Soc.*, **124**, 536–537.

15 (a) Nakagawa, T., Fujisawa, H. and Mukaiyama, T. (2003) *Chem. Lett.*, **32**, 696–607; (b) Nakagawa, T., Fujisawa, H., Nagaza, Y. and Mukaiyama, T. (2004) *Bull. Chem. Soc. Jpn.*, **77**, 1555–1567.

16 Kawano, Y., Fujisawa, H. and Mukaiyama, T. (2005) *Chem. Lett.*, **34**, 614–615.

17 Kobayashi, S., Hamada, T. and Manabe, K. (2002) *J. Am. Chem. Soc.*, **124**, 5640–5641.

18 (a) Hamada, T., Manabe, K. and Kobayashi, S. (2004) *J. Am. Chem. Soc.*, **126**, 7768–7769; (b) Hamada, T., Manabe, K. and Kobayashi, S. (2006) *Chem. Eur. J.*, **12**, 1205–1215.

19 Zhang, M. and Zhang, A.-Q. (2004) *Synth. Commun.*, **34**, 4531–4535.

20 Balalaie, S., Bararjanian, M., Hekmat, S. and Salehi, P. (2006) *Synth. Commun.*, **36**, 2549–2557.

21 (a) Ren, Z., Cao, W., Tong, W. and Jing, X. (2002) *Synth. Commun.*, **32**, 1947–1952; (b) Gupta, M. and Wakhloo, B.P. (2007) *Arkivoc*, 94–98.

22 Ren, Y. and Cai, C. (2007) *Catal. Lett.*, **118**, 134–138.

23 For a recent review, see: Ballini, R., Barboni, L., Fringuelli, F., Palmieri, A., Pizzo, F. and Vaccaro, L. (2007) *Green Chem.*, **9**, 823–838.

24 (a) Ballini, R. and Bosica, G. (1996) *Tetrahedron Lett.*, **37**, 8027–8030; (b) Ballini, R., Barboni, L., Bosica, G., Filippone, P. and Peretti, S. (2000) *Tetrahedron*, **56**, 4095–4099; (c) Giorgi, G., Miranda, S., López-Alvarado, P., Avendaño, C., Rodriguez, J. and Menéndez, J.C. (2005) *Org. Lett.*, **7**, 2197–2200.

25 (a) Bensa, D., Brunel, J.M., Buono, G. and Rodriguez, J. (2001) *Synlett*, 715–717; (b) Bensa, D. and Rodriguez, J. (2004) *Synth. Commun.*, **34**, 1515–1533.

26 For some recent examples, see: (a) Aggarwal, V.K., Emme, I. and Fulford, S.Y. (2003) *J. Org. Chem.*, **68**, 692–700; (b) Luo, S., Wang, P.G. and Cheng, J.-P. (2004) *J. Org. Chem.*, **69**, 555–558; (c) Faltin, C., Fleming, E.M. and Connon, S.J. (2004) *J. Org. Chem.*, **69**, 6496–6499; (d) Krishna, P.R., Kannan, V. and Reddy, P.V.N. (2004) *Adv. Synth. Catal.*, **346**, 603–606; (e) Zhao,

S. and Chen, Z. (2005) *Synth. Commun.*, **35**, 121–127.

27 Boyer, B., Hambardzoumian, A., Roque, J.-P. and Beylerian, N. (2002) *J. Chem. Soc., Perkin Trans.*, **2**, 1689–1691.

28 Straub, T.S. (1995) *Tetrahedron Lett.*, **36**, 663–664.

29 Wang, B., Kang, Y.-R., Yang, L.-M. and Suo, J.-S. (2003) *J. Mol. Catal. A*, **203**, 29–36.

30 For a review on use of polymers with main chain chirality, see: Pu, L. (1998) *Tetrahedron: Asymmetry*, **9**, 1457–1477.

31 Flisak, J.R., Gombatz, K.J., Holmes, M.M., Jarmas, A.A., Lantos, I., Mendelson, W.L., Novack, V.J., Remich, J.J. and Snyder, L. (1993) *J. Org. Chem.*, **58**, 6247.

32 These reactions are reviewed in: Pellissier, H. (2007) *Tetrahedron*, **63**, 1297–1330.

33 Wang, G.-W., Zhang, Z. and Dong, Y.-W. (2004) *Org. Process Res. Dev.*, **8**, 18–21.

34 Pande, A., Ganesan, K., Jain, A.K., Gupta, P.K. and Malhotra, R.C. (2005) *Org. Process Res. Dev.*, **9**, 133–136.

35 Cai, Y., Peng, Y. and Song, G. (2006) *Catal. Lett.*, **109**, 61–64.

36 Gao, G.-H., Lu, L., Zou, T., Gao, J.-B., Liu, Y. and He, M.-Y. (2007) *Chem. Res. Chin. Univ.*, **23**, 169–172.

37 For some recent reviews, see: (a) Hunter, S.E. and Savage, P.E. (2004) *Chem. Eng. Sci.*, **59**, 4903–4909; (b) Sato, M., Ikushima, Y., Hatakeda, K. and Zhang, R. (2006) *Anal. Sci.*, **22**, 1409–1416.

38 (a) Nolen, S.A., Liotta, C.L., Eckert, C.A. and Gläser, R. (2003) *Green Chem.*, **5**, 663–669; (b) Comisar, C.M. and Savage, P.E. (2004) *Green Chem.*, **6**, 227–231; (c) Lu, X., Li, Z. and Gao, F. *Ind. Eng. Chem. Res.*, (2006) **45**, 4145–4149.

39 (a) Bröll, D., Kaul, C., Krämer, A., Krammer, P., Richter, T., Jung, M., Vogel, H. and Zehner, P. (1999) *Angew. Chem. Int. Ed.*, **38**, 2998–3014; (b) Ikushima, Y., Hatakeda, K., Sato, O., Yokohama, T. and Arai, M. (2001) *Angew. Chem. Int. Ed.*, **40**, 210–213.

40 Jin, F., Zhou, Z., Enomoto, H., Moriya, T. and Higashijima, H. (2004) *Chem. Lett.*, **33**, 126–127.

41 Oka, H., Yamago, S., Yoshida, J. and Kajimoto, O. (2002) *Angew. Chem. Int. Ed.*, **41**, 623–625.

42 Comisar, C.M. and Savage, P.E. (2005) *Green Chem.*, **7**, 800–806.

43 Hunter, S.E., Felczak, C.A. and Savage, P.E. (2004) *Green Chem.*, **6**, 222–226.

44 For some recent reviews on organic reactions in water, see: (a) Li, C.-J. (1993) *Chem. Rev.*, **93**, 2023–2035;(b) Li, C.-J. and Chan, T.H. (1997) *Organic Reactions in Aqueous Media*, John Wiley & Sons, Inc., New York;(c) Grieco, P.A. (ed.) (1998) *Organic Synthesis in Water*, Thomson Science, Glasgow; (d) Lindstrom, U.M. (2002) *Chem. Rev.*, **102**, 2751–2772; (e) Li, C.-J. (2005) *Chem. Rev.*, **105**, 3095–3166; (f) Herrerías, C.I., Yao, X., Li, Z. and Li, C.-J. (2007) *Chem. Rev.*, **107**, 2546–2562; (g) Lindstrom, U.M. (ed.) (2007) *Organic Reactions in Water: Principles, Strategies and Applications*, Blackwell, Oxford.

45 (a) Anastas, P.T. and Warner, J.C. (1998) *Green Chemistry: Theory and Practice*, Oxford University Press, Oxford;(b) Clark, J. and Macquarrie, D. (2002) *Handbook of Green Chemistry and Technology*, Blackwell, Oxford.

46 For examples, see: (a) Climent, M.J., Corma, A., Iborra, S. and Velty, A. (2002) *J. Mol. Catal. A*, **182–183**, 327–342; (b) Bass, J.D., Anderson, S.L. and Katz, A. (2003) *Angew. Chem. Int. Ed.*, **42**, 5219–5222; (c) Kabashima, H., Tsuji, H., Shibuya, T. and Hattori, H. (2000) *J. Mol. Catal. A*, **155**, 23–29; (d) Doskocyl, E.J. (2005) *J. Phys. Chem. B*, **109**, 2315–2320; (e) Bailly, M.-L., Chizallet, C., Costentin, G., Kraff, J.-M., Lauron-Pernot, H. and Che, M. (2005) *J. Catal.*, **235**, 413–422.

47 Lan, P., Berta, D., Porco, J.A. Jr., South, M.S. and Parlow, J.J. (2003) *J. Org. Chem.*, **68**, 9678–9686.

48 Bigi, F., Chesini, L., Maggi, R. and Sartori, G. (1999) *J. Org. Chem.*, **64**, 1033–1035.

49 Maggi, R., Ballini, R., Sartori, G. and Sartorio, R. (2004) *Tetrahedron Lett.*, **45**, 2297–2299.

50 Jin, T.-S., Zhang, J.-S., Wang, A.-Q. and Li, T.-S. (2004) *Synth. Commun.*, **34**, 2611–2616.
51 Ebitani, K., Motokura, K., Mori, K., Mizugaki, T. and Kaneda, K. (2006) *J. Org. Chem.*, **71**, 5440–5447.
52 Tamami, B. and Fadavi, A. (2005) *Catal. Commun.*, **6**, 747–751.
53 (a) Isobe, K., Hoshi, T., Suzuki, T. and Hagiwara, H. (2005) *Mol. Diversity*, **9**, 317–320; (b) Hagiwara, H., Isobe, K., Numamae, A., Hoshi, T. and Suzuki, T. (2006) *Synlett*, 1601–1603; (c) Hagiwara, H., Inotsume, S., Fukushima, M., Hoshi, T. and Suzuki, T. (2006) *Chem. Lett.*, **35**, 926–927.
54 Mori, K., Oshiba, M., Hara, T., Mizugaki, T., Ebitani, K. and Kaneda, K. (2006) *New J. Chem.*, **30**, 44–52.
55 Chheda, J.N. and Dumesic, J.A. (2007) *Catal. Today*, **123**, 59–70.
56 Lakshmi-Kantam, M., Balasubrahmanyam, V., Shiva-Kumar, K.B., Venkanna, G.T. and Figueras, F. (2007) *Adv. Synth. Catal.*, **349**, 1887–1890.
57 For a review on oxidation with solid bases, see: Fraile, J.M., García, J.I. and Mayoral, J.A. (2000) *Catal. Today*, **57**, 3–16.
58 Cativiela, C., Figueras, F., Fraile, J.M., García, J.I. and Mayoral, J.A. (1995) *Tetrahedron Lett.*, **36**, 4125–4128.
59 Honma, T., Nakajo, M., Mizugaki, T., Ebitani, K. and Kaneda, K. (2002) *Tetrahedron Lett.*, **43**, 6229–6232.
60 Subba Rao, Y.V., De Vos, D.E. and Jacobs, P.A. (1997) *Angew. Chem. Int. Ed. Engl.*, **36**, 2661–2663.
61 Fraile, J.M., García, J.I., Mayoral, J.A., Sebti, S. and Tahir, R. (2001) *Green Chem.*, **3**, 271–274.
62 Kirm, I., Medina, F., Rodríguez, X., Cesteros, Y., Salagre, P. and Sueiras, J. (2004) *Appl. Catal. A*, **272**, 175–185.
63 Moreau, C., Lecomte, J. and Roux, A. (2006) *Catal. Commun.*, **7**, 941–944.

# 4
# Green Oxidation in Water
*Roger A. Sheldon*

## 4.1
## Introduction

The widespread replacement of traditional organic syntheses employing stoichiometric reagents, such as Brønsted and Lewis acids and bases and oxidants and reductants, is the key to the development of a green and sustainable (fine) chemical industry [1]. A wide variety of industrially important catalytic processes – hydrogenation, carbonylation, hydroformylation, olefin metathesis, polymerization and telomerization – can be effectively performed in an aqueous medium [2]. They involve transition metals in low oxidation states coordinated to soft ligands, such as phosphines, as the catalytically active species and organometallic compounds as reactive intermediates. Performing such reactions in aqueous–organic biphasic media is simply a matter of replacing the ligands used in organic media with hydrophilic equivalents, such as sulfonated triarylphosphines, that are highly water soluble. Catalytic oxidations, in contrast, generally involve transition metals in high oxidation states as the active species, and relatively simple, hard ligands, such as carboxylate. Reactive intermediates tend to be coordination complexes rather than organometallic species. Furthermore, water coordinates strongly to the hard metal center, thereby suppressing coordination of a less polar hydrocarbon substrate, resulting in inhibition of the catalyst or deactivation by hydrolysis in aqueous media. On the other hand, coordination of nitrogen- and/or oxygen-containing ligands can lead to the generation of more active oxidants by promoting the formation of high oxidation states. For example, in heme-dependent oxygenases and peroxidases, the formation of active high-valent oxoiron complexes is favored by coordination to a macrocyclic porphyrin ligand in the active site.

Because a large variety of oxidation processes can be performed in water, it is essential that we first define our frame of reference for this chapter. Enzymatic oxidation processes are generally performed in water but are beyond the scope of this chapter and will not be covered in any detail. We shall also focus on the use of the green, inexpensive, and readily available oxidants – molecular oxygen and hydrogen peroxide – although other oxidants, for example hypochlorite, receive a cursory mention. It is worth mentioning in this context that there are many shades of green,

*Handbook of Green Chemistry, Volume 5: Reactions in Water.* Edited by Chao-Jun Li
Copyright © 2010 WILEY-VCH Verlag GmbH & Co. KGaA, Weinheim
ISBN: 978-3-527-31591-7

and if your starting point is a stoichiometric oxidation with a hexavalent chromium reagent then an oxidation with hypochlorite looks rather green.

The aerobic oxidation of water-soluble alcohols, diols, and carbohydrates over heterogeneous noble metal catalysts (Pt, Pd, Ru) in aqueous media has been extensively studied and has a long history dating back to the nineteenth century. Indeed, it was the first catalytic process to be studied. Since this work has been extensively reviewed elsewhere [3], we shall largely exclude it from our discussion. However, there is currently much interest in the use of metal nanoparticles or nanoclusters as catalysts [4] in a variety of reactions, including oxidations. These may be considered as quasi-homogeneous catalysts or intermediate species at the interface of homogeneous and heterogeneous catalysis. Furthermore, the current focus of attention on the replacement of petroleum hydrocarbon feedstocks, derived from fossil resources, by carbohydrates derived from renewable raw materials [5] is stimulating a revival in interest in catalytic oxidations of carbohydrate feedstocks and water is the solvent of choice for these reactions.

If we consider the oxidation of sparingly soluble substrates in an aqueous biphasic medium, we can distinguish two different approaches, based on whether the catalytic reaction takes place in the organic or the water phase. In the first category the substrate is dissolved in, or forms itself, an organic phase while the oxidant, and possibly also the catalyst, resides in the aqueous phase. A phase-transfer agent is employed to transfer the catalyst and/or oxidant to the organic phase where the reaction takes place. Many catalytic oxidations with water-soluble oxidants, such as hydrogen peroxide, hypochlorite, and persulfate, are examples of this category. In the second category, the substrate resides in a separate organic phase and the catalyst and oxidant are dissolved in the water phase where the reaction takes place. The product is separated as the organic phase and the catalyst, contained in the aqueous phase, is easily recovered and recycled.

## 4.2
### Water-soluble Ligands

Some examples of water-soluble ligands that have been used in catalytic oxidations with oxygen or hydrogen peroxide are shown in Scheme 4.1. Much of this research falls into the category of biomimetic oxidations and, hence, water-soluble derivatives of porphyrins **1** and the structurally related phthalocyanines **2** and other tetradentate nitrogen ligands, such as the tetraamido macrocyclic ligand (TAML, **3**), have been widely used. As we shall see later, water-soluble metal complexes of bidentate nitrogen ligands such as sulfonated phenanthrolines (**4**), have also become popular as catalysts for the aerobic oxidation of alcohols.

## 4.3
### Oxidations Catalyzed by Metalloporphyrins and Metallophthalocyanines

Oxidoreductases – oxygenases, peroxidases, and oxidases – are implicated in a wide variety of *in vivo* degradation processes of biopolymers and xenobiotics. It is not

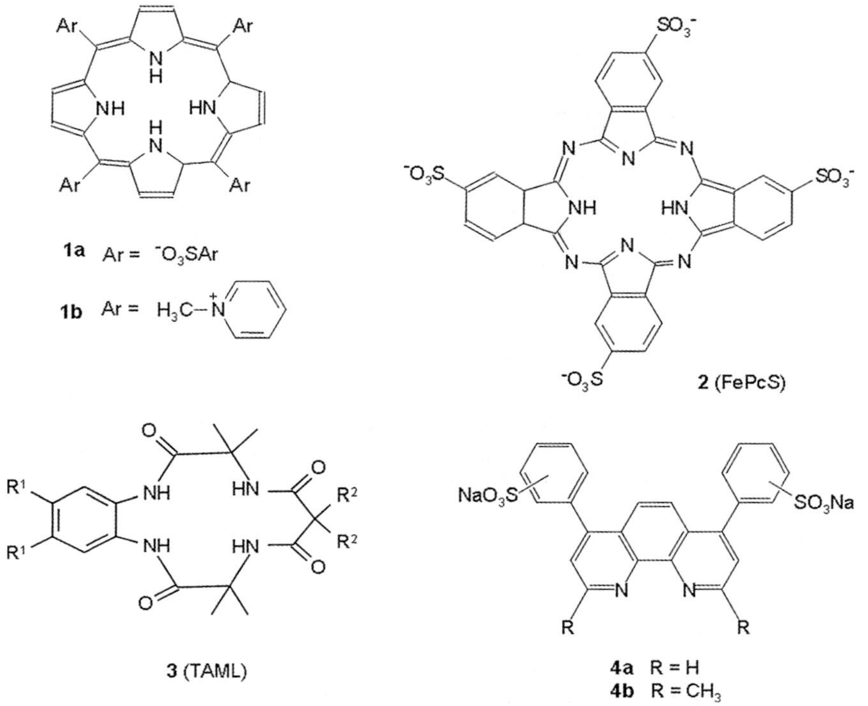

**Scheme 4.1** Water-soluble ligands.

surprising, therefore, that metal complexes of porphyrins and phthalocyanines have been investigated as biomimetic catalysts for the modification and/or degradation of a variety of biopolymers and organic pollutants. For example, metal complexes of water-soluble metalloporphyrins [6] and metallophthalocyanines [7] have been studied as biomimetic catalysts for the aerobic oxidation of phenolic substrates as model compounds for lignin, the goal being an environmentally friendly process for delignification of wood pulp to produce cellulose for paper manufacture. Conventional processes involve the use of $Cl_2$ or $ClO_2$ as oxidants and produce effluents containing chlorinated phenols. Water-soluble polyoxometalates such as $PV_2Mo_{10}O_{40}^{5-}$ have also been used as catalysts for aerobic delignification [8]. FePcS (2) has also been successfully applied as a catalyst for the oxidative destruction of recalcitrant chlorinated phenol pollutants, such as 2,4,6-trichlorophenol, in waste water, using $H_2O_2$ as the primary oxidant [9]. Similarly, Fe and Mn complexes of **1a** and **1b** catalyzed the oxidation of phenols with $KHSO_5$ as the primary oxidant [10, 11].

More recently, Sorokin and co-workers showed that FePcS is an extremely effective catalyst for the oxidation of starch, simultaneously at the $C_6$ primary alcohol group and via cleavage of the $C_2$–$C_3$ vicinal diol (Scheme 4.2), using hydrogen peroxide in an aqueous medium [12, 13]. Selective oxidation was observed with FePcS loadings as low as 0.003–0.016 mol%. In contrast, the same amount of $FeSO_4$ gave no detectable oxidation. Hydrophilic starches, obtained by partial oxidation, are commercially interesting products with many potential applications in the paper and textile industries and as water superabsorbents. Current methods for achieving such oxidations involve the use of stoichiometric amounts of oxidants such as hypochlorite, $N_2O_4$, or periodate, producing copious quantities of inorganic waste [13]. The combination of a green oxidant ($H_2O_2$) with very low loadings of a relatively inexpensive iron-based catalyst for the one-step modification of starch affords obvious economic and environmental benefits.

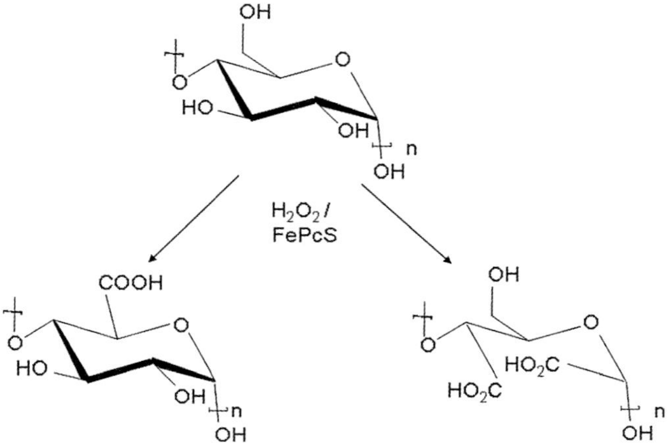

**Scheme 4.2** Catalytic oxidation of starch with hydrogen peroxide.

A major issue associated with the use of porphyrins and, to a lesser extent, phthalocyanines is their susceptibility towards oxidative degradation. Hence there is a need for macrocyclic ligands which stabilize higher oxidation states of, for example, iron and are oxidatively stable. Collins and co-workers [14, 15] developed a series of iron(III) complexes of tetraamido macrocyclic ligands (TAMLs) of general structure **3**, with greatly enhanced stabilities towards oxidative and hydrolytic degradation. Iron complexes of TAMLs are efficient activators of aqueous hydrogen peroxide, over a broad pH range, with a wide variety of potential applications, for example, to replace chlorine bleaching in the pulp and paper industry and for use in waste water treatment. Applications in organic synthesis have, as yet, not been explored.

## 4.4
### Epoxidation and Dihydroxylation of Olefins in Aqueous Media

The epoxidation and the related vicinal dihydroxylation of olefins are reactions of great industrial importance and commercially available 30% aqueous hydrogen peroxide provides obvious environmental and economic benefits compared with classical methods involving hypochlorite or percarboxylic acids or metal-catalyzed epoxidations with alkyl hydroperoxides. Both of the above-mentioned strategies have been employed, using a phase-transfer agent to transfer the catalyst and the oxidant to the organic phase or using water-soluble ligands to promote the oxidation of the olefin substrate in the aqueous phase [16].

The use of tungstate, in the presence of phosphate and a tetraalkylammonium salt as a phase-transfer agent, for the epoxidation of olefins with 30% aqueous hydrogen peroxide, in a biphasic dichloroethane–water medium, was first described by Venturello et al. in 1983 [17]. Ishii et al. [18] subsequently reported the use of a heteropolytungstate, $H_3PW_{12}O_{40}$, in an aqueous biphasic system under phase-transfer conditions for the epoxidation of olefins with hydrogen peroxide. Later, it was shown by Bregeault and co-workers [19, 20] that both systems involve the same peroxotungstate complex, $(R_4N)_3PO_4[W(O)(O_2)_2]_4$, as the catalytically active species.

Noyori and co-workers [21, 22] reported a significant improvement of the original system. A phase-transfer agent, comprising a lipophilic tetraalkylammonium cation and a hydrogensulfate anion ($HSO_4^-$), together with catalytic amounts of $H_2NCH_2PO_3H_2$ and sodium tungstate, proved to be an effective combination for the epoxidation of olefins with hydrogen peroxide in toluene–water or in the absence of an organic solvent (Scheme 4.3).

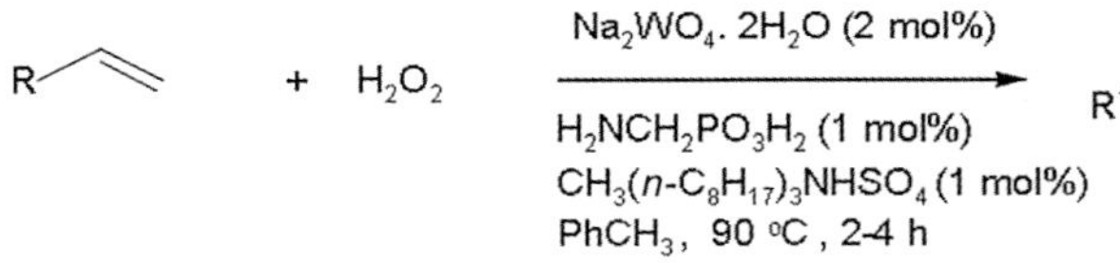

| R | Conv.(%) | Yield(%) |
|---|---|---|
| $C_6H_{13}$ | 96 | 94 |
| $C_8H_{17}$ | 99 | 99 |
| $C_{10}H_{21}$ | 98 | 97 |
| $C_6H_5$ | 52 | 3 |

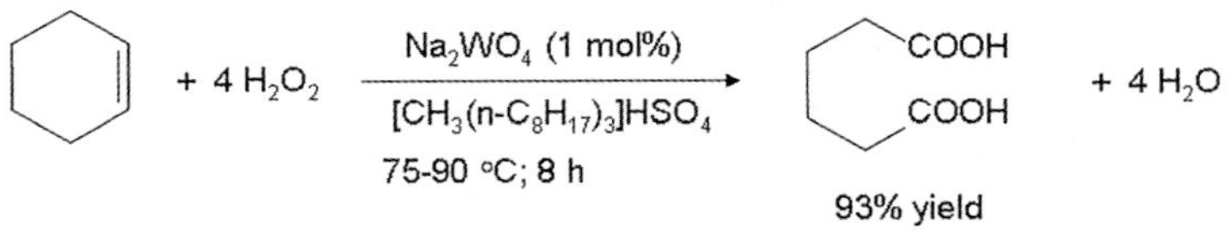

**Scheme 4.3** Solvent- and halide-free epoxidation of olefins with aqueous $H_2O_2$.

The same system was used for the direct oxidation of cyclohexene to adipic acid (see Scheme 4.3), by oxidation with 4 equiv. of 30% aqueous hydrogen peroxide, via the initial formation of cyclohexene oxide [23]. The cost of 4 equiv. of hydrogen peroxide precludes application in adipic acid manufacture, but the methodology obviously has broad synthetic utility in the oxidative cleavage of cyclic olefins [23].

Reedijk and co-workers [24] reported that a similar system, comprising sodium tungstate, tungstic acid, and chloroacetic acid together with methyltrioctylammonium chloride as phase-transfer agent, was effective for the epoxidation of olefins with 50% aqueous hydrogen peroxide. Similarly, Xi et al. [25] showed that a cetylpyridinium heteropolytungstate, $[C_5H_5NC_{16}H_{33}]_3PO_4[WO_4]_3$, catalyzed the epoxidation of propylene and other olefins with hydrogen peroxide in an aqueous biphasic system via so-called *reaction-controlled phase-transfer catalysis*. The catalyst dissolves in the reaction medium through the *in situ* formation of a soluble peroxo complex, $[C_5H_5NC_{16}H_{33}]_3PO_4[W(O)_2(O_2)]_4$. When the reaction is complete, the original catalyst precipitates from the solution and can be filtered and recycled.

Hage et al. followed the second strategy and used a water-soluble manganese complex of 1,4,7-trimethyl-1,4,7-triazacyclononane (tmtacn), which was originally developed as a highly effective catalyst for the low-temperature bleaching of stains [26], as a catalyst for the selective epoxidation of olefins with $H_2O_2$ in aqueous MeOH or water alone (Scheme 4.4) [26]. However, large amounts of $H_2O_2$ (10 equiv. or more) were required, indicating that considerable non-productive dismutation of $H_2O_2$ occurs. Subsequently, it was shown that the catalase-type dismutation of the hydrogen peroxide could be largely suppressed by the addition of oxalate [27], ascorbic acid [28], or glyoxylic acid methyl ester hemiacetal [29] as a cocatalyst, or by anchoring the ligand to a solid support [30].

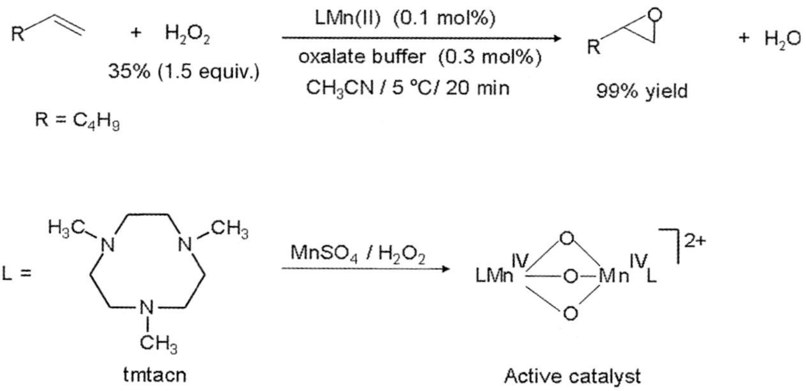

$$R\diagup\!\!\!\diagdown + H_2O_2 \xrightarrow[\text{CH}_3\text{CN / 5 °C/ 20 min}]{\text{LMn(II) (0.1 mol%)}\atop \text{oxalate buffer (0.3 mol%)}} R\triangleleft + H_2O$$

35% (1.5 equiv.)    99% yield

$R = C_4H_9$

$$L = \underset{\underset{CH_3}{|}}{\underset{N}{\bigg\langle}}\overset{H_3C\diagdown N\diagup\diagdown N\diagup CH_3}{\phantom{X}}\bigg\rangle \quad \xrightarrow{MnSO_4 \,/\, H_2O_2} \quad \left[LMn\overset{IV}{\diagup}\overset{O}{\underset{O}{\diagdown}}\overset{IV}{\diagdown}Mn\,L\right]^{2+}$$

tmtacn                                          Active catalyst

**Scheme 4.4** Manganese-catalyzed epoxidation with $H_2O_2$ in water.

Interestingly, the corresponding *cis*-diols were observed as byproducts in many cases. It was subsequently found [31] that the formation of the *cis*-diol was favored by the addition of catalytic quantities of carboxylic acids. A subsequent detailed mechanistic study [32] showed that the active catalyst is a dinuclear, $Mn^{III}_2(\mu\text{-}RCO_2)_2(\text{tmtacn})_2$ species, formed *in situ* by reaction of the catalyst precursor, $[Mn^{IV}_2(\mu\text{-}O)_3(\text{tmtacn})_2](PF_6)_2$, with $H_2O_2$ and the carboxylic acid (Scheme 4.5). The results of, *inter alia*, oxygen labeling experiments were consistent with a mechanism involving the formation of the epoxide and *cis*-diol from a common intermediate and a hydroperoxomanganese(III) species as active oxidant (see Scheme 4.5). The activity and selectivity for epoxidation versus *cis*-hydroxylation are dependent on both electronic and steric properties of the carboxylic acid cocatalyst. Formation of the *cis*-diol is favored with sterically hindered acids and the highest turnover number (>2000) for *cis*-hydroxylation was observed with 2,6-dichlorobenzoic acid [32].

Moderate enantioselectivities were observed [33] when BOC-protected amino acids were used as chiral carboxylic acid ligands, which suggests that further ligand tuning could lead to the development of effective methodologies for asymmetric cis-hydroxylation of olefins.

Scheme 4.5 Mechanism of epoxidation and cis-hydroxylation.

Burgess and co-workers [34, 35] showed that simple manganese(II) salts, such as MnSO$_4$, catalyze the epoxidation of olefins with 30% H$_2$O$_2$ in aqueous dimethylformamide or *tert*-butanol, in the presence of sodium hydrogencarbonate. The latter is an essential component of the system because it reacts with the hydrogen peroxide to form percarbonate, HCO$_4^-$ (Scheme 4.6). It was proposed that the percarbonate oxidizes the manganese to an Mn(IV) species, which is the active oxidant. Here again, additives such as sodium acetate or salicylic acid had a rate- and selectivity-enhancing effect [35].

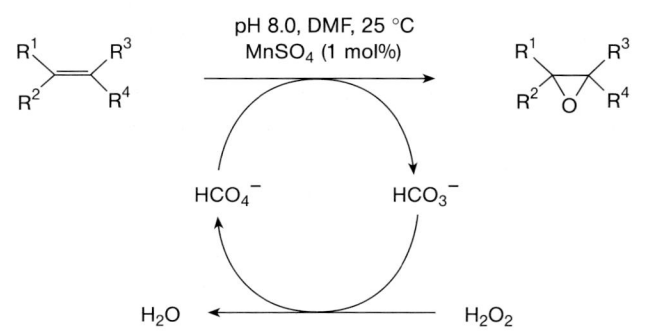

Scheme 4.6 MnSO$_4$-catalyzed epoxidation with hydrogen peroxide–hydrogencarbonate.

An interesting elaboration on this theme is the combination of the $MnSO_4$–hydrogencarbonate system with *in situ* generation of hydrogen peroxide by glucose oxidase-catalyzed oxidation of glucose to afford a chemoenzymatic epoxidation of olefins in an aqueous medium, reported by Chan and co-workers [36]. Lipophilic olefins could be epoxidized in a two-phase system by adding the surfactant sodium dodecyl sulfate. Immobilization of the glucose oxidase by anchoring to silica gel permitted recycling of the enzyme eight times with no significant loss of activity.

Recently, there has been a flourish of interest in the rational design of novel, non-natural enzymatic transformations which have become collectively grouped under the term enzyme (catalytic) promiscuity [37]. In a further elaboration of the $MnSO_4$–hydrogencarbonate system, Okrasa and Kazlauskas [38] exchanged the zinc atom in carbonic anhydrase with manganese to obtain a semi-synthetic enzyme which, in the presence of hydrogencarbonate, catalyzed the epoxidation of olefins with hydrogen peroxide.

Other coordination complexes of the first-row transition elements, manganese and iron, such as porphyrin and salen complexes, also catalyze epoxidations with aqueous hydrogen peroxide [16], but these ligands are prone to rapid oxidative degradation and have limited utility. Ligands containing pyridine and amine coordinating groups fare better under oxidizing conditions. Iron complexes of the pyridylamine ligands 5 and 6 (Scheme 4.7), for example, have been investigated by the groups of Que [39] and Jacobsen [40], respectively, and exhibit moderate olefin epoxidation and/or dihydroxylation activities with aqueous hydrogen peroxide. It should be noted, however, that the reactions are performed in the presence of an organic solvent, generally acetonitrile, and are not strictly examples of catalytic oxidations in water. The activities of these catalysts are also influenced by additives; for example, the iron complex of 6 gave efficient epoxidation in the presence of acetic acid [40]. High-valent dinuclear iron oxo species are implicated as the active oxidants in these reactions, analogous to the putative intermediates in epoxidations mediated by iron-dependent monooxygenases or peroxidases. As with the manganese-catalyzed epoxidations described above, the active catalytic species may also contain bridging carboxylate ligands.

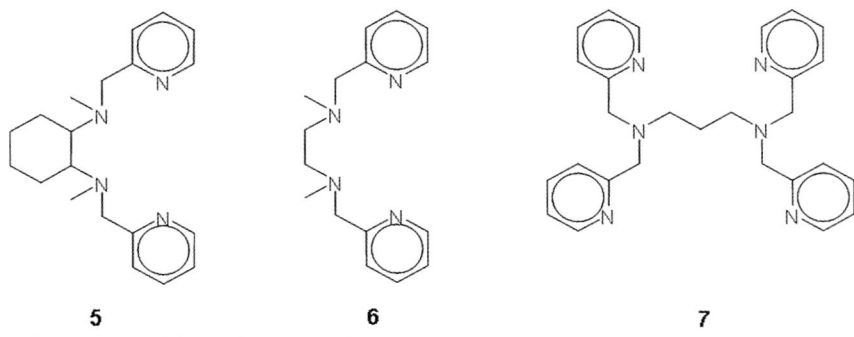

**Scheme 4.7** Pyridylamine ligands used in combination with Fe or Mn.

Feringa and co-workers [41] showed that manganese complexes of the pyridylamine ligand **7** were active epoxidation catalysts with 30% hydrogen peroxide, albeit in acetone as solvent. A dinuclear manganese oxo complex was implicated as the active oxidant, analogous to the iron complexes discussed above. We note, however, that none of these catalysts exhibits the high activities observed with the highly active manganese complex of tmtacn (see above).

Beller and co-workers [42, 43] showed that the ruthenium complex **8**, containing a pyridine-2,6-dicarboxylate and a terpyridine ligand, [Ru(terpy)(pydic)], is a remarkably effective epoxidation catalyst for a variety of olefins (see Scheme 4.8) with 3 equiv. of 30% hydrogen peroxide at very low catalyst loadings (0.005 mol%). A tertiary alcohol such as *tert*-amyl alcohol was used as a cosolvent. Based on its high activity and broad scope, this system appears to have considerable synthetic utility. An asymmetric variant of this reaction was developed by replacing the terpy by chiral pybox ligands (see Scheme 4.8) [44].

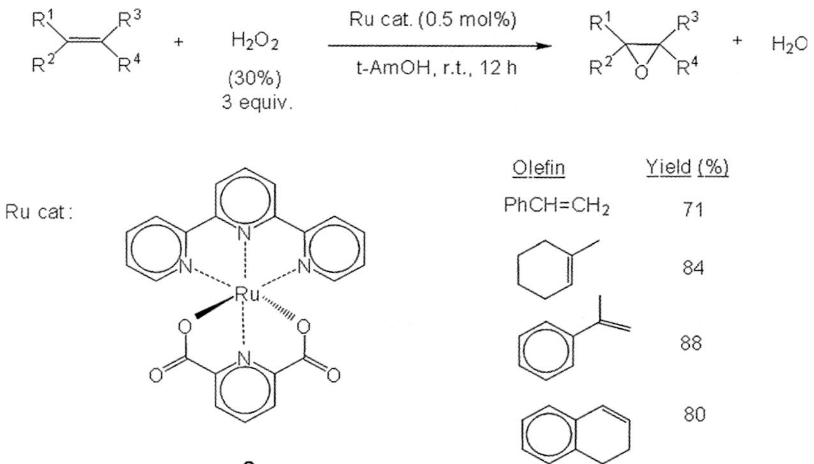

**Scheme 4.8** Ruthenium-catalyzed epoxidations with 30% $H_2O_2$.

More recently, Beller and co-workers turned their attention to the use of iron(III) complexes of pyridine-2,6-dicarboxylic acid ($H_2$pydic) as epoxidation catalysts. They found [45] that a combination of $H_2$pydic, $FeCl_3 \cdot 6H_2O$ and an organic base (a combination that is reminiscent of Barton and Doller's GoAgg system [46]) afforded an effective catalyst for epoxidations with 30% aqueous hydrogen peroxide in *tert*-amyl alcohol (Scheme 4.9). More recently, an asymmetric variant of this system was designed [47] by using the chiral amine ligand **9** of the type popularized by Noyori and co-workers [48] for asymmetric reductions of ketones. The enantioselectivities were only moderate with most olefins tried and this system needs to be improved with respect to enantioselectivity and scope, but it could represent an important step forward in the quest for cost-effective and environmentally attractive asymmetric epoxidation with aqueous hydrogen peroxide.

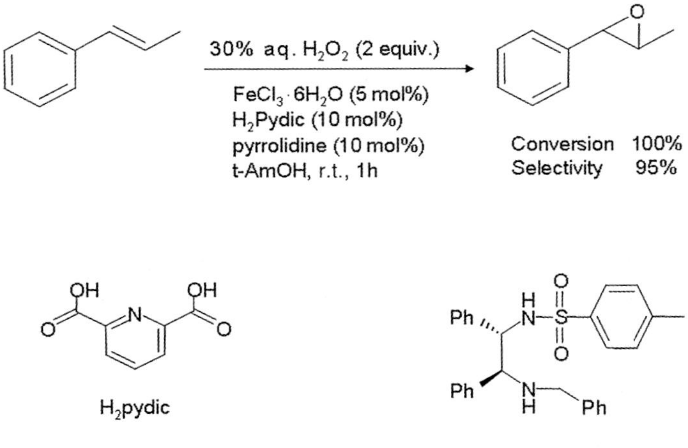

**Scheme 4.9** Iron-catalyzed epoxidations with 30% $H_2O_2$.

Some of the systems discussed above (e.g. Fe and Mn) give the *cis*-1,2-diol under certain conditions, via a concerted mechanism, whereas others (e.g. the tungstate-based systems) can give the *trans*-1,2-diol via acid-catalyzed ring opening of an initially formed epoxide. Sato and co-workers [49], on the other hand, reported a simple catalytic organic solvent- and metal-free system for the oxidation of olefins to the corresponding *trans*-1,2-diols, using 30% hydrogen peroxide (see Scheme 4.10). The catalyst is a resin-supported sulfonic acid, such as Amberlyst 15, Nafion, or the related Nafion–silica composites, and could be recovered by simple filtration and recycled five times without loss of activity.

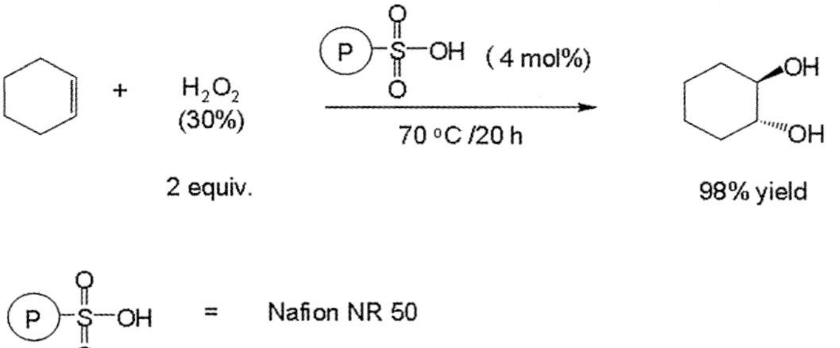

**Scheme 4.10** Olefin dihydroxylation with 30% $H_2O_2$ over Nafion resin.

Similarly, metal-substituted zeolites such as titanium silicalite (TS-1) [50] and titanium beta [51] are recyclable heterogeneous catalysts for the epoxidation and/or dihydroxylation of olefins with aqueous hydrogen peroxide, usually in the presence of an organic cosolvent, such as methanol or *tert*-butyl alcohol, but the reactions can, in principle, be performed in the absence of an added organic solvent. The TS-1-catalyzed epoxidation of propylene with aqueous $H_2O_2$ is currently being scaled up to the industrial scale in a collaboration between BASF, Dow and Solvay [52]. A serious limitation of these molecular sieve catalysts in organic synthesis is their narrow scope, limited to substrates which are able to access their micropores.

## 4.5 Alcohol Oxidations in Aqueous Media

The oxidations of alcohols to the corresponding carbonyl compounds are pivotal reactions in organic synthesis and catalytic methodologies employing dioxygen (air) or hydrogen peroxide as the terminal oxidant are particularly attractive from both economic and environmental viewpoints. If the reaction can be performed in an aqueous medium, thus avoiding the use of volatile organic solvents, this is an added benefit.

The oxidation of alcohols with aqueous hydrogen peroxide using a tungstate catalyst, in the presence of a tetraalkylammonium salt as a phase-transfer agent, in a biphasic system composed of water and 1,2-dichloroethane, was first reported by DiFuria and co-workers in 1986 [53]. As with the analogous tungsten-based epoxidation system described above, Noyori and co-workers [22, 54] substantially optimized this methodology to afford an extremely effective chloride- and organic solvent-free system. A combination of 0.002 mol% sodium tungstate and 0.002 mol% of the phase-transfer agent methyltrioctylammonium hydrogensulfate, $[CH_3\ (n\text{-}C_8H_{17})_3N]^+HSO_4^-$, was an effective catalyst for the selective oxidation of alcohols with 1.1 equiv. of 30% $H_2O_2$ at 90 °C in an organic solvent-free medium (Scheme 4.11). As in the analogous epoxidations (see above) the combination of the lipophilic cation with the hydrogensulfate anion is important for activity. Substrate catalyst ratios as high as 400 000 could be used, affording turnover numbers up to 180 000. A wide variety of secondary alcohols afforded the corresponding ketones in high yields. Olefinic alcohols

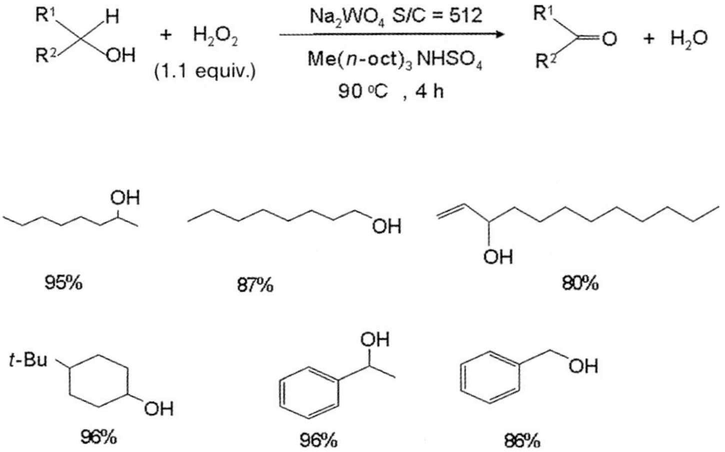


Scheme 4.11 Chloride- and organic solvent-free oxidation of alcohols with aqueous $H_2O_2$.

underwent chemoselective oxidation to the corresponding unsaturated ketones. Primary alcohols gave the corresponding carboxylic acid via further oxidation of the hydrate of the intermediate aldehyde. High yields of carboxylic acids were obtained from a variety of primary alcohols using 2.5 equiv. of hydrogen peroxide. The reactions involve a tetraalkylammonium pertungstate species as the active oxidant.

Neumann and co-workers [55] described the use of a hydrolytically and oxidatively stable sandwich-type tungsten-containing polyoxometalate, $Na_{12}[(WZn_3(H_2O)_2]$ $[(ZnW_9O_{34})_2]$, as a catalyst for the oxidation of alcohols with hydrogen peroxide in an aqueous biphasic medium without an added organic solvent. It was not necessary to prepare the catalyst beforehand. It could be assembled *in situ* by mixing sodium tungstate, zinc nitrate, and nitric acid in water.

The palladium(II)-catalyzed aerobic oxidation of alcohols has been extensively investigated [56]. A general problem encountered in palladium-catalyzed aerobic oxidations is the sluggish reoxidation of Pd(0) to Pd(II), which results in the agglomeration of the Pd(0) particles to palladium black and accompanying deactivation of the catalyst. In the classical Wacker process for the oxidation of ethylene to acetaldehyde, in an aqueous medium, this problem is circumvented by adding copper(II) [57], which reoxidizes the Pd(0) to Pd(II) with concomitant formation of Cu(I). The latter is reoxidized by dioxygen to complete the catalytic cycle.

Substituted olefins afford the corresponding ketone and the analogous oxidation of propylene to acetone has been developed to an industrial scale. The Wacker oxidation of higher terminal olefins, in contrast, is fraught with problems: lower rates and complex product mixtures as a result of competing Pd catalyzed isomerization of the olefin substrate and the formation of chlorinated byproducts owing to the high chloride concentrations used. Various approaches have been examined to circumvent these problems, such as the use of an organic cosolvent, phase-transfer agents, or microemulsion systems, with varying degrees of success [57].

As a continuation of our previous work [58] with palladium complexes of sulfonated phosphines as catalysts for carbonylations in aqueous media, we investigated the use of Pd(II) complexes of analogous water-soluble diamine ligands in the expectation that they could stabilize a transient Pd(0) species under oxidizing conditions, and prevent the formation of palladium black. The commercially available sulfonated bathophenanthroline formed a water-soluble complex when mixed with an aqueous solution of palladium(II) acetate. This complex was shown to catalyze the selective aerobic oxidation of terminal olefins, to the corresponding ketones, in a chloride- and organic solvent-free aqueous medium (Scheme 4.12) [59].

We subsequently found [60–62] that the Pd(II) complex of sulfonated bathophenanthroline (4, R = H) and related water-soluble diamine ligands are stable, recyclable catalysts for the aerobic oxidation of alcohols in a two-phase aqueous–organic medium where the organic phase consists of the alcohol substrate and the carbonyl product (Scheme 4.13). Reactions were generally complete in 5 h at 100 °C/30 bar air with as little as 0.25 mol% catalyst. No organic solvent is required (unless the substrate is a solid) and the product is easily recovered by phase separation. The catalyst is stable and remains in the aqueous phase, facilitating recycling to the next batch.

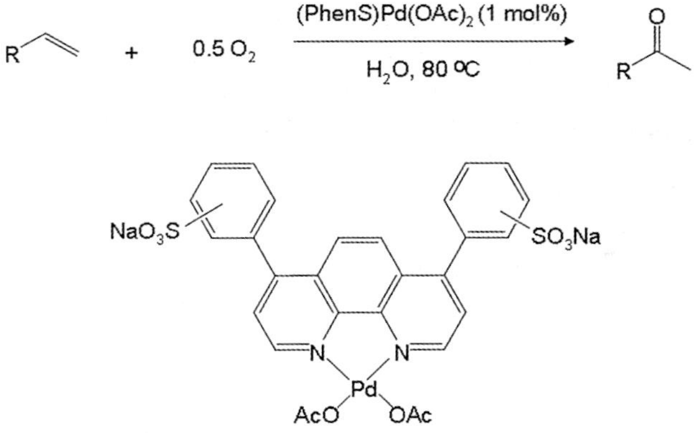

## 4.5 Alcohol Oxidations in Aqueous Media

**Scheme 4.12** Palladium-catalyzed aerobic oxidation of terminal olefins in water.

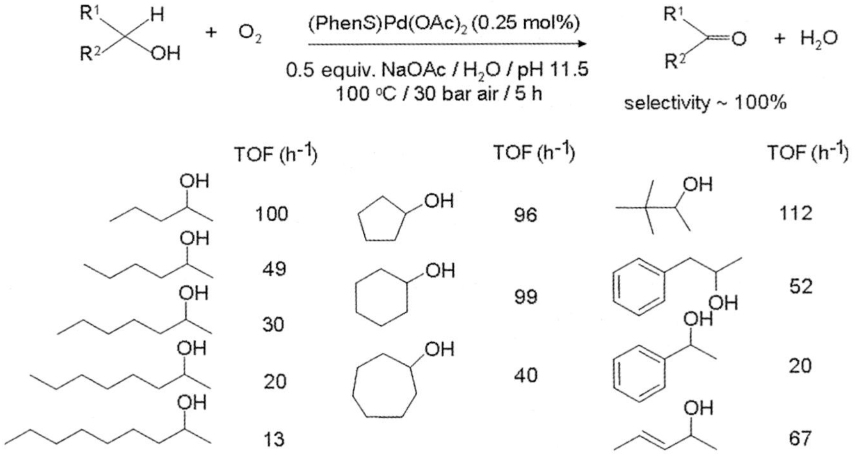

**Scheme 4.13** Palladium-catalyzed aerobic oxidation of secondary alcohols in water.

A wide range of primary and secondary alcohols were oxidized with turnover frequencies (TOFs) ranging from 10 to 100 h$^{-1}$, depending on the structure and the solubility of the alcohol in water. The alcohol must be at least sparingly soluble in water as the reaction occurs in the water phase. Secondary alcohols afforded the corresponding ketones with >99% selectivity in virtually all cases studied. Primary alcohols afforded the corresponding carboxylic acids via further oxidation of the initially formed aldehyde; for example, 1-hexanol afforded 1-hexanoic acid in 95% yield. It is important to note that this was achieved without the necessity to neutralize the carboxylic acid product with 1 equiv. of base. When the reaction was performed in the presence of 1 mol% of the stable free radical TEMPO (2,2,6,6-tetramethylpiperidinoxyl), over-oxidation was suppressed and the aldehyde was obtained in high yield; for example, 1-hexanol afforded hexanal in 97% yield.

A plausible catalytic cycle [61], consistent with the observed half-order in palladium, is depicted in Scheme 4.14. The active catalyst is formed by initial dissociation of a hydroxyl-bridged palladium(II) dimer. Coordination of the alcohol substrate and β-hydrogen elimination affords the carbonyl product and palladium(0), which is re-oxidized to palladium(II) by dioxygen. Further evidence in support of this mechanism has been reported by Stahl et al. [63].

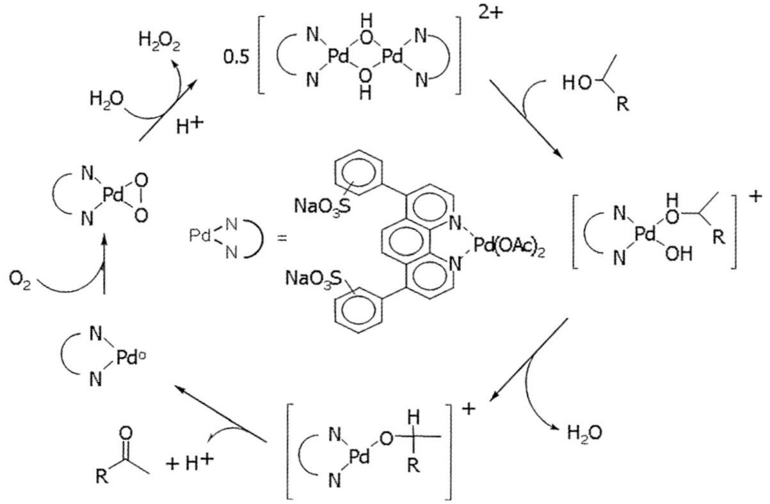

**Scheme 4.14** Mechanism of Pd-catalyzed aerobic oxidation of alcohols.

Compared with most existing systems for the aerobic oxidation of alcohols, the Pd–bathophenanthroline system is an order of magnitude more reactive, requires no organic solvent, involves simple product isolation and catalyst recycling, and has broad scope in organic synthesis. A shortcoming is the requirement that the alcohol substrate should be at least sparingly soluble in water. A second and more general limitation, from which many catalyst systems seem to suffer, is the low tolerance of (coordinating) functional groups in the solvent or the substrate. The bathophenanthroline–Pd(OAc)$_2$ system tolerated only a single ether functionality, and many other functional groups, for example containing N or S as heteroatoms, which coordinate more strongly to palladium, were not tolerated. With a view to obtaining superior systems with higher activities and better functional group tolerance and broader substrate scope, we studied electronic [64] and steric [65] effects of substituents in the phenanthroline ligands on the rates and substrate scope of these reactions. Electron-donating substituents in the alcohol substrate and electron-withdrawing substituents in the ligand increased the rate in accordance with the proposed mechanism. Buffin et al. [66] observed similar electronic effects in a study of the structurally related Pd(II)–biquinoline complexes as catalysts for the aerobic oxidation of alcohols in water.

We further reasoned that substitution at the 2- and 9-positions in the phenanthroline ring would create steric crowding in the hydroxyl-bridged dimer and favor its

## 4.5 Alcohol Oxidations in Aqueous Media

dissociation to the catalytically active monomer and, hence, increase the overall activity. This indeed proved to be the case: the bathocuproin sulfonate (4, R = CH$_3$) and neocuproin complexes of Pd(II) exhibited TOFs of 49, 150, and 1800 h$^{-1}$ in the oxidation of 1-hexanol, although the reaction with the neocuproin complex was performed in 50 : 50 v/v DMSO–water whereas the reactions with bathophenanthroline and bathocuproin sulfonates were performed in water. Nonetheless, the Pd(II)–neocuproin-catalyzed oxidations could be performed at low catalyst loadings (0.1 mol%) affording TOFs of >1500 h$^{-1}$) in 1 : 1 DMSO–water or ethylene carbonate–water mixtures and tolerating a wide variety of functional groups in the alcohol substrate (Scheme 4.15).

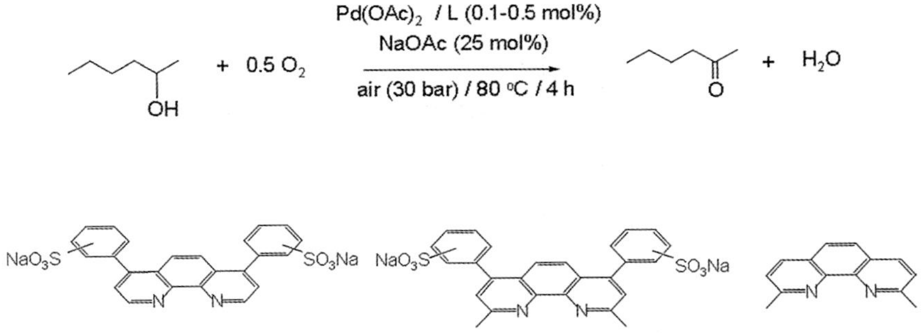

| L | PhenS | BathocuproinS | Neocuproin |
|---|---|---|---|
| TOF (h$^{-1}$) | 50 | 150 | 1800 |
| Solvent | H$_2$O | H$_2$O | H$_2$O / DMSO (1:1) |

**Scheme 4.15** Comparison of different ligands in Pd-catalyzed oxidations of alcohols.

However, a more detailed examination of the results obtained with the Pd(II)–bathophenanthroline and Pd(II)–neocuproin complexes revealed a remarkable difference in the oxidation of the unsaturated alcohol substrate shown in Scheme 4.16: with the former the major product was derived from oxidation of the olefinic double bond whereas the latter afforded >99% selective oxidation of the alcohol moiety. This suggested that we were concerned with totally different types of catalyst. Indeed, further investigation revealed that the Pd(II)–neocuproin complex dissociates completely to afford Pd nanoclusters, which are the actual catalyst [67].

Moiseev and co-workers [68] had previously shown that giant Pd clusters (nowadays known as Pd nanoclusters) are good catalysts for the oxidation of alcohol moieties and selectively oxidize allylic C–H bonds in olefins. More recently, supported Pd-on-hydroxyapatite [69], Au-on-CeO$_2$ [70], Au-on-Mg$_2$AlO$_4$ [71], and Pd–Au-on-TiO$_2$ [72] nanoclusters have been shown to be excellent catalysts for alcohol oxidations. The reactions were performed successfully in organic solvents or solvent-free and also in water. For example, Christensen et al. reported the aerobic oxidation of aqueous (bio)ethanol to acetic acid over Au-on-Mg$_2$AlO$_4$ [71]. Interestingly, when the oxidation is performed in methanol, the methyl ester of the corresponding carboxylic acid is obtained; for example, the renewable raw materials

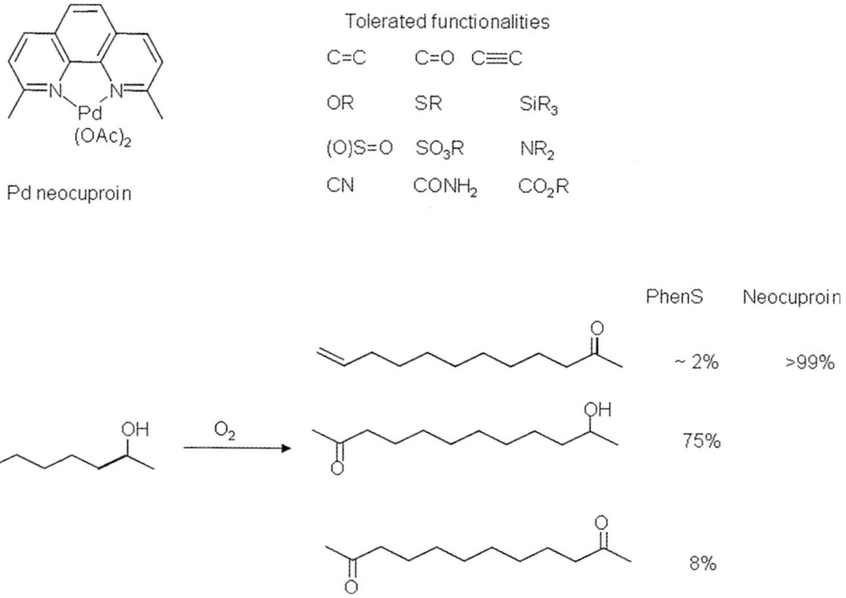

**Scheme 4.16** Neocuproin–Pd(OAc)$_2$-catalyzed oxidation of alcohols.

furfural and hydroxymethylfurfural gave methyl furoate and the dimethyl ester of furan-1,4-dicarboxylic acid [73]. Prati and Rossi [74] were pioneers in the use of Au nanoclusters as catalysts for the aerobic oxidation of alcohol moieties in aqueous media. They showed that Au nanoclusters are excellent catalysts for the aerobic oxidation of carbohydrates, such as glucose to gluconic acid [74].

Alternatively, mono- and bimetallic nanoclusters stabilized and immobilized in polymer matrices have been introduced as "quasi-homogeneous catalysts" for aerobic alcohol oxidations in water. For example, Au and Pd nanoclusters embedded in microgels comprising crosslinked N,N-dimethylacrylamide-based polymers [75], Au dispersed in the stabilizing hydrophilic poly(N-vinyl-2-pyrrolidone) [76], Au incarcerated in a polystyrene matrix [77], and Pd [78] or Pt [79] nanoclusters dispersed in an amphiphilic polystyrene–poly(ethylene glycol) resin have been described. In a further elaboration, the occlusion of preformed, polyvinylpyrrolidone-stabilized Au–Pd nanoclusters in a porous polyimide membrane was recently reported [80]. The resulting heterogeneous catalyst was tested in aerobic alcohol oxidations in dimethylformamide or under solvent-free conditions but not in water.

Another interesting example of a "quasi-homogeneous catalyst" is the recyclable, thermoresponsive catalyst in a micellar-type system reported by Ikegami and co-workers [81]. A polyacrylamide-based copolymer containing pendant tetraalkylammonium cations and a polyoxometalate, $PW_{12}O_{40}^{3-}$, as the counteranion was used as a catalyst for the oxidation of alcohols with hydrogen peroxide in water (Scheme 4.17). At room temperature, the substrate and the aqueous hydrogen peroxide, containing the catalyst, formed distinct separate phases. When the mixture was heated to 90 °C, a stable emulsion was formed, in which the reaction took place

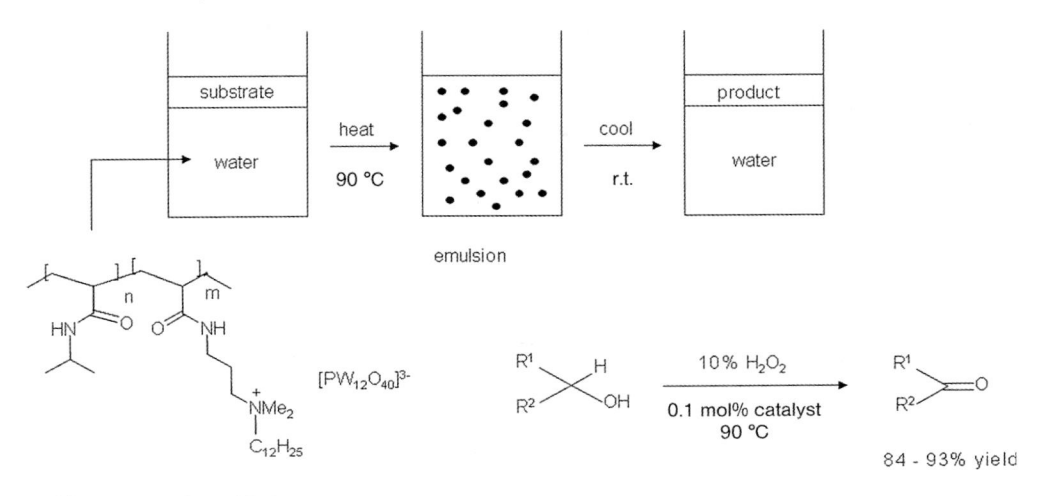

**Scheme 4.17** Oxidation of alcohols with hydrogen peroxide with a thermoresponsive catalyst in a micellar system.

As with the epoxidations and vicinal dihydroxylations discussed in the preceding section, water-soluble complexes of ruthenium, or even better the less expensive and more environmentally acceptable first-row elements iron and manganese, would be expected to catalyze the oxidation of alcohols. Indeed, it has been shown [82] that the water-soluble dinuclear ruthenium(II) complex $Ru_2(\mu\text{-}OAc)_3(\mu\text{-}CO_3)$ catalyzes the aerobic oxidation of a variety of alcohols in water at 80 °C and ambient pressure but at high catalyst loadings (10 mol% Ru). A marked preference for primary versus secondary alcohol moieties was observed. The use of manganosilicate molecular sieves as heterogeneous catalysts for the oxidation of alcohols with peroxodisulfate as the terminal oxidant has been described [83]. Obviously, it would be more interesting if this could be performed with $H_2O_2$ or $O_2$ as the terminal oxidant.

Hypervalent iodine compounds, in stoichiometric amounts, are also known to oxidize alcohols and the use of iodosylbenzene or a polymer-supported iodine(III) reagent, in combination with KBr as a cocatalyst, for the oxidation of alcohols in water has been described [84]. More recently, a related catalytic system, consisting of $PhIO_2$ (2 mol%), $Br_2$ (2 mol%), and $NaNO_2$ (1 mol%), for the aerobic oxidation of alcohols in water at 55 °C has been described [85].

The stable free radical TEMPO is an example of an organocatalyst that is effective in the oxidation of a broad range of alcohols [86], including simple carbohydrates [87] and polysaccharides [88], using hypochlorite (household bleach) as the terminal oxidant (Scheme 4.18). The stoichiometric oxidation of primary alcohols, to the corresponding aldehydes, by the oxoammonium cation, derived from one-electron oxidation of TEMPO, was first reported by Golubev et al. in 1965 [89]. The reaction is rendered catalytic in TEMPO by using single oxygen donors such as m-chloroperbenzoic acid [90], persulfate(oxone) [91], periodic acid ($H_5IO_6$) [92], and sodium hypochlorite [93] to generate the oxoammonium cation *in situ*. In particular, the TEMPO–hypochlorite protocol, using 1 mol% TEMPO in combination with 10 mol% sodium bromide as cocatalyst, in dichloromethane–water at pH 9 and 0 °C has been widely applied in organic synthesis. The method was first described in 1987 by Montanari and co-workers, who used 4-methoxy-TEMPO as the catalyst [93]. The catalytic cycle involves alternating oxidation of the alcohol by the oxoammonium cation and regeneration of the latter by reaction of the TEMPOH with the primary oxidant (hypochlorite). Hence TEMPO is the catalyst precursor, which is presumably oxidized by bromine or chlorine (Scheme 4.18) to the oxoammonium cation, which enters the catalytic cycle.

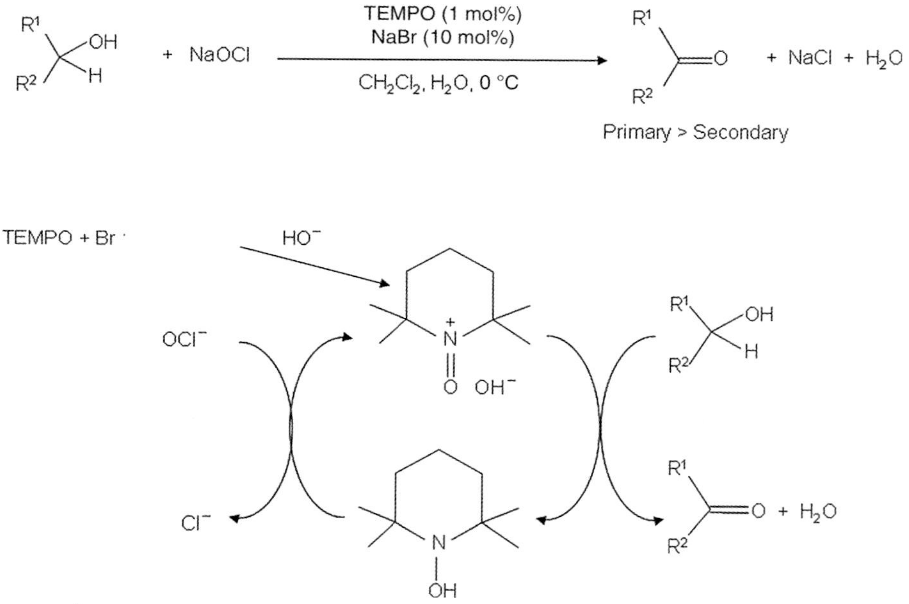

**Scheme 4.18** TEMPO-catalyzed oxidation of alcohols with NaOCl.

The Montanari protocol, although widely applicable, suffers from several environmental and/or economic drawbacks. It is not waste free because at least 1 equiv. of sodium chloride is produced per molecule of alcohol oxidized, and the use of hypochlorite as oxidant can also lead to the formation of chlorinated by-products. Other shortcomings are the use of 10 mol% bromide as a cocatalyst and dichloromethane as a solvent. Furthermore, although only 1 mol% is used, TEMPO is rather expensive, which means that efficient recycling is an important issue. Some of these problems were circumvented by replacing the TEMPO with a recyclable oligomeric TEMPO, referred to as PIPO (polymer-immobilized piperidinyloxyl), derived from the commercially available antioxidant and light stabilizer Chimassorb 944, an oligomeric sterically hindered amine (Scheme 4.19). PIPO proved to be a very effective and recyclable catalyst for the oxidation of alcohols, including a wide variety of carbohydrates, with hypochlorite in a bromide- and chlorinated hydrocarbon-free system [94–97]. The reaction is performed with 1 mol% of PIPO and 1.25 equiv. of NaOCl in water as the sole solvent or in a water–$n$-hexane mixture.

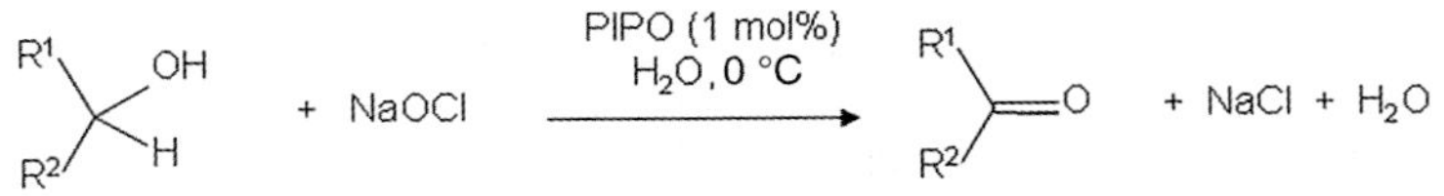

| substrate | time (min) | conv. (%) | sel. (%) |
|---|---|---|---|
| 1-octanol | 45 | 90 | 50 |
|  | 45 | 80 | 94[a] |
| 1-hexanol | 45 | 89 | 95[a] |
| 2-octanol | 45 | 99 | >99 |
| cyclooctanol | 45 | 100 | >99 |
| benzyl alcohol | 30 | 100 | >99 |
| 1-phenylethanol | 30 | 100 | >99 |

[a] $n$-hexane as solvent

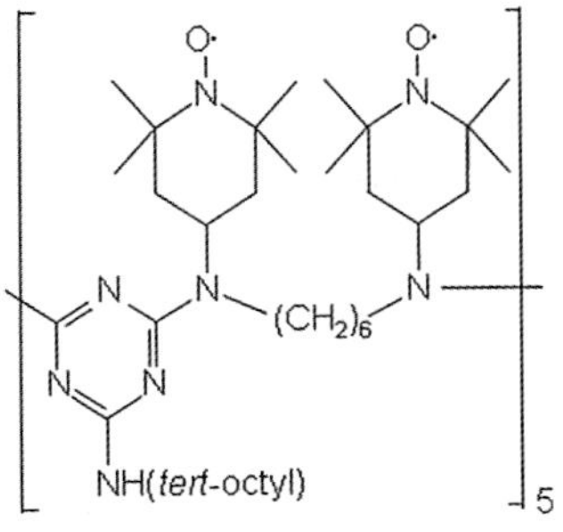

Rel. rates:
- $PhCH_2OH$ / $PhCH(OH)CH_3$  95/4
- 1-octanol / 2-octanol  86/1

**Scheme 4.19** PIPO catalyzed oxidation of alcohols with NaOCl.

Recently, a so-called ion-supported TEMPO was synthesized by building a TEMPO moiety into the side chain of a dialkylimidazolium salt (Scheme 4.20). The resulting material catalyzed the oxidation of alcohols with NaOCl or $I_2$ in water or an ionic liquid–water mixture [98].

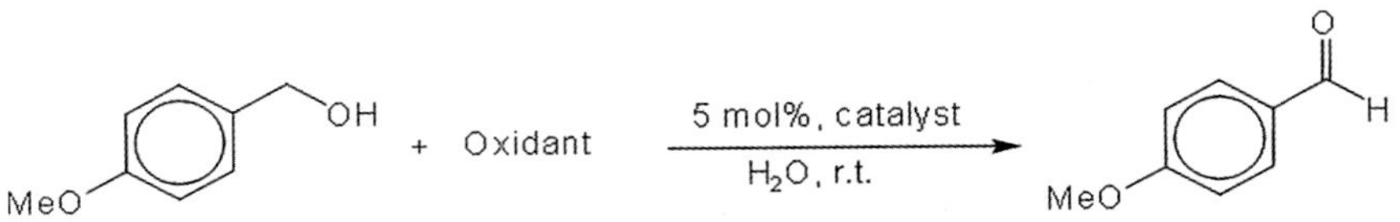

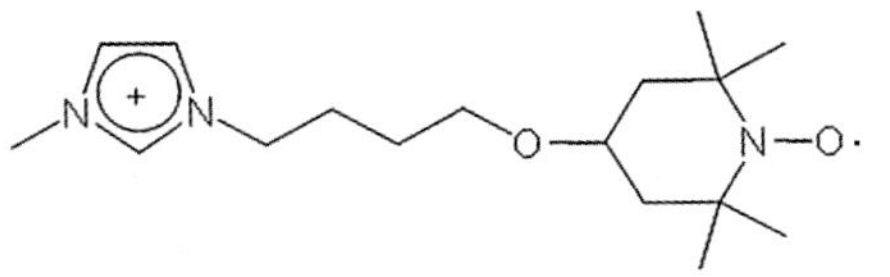

| Oxidant | Time (min) | Yield (%) |
|---|---|---|
| $I_2$ | 40 | 98 |
| NaOCl | 3 | 96 |

**Scheme 4.20** Ion-supported TEMPO as oxidation catalyst.

Although these recyclable systems offer many economic and environmental benefits, they still suffer from the disadvantage of requiring hypochlorite as a stoichiometric oxidant. Its industrial potential would be significantly enhanced if the latter could be replaced by dioxygen or hydrogen peroxide. Copper complexes of bipyridine ligands, in combination with TEMPO, have been shown to catalyze the aerobic oxidation of alcohols in aqueous acetonitrile or dimethylformamide [99, 100]. Furthermore, a copper-dependent oxidase, laccase (EC 1.10.3.2), in combination with TEMPO as a cocatalyst (or so-called mediator) was shown by Galli and co-workers to catalyze the aerobic oxidation of primary benzylic alcohols [101]. The laccase–TEMPO-catalyzed selective aerobic oxidation of the primary alcohol moiety in carbohydrates had been previously reported in two patents [102, 103]. It is generally believed that these reactions involve one-electron oxidation of the TEMPO by the oxidized form of the laccase to afford the oxoammonium cation, followed by reoxidation of the reduced form of laccase by dioxygen [95].

Laccases are extracellular enzymes that are secreted by white rot fungi and play an important role in the delignification of lignocellulose, the major constituent of wood, by these microorganisms [104]. There is currently considerable commercial interest in laccases for application in pulp bleaching (as a replacement for chlorine) in paper manufacture and remediation of phenol-containing waste streams. The selective oxidation of the primary alcohol groups in starch, to the corresponding carboxyl functionalities, is another reaction of commercial importance (see earlier). It has been shown [98, 99] that laccase, in combination with TEMPO, or derivatives thereof, is able to catalyze the aerobic oxidation of the primary alcohol moieties in starch, but the relatively high enzyme costs are an obstacle to commercialization. Inefficient laccase use is a result of its instability towards the oxidizing reaction conditions. We have recently shown that the stability of the laccase under reaction conditions can be improved by immobilization as a cross-linked enzyme aggregate (CLEA) [105]. Indeed, laccase CLEA has broad potential as a catalyst for a variety of applications.

The combinations of TEMPO with $NaNO_2$ and 1,3-dibromo-5,5-dimethylhydantoin [106] or $FeCl_3$ [107] as cocatalysts have also been shown to catalyze the aerobic oxidation of alcohols in water at 80 °C.

## 4.6
### Aldehyde and Ketone Oxidations in Water

The Baeyer–Villiger oxidation of aldehydes and ketones to the corresponding esters or lactones (Scheme 4.21) is a widely applied reaction in organic synthesis. Traditionally, it is performed with an organic peracid such as peracetic acid. However, the use of a peracid results in the formation of 1 equiv. of the corresponding carboxylic acid salt as waste, which has to be recycled or disposed of. Moreover, organic peracids are expensive and/or hazardous (because of shock sensitivity), which limits their commercial application. The transport and storage of peracetic acid, for example, have been severely restricted, making its use prohibitive.

Consequently, increasing attention is focused on the development of procedures deploying aqueous hydrogen peroxide as the primary oxidant, preferably in water as the sole solvent [108].

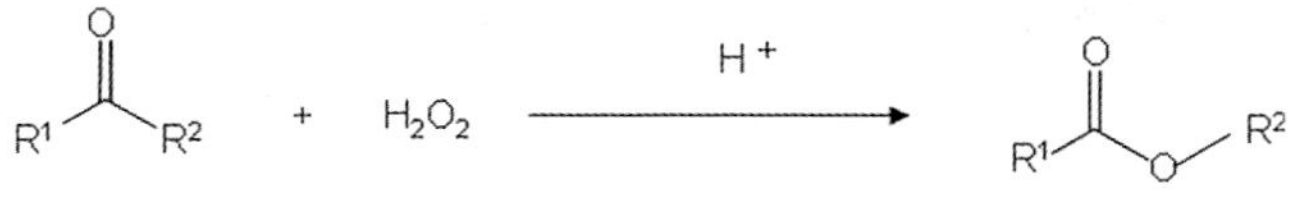

$$R^1-CO-R^2 + H_2O_2 \xrightarrow{H^+} R^1-CO-O-R^2$$

$R^1$, $R^2$ = alkyl, H
**Scheme 4.21** Baeyer–Villiger oxidation of a ketone.

In their studies of the tungstate-catalyzed oxidation of primary alcohols to the corresponding carboxylic acids, via the corresponding aldehydes (see above), Noyori and co-workers discovered that aldehydes could be selectively converted to carboxylic acids by reaction with aqueous hydrogen peroxide in an aqueous–organic biphasic system, in the presence of the acidic, lipophilic phase-transfer catalyst methyltrioctylammonium hydrogensulfate, $[CH_3(n\text{-}C_8H_{17})_3N]^+ HSO_4^-$, without the need for an organic solvent, halide, or even a metal catalyst (Scheme 4.22) [22, 109]. The reactions proceeded via the formation of the aldehyde perhydrate, by addition of hydrogen peroxide to the carbonyl group, which undergoes Brønsted acid-catalyzed rearrangement to the ester (lactone) product and water.

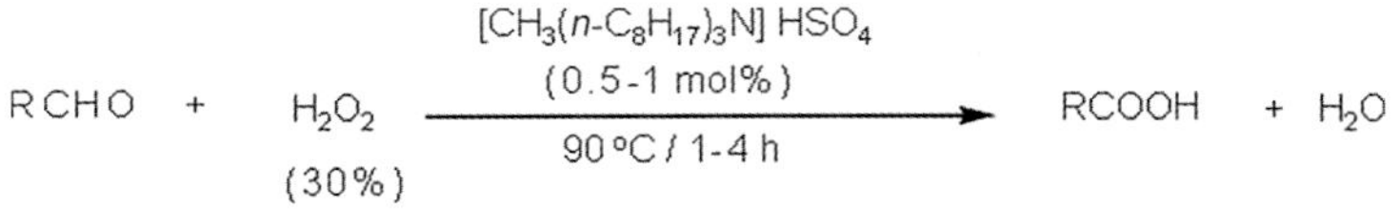

$$RCHO + H_2O_2 \xrightarrow[90\,°C\,/\,1\text{-}4\,h]{[CH_3(n\text{-}C_8H_{17})_3N]\,HSO_4\ (0.5\text{-}1\ \text{mol}\%)} RCOOH + H_2O$$
(30%)

| R | $H_2O_2$ (equiv.) | Yield RCOOH (%) |
|---|---|---|
| $n\text{-}C_7H_{15}$ | 1.1 | 85 |
| $PhCH_2CH_2$ | 1.1 | 78 |
| $C_6H_5$ | 2.5 | 85 |
| $4\text{-}NO_2C_6H_4$ | 2.5 | 93 |
| $4\text{-}MeOC_6H_4$ | 2.5 | 9 |

**Scheme 4.22** Catalytic oxidation of aldehydes with aqueous $H_2O_2$.

Cyclic ketones were oxidized to the corresponding dicarboxylic acids, via initial Baeyer–Villiger reaction, using hydrogen peroxide in the presence of a catalyst comprising 1 mol% tungstic acid, $H_2WO_4$, under organic solvent- and halide-free conditions (Scheme 4.23) [110]. For example, cyclohexanone afforded adipic acid in 91% isolated yield using 3.3 equiv. of hydrogen peroxide. The acidic nature of the catalyst is crucial because when $Na_2WO_4$ was used no reaction took place. Since analogous systems can also oxidize alcohols to ketones (see above), reaction of cyclohexanol with 4.4 equiv. of hydrogen peroxide afforded adipic acid in 87% isolated yield.

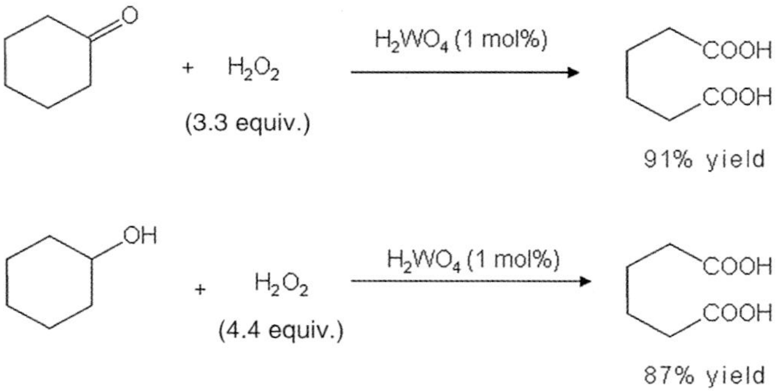

## 4 Green Oxidation in Water

**Scheme 4.23** Catalytic oxidation of cyclohexanol/cyclohexanone with $H_2O_2$.

Cyclohexanone + $H_2O_2$ (3.3 equiv.) → [$H_2WO_4$ (1 mol%)] → adipic acid (COOH, COOH), 91% yield

Cyclohexanol + $H_2O_2$ (4.4 equiv.) → [$H_2WO_4$ (1 mol%)] → adipic acid (COOH, COOH), 87% yield

Corma et al. [111] reported that Sn-beta containing tetrahedrally coordinated Sn(IV) in the zeolite framework is a highly active and selective heterogeneous catalyst for the Baeyer–Villiger oxidation of ketones and aldehydes with aqueous hydrogen peroxide. The Sn-substituted mesoporous silica Sn-MCM-41 was subsequently shown to exhibit similar activity [112]. The reactions are not strictly oxidations in water as they were performed in a water-miscible organic solvent, such as dioxane, and it was not clear if they could be successfully carried out in water alone. Evidence has been presented [113] to support a mechanism involving simultaneous coordination of the carbonyl group of the substrate and the hydrogen peroxide to the catalyst.

## 4.7
## Sulfoxidations in Water

Several pharmaceuticals contain a sulfoxide or sulfone moiety and the oxidation of dialkyl sulfides to the corresponding sulfoxides (or sulfones), including the corresponding enantioselective oxidation of prochiral sulfides, are important reactions in organic synthesis [114]. Noyori and co-workers found that aromatic and aliphatic sulfides are oxidized to the corresponding sulfones, in high yields, using 30% aqueous $H_2O_2$ in an aqueous biphasic system, in the absence of an organic solvent (Scheme 4.24) [22, 115]. The catalyst consisted of sodium tungstate, phenylphosphonic acid ($PhPO_3H_2$), and a tetraalkylammonium hydrogensulfate as a phase-transfer agent. Using a slight excess of $H_2O_2$, smooth oxidation to the sulfone was observed at 50 °C with a substrate-to-catalyst ratio of 1000–5000 (Scheme 4.24). Olefinic double bonds and primary and secondary alcohol functionalities remained intact under these conditions. Alternatively, the sulfoxide could be obtained, in high yield, by performing the reaction in the absence of the tungstate or at lower temperatures, such as 0 °C. It was suggested [115] that the function of the phenylphosphonic acid cocatalyst is to increase the reactivity of peroxo ligands by coordination to the W(VI).

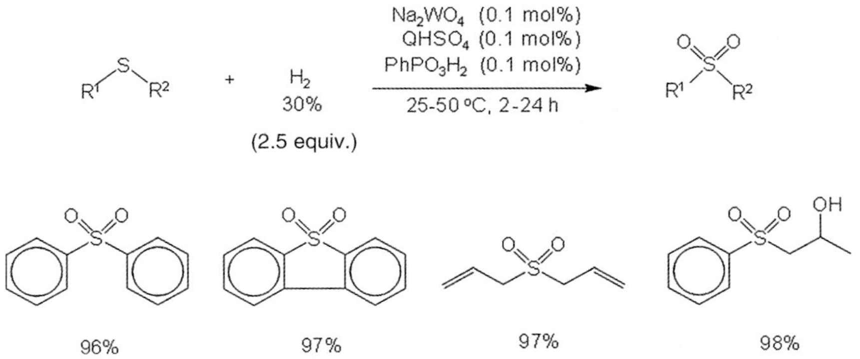

**Scheme 4.24** Catalytic sulfoxidations with $H_2O_2$.

Reaction conditions: $Na_2WO_4$ (0.1 mol%), $QHSO_4$ (0.1 mol%), $PhPO_3H_2$ (0.1 mol%), $H_2O_2$ 30% (2.5 equiv.), 25–50 °C, 2–24 h.

Products: 96%, 97%, 97%, 98%.

Enantioselective sulfoxidation of prochiral sulfides is also of industrial interest, for example in the synthesis of the anti-ulcer drug esomeprazole, an enantiomerically pure sulfoxide [114, 116]. The most practical method which is used in the industrial synthesis of esomeprazole involves titanium-catalyzed oxidation with an alkyl hydroperoxide, and a dialkyltartrate as chiral ligand, in an organic solvent such as dichloromethane [114, 116]. A variety of oxidoreductases are known to catalyze the enantioselective oxidation of prochiral sulfides, usually as whole-cell biotransformations in aqueous media, but no simple metal complexes have been shown to be effective in water and the development of practical systems employing aqueous hydrogen peroxide as the primary oxidant is still an important challenge. In this context, it is worth mentioning the enantioselective sulfoxidation of prochiral sulfoxides catalyzed by the semi-synthetic peroxidase vanadium phytase [117] in an aqueous medium.

## 4.8
## Conclusion

Important advances have been and are still being made in the design of effective catalytic systems for oxidations with the green oxidants dioxygen and hydrogen peroxide in an aqueous mono- or biphasic system, in the absence of organic solvents. Various practical systems have been reported, based on metal complexes of water-soluble, oxidatively stable ligands or using a phase-transfer agent to transport the catalyst and/or active oxidant to the organic phase. There is a marked trend towards the use of complexes of inexpensive and more environmentally acceptable first-row elements (Fe, Mn, and Cu) rather than noble metal complexes as catalysts. Methods based on stable N-oxy radicals as organocatalysts for alcohol oxidations are becoming increasingly popular, largely because they are "heavy metal free." The last few years have also witnessed the ascendance of noble metal (particularly Au) nanoclusters as catalysts for aerobic oxidations of, inter alia, simple alcohols and polyols.

These methodologies constitute green alternatives – clean oxidants, no need for organic solvents, facile product separation and catalyst recycling – for traditional oxidations. In the future, we expect that they will be further applied in industrial organic synthesis.

## References

1 Sheldon, R.A., Arends, I.W.C.E. and Hanefeld, U. (2007) *Green Chemistry and Catalysis*, Wiley-VCH Verlag GmbH, Weinheim.
2 Lindström, U.M. (2007) *Organic Reactions in Water*, Blackwell; Cornils, B. and Herrmann, W. (2006) *Aqueous Phase Organometallic Chemistry*, 2nd edn, Wiley-VCH Verlag GmbH, Oxford, Weinheim.
3 For example, see: Besson, M. and Gallezot, P. (2001) In: *Fine Chemicals Through Heterogeneous Catalysis* (eds R.A. Sheldon and H. van Bekkum), Wiley-VCH Verlag GmbH, Weinheim, pp. 491–506, and 507–518.
4 For a recent review of catalysis by gold nanoparticles, see: Hashmi, A.S.K. and Hutchings, G.J. (2006) *Angew. Chem. Int. Ed.*, **45**, 7896–7936.
5 Christensen, C.H., Rass-Hansen, J., Marsden, C.C., Taarning, E. and Egeblad, K. (2008) *ChemSusChem*, **1**, 283–289; Clark, J.H. (2007) *J. Chem. Technol. Biotechnol.*, **82**, 603–609; Lange, J.P. (2007) *Biofuels Bioprod. Bioref.*, **1**, 39–48; Dale, B.E. (2007) *Biofuels Bioprod. Bioref.*, **1**, 24–38; Dale, B.E. (2003) *J. Chem. Technol. Biotechnol.*, **78**, 1093–1103.
6 Watson, P.A., Wright, L.J. and Fullerton, T.J. (1993) *J. Wood Chem. Technol.*, **13**, 371, 391, 411.
7 Hampton, K.W. and Ford, W.T. (1996) *J. Mol. Catal. A: Chemical*, **113**, 167.
8 Weinstock, I.A., Atilla, R.H., Reiner, R.S., Moen, M.A., Hammel, K.E., Houtman, C.J., Hill, C.L. and Harrup, M.K. (1997) *J. Mol. Catal. A: Chemical*, **116**, 59.
9 Sorokin, A. Séris, J.-L. and Meunier, B. (1995) *Science*, **268**, 1163; Sorokin, A. DeSuzzoni-Dezard, S. Poullain, D. Noel, J.P. and Meunier, B. (1996) *J. Am. Chem. Soc.*, **118**, 7410–7411; Meunier, B. and Sorokin, A. (1997) *Acc. Chem. Res.*, **30**, 470–476.
10 Shakla, R.S. Robert, A. and Meunier, B. (1996) *J. Mol. Catal. A: Chemical*, **113**, 45.
11 Kamp, N.W.J. and Lindsay Smith, J.R. (1996) *J. Mol. Catal. A: Chemical*, **113**, 131.
12 Kachkarova-Sorokina, S.L. Gallezot, P. and Sorokin, A.B. (2004) *Chem. Commun.*, 2844–2845.
13 Sorokin, A.B. Kachkarova-Sorokina, S.L. Donze, C. Pinel, C. and Gallezot, P. (2004) *Top. Catal.*, **27**, 67–76.
14 Collins, T.J. (2002) *Acc. Chem. Res.*, **35**, 782–790; Banerjee, D. Markley, A.L. Yano, T. Ghosh, A. Berget, P.B. Minkley, E.G. Khetan, S.K. and Collins, T.J. (2006) *Angew. Chem. Int. Ed.*, **45**, 3974–3977; Chahbane, N. Popescue, D.L. Mitchell, D.A. Chanda, A. Lenoir, D. Ryabov, A.D. Schramm, K.W. and Collins, T.J. (2007) *Green Chem.*, **9**, 49–57.
15 Collins, T.J. (1994) *Acc. Chem. Res.*, **27**, 279–285.
16 For a review of metal-catalyzed epoxidations with hydrogen peroxide, see: Lane, B.S. and Burgess, K. (2003) *Chem. Rev.*, **103**, 2437–2473; see also: Bregeault, J.M. (2003) *Dalton Trans.*, 3289–3302.
17 Venturello, G., Alneri, E. and Ricci, M. (1983) *J. Org. Chem.*, **48**, 3831–3833.
18 Ishii, Y., Yamawaki, K., Ura, T., Yamada, H., Yoshiba, T. and Ogawa, M. (1988) *J. Org. Chem.*, **53**, 3587.
19 Aubry, C., Chottard, G., Platzer, N., Bregault, J.M., Thouvenot, R., Chauveau,

F., Huet, C. and Ledon, H. (1991) *Inorg. Chem.*, **30**, 4409.

20 For recent reviews, see: Mahha, Y., Salles, L., Piquemal, J.Y., Briot, E., Atlamsami, A. and Bregeault, J.M. (2007) *J. Catal.*, **249**, 338–348; Bregeault, J.M., Vennat, M., Salles, L., Piquemal, J.Y., Mahha, Y., Briot, E., Bakala, P.C., Atlasami, A. and Thouvenot, R. (2006) *J. Mol. Catal. A: Chemical*, **250**, 177–189.

21 Sato, K., Aoki, M., Ogawa, M., Hashimoto, T. and Noyori, R. (1996) *J. Org. Chem.*, **61**, 8310–8311.

22 Noyori, R., Aoki, M. and Sato, K. (2003) *Chem. Commun.*, 1977–1986.

23 Sato, K., Aoki, M. and Noyori, R. (1998) *Science*, **281**, 1646–1647.

24 Maheswari, P.U., de Hoog, P., Hage, R., Gamez, P. and Reedijk, J. (2005) *Adv. Synth. Catal.*, **347**, 1759–1764.

25 Xi, Z., Zhou, N., Sun, Y. and Li, K. (2001) *Science*, **292**, 1139–1141.

26 Hage, R. and Lienke, A. (2006) *Angew. Chem. Int. Ed.*, **45**, 206–222 and references cited therein; Hage, R. (1996) *Recl. Trav. Chim. Pays-Bas*, **115**, 385.

27 De Vos, D.E., Sels, B.F., Reynaers, M., Subba Rao, Y.V. and Jacobs, P.A. (1998) *Tetrahedron Lett.*, **39**, 3221.

28 Berkessel, A. and Sklorz, C.A. (1999) *Tetrahedron Lett.*, **40**, 7965.

29 De Vos, D.E., De Wildeman, S., Sels, B.F., Grobet, P.J. and Jacobs, P.A. (1999) *Angew. Chem. Int. Ed.*, **38**, 980.

30 Brinksma, J., Schmieder, L., van Vliet, G., Boaron, R., Hage, R., De Vos, D.E., Alsters, P.L. and Feringa, B.L. (2002) *Tetrahedron Lett.*, **43**, 2619.

31 Hage, R. and Lienke, A. (2006) *J. Mol. Catal. A: Chemical*, **251**, 150–158; de Boer, J.W., Brinksma, J., Browne, W.R., Meetsma, A., Alsters, P.L., Hage, R. and Feringa, B. (2005) *J. Am. Chem. Soc.*, **127**, 7990.

32 de Boer, J.W., Browne, W.R., Brinksma, J., Alsters, P.L., Hage, R. and Feringa, B.L. (2007) *Inorg. Chem.*, **46**, 6353–6372.

33 de Boer, J.W., Browne, W.R., Harutyunyan, S.R., Bini, L., Tiemersma-Wegman, T.D., Alsters, P.L., Hage, R. and Feringa, B. (2008) *Chem. Commun.*, 3747–3749.

34 Lane, B.S. and Burgess, K. (2001) *J. Am. Chem. Soc.*, **123**, 2933.

35 Lane, B.S., Vogt, M., De Rose, V.J. and Burgess, K. (2002) *J. Am. Chem. Soc.*, **124**, 11947.

36 Tong, K.-H., Wong, K.-Y. and Chan, T.H. (2005) *Tetrahedron*, **61**, 6009–6014.

37 For leading references, see: Kazlauskas, R. (2005) *Curr. Opin. Chem. Biol.*, **9**, 195; Taglieber, A., Höbenreich, H., Carballeira, J.D., Mondiere, R.J.G. and Reetz, M.T. (2007) *Angew. Chem. Int. Ed.*, **46**, 8597; Reetz, M.T., Mondiere, R. and Carballeira, J.D. (2007) *Tetrahedron Lett.*, **48**, 1679; Hult, K. and Berglund, P. (2007) *Trends Biotechnol.*, **25**, 231.

38 Okrasa, K. and Kazlauskas, R.J. (2006) *Chem. Eur. J.*, **12**, 1587.

39 Chen, K., Costas, M., Kim, J., Tipton, A.K. and Que, L. (2002) *J. Am. Chem. Soc.*, **124**, 3026; Bassan, A., Blomberg, M.R.A., Siegbahn, P.E.M. and Que, L. (2002) *J. Am. Chem. Soc.*, **124**, **11056**, and references cited therein.

40 White, M.C., Doyle, A.G. and Jacobsen, E.N. (2001) *J. Am. Chem. Soc.*, **123**, 7194.

41 Brinksma, J., Hage, R., Kerschner, J. and Feringa, B.L. (2000) *Chem. Commun.*, 537.

42 Tse, M.K., Klawonn, M., Bhor, S., Döbler, C., Anilkumar, G., Hugl, H., Maegerlein, W. and Beller, M. (2005) *Org. Lett.*, **7**, 987–990.

43 For a review, see: Mägerlein, W., Dreisbach, C., Hugl, H., Tse, M.K., Klawonn, M., Bhor, S. and Beller, M. (2007) *Catal. Today*, **121**, 140–150.

44 Tse, M.K., Döbler, C., Bhor, S., Klawonn, M., Mägerlein, W., Hugl, H. and Beller, M. (2004) *Angew. Chem. Int. Ed.*, **43**, 5255; Tse, M.K., Bhor, S., Klawonn, M., Anilkumar, G., Jiao, H., Spannenberg, A., Döbler, C., Mägerlein, W., Hugl, H. and Beller, M. (2006) *Chem. Eur. J.*, **12**, 1855;

Anilkumar, G., Bhor, S., Tse, M.K., Klawonn, M., Bitterlich, B. and Beller, M. (2005) *Tetrahedron: Asymmetry*, **16**, 3536–3561.

45 Anilkumar, G., Bitterlich, B., Gelalcha, F.G., Tse, M.K. and Beller, M. (2007) *Chem. Commun.*, 289–291.

46 Barton, D.H.R. and Doller, D. (1992) *Acc. Chem. Res.*, **25**, 504–512.

47 Gelalcha, F.G., Bitterlich, B., Anilkumar, G., Tse, M.K. and Beller, M. (2007) *Angew. Chem. Int. Ed.*, **46**, 7293–7296.

48 Uematsu, N., Fujii, A., Hashiguchi, S., Ikariya, T. and Noyori, R. (1996) *J. Am. Chem. Soc.*, **118**, 4916.

49 Usui, Y., Sato, K. and Tanaka, M. (2003) *Angew. Chem. Int. Ed.*, **42**, 5623–5625.

50 Sheldon, R.A. and van Vliet, M.C.A. (2001) In: *Heterogeneous Catalysis for Fine Chemicals* (eds R.A. Sheldon and H. van Bekkum), Wiley-VCH Verlag GmbH, Weinheim, pp. 473–490 and references cited therein.

51 Sato, T., Dakka, J. and Sheldon, R.A. *Stud. Surf. Sci. Catal.*, (1994) **84**, 1853–1860; Corma, A. Navarro, M.T. and Perez-Pariente, J. (1994) *Chem. Commun.*, 147–148.

52 www.chemicals-technology.com/projects/BASF-HPPO.

53 Bortolini, O., Conte, V., DiFuria, F. and Modena, G. (1986) *J. Org. Chem.*, **51**, 2661.

54 Sato, K., Aoki, M., Takagi, J. and Noyori, R. (1997) *J. Am. Chem. Soc.*, **119**, 12386; Sato, K., Takagi, J., Aoki, M. and Noyori, R. (1998) *Tetrahedron Lett.*, **39**, 7549; Sato, K., Aoki, M., Takagi, J., Zimmermann, K. and Noyori, R. (1999) *Bull. Chem. Soc. Jpn.*, **72**, 2287; for an analogous system see: Shi, X.Y. and Wei, J.F. (2005) *J. Mol. Catal. A: Chemical*, **229**, 13–17.

55 Sloboda-Rozner, D., Alsters, P.L. and Neumann, R. (2003) *J. Am. Chem. Soc.*, **125**, 5280–5281; Sloboda-Rozner, D., White, P., Alsters, P. and Neumann, R. (2004) *Adv. Synth. Catal.*, **346**, 339–345; see also: Haimov, A. and Neumann, R. (2006) *J. Am. Chem. Soc.*, **128**, 15697–15700.

56 For reviews, see: Muzart, J. and Tetrahedron, (2003) **59**, 5789; Stahl, S.S. (2004) *Angew. Chem. Int. Ed.*, **43**, 3400.

57 For recent reviews of Wacker oxidations, see: Hintermann, L. (2004) In: *Transition Metals in Organic Synthesis* (eds M. Beller and C. Bolm), Wiley-VCH Verlag GmbH, Weinheim, pp. 379–388; Monflier, E. and Mortreux, A. (2004) In: *Aqueous Phase Organometallic Chemistry* (eds B. Cornils and W. Herrmann), 2nd edn, Wiley-VCH Verlag GmbH, Weinheim, pp. 481–487.

58 Papadogianakis, G., Verspui, G., Maat, L. and Sheldon, R.A. (1997) *Catal. Lett.*, **47**, 43.

59 ten Brink, G.-J., Arends, I.W.C.E., Papadogianakis, G. and Sheldon, R.A. (1998) *Chem. Commun.*, 2359–2360; ten Brink, G.-J., Papadogianakis, G., Arends, I.W.C.E. and Sheldon, R.A. (2000) *Appl. Catal. A*, **435**, 194–195.

60 ten Brink, G.J., Arends, I.W.C.E. and Sheldon, R.A. (2000) *Science*, **287**, 1636.

61 ten Brink, G.J., Arends I.W.C.E. and Sheldon, R.A. (2002) *Adv. Synth. Catal.*, **344**, 355.

62 Sheldon, R.A., Arends, I.W.C.E. and ten Brink, G.J. (2002) *Acc. Chem. Res.*, **35**, 774.

63 Stahl, S.S., Thorman, J.L., Nelson, R.C. and Kozee, M.A. (2001) *J. Am. Chem. Soc.*, **23**, 7188–7189; see also: Steinhoff, B.A. Stahl, S.S. (2002) *Org. Lett.*, **4**, 4179; Steinhoff, B.A., Fix, S.R. and Stahl, S.S. (2002) *J. Am. Chem. Soc.*, **124**, 766; Steinhoff, B.A., King, A.E. and Stahl, S.S. (2006) *J. Org. Chem.*, **71**, 1861–1868; Conley, N.R., Labios, L.A.B., Pearson, D.M., McCrory, C.M. and Waymouth, R.M. (2007) *Organometallics*, **26**, 5447.

64 ten Brink, G.J., Arends, I.W.C.E., Hoogenraad, M., Verspui, G. and Sheldon, R.A. (2003) *Adv. Synth. Catal.*, **345**, 497.

65 ten Brink, G.J., Arends, I.W.C.E., Hoogenraad, M., Verspui, G. and Sheldon, R.A. (2003) *Adv. Synth. Catal.*, **345**, 1341.

66 Buffin, B.P., Belitz, N.L. and Verbeke, S.L. (2008) *J. Mol. Catal. A: Chemical*, **284**, 149–154; Buffin, B.P., Clarkson, J.P., Belitz, N.L. and Kundu, A. (2005) *J. Mol. Catal. A: Chemical*, **225**, 111–116.
67 In preparation.
68 Kovtun, G., Kameneva, T., Hladyi, S., Starchevsky, M., Pazdersky, Y., Stolarov, I., Vargaftik, M. and Moiseev, I.I. (2002) *Adv. Synth. Catal.*, **344**, (9); 957–964; Moiseev, I.I. and Vargaftik, M.N. (1998) *New J. Chem.*, **22**, 1217–1227; Pasichnyk, P.I., Starchevsky, M.K., Pazdersky, Y.A., Zagorodnikov, V.P., Vargaftik, M.N. and Moiseev, I.I. (1994) *Mendeleev Commun.*, 1–2; Moiseev, I.I., Stromnova, T.A. and Vargaftik, M.N. (1994) *J. Mol. Catal.*, **86**, 71–94.
69 Mori, K., Hara, T., Mizugaki, T., Ebitani, K. and Kaneda, K. (2004) *J. Am. Chem. Soc.*, **126**, 10657–10666.
70 Abad, A., Concepcion, P., Corma, A. and Garcia, H. (2005) *Angew. Chem. Int. Ed.*, **44**, 4066–4069.
71 Christensen, C.H., Jørgensen, B., Rass-Hansen, J., Egeblad, K., Madfsen, R., Klitgaard, S.K., Hansen, S.M., Hansen, M.R., Andersen, H.C. and Riisager, A. (2006) *Angew. Chem. Int. Ed.*, **45**, 4648–4651.
72 Enache, D.I., Edwards, J.K., Landon, P., Solsona-Espriu, B., Carley, A.F., Herzing, A.A., Watanabe, M., Kiely, C.J., Knight, D.W. and Hutchings, G.J. (2006) *Science*, **311**, 362–365.
73 Taarning, E., Nielsen, I.S., Egeblad, K., Madsen, R. and Christensen, C.H. (2008) *ChemSusChem*, **1**, 75–78; see also Taarning, E., Madsen, A.T., Marchetti, J.M., Egeblad, K. and Christensen, C.H. (2008) *Green Chem.*, **10**, 408–414.
74 Prati, L. and Rossi, M. (1997) *Stud. Surf.Sci.Catal.*, **110**, 509; Prati, L. and Rossi, M. (1998) *J. Catal.*, **176**, 552; Bianchi, C., Porta, F., Prati, L. and Rossi, M. (2000) *Top. Catal.*, **13**, 231; Comotti, M., Della Pina, C., Matarrese, R., Rossi, M. and Siani, A. (2005) *Appl. Catal. A: General*, **291**, 204–209; Prati, L. and Porta, F.

(2005) *Appl. Catal. A: General*, **291**, 199–203.
75 Biffis, A. and Minati, L. (2005) *J. Catal.*, **236**, 405–409; Biffis, A., Cunial, S., Spontoni, P. and Prati, L. (2007) *J. Catal.*, **251**, 1–6.
76 Tsunoyama, H., Sakurai, H., Negishi, Y. and Tsukuda, T. (2005) *J. Am. Chem. Soc.*, **127**, 9374–9375.
77 Miyamura, H., Matsubara, R., Miyazaki, Y. and Kobayashi, S. (2007) *Angew. Chem. Int. Ed.*, **46**, 4151–4154.
78 Uozumi, Y. and Nakao, R. (2003) *Angew. Chem. Int. Ed.*, **42**, 194–197.
79 Yamada, Y.M.A., Arakawa, T., Hocke, H. and Uozumi, Y. (2007) *Angew. Chem. Int. Ed.*, **46**, 704–706.
80 Mertens, P.G.N., Vandezande, P., Ye, X., Poelman, H., De Vos, D.E. and Vankelecom, I.F.J. (2008) *Adv. Synth. Catal.*, **350**, 1241–1247.
81 Hamamoto, H., Suzuki, Y., Yamada, Y.M.A., Tabata, H., Takahashi, H. and Ikegami, S. (2005) *Angew. Chem Int.*, **44**, 4536.
82 Komiya, N., Nakae, T., Sato, H. and Naota, T. (2006) *Chem. Commun.*, 4829–4831.
83 Manyar, H.G., Chaure, G.S. and Kumar, A. (2006) *Green Chem.*, **8**, 344–348.
84 Tohma, H., Maegawa, T., Takizawa, S. and Kita, Y. (2002) *Adv. Synth. Catal.*, **344**, 328–337.
85 Liu, Z.-L., Yang, L., Mu, R., Liu, Z.Q., Yang, Z. and Liu, Z.G. (2005) *Adv. Synth. Catal.*, **347**, 1333.
86 de Nooy, A.E.J., Besemer, A.C. and van Bekkum, H. (1996) *Synthesis*, 1153–1174.
87 Bragd, P.L., Besemer, A.C. and van Bekkum, H. (2001) *J. Mol. Catal. A: Chemical*, **170**, 35–42.
88 Bragd, P.L., van Bekkum, H. and Besemer, A.C. (2004) *Top. Catal.*, **27**, 49–66.
89 Golubev, V.A., Rozantsev, E.G. and Neiman, M.B. (1965) *Bull. Acad. Sci. USSR, Chem. Ser.*, **14**, 1898.
90 Cella, J.A., Kelley, J.A. and Kenehan, E.F. (1975) *J. Org. Chem.*, **40**, 1850; Rychovsky,

S.D. and Vaidyanathan, R. (1999) *J. Org. Chem.*, **64**, 310.
91 Bolm, C., Magnus, A.S. and Hildebrand, J.P. (2000) *Org. Lett.*, **2**, 1173.
92 Kim, S.S. and Nehru, K. (2002) *Synlett.*, 616.
93 Anelli, L. Biffi, C. Montanari, F. and Quici, S. (1987) *J. Org. Chem.*, **52**, 2559; see also: Anelli, P., Banfi, S., Montanari, F. and Quici, S. (1989) *J. Org. Chem.*, **54**, 2970; Anelli, P.L., Montanari, F. and Quici, S. (1990) *Org. Synth.*, **69**, 212.
94 Dijksman, A., Arends, I.W.C.E. and Sheldon, R.A. (2000) *Chem. Commun.*, 271.
95 Dijksman, A., Arends, I.W.C.E. and Sheldon, R.A. (2001) *Synlett*, 102.
96 Dijksman, A., Arends, I.W.C.E. and Sheldon, R.A. (2001) In: *Supported Catalysts and Their Applications* (eds D.C. Sherrington and A.P. Kybett), Royal Society of Chemistry, Cambridge, pp. 118–124.
97 For other strategies for immobilization of TEMPO, see: Bolm, C. and Fey, T. (1999) *Chem. Commun.*, 1795; Fey, T., Fischer, H., Bachmann, S., Albert, K. and Bolm, C. (2001) *J. Org. Chem.*, **66**, 8154; Brunel, D., Lentz, P., Sutra, P., Deroide, B., Fajula, F. and Nagy, J.B. (1999) *Stud. Surf. Sci. Catal.*, **25**, 237; Verhoef, M.J., Peters, J.A. and van Bekkum, H. (1999) *Stud. Surf. Sci. Catal.*, **125**, 465; Brunel, D., Fajula, F., Nagy, J., Deroide, B., Verhoef, M., Veum, L., Peters, J. and van Bekkum, H. (2001) *Appl. Catal. A*, **213**, 73; Gilhespy, M., Lok, M. and Baucherel, X. (2005) *Chem. Commun.*, 2871; Ferreira, P., Hayer, W., Phillips, E., Rippon, D. and Tsang, S.C. (2004) *Green Chem.*, **6**, 310; Ciriminna, R., Blum, J., Avnis, D. and Pagliaro, M. (2000) *Chem. Commun.*, 1441; Ciriminna, R., Bolm, C., Fey, T. and Pagliaro, M. (2002) *Adv. Synth. Catal.*, **344**, 159; Luo, J., Pardin, C., Dell, W.D. and Zhu, X.X. (2007) *Chem. Commun.*, 2136–2138.
98 Qian, W., Jin, E., Bao, W. and Zhang, Y. (2006) *Tetrahedron*, **62**, 556–562; see also: Wu, X., Ma, L., Ding, M. and Gao, L. (2005) *Synlett.*, **4**, 607–610.
99 Sheldon, R.A. and Arends, I.W.C.E. (2004) *Adv. Synth. Catal.*, **346**, 1051–1107. and references cited therein.
100 Gamez, P., Arends, I.W.C.E., Reedijk, J. and Sheldon, R.A. (2003) *Chem. Commun.*, 2414; Gamez, P., Arends, I.W.C.E., Sheldon, R.A. and Reedijk, J. (2004) *Adv. Synth. Catal.*, **346**, 805–811.
101 Fabbrini, M., Galli, C., Gentili, P. and Macchitella, D. (2001) *Tetrahedron Lett.*, **42**, 7551.
102 Viikari, L., Niku-Paavola, M.L., Buchert, J., Forssell, P., Teleman, A. and Kruus, K. (1999) Patent WO 992324,0.
103 Jetten, J.M., van den Dool, R.T.M., van Hartingsveldt, W. and van Wandelen, M.T.R. (2000) Patent WO 00/50621.
104 Rochefort, D., Leech, D. and Bourbonnais, R. (2004) *Green Chem.*, 614.
105 For recent reviews, see: Sheldon, R.A. (2007) *Biochem. Soc. Trans.*, **35**, 1583–1587; Sheldon, R.A., Sorgedrager, M. and Janssen, M.H.A. (2007) *Chim. Oggi (Chem. Today)*, **25**, 1 48–52.
106 Liu, R., Dong, C., Liang, X., Wang, X. and Hu, X. (2005) *J. Org. Chem.*, **70**, 729–731.
107 Wang, N., Liu, R., Chen, J. and Liang, X. (2005) *Chem. Commun.*, 5322–5324.
108 For a review, see: ten Brink, G.J., Arends, I.W.C.E. and Sheldon, R.A. (2004) *Chem. Rev.*, **104**, 4105–4123.
109 Sato, K., Hyodo, M., Takagi, J., Aoki, M. and Noyori, R. (2000) *Tetrahedron Lett.*, **41**, 1439.
110 Usui, Y. and Sato, K. (2003) *Green Chem.*, **5**, 373–375.
111 Corma, A., Nemeth, L.T., Renz, M. and Valencia, S. (2001) *Nature*, **412**, 423–425.
112 Corma, A. Navarro, M.T. Nemeth, L. and Renz, M. (2001) *Chem. Commun.*, 2190–2191.
113 Boronat, M., Corma, A., Renz, M., Sastre, G. and Viruela, P.M. (2005) *Chem. Eur. J.*, **11**, 6905–6915.

114 Legros, J., Dehli, J.R. and Bolm, C. (2005) *Adv. Synth. Catal.*, **347**, 19–31, and references cited therein.
115 Sato, K., Hyodo, M., Aoki, M., Zheng, X.-Q. and Noyori, R. (2001) *Tetrahedron*, **57**, 2469.
116 Federsel, H.-J. and Larrson, M. (2004) in *Asymmetric Catalysis on Industrial Scale; Challenges, Approaches and Solutions* (eds H.-U. Blaser and E. Schmidt), Wiley-VCH Verlag GmbH, Weinheim, p. 413.
117 van de Velde, F., Könemann, L., van Rantwijk, F. and Sheldon, R.A. (1998) *Chem. Commun.*, 1891–1892; van de Velde, F., Arends, I.W.C.E. and Sheldon, R.A. (2000) *J. Inorg. Biochem.*, **80**, 81–89.

# 5
# Green Reduction in Water
*Xiaofeng Wu and Jianliang Xiao*

## 5.1
## Introduction

Reduction reactions are one of the most frequently encountered transformations in chemical synthesis. They can be effected with metals, hydrides, enzymes, catalytic hydrogenation and transfer hydrogenation. However, it is the last category of reduction that has gained far more prominence in various areas of synthetic chemistry over the past a few decades. Hydrogenation and transfer hydrogenation are catalyzed by both homogeneous and heterogeneous catalysts [1–3]. Whereas the latter are easy to use, recyclable, and long-living, the former offer better activity and selectivity under mild conditions. In this context, it is not surprising that heterogeneous hydrogenation has been widely used in the production of commodity chemicals such as ammonia, methanol, cyclohexane, and fatty acids, while homogeneous hydrogenation has found dominant applications in the synthesis of functionalized compounds [1–3]. It is this latter area that forms the focal point of this chapter. Our emphasis will be placed on reactions catalyzed by soluble molecular catalysts, that is, transition metal complexes; however, examples of heterogeneous catalysis will be presented whenever appropriate.

Hydrogenation, that is, reduction using $H_2$ under catalysis, is probably the most widely studied reaction in aqueous media [4–10]. In the 1960s and 1970s, simple water-soluble metal salts such as $[Co(CN)_5]^{3-}$ and $RhCl_3$ were studied for the hydrogenation of olefins in water. However, aqueous-phase hydrogenation did not attract much attention until the introduction of water-soluble phosphines as ligands for rhodium-catalyzed hydrogenation and hydroformylation in the mid-1970s and the wider awareness of the product–catalyst separation issue facing homogeneous catalysis from the 1990s onwards [6]. Today, almost all of the common functional groups in synthetic organic chemistry have been hydrogenated in water.

Transfer hydrogenation uses hydrogen sources other than $H_2$. The initial studies with water-soluble catalysts appeared in the late 1980s, and in spite of the well-documented studies of aqueous-phase hydrogenation, this area had been less developed until recently [11]. As a tool in synthesis, transfer hydrogenation is

*Handbook of Green Chemistry, Volume 5: Reactions in Water.* Edited by Chao-Jun Li
Copyright © 2010 WILEY-VCH Verlag GmbH & Co. KGaA, Weinheim
ISBN: 978-3-527-31591-7

complementary to hydrogenation; it requires neither the hazardous hydrogen gas nor pressure vessels, and it is easy to execute. Furthermore, there are a number of chemicals that are readily available and can be used as hydrogen donors, the most popular being 2-propanol and formic acid. Still further, it may allow for reductions that cannot be effected under the conditions of hydrogenation.

One of the major incentives for the development of aqueous-phase hydrogenation and transfer hydrogenation chemistry is to facilitate catalyst–product separation. Since water is a highly polar and protic solvent ($\varepsilon = 78$, $E_T^N = 1$), most common organic compounds are insoluble or only sparsely soluble in water [12]. This means that a product can be easily separated from the solvent by simple phase separation or extraction, provided that the catalyst is preferentially made water soluble. A number of water-soluble metal catalysts are now readily accessible; these include those containing hydrophilic ligands and those bonding with water. There is another significant advantage on offer when water is used as solvent: being inexpensive, readily available, non-toxic, non-flammable, and eco-benign, water is a natural choice for "greening" chemistry.

However, the use of water also presents challenges. The insolubility of many organic compounds in water implies possible diffusion control if the reduction occurs in the aqueous phase. The same is true with hydrogen, which has a lower solubility in water (0.81 mM at 20 °C) than in common organic solvents. The immiscibility of a compound with water could be exploited, however, to benefit a reaction; the hydrophobic interaction could drive a reaction to occur "on water," where stronger hydrogen-bonding interactions via the surface –OH groups may lead to faster reaction rates [13, 14]. Furthermore, although it appears that hydrogenation and transfer hydrogenation in water operate with mechanisms similar to those in organic solvents, various studies have now shown that water is not an innocent spectator. It may interact with intermediates and transition states, particularly when these bear hydrogen-bond donor or acceptor functionalities, and it may react with a hydride or dihydrogen species and participate in an acid–base equilibrium with the catalyst [6]. The role of water will be briefly addressed towards the end of this chapter.

The unique properties of water have attracted a great deal of interest in its potential applications in catalysis, particularly in hydrogenation. Over the past decade or so, a number of review articles have been published on hydrogenation, including transfer hydrogenation in water, covering the literature up to the early 2000s [4–11, 15–18]. We have therefore restricted the coverage of this chapter to the progress made mainly in the past 5 years.

## 5.2
### Water-soluble Ligands

A metal complex catalyst can be made water soluble by modification of its ligands such that they become sufficiently hydrophilic. This is usually done by attaching ionic or hydrogen-bonding groups to a ligand. Typical hydrophilic structural elements are shown in Table 5.1, the sulfonated and ammonium tags being most popular.

Table 5.1 Polar groups used to confer water solubility on ligands.

| Polar group | Example |
| --- | --- |
| Sulfonate | $-SO_3H$, $-SO_3Na$ |
| Ammonium | $-NR_3^+$, $-NH_2$ |
| Carboxylate | $-COOH$, $-COONa$ |
| Carbohydrate | $-O_2C_{12}H_{14}O_3(OH)_6$ |
| Phosphonium and phosphonate | $-PR_3^+$, $-P(O)(OR)_2$, $-P(O)(ONa)_2$ |
| Hydroxy and polyether | $-OH$, $-(CH_2CH_2O)_nH$ |
| Guadinium | $-NHC(=NH)NH_2$ |

However, an ionic metal complex can be soluble in water without calling for a special ligand. Still further, there are complexes which are soluble in water by bonding to water. Examples of water-soluble ligands are briefly presented below; Section 5.4 shows examples of water-soluble metal complexes.

## 5.2.1
## Water-soluble Achiral Ligands

Phosphines have been the most widely used ligands for reduction in water, and representative water-soluble achiral ligands (1–6) are illustrated. The most intensively investigated water-soluble phosphine ligands for hydrogenation and transfer hydrogenation in aqueous media are (3-sulfonatophenyl)diphenylphosphine (TPPMS, **1**) and tris(3-sulfonatophenyl)phosphine (TPPTS, **2**) [7, 8, 19–22]. Following on from the successful demonstration of these phosphines in aqueous-phase reduction, a wide range of water-soluble ligands have been investigated and applied to various reduction reactions. Nitrogen-containing ligands have also been explored, achieving water-solubility by a similar strategy; typical examples include diamines, imines, and pyridines (**5**, **6**). Water-soluble catalysts are often synthesized by *in situ* reaction of a water-soluble ligand with a metal precursor complex.

## 5.2.2
### Water-soluble Chiral Ligands

Water-soluble chiral ligands can be prepared via similar chemistry, namely by the introduction of a polar functional group on a normal chiral ligand. Selected examples (**7–22**) are illustrated. Phosphorus-containing ligands have been the most widely investigated. However, nitrogen-containing water-soluble ligands have found wide applications in transfer hydrogenation in aqueous media since a seminal paper on asymmetric transfer hydrogenation using TsDPEN [TsDPEN = N-(p-toluene-sulfonyl)-1,2-diphenylethylenediamine] by Noyori and co-workers in 1995 [23–26]. Water-soluble analogs have since been prepared (**19–22**) (Ts = p-CH$_3$C$_6$H$_4$SO$_2$–).

## 5.3
### Hydrogenation in Water

As introduced at the beginning, hydrogenation is probably the most widely studied reaction in aqueous media. This is a totally atom-economic reaction, and when a water-soluble catalyst is chosen, there is in principle no waste to generate following product separation. A wide variety of substrates have been hydrogenated, with the stage dominated by achiral reactions.

## 5.3.1 Achiral Hydrogenation

### 5.3.1.1 Hydrogenation of Olefins

Achiral hydrogenation of simple olefins provides the earliest examples of aqueous-phase hydrogenation. The catalysts used were almost exclusively rhodium and ruthenium complexes containing water-soluble phosphines [4–11, 15–18]. A recent example is seen in the water-soluble $RuCl_2(TPPTS)_3$, which catalyzed the hydrogenation of unsaturated hydrocarbons, such as 1-alkenes, styrene, cyclooctenes, and even benzene, in water under 10 bar $H_2$ at 150 °C, affording moderate to high conversions [27]. As expected, aliphatic unsaturated hydrocarbons were more easily hydrogenated than aromatic compounds. Ruthenium carbonyl complexes bearing TPPMS (**1**), such as $Ru(CO)_3(TPPMS)_2$ and $RuH_2(CO)(TPPMS)_3$, have also been shown to be efficient, recyclable catalysts for the hydrogenation of these olefins in a water–*n*-heptane (1 : 1) mixture [28, 29].

Hydrogenation of functionalized olefins is seen in an early study. As shown in Scheme 5.1, various olefins were chemoselectively hydrogenated at the C=C double bonds by using $RhCl_3$ in the presence of TPPTS (**2**) at room temperature and 1 bar $H_2$. In the case of dienes, the less hindered C=C bond was first hydrogenated and the reaction could be terminated at the monoene stage [30].

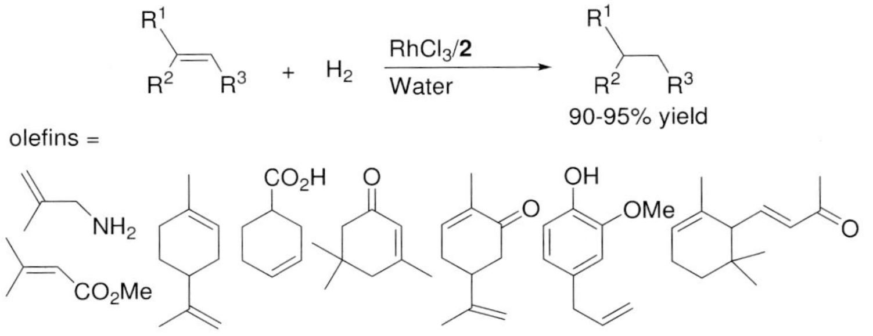


**Scheme 5.1**

Functionalized olefins that are water soluble present a problem for product separation, particularly when the catalyst is also water-soluble. This has been addressed by using supercritical $CO_2$–water biphasic catalysis, in which a "$CO_2$-philic," rather than hydrophilic, catalyst resides in the supercritical fluid, while the substrate and product dissolve in and are removed by water [31]. Itaconic acid, α-acetamidocinnamic acid, fumaric acid, and methyl acetamidoacrylate were hydrogenated this way with $[Rh(COD)_2][BF_4]$ and a $CO_2$-philic phosphine, $P(3\text{-}C_6H_4CH_2CH_2C_6F_{13})_3$, furnishing >99% conversions at 30 bar $H_2$ and 56 °C. The catalyst-containing $CO_2$ phase was reusable, with rhodium leaching at the ppm level in each recycle run when an excess of ligand was employed.

Selective hydrogenation of α,β-unsaturated aldehydes has been extensively studied in water, and good control of the selectivity has been established. Rh and Ru complexes are generally the catalysts of choice in this transformation and often, but not always, the former favor hydrogenation at C=C double bonds whereas the latter prefer C=O bond reduction. For example, as shown in Scheme 5.2 (S/C = substrate/catalyst molar ratio), α,β-unsaturated aldehydes were hydrogenated in a water–toluene (1:1) mixture, at the C=C bond by Rh-TPPTS with up to 93% conversion and 97% selectivity, and at the C=O moiety by Ru-TPPTS with up to 100% conversion and 99% selectivity. The catalyst could be recycled to yield even a slightly higher activity and selectivity [32].

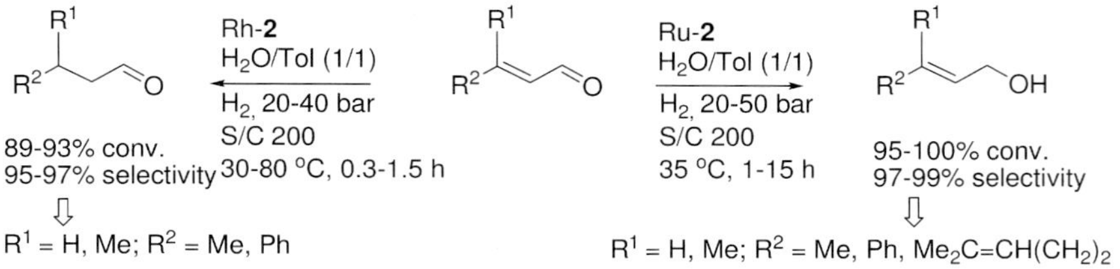

Scheme 5.2

The water-soluble tetranuclear complex $Rh_4(O_2CPr)_4Cl_4(MeCN)_4$ (Pr = n-propyl) was reported to catalyze selectively the hydrogenation of α,β-unsaturated alcohol, ketone, nitrile, carboxylic acid, and amide substrates at the C=C bond under 1 bar $H_2$ at room temperature [33]. Likewise, the water-soluble Ru(II), Rh(I) and Rh(III) complexes of N-methyl-PTA (PTA-Me; PTA = 1,3,5-triaza-7-phosphaadamantane, **3**), such as $RuI_4(PTA-Me)_2$, $[RuI_2(PTA-Me)_3(H_2O)][I_3]$ and $[Rh_4(PTA-Me)_2][I]$, were shown to be active catalysts for the hydrogenation of cinnamaldehyde at the C=C bond with Rh but at the C=O moiety with Ru in a $H_2O$–toluene or $H_2O$–chlorobenzene biphasic mixture [34]. For instance, cinnamaldehyde was converted to $PhCH_2CH_2CHO$ in 95% conversion with 84% selectivity and a turnover frequency (TOF) of $190\,h^{-1}$ by using $RuI_4(PTA-Me)_2$ in $H_2O$–toluene. However, several PTA-containing half-sandwich Ru(II) complexes, such as $CpRuH(PTA)_2$ and $[CpRu(MeCN)(PTA)_2][PF_6]$, were shown to be selective towards C=C saturation, albeit with low activities [35–37]. Unlike the TPPTS and other PTA complexes mentioned, hydrogenation with the half-sandwich catalysts may proceed via an ionic mechanism, involving transfer of $H_2$ as $H^+$ and $H^-$ and no coordination of the substrate (Scheme 5.3) [37]. This is reminiscent of Noyori and co-workers' metal–ligand bifunctional mechanism put forward for the Ru(II)(diphosphine)(diamine)-catalyzed hydrogenation of ketones [38, 39]. The 1,4-addition that involves activation of the double bond via N–H–O hydrogen bonding explains why the C=C bond is selectively reduced.

Half-sandwich complexes with an ionic tag have also been explored for aqueous hydrogenation. The imidazolium-functionalized complexes **23** and **24** are active for the hydrogenation of styrene in water–cyclohexane (1:2) [40]. The reaction proceeded readily under the conditions of S/C = 1000 and 40 bar $H_2$ at 80 °C, with the catalyst

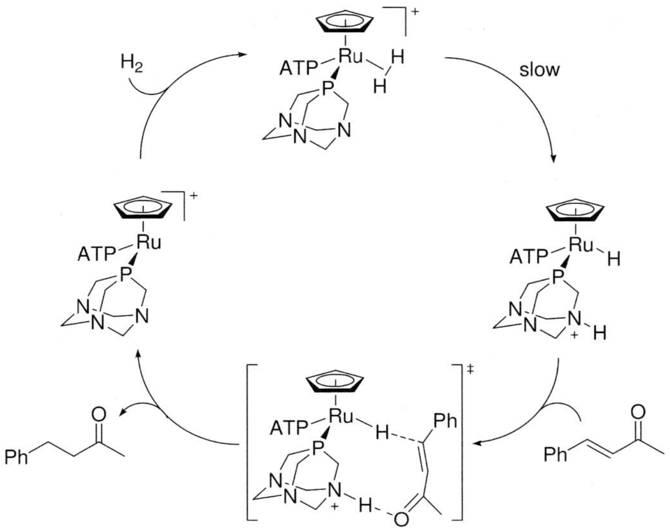

## 5.3 Hydrogenation in Water | 111

**Scheme 5.3**

being reusable, although slightly decreased activity was noted in recycle runs. Catalysts of this type have also found use in catalysis in ionic liquids.

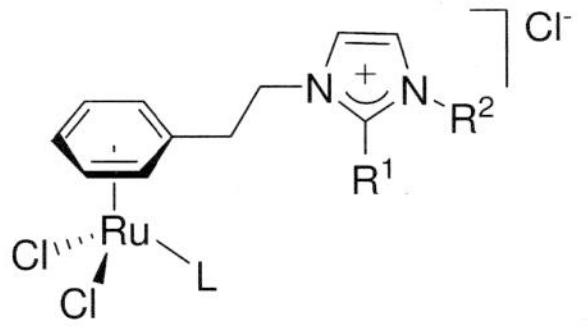

**23**: $R^1 = R^2 = Me$, $L = PPh_3$
**24**: $R^1 = H$, $R^2 = Bu$, $L = PPh_3$

Recent studies have revealed that the selectivity pattern of the M-TPPTS catalysts in the hydrogenation of unsaturated carbonyls is affected by a variety of parameters, such as the $H_2$ pressure, temperature, catalyst concentration, ligand-to-metal ratio, substrate concentration, and solution pH [41–43]. Significantly, these parameters could be maneuvered to alter the normal selectivity patterns. For instance, in the case of the Rh-TPPTS-catalyzed process, a higher $H_2$ pressure shifts the hydrogenation towards the C=O bond, whereas a large excess of ligand **2** favors saturation of the C=C bond [42], and although the Ru-TPPTS catalyst is selective towards C=O saturation at high catalyst concentration, it favors hydrogenation of the C=C bond when the catalyst concentration is low [44]. Solution pH has also been found to be important. Thus, selective hydrogenation of *trans*-cinnamaldehyde at the C=C bond is achievable at a low pH (<5) with 1 bar of $H_2$; but the selectivity favors the C=O bond when the pH is increased to >7 [43]. On the other hand, selective saturation of the C=O bond could also be achieved at a low pH, but at a higher $H_2$ pressure of 8 bar. These seemingly conflicting observations result from differing active catalytic species being involved in the reaction, for example $RuClH(TPPMS)_3$ and $RuH_2(TPPMS)_4$, the proportion of which varies with the solution pH and $H_2$ pressure. The former is selective towards the C=C bond and the latter towards the carbonyl [45, 46].

More recently, a DFT investigation of the Ru-catalyzed selective hydrogenation of α,β-unsaturated aldehydes in aqueous–organic biphasic media showed that the favored C=O hydrogenation under basic conditions is due to the presence of water, which forms a hydrogen bond with the aldehyde, thus facilitating C=O reduction [47]. However, a similar mechanistic study [48] revealed that the selective reduction of the C=C bond by $RuClH(PR_3)_3$ is due to a lower barrier of C=C versus C=O insertion into the Ru–H bond, whereas the reduction of the C=O bond, instead of C=C, with $RuH_2(TPPMS)_4$ stems from the energy difference in the subsequent step of protonation [49]. It is worth noting that these studies show that, being involved in various steps of the catalytic cycle, water as solvent is not an innocent spectator in the hydrogenation (also see Section 5.5).

Hydrogenation of unsaturated polymers in water is also possible. Such reactions have been successfully carried out in conventional organic solvents [50] and other media, such as ionic liquids [51, 52]. Rh-TPPTS complexes were recently demonstrated to catalyze efficient hydrogenation of poly(1,4-butadiene)-*block*-poly(ethylene oxide) (PB-*b*-PEO) in water, as shown in Scheme 5.4 [53]. The reaction was homogeneous and proceeded in PB-*b*-PEO/DTAC nanomicelles (DTAC = dodecyltrimethylammonium chloride, a cationic surfactant), affording high conversions with high catalytic activities (TOF > 840 $h^{-1}$). The catalyst could be recycled, maintaining high catalytic activity in a consecutive run even at a rhodium concentration of only 1 ppm in water.

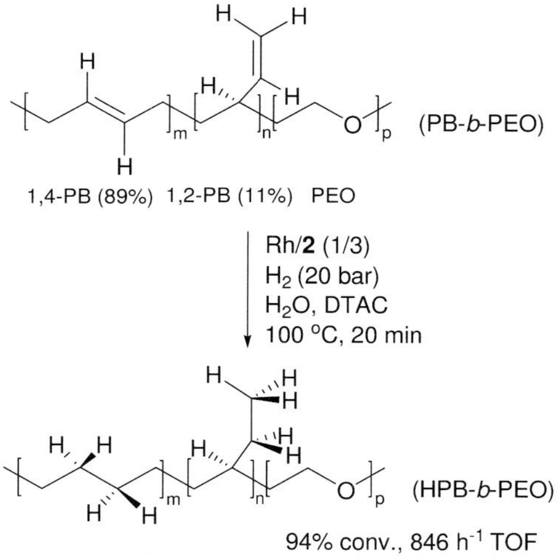

**Scheme 5.4**

Higher olefins, such as 1-octene, can be difficult to hydrogenate due to reduced solubility in water. An amphiphilic Ru nanoparticle catalyst stabilized by the water-soluble poly(*N*-vinyl-2-pyrrolidone) (PVP) demonstrated a high activity in the hydrogenation of $CH_2=CH(C_nH_{2n+1})$ ($n=4$–9) and cyclohexene in a water–decane mixture, with TOF as high as 23 000 h$^{-1}$ being reached at 40 bar H$_2$ and 80 °C [54]. The amphiphilic nature of the catalyst leads to an enhanced concentration of substrate around the polymer-trapped ruthenium and thus high hydrogenation rates. Other nanoparticle catalysts have also been shown to be active catalyst precursors for heterogeneous hydrogenations in aqueous phase [55–60].

### 5.3.1.2 Hydrogenation of Carbonyl Compounds

In addition to the aforementioned unsaturated carbonyl compounds, simple ketones and aldehydes have also been reduced in aqueous solutions. The water-soluble, half-sandwich iridium complex [Cp*Ir(H$_2$O)$_3$]$^{2+}$ (also see the structures in Section 5.4) was shown to be active for the hydrogenation of carbonyl compounds and alkenes in water under mild reaction conditions (1–7 bar H$_2$, 25 °C) [61]. An isoelectronic ruthenium complex, RuCl$_2$L(p-cymene) (L = 1-butyl-3-methylimidazol-2-ylidene), showed good activity in the hydrogenation of acetone, acetophenone, and propanal in water at 10 bar H$_2$ and 80 °C [62]. Remarkably, replacing the carbene ligand with phosphine resulted in no hydrogenation. Similarly to the PTA catalysts mentioned in Section 5.3.1.1, the Ru–carbene complex catalyzes the preferential hydrogenation of C=C bonds in α,β-unsaturated compounds.

Recently, a diamine-ligated half-sandwich iridium complex (**25**) was shown to catalyze the hydrogenation of a wide range of aldehydes in neat water (Scheme 5.5) [63]. Aromatic, aliphatic, heterocyclic, and α,β-unsaturated aldehydes

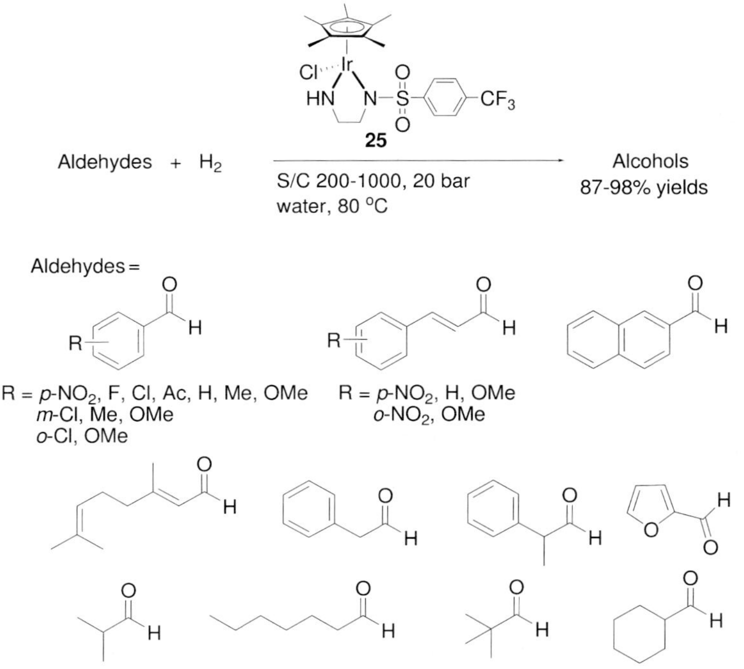

**Scheme 5.5**

were all viable substrates, and in the case where C=C or ketone C=O bonds co-exist, only the formyl group was saturated. Interestingly, this catalyst is also highly active in transfer hydrogenation of aldehydes with formate in water (see Section 5.4.2).

As with the catalysts bearing water-soluble phosphines, the phosphine-free half-sandwich complexes also display pH dependence in hydrogenation. For instance, $[Cp^*Ir(H_2O)_3]^{2+}$ was active in the pH range from −1 to 4 [61, 62]. This dependence can again be traced to the variation of catalytic species with pH (see Section 5.5.2). Similar reports were presented recently, suggesting that the active catalyst species [64] or the stereoselectivity of the catalyst [65] could be controlled by the pH of the aqueous phase.

Apart from the soluble metal catalysts, nanoparticles have also shown promise in water. The aforementioned PVP-stabilized Ru nanoparticles allow for rapid hydrogenation of ketones [54]. Similarly, Ru nanoparticles immobilized on the water-soluble polymer polyorganophosphazenes ($-[N=PR_2]_n-$) were found to be active for the hydrogenation of unsaturated ketones or aromatic compounds such as pyruvic acid and p-aminomethylbenzoic acid in water [66]. For example, as shown in Scheme 5.6, pyruvic acid was completely reduced to lactic acid by Ru on PDMP (PDMP = polydimethylphosphazene) with 100% selectivity and an SA of 14.3 (SA = specific activity: moles of converted substrate per gram-atom of ruthenium per hour) under mild conditions; no catalyst deactivation was observed in the recycle runs. Likewise, methyl pyruvate was hydrogenated under the catalysis of Pt nanoparticles supported on poly(diallyldimethylammonium chloride) (PDDA) (Scheme 5.6) (TON = turnover number); the catalyst also allowed for the hydrogenation of aliphatic and aromatic aldehydes under the same conditions in water [59, 67].

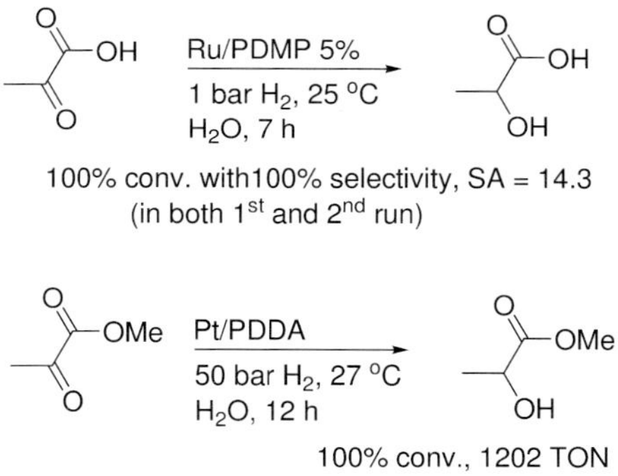

**Scheme 5.6**

### 5.3.1.3 Hydrogenation of Aromatic Rings

Aromatic rings are difficult to saturate and the hydrogenation is generally effected with heterogeneous catalysts. A number of soluble metal complexes were previously reported to catalyze arene hydrogenation in water [60, 68–74]. In most cases, however, decomposition into heterogeneous metal particles took place, which catalyzed the hydrogenation. In the case of half-sandwich Ru(II)–cymene complexes, the decomposition was shown to be a function of solution pH; the decomposition accelerated at higher pH, correlating with an increasing hydrogenation activity [72, 75].

Water is a viable alternative to the frequently used polar organic solvents. The Ru–PVP catalysts mentioned in Section 5.3.1.1 permit rapid, complete hydrogenation of simple benzenes, styrene, anisole, and benzonate, with TOFs as high as 45 000 h$^{-1}$ being reported at 40 bar $H_2$ and 80 °C in a water–cyclohexane mixture [76]. Similarly, colloidal Rh suspension stabilized by highly water-soluble $N,N'$-dimethyl-$N$-cetyl-$N$-(2-hydroxyethyl)ammonium bromide or chloride (HEA16X, X = Br or Cl) catalyzes the hydrogenation of $N$- and $O$-heteroaromatic compounds in water under mild conditions (Scheme 5.7) [77]. The catalyst could be reused without losing activity. However, no catalytic activities were detected for sulfur compounds such as thiophene.

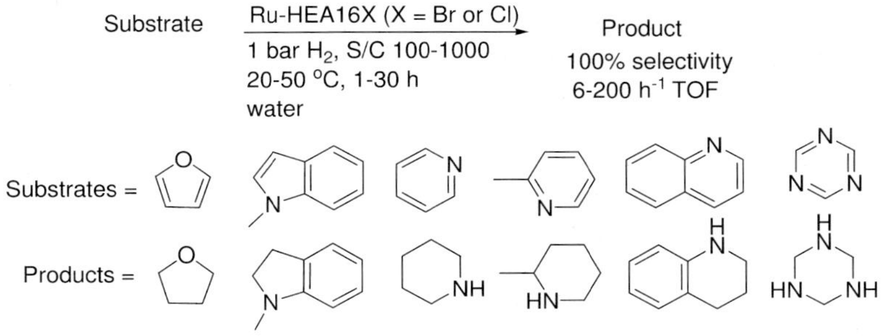

**Scheme 5.7**

An even simpler example is seen in the commercial Rh/C catalyst, which catalyzes the hydrogenation of a range of aromatic rings under mild conditions in water [78]. Selected examples are provided in Scheme 5.8. The rhodium catalysts in both Schemes 5.7 and 5.8 allow for efficient reduction of pyridine derivatives, which are known to poison platinum metal catalysts.

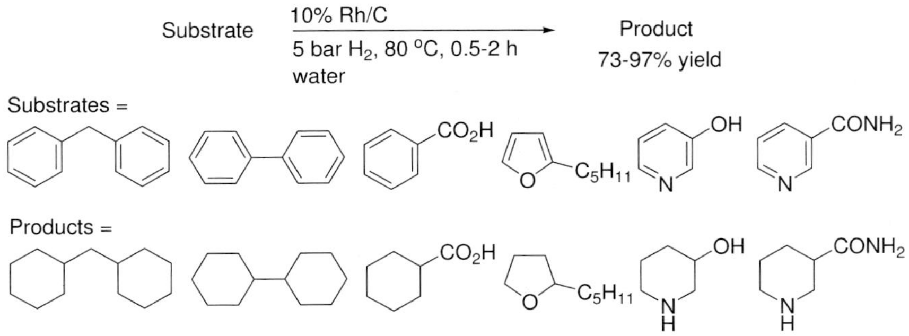

**Scheme 5.8**

### 5.3.1.4 Hydrogenation of Other Organic Groups

In addition to those discussed, compounds such as imines, nitro compounds, and nitriles have also been hydrogenated in water. The examples are far fewer, however, and most are concerned with heterogeneous catalysts. Benzonitrile was cleanly hydrogenated to benzylamine with Pd/C in the presence of $NaH_2PO_4$ in an $H_2O$–$CH_2Cl_2$ mixture, affording 95% selectivity and 90% isolated yield under mild reaction conditions (30 °C, 6 bar $H_2$) [60]. The PDDA-supported Pt nanoparticles mentioned in Section 5.3.1.2 permit efficient hydrogenation of chloronitrobenzenes at 50 bar $H_2$ and 27 °C, affording a higher TON in neat water than in a water–toluene mixture [67]. Somehow similarly, addition of water accelerates significantly the hydrogenation of p-chloronitrobenzene catalyzed by various supported metal catalysts, $M/SiO_2$ (M=Ru, Ni, Co, Fe), in ethanol [79]. Nitrobenzene, along with benzaldehyde and cyclohexanone, has been reduced in high-temperature pressurized water, using $H_2$ generated *in situ* from formate and thus providing a "gasless" approach to hydrogenation [80].

Imines have been reduced with β-cyclodextrin (β-CD)-modified Pd nanoparticles in water [81]. β-CD is capable of transferring hydrophobic molecules into water by hosting the molecules inside its cavity, thereby facilitating the hydrogenation of water-insoluble substrates in water. A range of aldimines were hydrogenated; examples are shown in Scheme 5.9. The catalyst is also effective in reducing various α,β-unsaturated ketones and aldehydes under similar conditions. In the case of isophorone hydrogenation, the presence of β-CD leads to a 250-fold increase in TOF.

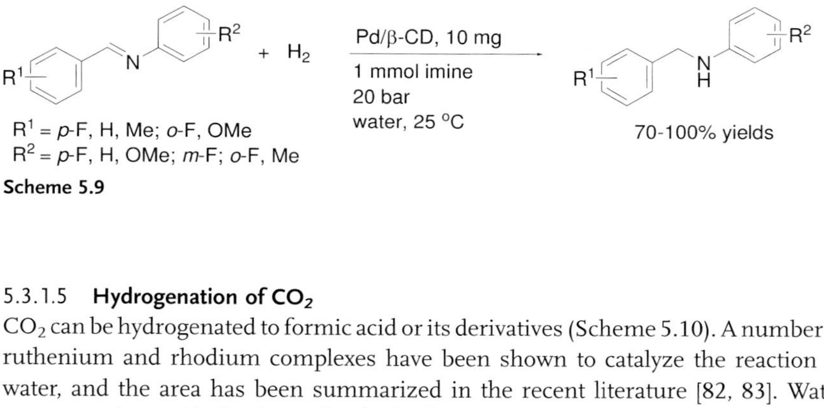

$R^1$ = p-F, H, Me; o-F, OMe
$R^2$ = p-F, H, OMe; m-F; o-F, Me

Pd/β-CD, 10 mg
1 mmol imine
20 bar
water, 25 °C

70–100% yields

**Scheme 5.9**

### 5.3.1.5 Hydrogenation of $CO_2$

$CO_2$ can be hydrogenated to formic acid or its derivatives (Scheme 5.10). A number of ruthenium and rhodium complexes have been shown to catalyze the reaction in water, and the area has been summarized in the recent literature [82, 83]. Water appears to be an ideal solvent for the hydrogenation. In the gas phase, the hydrogenation is endergonic with $\Delta G°_{298} = 33$ kJ $mol^{-1}$, whereas in the aqueous solution, the standard free energy becomes $-4$ kJ $mol^{-1}$. In the case of the latter, however, the hydrogenation is more complicated, as $CO_2$ is in equilibrium with hydrogencarbonate ($pK_1 = 6.35$) and carbonate ($pK_2 = 10.33$), both of which can be hydrogenated to formate. In addition, the product formic acid ionizes in water ($pK_a = 3.6$). Therefore, as might be expected, $CO_2$ hydrogenation in water is pH dependent. Most of the hydrogenation reactions have been run under basic conditions, enhancing the possibility of hydrogenating hydrogencarbonate instead of $CO_2$.

Rhodium and ruthenium complexes dominate the scene, most containing phosphine ligands. Typical examples are $RhCl(TPPTS)_3$ and $[RuCl_2(TPPMS)_2]_2$. In the

## 5.3 Hydrogenation in Water

**Scheme 5.10**

$$CO_2 + H_2 \rightleftharpoons HCOOH$$

$$CO_2 + H_2O \rightleftharpoons H_2CO_3 \xrightleftharpoons{pK_1} HCO_3^- + H^+ \xrightleftharpoons{pK_2} CO_3^{2-} + H^+$$

$$HCOOH \xrightleftharpoons{pK_a} HCO_2^- + H^+$$

hydrogenation reaction in water using RhCl(TPPTS)$_3$, a TOF of 7260 h$^{-1}$ was observed at a total pressure of 40 bar (H$_2$:CO$_2$ = 1:1) and 81 °C in the presence of NHMe$_2$ [84], while the Ru(II) catalyst afforded a TOF of 9600 h$^{-1}$ at a total pressure of 95 bar (H$_2$:CO$_2 \approx$ 2:1) and 80 °C in the presence of NaHCO$_3$ [85].

Half-sandwich metal complexes containing bipyridine (bipy)-type ligands have also been shown to be active [82, 83, 86–88]. An interesting recent example is seen in an Ir(III) complex (**26**) bearing the dihydroxy-bipy ligand **6**, which afforded a TON of 190 000 in 57 h with an initial TOF of 42 000 h$^{-1}$ at a total pressure of 60 bar (H$_2$:CO$_2$ = 1:1) and 120 °C in an aqueous KOH (1 M) solution (Scheme 5.11) [88]. The analogous rhodium and ruthenium catalysts were less active. A kinetic study of closely related Ru(II) and Ir(III) complexes suggests that the hydrogenation is rate limited by hydride formation in the case of the former but by hydride transfer to CO$_2$ in the case of the latter [87].

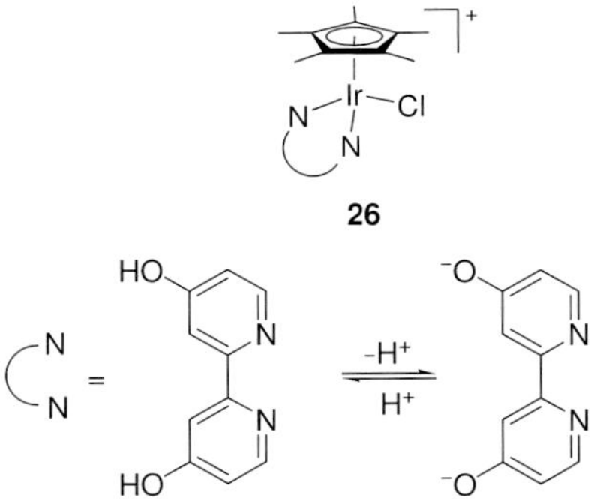

**Scheme 5.11**

The high catalytic activity stems from the electron-rich nature of the ligand **6**, which is deprotonated during the reaction (Scheme 5.11); the resulting oxyanion is a much stronger electron donor than the hydroxyl group ($\sigma_p^+ = -2.3$ vs $-0.91$). There indeed exists a correlation between the initial TOF and $\sigma_p^+$, and this is consistent with previous theoretical and experimental studies, which suggest that electron-rich ligands facilitate the hydrogenation [89, 90]. The equilibrium shown in Scheme 5.11 also confers tunable solubility on the catalyst: on the completion of hydrogenation, the solution turns acidic, rendering the ligand neutral and thus triggering the precipitation of the catalyst. Indeed, the catalyst was shown to be recyclable by filtration, with iridium leaching <1 ppm in each run. The product HCO$_2$K was separated from the catalyst-free aqueous solution by evaporation of H$_2$O.

## 5.3.2
## Asymmetric Hydrogenation

Water is attractive for enantioselective hydrogenation, as it can "green" the processes of fine chemicals and pharmaceutical synthesis and permit easy separation and recycling of expensive chiral catalysts including the metals and ligands. In many cases, however, chiral catalysts in water afforded lower enantioselectivities and/or activities than in organic solvents, especially in the early days [5, 15, 16]. This is partly due to factors such as catalyst modification, altered solvation, diffusion control, and reduced substrate and catalyst solubility. As with achiral hydrogenation, aqueous-phase asymmetric hydrogenation usually necessitates the use of water-soluble ligands. However, these are generally more difficult to access than achiral ligands, explaining in part why asymmetric hydrogenation reactions in water are very limited in terms of both the type and scope of substrates.

### 5.3.2.1 Asymmetric Hydrogenation of Olefins

Most reports in this area are concerned with some standard substrates. Methyl 2-acetamidoacrylate was hydrogenated with the water-soluble, hydroxy-functionalized complex [RhL(NBD)][SbF$_6$] (**27**) (NBD = norbornadiene, L = **14**), giving rise to 100% conversion and >99% *ee* in water at about 3 bar H$_2$ at room temperature in 5–7 h (Scheme 5.12) [91]. The catalyst was stable, being recyclable up to four times without losing any activity and enantioselectivity. The hydroxyl ligand **14** has the backbone of Burk's well-established DuPhos ligands [92]. A similar catalyst **28** bearing the ligand **15** was explored for the hydrogenation of itaconic acid in water in the presence of a small amount of MeOH, affording excellent *ee*s (Scheme 5.13) [93]. With no erosion in enantioselectivity, these results demonstrate the potential of this type of ligand in aqueous-phase asymmetric hydrogenation.

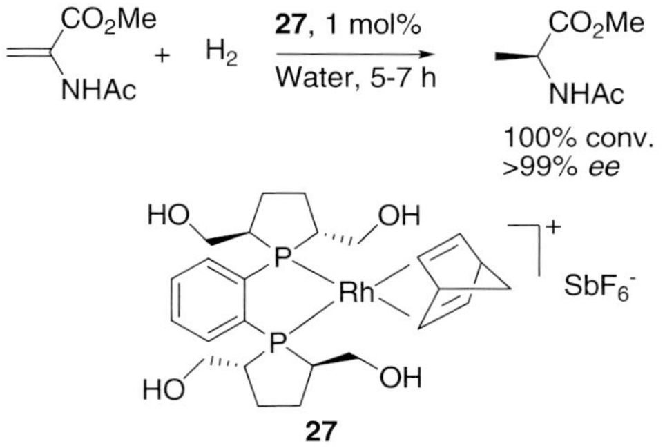

**Scheme 5.12**

2-Acetamidoacrylic acid can be directly hydrogenated in water with the catalyst **27** [94]. In a more recent study, monodentate phosphorus ligands were also shown to be effective. The phosphoramidite **29** derived from (*S*)-BINOL and the related, poly(ethylene glycol)-functionalized **30** allow for Rh-catalyzed enantioselec-

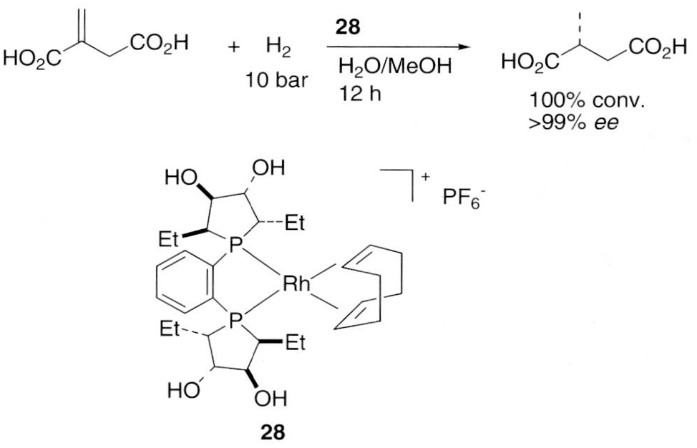

## Scheme 5.13

tive hydrogenation, affording *ees* up to 95% [95]. Whereas **29** led to faster catalysis in CH$_2$Cl$_2$ (TOF 400 vs 133 h$^{-1}$), the water-soluble **30** was far more effective in a polar MeOH–H$_2$O mixture (TOF 1200 vs 20 h$^{-1}$). In neat water, however, the catalysts derived from both ligands showed decreased activity and enantioselectivity, although **30** was still considerably more effective than **29** (82 vs 16% *ee*; TOF 55 vs 20 h$^{-1}$), illustrating a typical problem for aqueous-phase asymmetric hydrogenation.

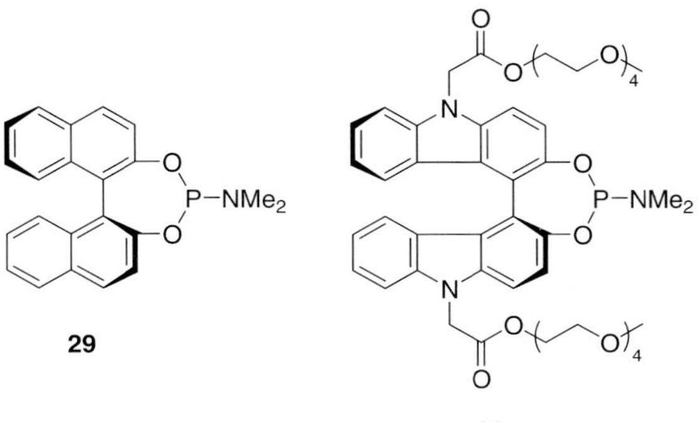

Asymmetric hydrogenation coupled with enzymatic hydrolysis has been explored for the direct synthesis of amino acids in water. In one such example, the catalyst Rh–**29**, immobilized on an oxide surface via ionic interactions, was used to catalyze the hydrogenation of methyl 2-acetamidoacrylate in water, affording methyl *N*-acetylalanate in 95% *ee* at 5 bar H$_2$ [96]. Following filtration of the catalyst, an aminoacylase, such as *Aspergillus melleus* (AM), was introduced to catalyze the hydrolysis in a phosphate-buffered solution, leading to the formation of L-alanine with an increased *ee* of >98% (Scheme 5.14). It is noted that the hydrolysis took place first at the ester group. The two-step reaction could also be performed in a one-pot fashion, circumventing the need for filtration.

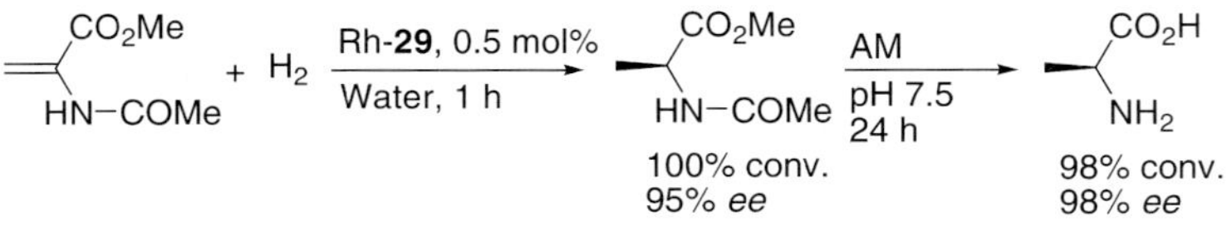

**Scheme 5.14**

In some instances, the low rates and enantioselectivities encountered in hydrogenation in water can be improved by the addition of amphiphiles [7, 97, 98]. An early example is the asymmetric hydrogenation of phosphonates to give α-aminophosphonic acids using a water-insoluble Rh–BPPM (**31**) catalyst (Scheme 5.15) [99]. Various phosphonates were readily reduced with *ee*s in the range 96–99% in the presence of a surfactant, sodium dodecyl sulfate (SDS). The introduction of SDS resulted in both a higher enantioselectivity and a much improved reaction rate, at least partly due to increased catalyst and substrate solubility in water.

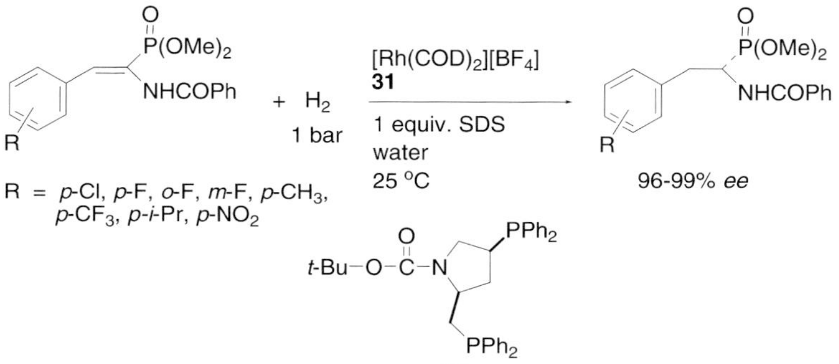

**Scheme 5.15**

SDS has also been shown to influence asymmetric hydrogenation of dehydroamino acids with water-soluble catalysts. This is seen in the hydrogenation of methyl (Z)-α-acetamidocinnamate by a rhodium catalyst containing a trehalose-derived phosphinite ligand, which furnished a faster reaction and a higher *ee* value of >99% in the presence of SDS (10–200%) at 5 bar of $H_2$ and room temperature. A number of other enamides were also fully hydrogenated within 1 h with excellent *ee*s under such conditions [100]. Although the precise role of amphiphiles remains speculative, it appears that the formation of micelles is important [101].

Asymmetric hydrogenation has recently been shown to be feasible in mixtures of water and imidazolium ionic liquids. One benefit of using such mixed solvents is that common chiral ligands can be directly employed without modification, since their metal complexes are generally soluble in ionic liquids. Surprisingly somehow, the mixed solvents can also confer better catalyst performance [102, 103]. Thus, methyl 2-acetamidoacrylate was hydrogenated under the catalysis of Rh–EtDuPhos in a [bmim][$PF_6$]–$H_2O$ mixture (1 : 1 v/v) at 20 °C and 5 bar of $H_2$, furnishing a 68% conversion and 96% *ee* in a 20 min reaction time. The catalyst-containing ionic liquid phase could be reused after extracting the product. In contrast, in [bmim][$PF_6$] without water, there was no reaction at all. The role of water was ascribed to helping create a well-mixed "emulsion-like" system [102].

### 5.3.2.2 Asymmetric Hydrogenation of Carbonyl and Related Compounds

Asymmetric hydrogenation of ketones provides synthetically important chiral alcohols, but it has been even less studied in water. In recent examples, β-keto esters were reduced in excellent ees (>97%) in water with Ru(II) complexes containing the 4,4'- and 5,5'-diaminomethyl-BINAP ligands **32** and **33** (Scheme 5.16) [104]. The catalyst could be recycled up to eight times without loss of reactivity or enantioselectivity. Further studies showed that the same ligands could also be used in the Ru-catalyzed hydrogenation of ethyl trifluoroacetoacetate in an acidic aqueous medium (1.0 ml of water, 0.125 ml of acetic acid and 0.125 ml of trifluoroacetic acid) to give about 70% ee, one of the best enantioselectivities obtained for the reduction of this substrate with Ru–BINAP or its derivatives [105]. The improved selectivity may be due to acid-facilitated equilibration of keto–enol hydrate involved in the catalytic cycle.

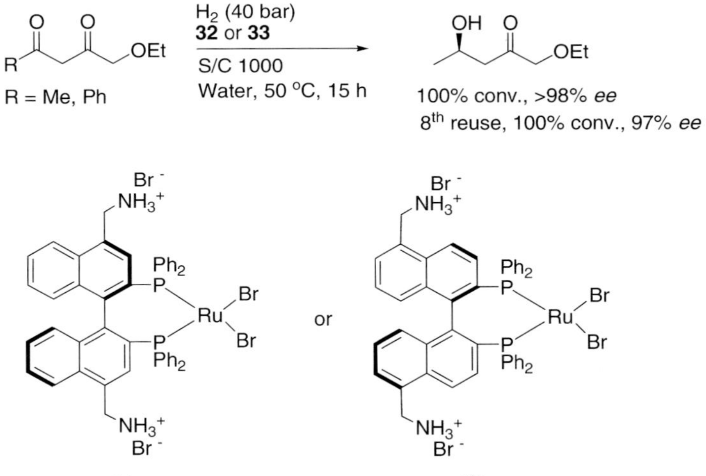

**Scheme 5.16**

Water has been shown to improve the catalyst performance in common solvents. Thus, in the Ru–(R)-BINAP-catalyzed asymmetric hydrogenation of methyl acetoacetate in methanol, the catalytic activity and selectivity were both enhanced by the addition of 3 wt% of water (TOF from 98 to 594 $h^{-1}$ and selectivity from 77 to 99.9%) [106]. Water was considered to restrict acetal formation in the initial stage of the hydrogenation. However, addition of more than 5 wt% of water caused a decrease in both the TOF and ee.

When dealing with highly polar substrates, water can be an ideal solvent. An example is the diastereoselective hydrogenation of folic acid disodium salt. This is a difficult reaction, involving the enantioselective saturation of a pyrazine ring. Amongst a series of water-soluble diphosphine ligands, a modified Josiphos (**34**) was found to be suitable for the rhodium-catalyzed hydrogenation in water at pH 7, giving up to 49% *de* for L-tetrahydrofolic acid, a pharmaceutically relevant intermediate, with 97% conversion at 30 °C after 12 h of reaction (Scheme 5.17) [107]. At a higher temperature, TOF up to 334 h$^{-1}$ and TON up to 2800 could be obtained, although the *de* was lowered. The TOF and TON were considered to be technically viable; but the diastereoselectivity was still too low.

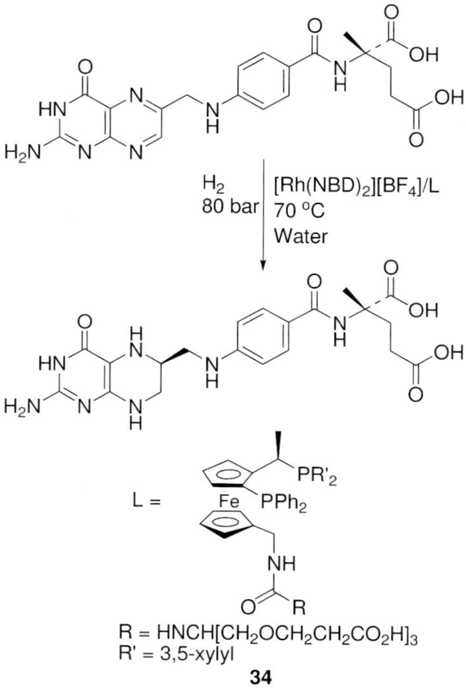

**Scheme 5.17**

Heterogeneous catalysts have also been explored in aqueous-phase asymmetric hydrogenation. For instance, Ru/C could be used to reduce the amino acid L-alanine to L-alaninol in >90% yield with >97% selectivity and 99% *ee* under 70 bar H$_2$ at 100 °C in an acidic aqueous phase [108]. A kinetic study predicted that the acidified solution was necessary to give high conversions. Under such conditions, the amino acid would be protonated and so readily hydrogenated [109]. Surfactant-stabilized Pt(0) nanoparticles, modified with (−)-cinchonidine, was shown to catalyze efficiently the asymmetric hydrogenation of ethyl pyruvate in water at 25 °C under 40 bar of H$_2$, giving rise to a complete reaction with *ee*s up to 55% in a 1 h reaction time. Both the conversion and *ee* were higher than those without using the surfactant [98].

## 5.4 Transfer Hydrogenation in Water

Transfer hydrogenation has often been performed in 2-propanol and azeotropic formic acid and triethylamine mixture [17, 21, 110], which act as the solvent and hydrogen source. Although formic acid and its salts are viable hydrogen sources and soluble in water, and aqueous formate has been used by enzymes for reduction reactions for millions of years, only in recent years has asymmetric transfer hydrogenation in water attracted significant attention. To some extent this reflects the relatively limited research into transfer hydrogenation undertaken in past decades. In the case of aldehydes, this is also partly due to concern over possible decarbonylation of the substrates and poisoning of the catalyst by the resulting CO [111–117].

Examples of chiral and achiral half-sandwich metal complexes that have recently found successful applications in aqueous-phase transfer hydrogenation are illustrated below. In contrast to many other metal catalysts, these complexes catalyze the transfer hydrogenation with no need for ligand modifications. Their water-solubility derives from their capability to coordinate with water. In the case of the chlorides, aquation gives rise to water solubility. These complexes exhibit varying solubilities in water. For example, **38** ($SO_4^{2-}$ being the counteranion) has a solubility of 136 mg ml$^{-1}$ (pH 3, 25 °C), and the related Cp*–Ir(III) complexes display solubilities up to 760 mg ml$^{-1}$ [61, 87]. It is noted, however, that most of these complexes show only limited solubility in water; this can be considerably enhanced when the ligands are made water soluble, for example **19–21**.

**48**
a: M = Rh, b: M = Ir

**49**
a: M = Rh, b: M = Ir

**50**
a: M = Rh, b: M = Ir
i: R = Ts, ii: R = TsCF$_3$

**51**
a: M = Rh, b: M = Ir

**52**

**53**

**54**

**55**

**56**

**57**

**58**

**59**

### 5.4.1
### Achiral Transfer Hydrogenation of Carbonyl Compounds

Organometallic catalysis in aqueous media has attracted interest since the 1970s [20, 22, 118–124]. In spite of the well-documented studies of aqueous-phase hydrogenation, transfer hydrogenation in water had been less developed until a few years ago. In the 1980s, aqueous–organic biphasic transfer hydrogenation of C=C and C=O double bonds with formate was reported [119, 120, 124]. Up to 76% conversion was obtained for aldehyde reduction with RuCl$_2$(PPh$_3$)$_3$ in 30 min at 90 °C; the reduction was less effective for ketones, however [124].

Transition metal-catalyzed transfer hydrogenation of aldehydes in neat water was first carried out by Joo and co-workers [20, 123]. Unsaturated aldehydes were reduced to unsaturated alcohol by HCOONa with a ruthenium catalyst bearing the water-soluble TPPMS (**1**) (Scheme 5.18).

The reduction was efficient, with most reactions completed in a few hours, including those involving multi-substituted aromatic aldehydes. For example, 2,6-dichlorobenzaldehyde was converted into the corresponding alcohol in 1.5 h without hydrodechlorination occurring, and the reduction of α,β-unsaturated aldehydes was chemoselective, only furnishing unsaturated alcohol as the product. However, there was no reaction for substrates containing an OH group, such as 2-hydroxybenzaldehyde. Among the various catalysts tested, Ru(II)–**1** was found to be the most efficient [20, 22, 123]. Subsequent work demonstrated the transfer hydrogenation and also asymmetric transfer hydrogenation of unsaturated carboxylic acids to saturated carboxylic acids by formate in water, using a rhodium catalyst containing water-soluble phosphines [121].

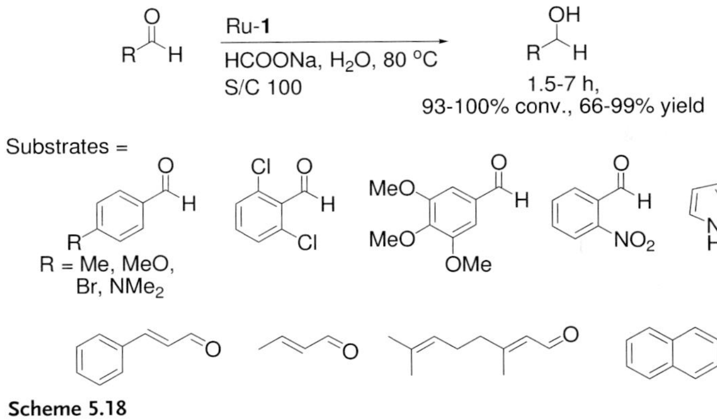

## 5.4 Transfer Hydrogenation in Water

Scheme 5.18

More recent research has revealed that ketones and aldehydes can be reduced by HCOONa or HCOOH in water with water-soluble half-sandwich Ru(II) and Ir(III) complexes **35–39** [61, 125–129]. The reduction was shown to be dependent on the solution pH, an important finding reminiscent of that in aqueous hydrogenation reactions [22]. Both water-soluble and -insoluble substrates were reduced, and in the favored pH window, TOFs up to 525 $h^{-1}$ were obtained with the Ir(III) catalyst **39a** and 153 $h^{-1}$ with the Ru(II) catalyst **38** (Table 5.2) [126, 127, 129]. A water-soluble molybdocene monohydride, $Cp_2Mo(H)OTf$ (**36**), was also found to catalyze the transfer hydrogenation of ketones and aldehydes in water, again with pH-dependent character-

**Table 5.2** Transfer hydrogenation of carbonyl compounds by HCOONa with **38**[a] and **39a**[b] in water.

| Substrate | Catalyst | Time (h) | Yield (%) | TOF[e] |
|---|---|---|---|---|
| Cyclohexanone | 38, pH 4.0 | 4 | 99 | 98 |
| 2-Butanone |  | 6 | 97 | 58 |
| Pyruvic acid |  | 4 | 99 | 96 |
| 4′-Acebenzsulf ass[c] |  | 3 | 98 | 103 |
| Acp[d] |  | 4 | 98 | 75 |
| 2-Trifluro-Acp |  | 4 | 99 | 153 |
| α-Tetralone |  | 13 | 97 | 21 |
| Cyclohexanone | 39a, pH 2.0 | 1 | 99 | 376 |
| Acp |  | 1 | 97 | 343 |
| 2-$CF_3$-Acp |  | 1 | 99 | 525 |
| Butanone |  | 4 | 99 | 150 |
| Pyruvic acid |  | 1 | 98 | 481 |
| 4′-$SO_3$Na-Acp |  | 1 | 99 | 419 |
| 1-Tetralone |  | 3 | 98 | 203 |

[a] 0.32 mmol ketones, 0.16 mol% **38**, 3 ml $H_2O$, 1.92 mol HCOONa, 70 °C, pH 4.0 [127].
[b] 0.32 mmol substrates, 0.5 mol% **39a**, 3 ml $H_2O$, 0.32 mol HCOOH, 70 °C, pH 2.0 [129].
[c] 4′-Acebenzsulf ass = 4′-acetylbenzenesulfonic acid sodium salt.
[d] Acp = acetophenone.
[e] Turnover frequency (mol $mol^{-1} h^{-1}$).

istics. Acetone could be converted into 2-propanol in about 8 h at 40 °C in water, and the reduction of benzaldehyde under the same conditions was instantaneous [130, 131].

A series of water-soluble ruthenium–arene complexes containing chelating 1,10-phenanthroline ligands have also been introduced (**40–43**) [132–138]. These complexes were found to catalyze transfer hydrogenation of ketones in aqueous solution using formic acid as hydrogen source, with **41** displaying a higher activity. For instance, TONs up to 164 were obtained in the reduction of acetophenone [132]. However, when the water-soluble ligand **1** was used instead of phenanthroline, similar half-sandwich Ru(II) complexes displayed much reduced activities on going from 2-propanol to water [138].

The half-sandwich catalysts above are unlikely to permit metal–ligand bifunctional catalysis [38, 39, 139]. This explains, to some extent, why the reduction rates are generally low. Accordingly, diamine ligands, having an $-NH_2$ functionality and so capable of activating a carbonyl substrate, were shown to be more effective. This is seen in the Ir(III)-catalyzed reduction of a wide range of aldehydes by HCOONa [140]. In particular, the catalyst **25** (Scheme 5.5) formed *in situ* from $[Cp^*IrCl_2]_2$ and the corresponding ligand, afforded TOFs of up to $1.3 \times 10^5 \, h^{-1}$ in the transfer hydrogenation of aldehydes in neat water. In contrast, when carried out in 2-propanol or azeotropic $HCOOH-NEt_3$ mixture, a much slower reduction resulted. The catalyst works for aromatic, α,β-unsaturated and aliphatic aldehydes and for those bearing functional groups such as halo, acetyl, alkenyl, and nitro groups, and is highly chemoselective towards the formyl group. For example, 4-acetylbenzaldehyde was reduced only to 4-acetylphenylethanol, and the reduction of 4-acetylcinnamaldehyde took place without affecting the ketone and olefin double bonds. Selected examples are presented in Scheme 5.19.

An interesting observation arising from the aldehyde reduction was that no reaction was detected with water-soluble substrates under the conditions employed. For example, the water-soluble 4-carboxybenzaldehyde or its sodium salt could not be reduced; but its ester analog, methyl 4-formylbenzoate, was reduced in a short time of 40 min with **25** at 80 °C at an S/C ratio of 5000. This suggests that the catalysis takes place on water rather than in water in these biphasic reactions, although other possibilities exist [116, 120]. Interestingly, **25** and its analogs also catalyze the hydrogenation of aldehydes in water, as discussed in Section 5.3.1.2 [63].

α,β-Unsaturated carbonyl compounds have been reduced at the C=C bonds in aqueous solution [141]. The reaction was catalyzed by $PdCl_2/SiO_2$ in an MeOH–$HCOOH-H_2O$ mixture (1:2:3) with microwave heating, affording the saturated carbonyls in moderate to excellent yields with high chemoselectivity.

## 5.4.2
### Asymmetric Transfer Hydrogenation

#### 5.4.2.1 Asymmetric Transfer Hydrogenation of Ketones

As with most other catalytic reactions using water as solvent, research into asymmetric transfer hydrogenation of ketones started with a search for water-soluble catalysts. As would be expected, this was achieved by synthesizing ligands that dissolve in water (**19–22**). However, recent studies have demonstrated that unmodi-

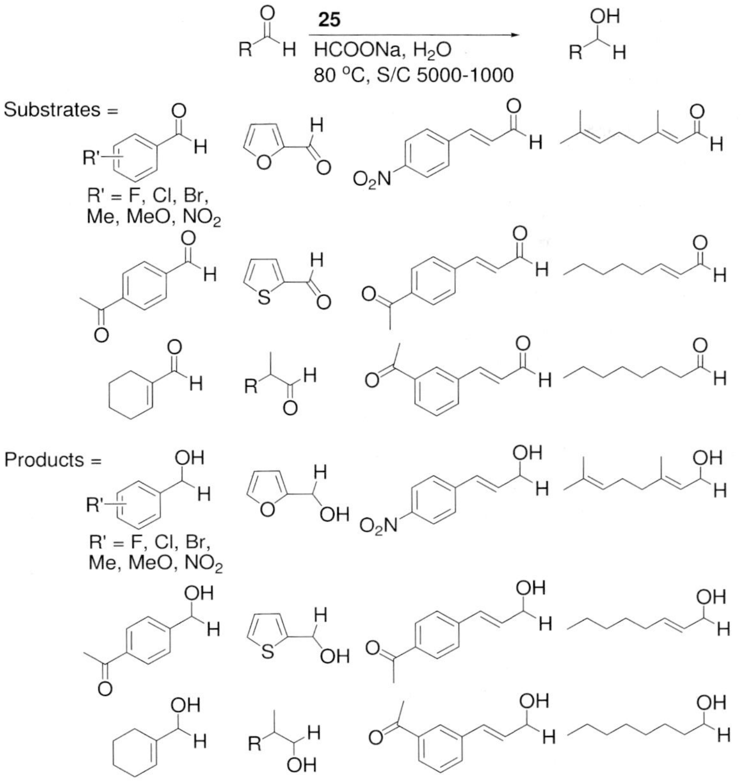

## Scheme 5.19

fied, water-insoluble ligands can deliver high activity and enantioselectivity for ketone reduction in water [10, 11, 133–135, 139, 142–196]. The structures above **44–59** show examples of catalysts containing water-insoluble ligands that are effective in aqueous-phase asymmetric transfer hydrogenation.

Williams and co-workers were the first to explore the asymmetric transfer hydrogenation of ketones with water-soluble Noyori–Ikariya-type catalysts in 2001 [191, 192]. The reduction was performed using catalysts containing a sulfonated TsDPEN or TsCYDN ligand (**19** or **21**) in 2-propanol with water added (up to 51% v/v). The catalyst was generated *in situ* by reacting the ligand with the chloro dimer, [RuCl$_2$(p-cymene)]$_2$ or [Cp*MCl$_2$]$_2$ (M=Rh, Ir). Although good to excellent *ee*s (up to 96%) were achieved, the reaction was generally sluggish under the chosen conditions. It was shown that the reaction went faster with increasing volume of water in the case of the Ir(III) catalysts. For example, the reduction of 3-fluoroacetophenone with the Ir–**21** catalyst gave an 82% conversion in 2.5 h when the water content was 34%, and a 94% conversion when the water level was increased to 51%; the enantioselectivities remained virtually unchanged, however, at 93–94% *ee*.

Asymmetric transfer hydrogenation of aromatic ketones by formate in neat water was demonstrated at about the same time, where a water-soluble catalyst was formed by combining [RuCl$_2$(p-cymene)]$_2$ with a (S)-proline amide ligand (**22b**) [190, 193]. The reduction was carried out with or without a surfactant, with better results obtained in its presence. As can be seen from Scheme 5.20, the reaction afforded good conversions with moderate to good *ee*s at 40 °C. These results represent the first examples of asymmetric transfer hydrogenation in water with no organic cosolvents. Water-soluble Rh(III)–Schiff base complexes were later reported to catalyze similar reductions in aqueous formate solution, affording moderate to good reaction rates and *ee*s [189].

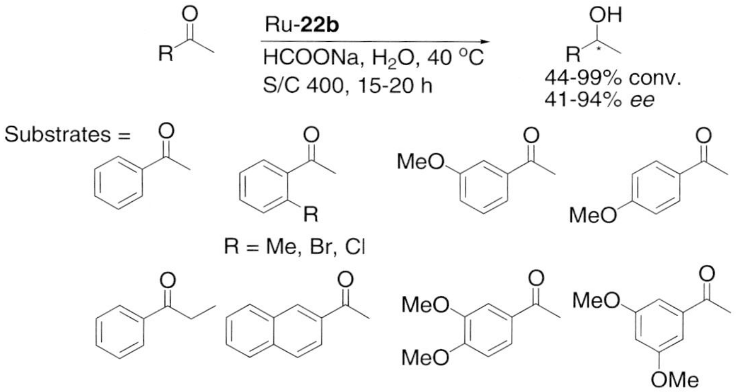

**Scheme 5.20**

A highly water-soluble ligand (**20a**) was developed around the same time [188]. Asymmetric transfer hydrogenation of ketones catalyzed by Ru–**20a** showed good activities and moderate to excellent enantioselectivities in the presence of a surfactant, SDS. Moreover, the catalyst, as it was designated, can be recycled twice without loss of enantioselectivity.

A common feature of these investigations, like those in other areas of aqueous-phase catalysis, is to make the catalysts soluble in water, and to circumvent the problem of low solubility of most organic substrates in water, surfactants are usually called upon. However, an investigation into the behavior of the water-soluble **60** and insoluble TsDPEN in ketone reduction by HCOONa in neat water revealed surprises [158, 197, 198]. While Ru–**60** was shown to be highly effective in neat water (see below), the TsDPEN-containing **44** was equally good. Thus, using the precatalyst generated by reacting TsDPEN with [RuCl$_2$(p-cymene)]$_2$, acetophenone was fully reduced to (R)-1-phenylethanol in 95% *ee* by HCOONa in water at 40 °C in a 1 h reaction time at an S/C of 100. In comparison, the reaction run in HCOOH–NEt$_3$ azeotropic mixture afforded a conversion of less than 2% in 1 h [158]. This initial finding has since been proved to be general, that is, water permits fast and enantioselective asymmetric reduction of unfunctionalized ketones by HCOONa with a range of metal–diamine catalysts. These catalysts can generally be prepared *in situ* by reacting the unmodified ligand with one of the aforementioned metal dimers without adding a base. They show varying solubilities in water, with those containing rhodium and iridium being more soluble than the ruthenium analogs. However, they display a much higher solubility in ketones and alcohols. Hence the reduction is often biphasic, with the catalysis probably taking place on water as mentioned earlier.

## 5.4 Transfer Hydrogenation in Water

**60**: MeO-(O)n-C6H4-CH(NH2)-CH(NHTs)-C6H4-(O)n-OMe

The performance of these catalysts in the reduction of the benchmark substrate acetophenone is shown in Table 5.3 [11, 18]. The monotosylated diamines, which have been shown to be successful ligands for asymmetric transfer hydrogenation of ketones in 2-propanol or HCOOH–NEt$_3$ azeotropic mixture, can all be applied to the reduction of acetophenone by HCOONa in water, with full conversions and up to 99% ees reached in short reaction times. In general, the reaction in water is faster than that in organic solvents, but with similar enantioselectivities. Under the given conditions, the Rh(III) catalysts appear to outperform both Ru(II) and Ir(III) in water in terms of catalytic activity and enantioselectivity, and the camphor-substituted **45** and **49** led to the best enantioselectivity (Table 5.3). It is noted that the reaction with the Rh–diamine catalysts can be carried out effectively in the open air without degassing and/or inert gas protection throughout, thus making the reduction easier to carry out than reactions catalyzed by most organometallic complexes (entries 21, 23, 29 and 32, Table 5.3).

**Table 5.3** Asymmetric transfer hydrogenation of acetophenone with various catalysts in aqueous media.

| Entry | Catalyst | [H]$^a$ | S/C$^b$ | Temperature (°C) | Time (h) | Conversion (%) | ee (%) | Ref. |
|---|---|---|---|---|---|---|---|---|
| 1 | Ru–20a | HCOONa | 100 | 40 | 24 | >99 | 95 | [188] |
| 2 | Rh–20a | HCOONa | 100 | 40 | 24 | 92 | 84 | [188] |
| 3 | Ir–20a | HCOONa | 100 | 40 | 24 | 10 | 58 | [188] |
| 4 | Ru–20b | HCOONa | 100 | 28 | 0.5 | 33 | 95 | [169] |
| 5 | Rh–20b | HCOONa | 100 | 28 | 0.5 | 97 | 97 | [169] |
| 6 | Ir–20b | HCOONa | 100 | 28 | 0.5 | 29 | 94 | [169] |
| 7 | Ru–22b | HCOONa | 400 | 40 | 18 | 98 | 69 | [190] |
| 8 | 44 | HCOONa | 100 | 40 | 1 | 99 | 95 | [158] |
| 9 | 44 | F–T$^c$ | 100 | 40 | 1.5 | >99 | 97 | [157] |
| 10 | 44 | F–T | 1000 | 40 | 9 | >99 | 96 | [157] |
| 11 | 44 | F–T | 5000 | 40 | 57 | 98 | 96 | [157] |
| 12 | 44 | F–T | 10000 | 40 | 110 | 98 | 94 | [157] |
| 13 | 44$^d$ | HCOONa | 100 | 40 | 3 | >99 | 96 | [167] |
| 14 | 45 | HCOONa | 100 | 40 | 2 | 99 | 97 | [176] |
| 15 | 45 | HCOONa | 1000 | 40 | 20 | 95 | 96 | [176] |
| 16 | 46a | HCOONa | 100 | 40 | 2 | 99 | 85 | [156] |
| 17 | 46b | HCOONa | 100 | 40 | 2.5 | >99 | 81 | [199] |
| 18 | 47 | HCOONa | 100 | 40 | 12 | 84 | 71 | [155] |
| 19 | 47 | HCOONa | 40 | 50 | – | 13 | 81 | [160] |
| 20 | 48a | HCOONa | 100 | 40 | 0.5 | 99 | 97 | [194] |
| 21 | 48a$^e$ | HCOONa | 100 | 40 | 0.5 | 99 | 97 | [194] |
| 22 | 48b | HCOONa | 100 | 40 | 3.5 | 99 | 93 | [194] |
| 23 | 48b$^e$ | HCOONa | 100 | 40 | 12 | 95 | 92 | [194] |
| 24 | 49a | HCOONa | 100 | 40 | 0.7 | 99 | 99 | [176] |
| 25 | 49a | HCOONa | 1000 | 40 | 20 | 89 | 99 | [176] |

(Continued)

Table 5.3 (Continued)

| Entry | Catalyst | [H]$^a$ | S/C$^b$ | Temperature (°C) | Time (h) | Conversion (%) | ee (%) | Ref. |
|---|---|---|---|---|---|---|---|---|
| 26 | 49b | HCOONa | 100 | 40 | 0.7 | 99 | 97 | [176] |
| 27 | 49b | HCOONa | 1000 | 40 | 2.5 | 97 | 98 | [176] |
| 28 | 50ai | HCOONa | 100 | 40 | 0.25 | >99 | 95 | [156] |
| 29 | 50ai$^e$ | HCOONa | 100 | 40 | 0.25 | 99 | 96 | [156] |
| 30 | 50bi | HCOONa | 100 | 40 | 3 | 99 | 93 | [156] |
| 31 | 50aii | HCOONa | 100 | 40 | 0.25 | >99 | 94 | [199] |
| 32 | 50aii$^e$ | HCOONa | 100 | 40 | 0.25 | >99 | 94 | [199] |
| 33 | 50bii | HCOONa | 100 | 40 | 1.5 | >99 | 92 | [199] |
| 34 | 51a | HCOONa | 100 | 40 | 20 | 92 | 55 | [155] |
| 35 | 51b | HCOONa | 100 | 40 | 5 | >99 | 27 | [155] |
| 36 | 52 | HCOONa | 200 | 28 | 3 | 100 | 96 | [175] |
| 37 | 54 | HCOONa | 100 | 60 | 2–5 | >99 | 93 | [134] |
| 38 | 55 | HCOONa | 100 | 60 | 2–5 | >99 | 44 | [134] |
| 39 | 57 | HCOONa | 100 | 40 | 0.5 | 100 | 93 | [170] |
| 40 | 57$^f$ | HCOONa | 100 | 40 | 0.5 | 100 | 94 | [170] |
| 41 | 57$^g$ | HCOONa | 100 | 40 | 0.5 | 100 | 94 | [170] |
| 42 | 58 | HCOONa | 20 | 30 | 12 | 100 | 67 | [159] |
| 43 | 59 | HCOONa | 20 | 30 | 40 | 100 | 84 | [159] |

$^a$[H] refers to hydrogen source.
$^b$S/C is substrate to catalyst molar ratio.
$^c$F–T = formic acid–triethylamine (mixtures at various molar ratios).
$^d$The reaction was carried out in PEG–H$_2$O.
$^e$The reaction was carried out in the open air without inert gas protection.
$^f$In the presence of the surfactant CTAB (cetyltrimethylammonium bromide).
$^g$In the presence of the surfactant SDS (sodium dodecyl sulfate).

Amino alcohol ligands were believed to be incompatible with formic acid as reductant in the past [117, 196]. Table 5.3 shows that the ligands **47** and **51** do catalyze the asymmetric transfer hydrogenation of acetophenone by HCOONa in water; however, the reduction rates and enantioselectivities were much lower than those obtained with the diamines. The metal complexes containing (−)-ephedrine yielded better results than others in terms of rates and/or ees, and in general the iridium catalysts exhibited higher activity [155, 160, 182]. Terpene-based amino alcohols and amino acids-ligated Cr(II) have also been explored in asymmetric transfer hydrogenation in water, but with only limited success [200, 201].

The diamine-enabled protocol has been applied to a range of ketones, and selected examples are shown [11, 18]. These substrates can be reduced efficiently with the catalysts **44–59** by HCOONa in water. Unfunctionalized aromatic ketones, heterocyclic ketones, and functionalized or multi-substituted ketones are all viable substrates with this reduction system. The reduction is generally easy to perform, affording the chiral alcohols with high ees in a short reaction time for most of the substrates. S/C ratios of 100–10 000 have been demonstrated to be feasible [157]. Of particular note is the Rh–diamine catalyst **50a**, which delivered high conversions for most of the ketones in a short reaction time, and in most cases the enantioselectivities were good to excellent, with ees reaching up to 99% and TOFs close to 4000 h$^{-1}$ in water in the open air [156]. The protocol works particularly well for some heteroaryl ketones, as shown by the

## 5.4 Transfer Hydrogenation in Water

examples illustrated. Thus, for instance, the reduction of 2-acetylfuran with **50ai** was complete within 5 min, yielding (R)-1-(2-furyl)ethanol in 99% ee.

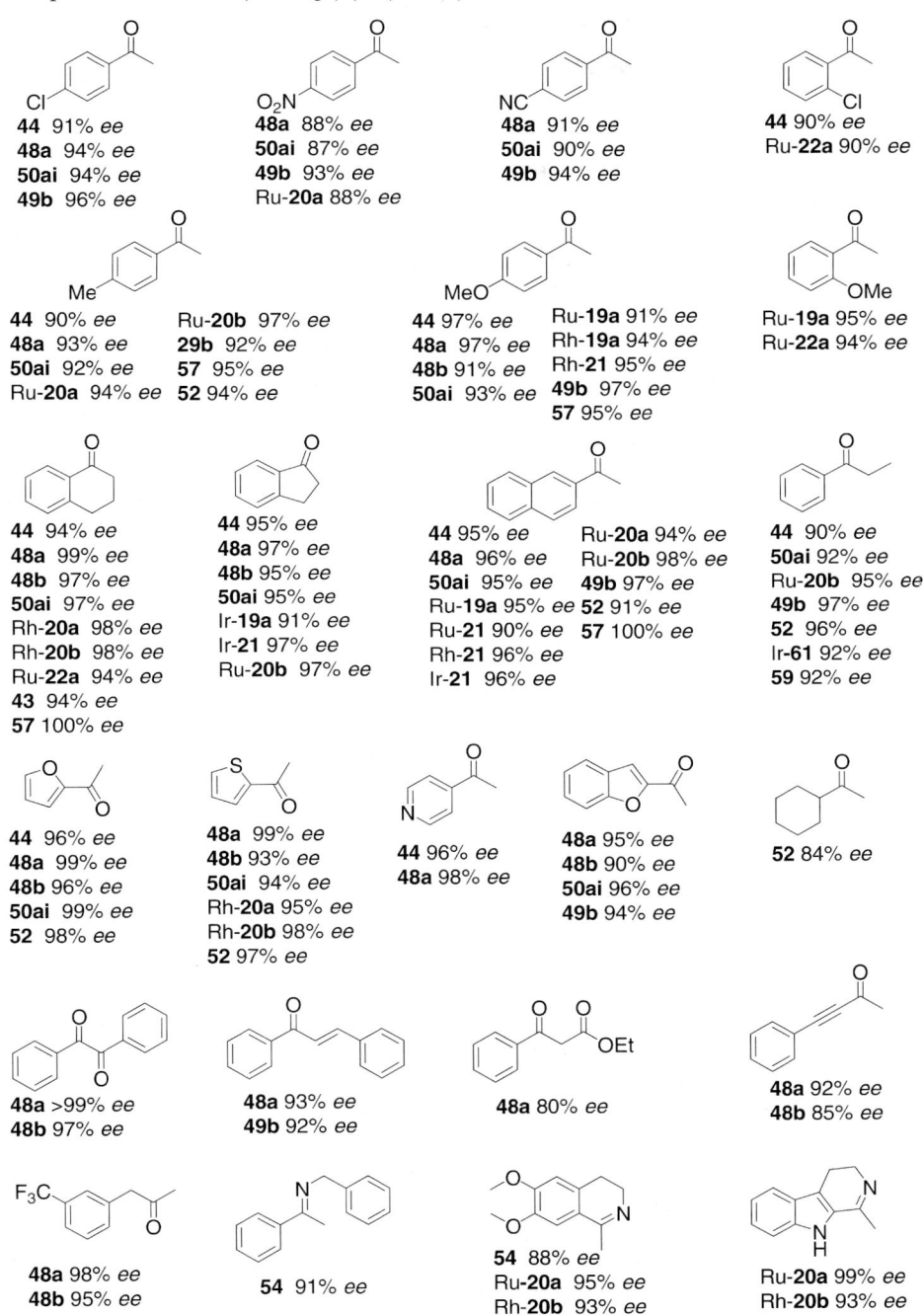

**4-Cl-acetophenone**
**44** 91% ee
**48a** 94% ee
**50ai** 94% ee
**49b** 96% ee

**4-O₂N-acetophenone**
**48a** 88% ee
**50ai** 87% ee
**49b** 93% ee
**Ru-20a** 88% ee

**4-NC-acetophenone**
**48a** 91% ee
**50ai** 90% ee
**49b** 94% ee

**2-Cl-acetophenone**
**44** 90% ee
**Ru-22a** 90% ee

**4-Me-acetophenone**
**44** 90% ee          **Ru-20b** 97% ee
**48a** 93% ee         **29b** 92% ee
**50ai** 92% ee        **57** 95% ee
**Ru-20a** 94% ee      **52** 94% ee

**4-MeO-acetophenone**
**44** 97% ee          **Ru-19a** 91% ee
**48a** 97% ee         **Rh-19a** 94% ee
**48b** 91% ee         **Rh-21** 95% ee
**50ai** 93% ee        **49b** 97% ee
                       **57** 95% ee

**2-OMe-acetophenone**
**Ru-19a** 95% ee
**Ru-22a** 94% ee

**α-tetralone**
**44** 94% ee
**48a** 99% ee
**48b** 97% ee
**50ai** 97% ee
**Rh-20a** 98% ee
**Rh-20b** 98% ee
**Ru-22a** 94% ee
**43** 94% ee
**57** 100% ee

**1-indanone**
**44** 95% ee
**48a** 97% ee
**48b** 95% ee
**50ai** 95% ee
**Ir-19a** 91% ee
**Ir-21** 97% ee
**Ru-20b** 97% ee

**2-acetylnaphthalene**
**44** 95% ee          **Ru-20a** 94% ee
**48a** 96% ee         **Ru-20b** 98% ee
**50ai** 95% ee        **49b** 97% ee
**Ru-19a** 95% ee      **52** 91% ee
**Ru-21** 90% ee       **57** 100% ee
**Rh-21** 96% ee
**Ir-21** 96% ee

**propiophenone**
**44** 90% ee
**50ai** 92% ee
**Ru-20b** 95% ee
**49b** 97% ee
**52** 96% ee
**Ir-61** 92% ee
**59** 92% ee

**2-acetylfuran**
**44** 96% ee
**48a** 99% ee
**48b** 96% ee
**50ai** 99% ee
**52** 98% ee

**2-acetylthiophene**
**48a** 99% ee
**48b** 93% ee
**50ai** 94% ee
**Rh-20a** 95% ee
**Rh-20b** 98% ee
**52** 97% ee

**4-acetylpyridine**
**44** 96% ee
**48a** 98% ee

**2-acetylbenzofuran**
**48a** 95% ee
**48b** 90% ee
**50ai** 96% ee
**49b** 94% ee

**cyclohexyl methyl ketone**
**52** 84% ee

**benzil**
**48a** >99% ee
**48b** 97% ee

**chalcone**
**48a** 93% ee
**49b** 92% ee

**ethyl benzoylacetate**
**48a** 80% ee

**4-phenyl-3-butyn-2-one**
**48a** 92% ee
**48b** 85% ee

**3-CF₃-phenylacetone**
**48a** 98% ee
**48b** 95% ee

**N-benzyl imine**
**54** 91% ee

**dihydroisoquinoline**
**54** 88% ee
**Ru-20a** 95% ee
**Rh-20b** 93% ee

**dihydro-β-carboline**
**Ru-20a** 99% ee
**Rh-20b** 93% ee

The water-soluble Ru(II) complexes **53** and **54** were recently synthesized and applied to the aqueous reduction of ketones, affording good yields and *ees* up to 93% [133–135]. The study reveals again the importance of tosylation of the diamine and substitution on the arene ring to both catalytic activity and enantioselectivity in water. These catalysts and Ru–**20** (see also Section 5.4.2.2) have also been extended into imine reduction in water. Selected examples are illustrated.

The PNNP ligand **61** (Scheme 5.21), which is highly effective in ruthenium-catalyzed asymmetric transfer hydrogenation of ketones in 2-propanol [202], could also be used for the aqueous-phase reduction when combined with [IrHCl$_2$(COD)]$_2$ [173, 177]. As shown in Scheme 5.21, the reduction of propiophenone was completed in 9 h with 88% *ee* at 60 °C at an S/C ratio of 100 in the presence of a phase-transfer catalyst. Moreover, the same reaction could be run without inert gas protection and at a higher S/C ratio of 8000. In the latter case, the reaction afforded 80% isolated yield with 85% *ee* in 101 h. The reduction with the analogous Schiff base ligand led to a much reduced enantioselectivity (34% *ee* for propiophenone), indicating again the importance of the N–H moiety to the reduction.

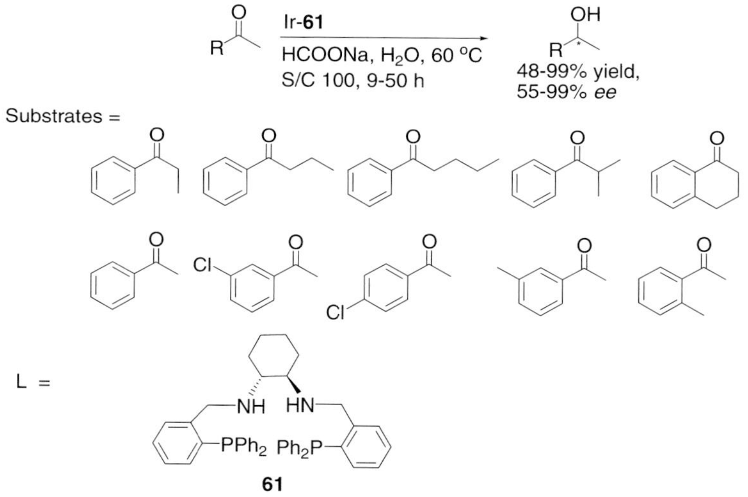

**Scheme 5.21**

The tethered complex **52** was shown to be an excellent catalyst in both organic and aqueous phase reduction of ketones [175]. Thus, acetophenone was reduced by HCOONa in water with **52** to give 100% conversion and 96% *ee* at 28 °C in 3 h, and in the case of 2-acetylfuran, the catalyst loading could be reduced to 0.01 mol% with an *ee* of 98% obtained. Remarkably, the catalyst even allowed for the reduction of aliphatic ketones in water, albeit with slightly lower *ees*.

#### 5.4.2.2 Asymmetric Transfer Hydrogenation of Imines

Asymmetric transfer hydrogenation of imines and related compounds in water has only recently been demonstrated (Scheme 5.22). Using the water-soluble ligand **20a**, imines and iminium salts could be smoothly reduced by HCOONa with Ru(II) catalysis in water in the presence of CTAB as a phase-transfer catalyst [174]. The reduction afforded moderate to excellent yields and *ee*s for both imines and iminiums. However, the catalyst failed to reduce acylic imines, which decomposed under the aqueous conditions.

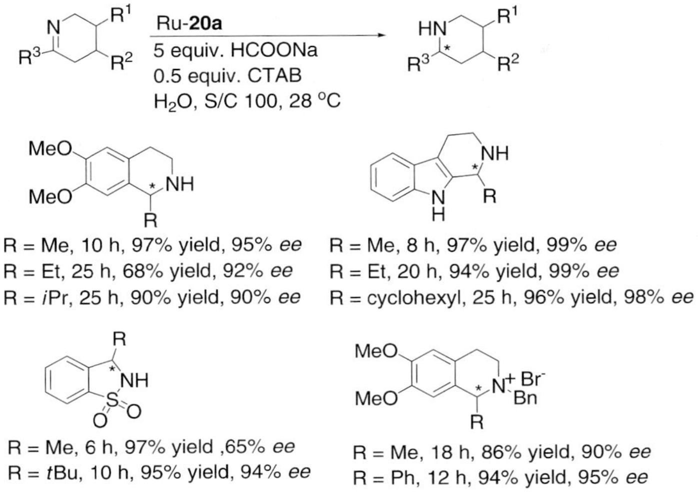

Ru-**20a**
5 equiv. HCOONa
0.5 equiv. CTAB
$H_2O$, S/C 100, 28 °C

R = Me, 10 h, 97% yield, 95% ee
R = Et, 25 h, 68% yield, 92% ee
R = *i*Pr, 25 h, 90% yield, 90% ee

R = Me, 8 h, 97% yield, 99% ee
R = Et, 20 h, 94% yield, 99% ee
R = cyclohexyl, 25 h, 96% yield, 98% ee

R = Me, 6 h, 97% yield, 65% ee
R = *t*Bu, 10 h, 95% yield, 94% ee

R = Me, 18 h, 86% yield, 90% ee
R = Ph, 12 h, 94% yield, 95% ee

**Scheme 5.22**

The water-soluble Ru(II)-arene complexes **43** and **53–55** were also shown to be effective for both ketone and imine reduction by HCOONa in water [134]. In the case of imines, both cyclic and acyclic substrates could be reduced by the catalysts, with *ee*s up to 91% obtained for acyclic and 88% for cyclic imines. Recently, a new water-soluble, aminated ligand **20b** was applied to the reduction of ketones and imines in water [169]. In comparison with the related Ru(II) and Ir(III) catalysts, Rh(III)–**20b** afforded the best performance in the transfer hydrogenation in terms of reaction rates and enantioselectivities. For instance, the reduction of acetophenone gave a 97% conversion with a 97% *ee* in 0.5 h at 28 °C and an S/C ratio of 100. The catalyst also worked well for cyclic imines, affording up to 93% *ee*.

### 5.4.2.3 Asymmetric Transfer Hydrogenation with Biomimetic Catalysts

Unlike organometallic catalysis, asymmetric transfer hydrogenation of ketones in aqueous media with enzymes and microorganisms is well documented [203–207]. Various aromatic and aliphatic ketones can be reduced stereoselectively using alcohol dehydrogenases, microorganisms, and whole microbial cells [203, 208, 209]. However, baker's yeast is by far the most widely used [203].

Aiming to broaden the substrate specificity of natural enzymes and discover new enzymes for novel transformations, artificial metalloenzymes, which integrate metal complexes with biocatalysts, have been explored in enantioselective catalysis since the 1970s [210–216]. The opportunity of combined chemogenetic optimization of artificial metalloenzymes, that is, chemically tuning the active metal centers while genetically modifying the host proteins, offers a new strategy to discover more efficient, enzyme-like catalysts [217]. Biotin displays a strong affinity for avidin and streptavidin, allowing biotinylated molecular metal catalysts to be incorporated into proteins and so giving rise to artificial metalloenzymes. Indeed, the incorporation of a biotinylated achiral 1,2-diamine–Ru(II) catalyst into a host protein, avidin or streptavidin, has recently been shown to afford a versatile artificial metalloenzyme capable of reducing ketones with formate in buffered aqueous solution [184, 217–219]. The reduction of aromatic ketones proceeded smoothly under optimized conditions, furnishing enantioselectivities of up to 97% ee. Selected chiral products are shown in Scheme 5.23 [184]. To identify the best metalloenzyme with matched active metal site and chiral protein pocket, the metal complex was modified by varying the arene ligand and the spacer group while the host protein was genetically optimized by point mutations. The results suggested that the catalytic activity of these artificial metalloenzymes is dependent on the localization of the biotinylated metal catalyst, with the properties of the $\eta^6$-bound arenes playing a critical role in the enantioselection.

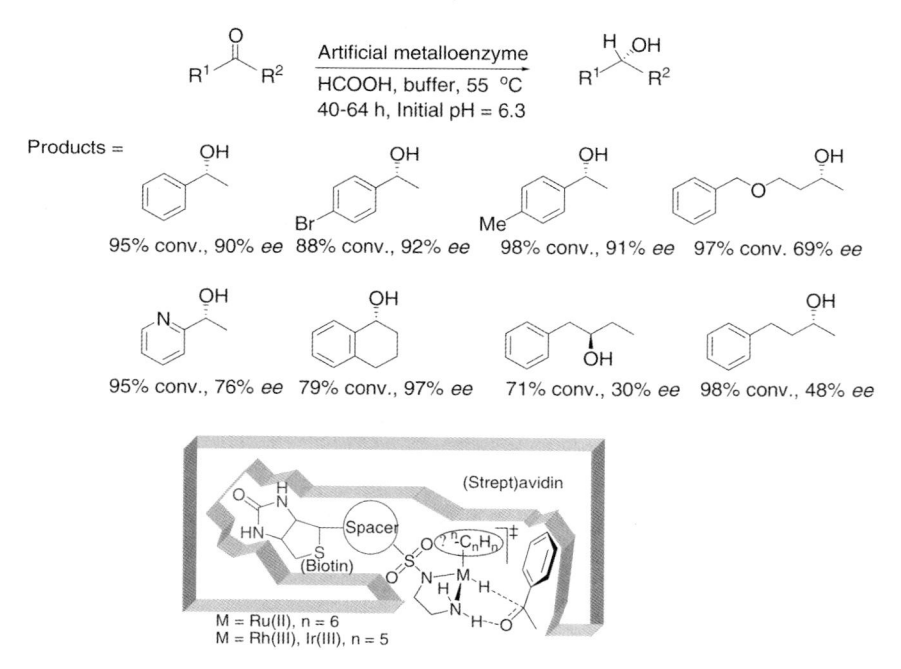

Scheme 5.23

In a twist to the approach above, β-cyclodextrin has been used to modify metal catalysts, enhancing the solubility of hydrophobic substrates in water. An Ru(II) complex so generated (**62**) served as an efficient catalyst in transfer hydrogenation in water using HCOONa as hydrogen source [220–222]. Non-conjugated ketones were reduced with high *ee*s and high yields, although the S/C ratios were low (Scheme 5.24). The β-cyclodextrin also appears to play an important role in the enantiocontrol through preorganization of the substrates in its hydrophobic cavity. Recently, the ruthenium unit has been attached to the secondary face of β-cyclodextrin (**63**); the catalyst allows for the reduction of challenging aliphatic ketones, impressively affording *ee*s up to 98% [222]. The enantioselection presumably arises from chiral relay from β-cyclodextrin to ruthenium, which changes the latter into a "chiral-at-metal" species.

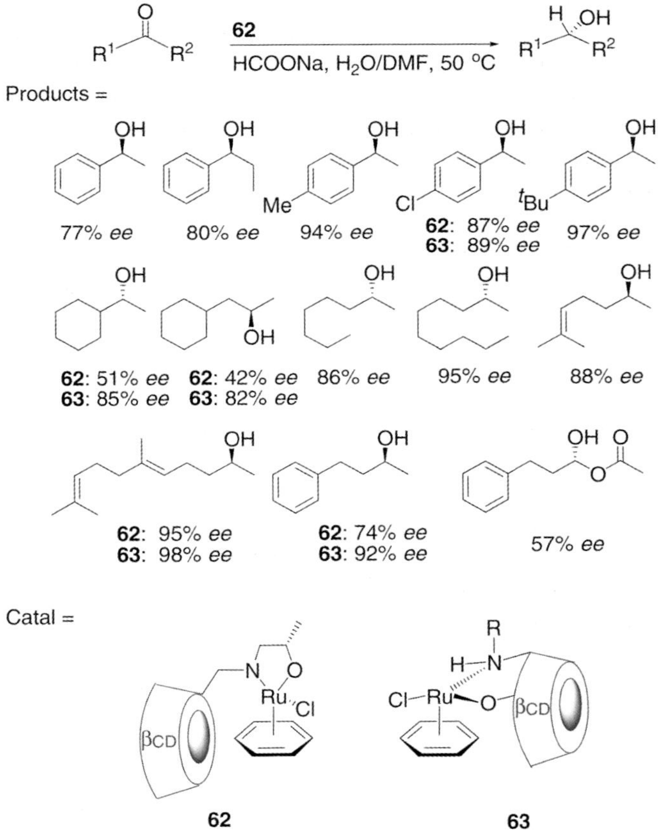

**Scheme 5.24**

## 5.4.3
### Water-facilitated Catalyst Separation and Recycle

Catalysts bearing water-soluble ligands generally allow easy catalyst–product separation in either hydrogenation or transfer hydrogenation reactions. However, being soluble in common organic solvents and insoluble in water, the unmodified chiral diamine ligands discussed above present a problem. This has been addressed in recent studies, which are briefly summarized here.

The half-sandwich catalysts in Sections 5.4.1–5.4.2 are often (partially) soluble in water but insoluble in non-polar solvents. In this case, the product can be extracted with a non-polar solvent such as diethyl ether without recourse to purpose-built water-soluble catalysts, and this has been demonstrated. A good example is the reduction catalyzed by Ru–**22b** mentioned earlier, which could be reused up to six times without compromising *ee*s [193]. Recently, a similar catalyst was also shown to be effective in the asymmetric transfer hydrogenation of aromatic ketones and recyclable in water [172].

For more practical and easier catalyst–product separation, highly water-soluble ligands or those that are supported on solid surfaces are desirable. As discussed previously, Ru–**20** has been shown to be an efficient catalyst for asymmetric transfer hydrogenation of ketones and imines in water. Given the hydrophilic nature of **20**, the catalyst could be readily separated from the product by simple decantation [188]. The PEG-supported **60** represents an example of water-soluble polymeric ligand. As with its non-supported counterpart, Ru–**60** is also highly effective for asymmetric transfer hydrogenation in water towards a wide range of aromatic ketones, with results comparable to those obtained with **44**. The advantage in using Ru–**60** is that the product can be easily separated from the catalyst-containing aqueous phase. To demonstrate its recyclability, the reduction of acetophenone by HCOONa with Ru–**60** in water was carried out, with the product extracted with diethyl ether [186]. An inductively coupled plasma (ICP) analysis showed that 0.4 mol% of ruthenium leached into the organic phase. Remarkably, the PEG-immobilized catalyst could be reused 14 times with no loss in enantioselectivity, demonstrating its excellent recyclability and lifetime under aqueous conditions. In contrast, when carried out in HCOOH–NEt$_3$ without water, recycling was possible only for two runs without the rates and *ee*s being eroded [197].

The polystyrene-supported diamines **64–65** have been combined with [RuCl$_2$(*p*-cymene)]$_2$ and [Cp*RhCl$_2$]$_2$ for asymmetric transfer hydrogenation [180]. Scheme 5.25 shows the reduction of ketones with Ru–**64** by HCOONa in water, affording excellent enantioselectivities. In these catalysts, the microenvironment within the polymer network appears to be important to stereoselection. The cross-linked polymer **65** has also proved to be efficient in the reduction in water, and the catalyst can be recycled five times, giving about the same *ee* values.

Inorganic supports have also been explored, as seen in the ligands **66–68**. When applied to Ru(II)-catalyzed asymmetric transfer hydrogenation of ketones, these ligands were effective in both organic and aqueous media, with **66** being the most efficient. Scheme 5.26 gives selected examples of ketone reduction by HCOONa catalyzed with Ru–**66** in water. Although taking a long time to complete in recycle runs even in the presence of a surfactant, the catalyst displayed excellent recyclability in terms of enantioselectivity – up to 11 recycles without loss of *ee*s [183, 185].

Dendrimer-supported ligands provide yet another example of recyclable catalysis. In the asymmetric transfer hydrogenation of ketones by HCOONa in water, the first-generation dendritic ligand **69** ligated to Rh(III) could be reused up to six times without enantioselectivity being eroded (Scheme 5.27) [178]. Remarkably, the S/C ratio could be increased to 10 000.

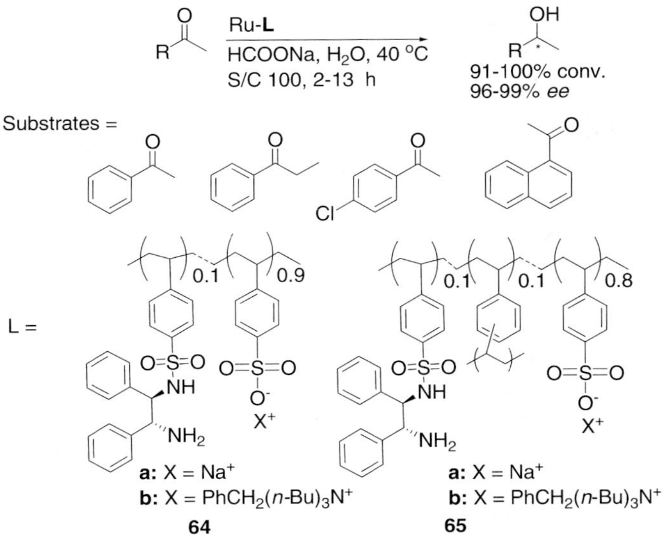

Scheme 5.25

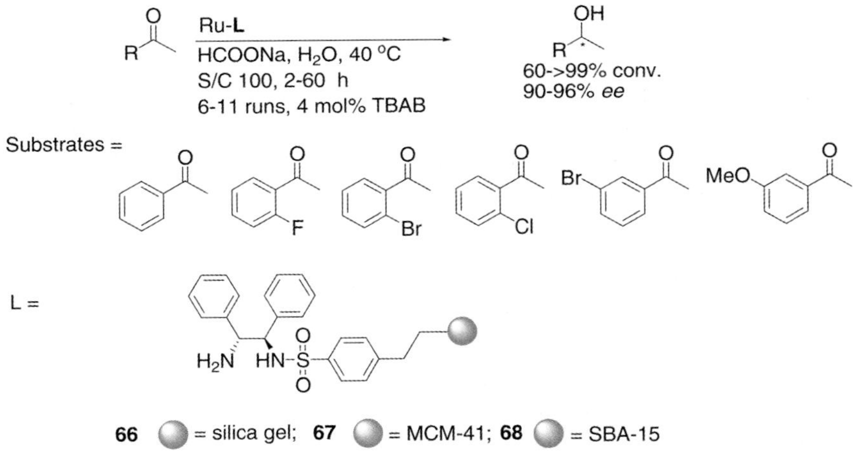

Scheme 5.26

## 5.5
## Role of Water

A number of examples have been presented which show that reduction in water is not only viable but can also gain in reaction rates and selectivities. In most cases, however, the role of water is not clear. In fact, this aspect of aqueous reduction has only received significant attention in recent years. What is clear now is that water is in general not an innocent spectator and it could play a role in every step of a catalytic cycle.

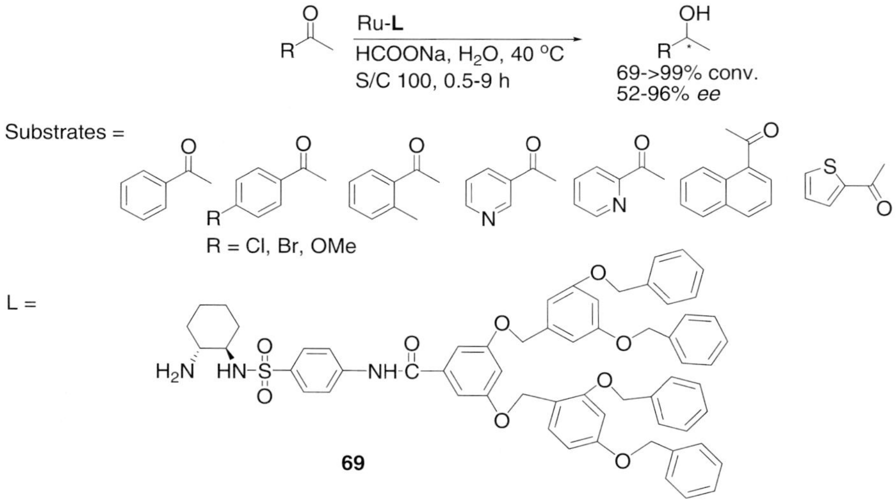

**Scheme 5.27**

## 5.5.1
### Coordination to Metals

Many metal complexes are known which contain coordinated water molecules, as seen in the previous sections. The water molecule may render the complex water soluble, stabilize it, and enhance or impede its reaction with a substrate. An illustrative example is found in some half-sandwich Ru(II), Rh(III), and Ir(III) chloro complexes, which can be used in both hydrogenation and transfer hydrogenation. These complexes readily undergo $I_a$-type aquation to give monoaqua dications in water (Scheme 5.28). In several instances, the aquation and anation equilibrium constants have been determined, which generally favor the formation of anation products [223–225]. In the case of [Ru(arene)(en)Cl][PF$_6$], the equilibrium constants are in the order of $K \approx 0.01$ M and do not vary significantly with the arene ligand [225]. The monoaqua complexes have been isolated and structurally characterized in a number of cases [129, 133, 225].

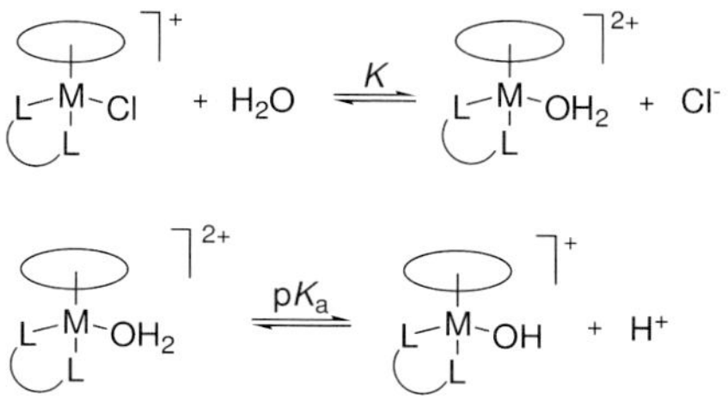

**Scheme 5.28**

The aqua complexes may undergo deprotonation in water, forming hydroxo species, which can slow catalysis by inhibiting the coordination of reductants such as $H_2$ or formate (Scheme 5.28). The $pK_a$ values of these half-sandwich complexes are ~7–9 and, interestingly, they do not appear to vary significantly with the central metal atom and its ligands [87, 224, 225]. However, the closely related triaqua complexes $[M(arene)(H_2O)_3]^{2+}$ are much more acidic, showing the importance of L^L in attenuating the electrophilic properties of M.

## 5.5.2
### Acid–Base Equilibrium

Water involvement in acid–base equilibria is ubiquitous. In hydrogenation reactions, water can act as a proton carrier, facilitating protonation of substrates and catalysts, or use its conjugate base to effect deprotonation. For instance, depending on the solution pH, a dehydroamino acid may exist in either neutral or ionized form in water. Consequently, the hydrogenation rate and/or selectivity may vary with the pH, as the mode of olefin coordination to catalysts could be affected [109, 226].

Recent DFT calculations revealed that hydride generation in hydrogenation reactions is facilitated by water [48, 49]. Thus, the model dihydrogen complex $RuCl(H_2)(OR)(PPh_3)$ is deprotonated by water molecules, leading to the corresponding hydride and $H_3O^+$ with a barrier of only ~16 kJ mol$^{-1}$ [48]. In a related complex, $RuH(OR)(PPh_3)(H_2O)$, external water was shown to act as an acid, protonating the coordinated alkoxide. In contrast, intramolecular reductive elimination to give HOR is more costly in terms of energy [49].

A significant consequence of the acid–base equilibrium is the effect of pH on hydrogenation. As mentioned earlier, the Ru–TPPMS catalyst exist mainly as RuClH(TPPMS)$_3$ at low pH but as $RuH_2(TPPMS)_4$ under basic conditions, and they show contrasting selectivities for α,β-unsaturated carbonyls [45, 46]. A good example illustrating the complex interplay of solution pH, concentration of reactants and catalysts, and hydride stability is found in the half-sandwich complex **39a**-catalyzed transfer hydrogenation of ketones in water. The reaction is highly pH dependent, showing a maximum rate at around pH 2 in the case of cyclohexanone. This pH dependence can be traced to the equilibria in Scheme 5.29 (L^L = bipy). In particular, below pH 1, the hydride is protonated, giving off $H_2$, and at pH > 7 the aqua complex is deprotonated, resulting in an inactive hydroxo species. It is noted, however, that the hydride itself is stable in a wide pH window of 1–9 [129].

A further example is seen in aqueous-phase asymmetric transfer hydrogenation. The catalysts **48a** and **48b** displayed pH windows of 5.5–10 and 6.5–8.5, respectively, for TOF > 50 h$^{-1}$ in the asymmetric transfer hydrogenation of acetophenone in water [194]. At lower pH values, not only did the TOF decrease, but also the enantioselectivities eroded. Various lines of evidence indicate that, apart from the effect of pH on [HCOO$^-$] (Scheme 5.29), the pH effect on both the reaction rate and enantioselectivity can be accounted for by the equilibrium shown in Scheme 5.30. The protonation of the chiral ligand explains why the catalyst behaves poorly at low pH values [154].

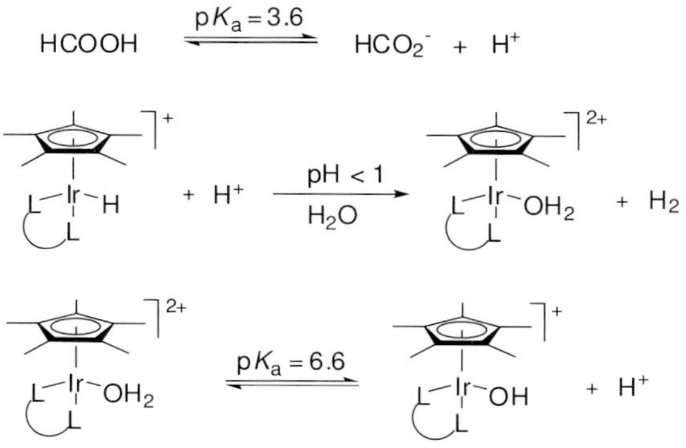

HCOOH $\rightleftharpoons$ (p$K_a$ = 3.6) $HCO_2^-$ + $H^+$

[Cp*Ir(L)H]$^+$ + H$^+$ $\xrightarrow{\text{pH} < 1, H_2O}$ [Cp*Ir(L)OH$_2$]$^{2+}$ + H$_2$

[Cp*Ir(L)OH$_2$]$^{2+}$ $\rightleftharpoons$ (p$K_a$ = 6.6) [Cp*Ir(L)OH]$^+$ + H$^+$

**Scheme 5.29**

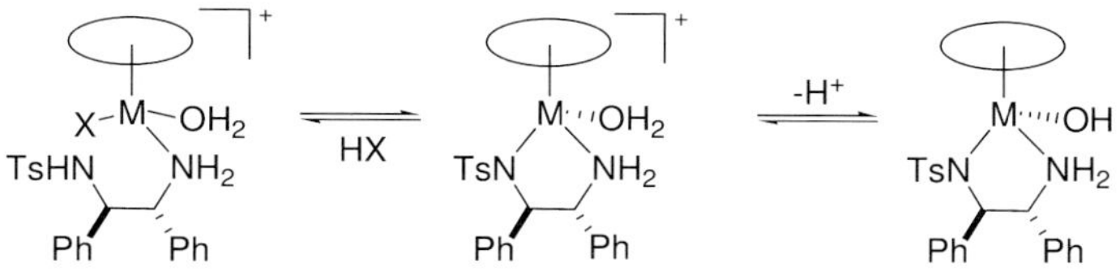

[X–M(OH$_2$)(TsHN)(NH$_2$)(Ph)(Ph)] $\rightleftharpoons$ HX [M(OH$_2$)(TsN)(NH$_2$)(Ph)(Ph)]$^+$ $\xrightarrow{-H^+}$ [M(OH)(TsN)(NH$_2$)(Ph)(Ph)]

**Scheme 5.30**

### 5.5.3
### H–D Exchange

One consequence of hydrogenation in water is H–D exchange. This has been studied both experimentally and theoretically. For instance, the water-soluble CpRuH(PTA)$_2$ catalyst, discussed in Section 5.3.1.1, readily undergoes H–D exchange in D$_2$O, having a $t_{1/2}$ of 127 min at 25 °C. Kinetic measurements show that this process is associated with an activation enthalpy $\Delta H^{\ne} = 68$ kJ mol$^{-1}$ and activation entropy $\Delta S^{\ne} = -94$ J K$^{-1}$ mol$^{-1}$, indicative of an associative mechanism. A possible process explaining the exchange involves protonation of the hydride to give a dihydrogen intermediate, deprotonation of which by the resulting hydroxide then leads to H–D exchange (Scheme 5.31) [227].

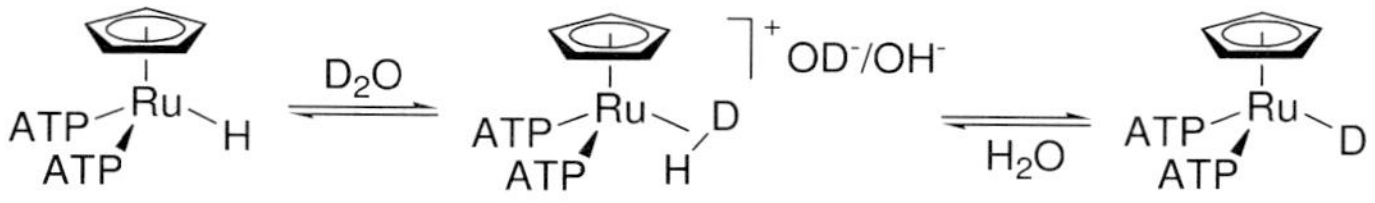

Cp(ATP)(ATP)Ru–H $\xrightarrow{D_2O}$ [Cp(ATP)(ATP)Ru(H)(D)]$^+$ $\xrightarrow{OD^-/OH^-, H_2O}$ Cp(ATP)(ATP)Ru–D

**Scheme 5.31**

Water-soluble phosphine complexes of rhodium and ruthenium are known to catalyze the H–D exchange between H$_2$ and D$_2$O. DFT calculations show that this is likely to occur again via a dihydrogen species formed by protonation of a metal hydride, and point to extensive H–D exchange in hydrogenation reactions in water [228]. In asymmetric transfer hydrogenation, RuCl(TsDPEN)(p-cymene) is also known to catalyze H–D exchange between DCOO⁻ and $H_2O$ [154].

## 5.5.4
### Participation in Transition States

Well-documented examples are available which show that water can accelerate chemical reactions. The acceleration may stem from hydrophobic effects or hydrogen bonding interactions, although the exact role of water remains to be defined in most cases. As mentioned earlier, there are examples of hydrogenation reactions which run faster in water than in common organic solvents. Recent DFT calculations have traced this to water participating in the transition states of hydride transfer.

The example of $CO_2$ hydrogenation catalyzed by $RuH_2(PMe_3)_4$ is illustrative. Calculations show that the reduction of $CO_2$ to formate with this complex in the presence of water involves an unusual mechanism (Scheme 5.32) [229, 230]. $CO_2$ does not interact directly with the metal center; instead, a nucleophilic attack by the hydride takes place, with a low activation barrier of only 14 kJ mol$^{-1}$. As is shown, this process is facilitated by hydrogen bonding between a coordinated $H_2O$ and the $CO_2$ oxygen [230]. A similar interaction is suggested for the aqueous-phase hydrogenation of α,β-unsaturated aldehydes with $RuH_2(PR_3)_4$ (Section 5.3.1.1). Furthermore, the presence of water suppresses the reverse reaction, that is, deinsertion of $CO_2$ from the resulting formate. In contrast, in the absence of water, the formate is formed by the usual $CO_2$ insertion into the Ru–H bond, the energy barrier of which is much higher, at 74 kJ mol$^{-1}$. In addition, deinsertion is easier in the absence of water, further reducing the reduction rate in aprotic solvents [230].

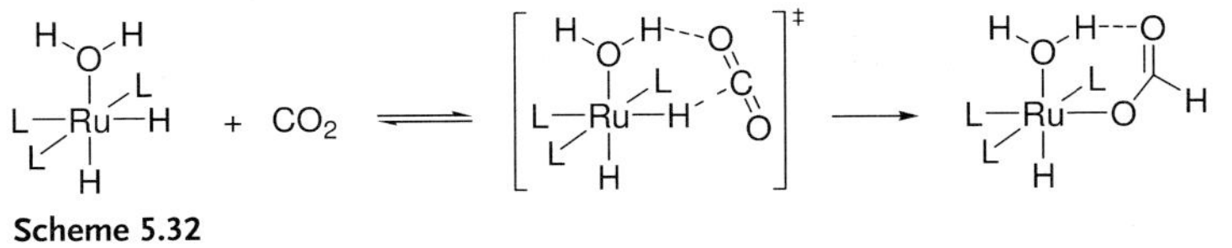

**Scheme 5.32**

Asymmetric transfer hydrogenation of ketones provides another example where water has been shown to confer faster reduction rates [158]. For instance, in the stoichiometric reduction of acetophenone by isolated Ru(II)–H, the rate in wet $CD_2Cl_2$ was six times that in dry $CD_2Cl_2$. DFT calculations revealed that water participates in the transition state of hydrogen transfer, stabilizing it by ∼16 kJ mol$^{-1}$ through hydrogen bonding with the ketone oxygen (Scheme 5.33) [154]. Of further interest is that the calculations, together with kinetic isotope measurements, suggest that the participation of water renders the hydrogen transfer process more stepwise than concerted [154, 231].

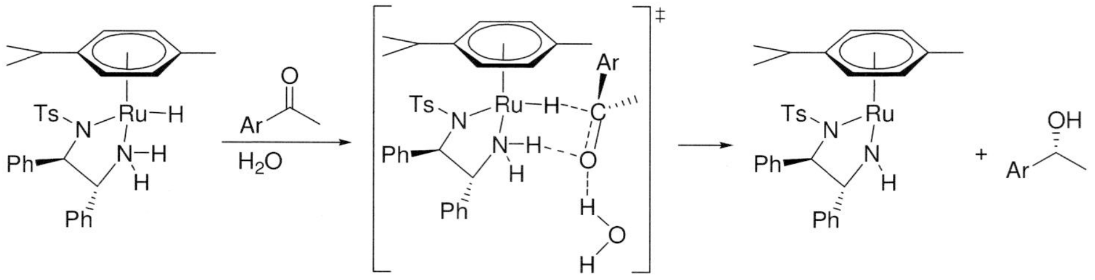

**Scheme 5.33**

## 5.6
## Conclusion

Aqueous-phase hydrogenation reactions have been extensively investigated over the past few decades. Numerous examples have been documented in the literature, as attested by this chapter, showing that the reduction in water is viable and, importantly, can gain from using water as solvent. Apart from making catalyst–product separation easy, water can confer enhanced activities and selectivities on a catalyst. These benefits often come with a price, however, namely that water-soluble ligands are usually necessary, and furthermore, their use may result in severe mass transfer problems. This said, the half-sandwich metal complexes in the preceding sections function remarkably well in asymmetric transfer hydrogenation, necessitating no ligand modifications. This brings up a fundamentally important issue, the role of water in catalytic hydrogenations or in any other reactions. In most cases, this is not clear or has been neglected. Although progress has been made, much remains to be done if more advantageous use of water is to be expected.

Water is the "greenest" solvent known to humans. It has contributed towards greening reduction reactions both commercially and in laboratories in the past. With the ever pressing environmental issues facing the chemical industries, water is expected to play an increasing role in one of the most important areas of organic synthesis, hydrogenation and transfer hydrogenation.

## References

1 Augustine, R.L. (1996) *Heterogeneous Catalysis for the Synthetic Chemist*, Marcel Dekker, New York.
2 de Vries, J.G., and Elsevier, S.J. (eds), (2006) *Handbook of Homogeneous Hydrogenation*, Wiley-VCH Verlag GmbH, Weinheim.
3 Ertl, G., Knözinger, H., Schüth, F. and Weitkamp, J. (eds), (2008) *Handbook of Heterogeneous Catalysis*, Wiley-VCH Verlag GmbH, Weinheim.
4 Li, C.J. and Chan, T.H. (1997) *Organic Reactions in Aqueous Media*, John Wiley & Sons Inc, New York.
5 Cornils, B. and Herrmann, W.A. (1998) *Aqueous-Phase Organometallic Catalysis: Concepts and Applications*, Wiley-VCH Verlag GmbH, Weinheim.
6 Joo, F. (2001) *Aqueous Organometallic Catalysis*, Kluwer, Dordrecht.
7 Dwars, T. and Oehme, G. (2002) *Adv. Synth. Catal.*, **344**, 239–260.

8 Sinou, D. (2002) *Adv. Synth. Catal.*, **344**, 221–237.
9 Breno, K.L., Ahmed, T.J., Pluth, M.D., Balzarek, C. and Tyler, D.R. (2006) *Coord. Chem. Rev.*, **250**, 1141–1151.
10 Liu, S.F. and Xiao, J.L. (2007) *J. Mol. Catal. A: Chemistry*, **270**, 1–43.
11 Wu, X.F. and Xiao, J.L. (2007) *Chem. Commun.*, 2449–2466.
12 Reichardt, C. (2004) *Solvents and Solvent Effects in Organic Chemistry*, Wiley-VCH Verlag GmbH, Weinheim.
13 Narayan, S., Muldoon, J., Finn, M.G., Fokin, V.V., Kolb, H.C. and Sharpless, K.B. (2005) *Angew. Chem. Int. Ed.*, **44**, 3275–3279.
14 Jung, Y. and Marcus, R.A. (2007) *J. Am. Chem. Soc.*, **129**, 5492–5502.
15 Herrmann, W.A. and Kohlpaintner, C.W. (1993) *Angew. Chem. Int. Ed.*, **32**, 1524–1544.
16 Benyei, A.C., Lehel, S. and Joo, F. (1997) *J. Mol. Catal. A: Chemistry*, **116**, 349–354.
17 Nomura, K. (1998) *J. Mol. Catal. A: Chemistry*, **130**, 1–28.
18 Wang, C., Wu, X.F. and Xiao, J.L. (2008) *Chem. Asian. J.*, **3**, 1750–1770.
19 Ahrland, S., Chatt, J., Davies, N.R. and Williams, A.A. (1958) *J. Chem. Soc.*, 276–288.
20 Joo, F. and Benyei, A. (1989) *J. Organomet. Chem.*, **363**, C19–C21.
21 Joo, F. and Katho, A. (1997) *J. Mol. Catal. A: Chemistry*, **116**, 3–26.
22 Joo, F. (2002) *Acc. Chem. Res.*, **35**, 738–745.
23 Hashiguchi, S., Fujii, A., Takehara, J., Ikariya, T. and Noyori, R. (1995) *J. Am. Chem. Soc.*, **117**, 7562–7563.
24 Fujii, A., Hashiguchi, S., Uematsu, N., Ikariya, T. and Noyori, R. (1996) *J. Am. Chem. Soc.*, **118**, 2521–2522.
25 Uematsu, N., Fujii, A., Hashiguchi, S., Ikariya, T. and Noyori, R. (1996) *J. Am. Chem. Soc.*, **118**, 4916–4917.
26 Noyori, R. and Hashiguchi, S. (1997) *Acc. Chem. Res.*, **30**, 97–102.
27 Parmar, D.U., Bhatt, S.D., Bajaj, H.C. and Jasra, R.V. (2003) *J. Mol. Catal. A: Chemistry*, **202**, 9–15.
28 Baricelli, P.J., Rodriguez, G., Rodriguez, A., Lujano, E. and Lopez-Linares, F. (2003) *Appl. Catal. A: General*, **239**, 25–34.
29 Baricelli, P.J., Izaguirre, L., Lopez, J., Lujano, E. and Lopez-Linares, F. (2004) *J. Mol. Catal. A: Chemistry*, **208**, 67–72.
30 Larpent, C., Dabard, R. and Patin, H. (1987) *Tetrahedron Lett.*, **28**, 2507–2510.
31 Burgemeister, K., Francio, G., Gego, V.H., Greiner, L., Hugl, H. and Leitner, W. (2007) *Chem. Eur. J.*, **13**, 2798–2804.
32 Grosselin, J.M., Mercier, C., Allmang, G. and Grass, F. (1991) *Organometallics*, **10**, 2126–2133.
33 Yang, Z.Y., Ebihara, M. and Kawamura, T. (2000) *J. Mol. Catal. A: Chemistry*, **158**, 509–514.
34 Smolenski, P., Pruchnik, F.P., Ciunik, Z. and Lis, T. (2003) *Inorg. Chem.*, **42**, 3318–3322.
35 Akbayeva, D.N., Gonsalvi, L., Oberhauser, W., Peruzzini, M., Vizza, F., Bruggeller, P., Romerosa, A., Sava, G. and Bergamo, A. (2003) *Chem. Commun.*, 264–265.
36 Bolano, S., Gonsalvi, L., Zanobini, F., Vizza, F., Bertolasi, V., Romerosa, A. and Peruzzini, M. (2004) *J. Mol. Catal. A: Chemistry*, **224**, 61–70.
37 Mebi, C.A. and Frost, B.J. (2005) *Organometallics*, **24**, 2339–2346.
38 Yamakawa, M., Ito, H. and Noyori, R. (2000) *J. Am. Chem. Soc.*, **122**, 1466–1478.
39 Noyori, R., Yamakawa, M. and Hashiguchi, S. (2001) *J. Org. Chem.*, **66**, 7931–7944.
40 Geldbach, T.J., Laurenczy, G., Scopelliti, R. and Dyson, P.J. (2006) *Organometallics*, **25**, 733–742.
41 Fujita, S.I., Sano, Y., Bhanage, B.M. and Arai, M. (2004) *J. Catal.*, **225**, 95–104.
42 Niuthitikul, K. and Winterbottom, M. (2004) *Chem. Eng. Sci.*, **59**, 5439–5447.
43 Papp, G., Elek, J., Nadasdi, L., Laurenczy, G. and Joo, F. (2003) *Adv. Synth. Catal.*, **345**, 172–174.

44 Nuithitikul, K. and Winterbottom, M. (2007) *Catal. Today*, **128**, 74–79.

45 Joo, F., Kovacs, J., Benyei, A.C. and Katho, A. (1998) *Catal. Today*, **42**, 441–448.

46 Joo, F., Kovacs, J., Benyei, A.C. and Katho, A. (1998) *Angew. Chem. Int. Ed.*, **37**, 969–970.

47 Joubert, J. and Delbecq, F. (2006) *Organometallics*, **25**, 854–861.

48 Kovacs, G., Ujaque, G., Lledos, A. and Joo, F. (2006) *Organometallics*, **25**, 862–872.

49 Rossin, A., Kovacs, G., Ujaque, G., Lledos, A. and Joo, F. (2006) *Organometallics*, **25**, 5010–5023.

50 Singha, N.K., Bhattacharjee, S. and Sivaram, S. (1997) *Rubber Chem. Technol.*, **70**, 309–367.

51 Wei, L., Jiang, J., Wang, Y. and Jin, Z. (2004) *J. Mol. Catal. A: Chemistry*, **221**, 47–50.

52 MacLeod, S. and Rosso, R.J. (2003) *Adv. Synth. Catal.*, **345**, 568–571.

53 Kotzabasakis, V., Georgopoulou, E., Pitsikalis, M., Hadjichristidis, N. and Papadogianakis, G. (2005) *J. Mol. Catal. A: Chemistry*, **231**, 93–101.

54 Lu, F., Liu, J. and Xu, J. (2007) *J. Mol. Catal. A: Chemistry*, **271**, 6–13.

55 Musolino, M.G., Cutrupi, C.M.S., Donato, A., Pietropaolo, D. and Pietropaolo, R. (2003) *J. Mol. Catal. A: Chemistry*, **195**, 147–157.

56 Semagina, N.V., Bykov, A.V., Sulman, E.M., Matveeva, V.G., Sidorov, S.N., Dubrovina, L.V., Valetsky, P.M., Kiselyova, O.I., Khokhlov, A.R., Stein, B. and Bronstein, L.M. (2004) *J. Mol. Catal. A: Chemistry*, **208**, 273–284.

57 Semagina, N., Joannet, E., Parra, S., Sulman, E., Renken, A. and Kiwi-Minsker, L. (2005) *Appl. Catal. A: General*, **280**, 141–147.

58 Drelinkiewicz, A., Laitinen, R., Kangas, R. and Pursiainen, J. (2005) *Appl. Catal. A: General*, **284**, 59–67.

59 Omota, F., Dimian, A.C. and Bliek, A. (2005) *Appl. Catal. A: General*, **294**, 121–130.

60 Hegedus, L. and Mathe, T. (2005) *Appl. Catal. A: General*, **296**, 209–215.

61 Makihara, N., Ogo, S. and Watanabe, Y. (2001) *Organometallics*, **20**, 497–500.

62 Csabai, P. and Joo, F. (2004) *Organometallics*, **23**, 5640–5643.

63 Wu, X., Corcoran, C., Yang, S. and Xiao, J. (2008) *ChemSuSChem*, **1**, 71–74.

64 Gulyas, H., Benyei, A.C. and Bakos, J. (2004) *Inorg. Chim. Acta*, **357**, 3094–3098.

65 Horvath, H.H. and Joo, F. (2005) *React. Kinet. Catal. Lett.*, **85**, 355–360.

66 Spitaleri, A., Pertici, P., Scalera, N., Vitulli, G., Hoang, M., Turney, T.W. and Gleria, M. (2003) *Inorg. Chim. Acta*, **352**, 61–71.

67 Maity, P., Basu, S., Bhaduri, S. and Lahiri, G.K. (2007) *Adv. Synth. Catal.*, **349**, 1955–1962.

68 Plasseraud, L. and SussFink, G. (1997) *J. Organomet. Chem.*, **539**, 163–170.

69 Fidalgo, E.G., Plasseraud, L. and Suss-Fink, G. (1998) *J. Mol. Catal. A: Chemistry*, **132**, 5–12.

70 Suss-Fink, G., Faure, M. and Ward, T.R. (2002) *Angew. Chem. Int. Ed.*, **41**, 99–101.

71 Faure, M., Vallina, A.T., Stoeckli-Evans, H. and Suss-Fink, G. (2001) *J. Organomet. Chem.*, **621**, 103–108.

72 Daguenet, C. and Dyson, P.J. (2003) *Catal. Commun.*, **4**, 153–157.

73 Daguenet, C., Scopelliti, R. and Dyson, P.J. (2004) *Organometallics*, **23**, 4849–4857.

74 Meister, G., Rheinwald, G., Stoecklievans, H. and Sussfink, G. (1994) *J. Chem. Soc., Dalton Trans.*, 3215–3223.

75 Zhang, L., Zhang, Y., Zhou, X.G., Li, R.X., Li, X.J., Tin, K.C. and Wong, N.B. (2006) *J. Mol. Catal. A: Chemistry*, **256**, 171–177.

76 Lu, F., Liu, J. and Xu, H. (2006) *Adv. Synth. Catal.*, **348**, 857–861.

77 Mevellec, V. and Roucoux, A. (2004) *Inorg. Chim. Acta*, **357**, 3099–3103.

78 Maegawa, T., Akashi, A. and Sajiki, H. (2006) *Synlett*, 1440–1442.

79 Ning, J., Xu, J., Liu, J., Miao, H., Ma, H., Chen, C., Li, X., Zhou, L. and

Yu, W. (2007) *Catal. Commun.*, **8**, 1763–1766.

80  Garcia-Verdugo, E., Liu, Z.M., Ramirez, E., Garcia-Serna, J., Fraga-Dubreuil, J., Hyde, J.R., Hamley, P.A. and Poliakoff, M. (2006) *Green Chem.*, **8**, 359–364.

81  Mhadgut, S.C., Palaniappan, K., Thimmaiah, M., Hackney, S.A., Torok, B. and Liu, J. (2005) *Chem. Commun.*, 3207–3209.

82  Himeda, Y. (2007) *Eur. J. Inorg. Chem.*, 3927–3941.

83  Jessop, P.G., Joo, F. and Tai, C.C. (2004) *Coord. Chem. Rev.*, **248**, 2425–2442.

84  Gassner, F. and Leitner, W. (1993) *J. Chem. Soc., Chem. Commun.*, 1465–1466.

85  Elek, J., Nadasdi, L., Papp, G., Laurenczy, G. and Joo, F. (2003) *Appl. Catal. A: General*, **255**, 59–67.

86  Hayashi, H., Ogo, S. and Fukuzumi, S. (2004) *Chem. Commun.*, 2714–2715.

87  Ogo, S., Kabe, R., Hayashi, H., Harada, R. and Fukuzumi, S. (2006) *Dalton Trans.*, 4657–4663.

88  Himeda, Y., Onozawa-Komatsuzaki, N., Sugihara, H. and Kasuga, K. (2007) *Organometallics*, **26**, 702–712.

89  Ohnishi, Y.Y., Matsunaga, T., Nakao, Y., Sato, H. and Sakaki, S. (2005) *J. Am. Chem. Soc.*, **127**, 4021–4032.

90  Tai, C.C., Pitts, J., Linehan, J.C., Main, A.D., Munshi, P. and Jessop, P.G. (2002) *Inorg. Chem.*, **41**, 1606–1614.

91  RajanBabu, T.V., Yan, Y.Y. and Shin, S.H. (2001) *J. Am. Chem. Soc.*, **123**, 10207–10213.

92  Burk, M.J. (2000) *Acc. Chem. Res.*, **33**, 363–372.

93  Li, W.G., Zhang, Z.G., Xiao, D.M. and Zhang, X.M. (2000) *J. Org. Chem.*, **65**, 3489–3496.

94  Holz, J., Heller, D., Sturmer, R. and Borner, A. (1999) *Tetrahedron Lett.*, **40**, 7059–7062.

95  Hoen, R., Leleu, S., Botman, P.N.M., Appelman, V.A.M., Feringa, B.L., Hiemstra, H., Minnaard, A.J. and van Maarseveen, J.H. (2006) *Org. Biomol. Chem.*, **4**, 613–615.

96  Simons, C., Hanefeld, U., Arends, I.W.C.E., Maschmeyer, T. and Sheldon, R.A. (2006) *Adv. Synth. Catal.*, **348**, 471–475.

97  Oehme, G., Grassert, I., paetzold, E., Fuhrmann, H., Dwars, T., Schmidt, U. and Iovel, I. (2003) *Kinet. Catal.*, **44**, 766.

98  Mevellec, V., Mattioda, C., Schulz, J., Rolland, J.P. and Roucoux, A. (2004) *J. Catal.*, **225**, 1–6.

99  Grassert, I., Schmidt, U., Ziegler, S., Fischer, C. and Oehme, G. (1998) *Tetrahedron: Asymmetry*, **9**, 4193–4202.

100  Yonehara, K., Ohe, K. and Uemura, S. (1999) *J. Org. Chem.*, **64**, 9381–9385.

101  Grassert, I., Kovacs, J., Fuhrmann, H. and Oehme, G. (2002) *Adv. Synth. Catal.*, **344**, 312–318.

102  Wolfson, A., Vankelecom, I.F.J. and Jacobs, P.A. (2005) *Tetrahedron Lett.*, **46**, 2513–2516.

103  Pugin, B., Studer, M., Kuesters, E., Sedelmeier, G. and Feng, X. (2004) *Adv. Synth. Catal.*, **346**, 1481–1486.

104  Berthod, M., Saluzzo, C., Mignani, G. and Lemaire, M. (2004) *Tetrahedron: Asymmetry*, **15**, 639–645.

105  Berthod, M., Mignani, G. and Lemaire, M. (2005) *J. Mol. Catal. A: Chemistry*, **233**, 105–110.

106  Bartek, L., Drobek, M., Kuzma, M., Kluson, P. and Cerveny, L. (2005) *Catal. Commun.*, **6**, 61–65.

107  Pugin, B., Groehn, V., Moser, R. and Blaser, H.U. (2006) *Tetrahedron: Asymmetry*, **17**, 544–549.

108  Jere, F.T., Miller, D.J. and Jackson, J.E. (2003) *Org. Lett.*, **5**, 527–530.

109  Jere, F.T., Jackson, J.E. and Miller, D.J. (2004) *Ind. Eng. Chem. Res.*, **43**, 3297–3303.

110  Wagner, K. (1970) *Angew. Chem. Int. Ed. Engl.*, **9**, 50.

111  Yang, J.W. and List, B. (2006) *Org. Lett.*, **8**, 5653–5655.

112  Yang, J.W., Fonseca, M.T.H. and List, B. (2004) *Angew. Chem. Int. Ed.*, **43**, 6660–6662.

113 Peris, E. and Crabtree, R.H. (2004) *Coord. Chem. Rev.*, **248**, 2239–2246.
114 Naskar, S. and Bhattacharjee, M. (2005) *J. Organomet. Chem.*, **690**, 5006–5010.
115 Miecznikowski, J.R. and Crabtree, R.H. (2004) *Polyhedron*, **23**, 2857–2872.
116 Miecznikowski, J.R. and Crabtree, R.H. (2004) *Organometallics*, **23**, 629–631.
117 Gladiali, S. and Alberico, E. (2006) *Chem. Soc. Rev.*, **35**, 226–236.
118 Joo, F., Csuhai, E., Quinn, P.J. and Vigh, L. (1988) *J. Mol. Catal.*, **49**, L1–L5.
119 Bar, R. and Sasson, Y. (1981) *Tetrahedron Lett.*, **22**, 1709–1710.
120 Bar, R., Bar, L.K., Sasson, Y. and Blum, J. (1985) *J. Mol. Catal.*, **33**, 161–177.
121 Sinou, D., Safi, M., Claver, C. and Masdeu, A. (1991) *J. Mol. Catal.*, **68**, L9–L12.
122 Zoran, A., Sasson, Y. and Blum, J. (1984) *J. Mol. Catal.*, **26**, 321–326.
123 Benyei, A. and Joo, F. (1990) *J. Mol. Catal.*, **58**, 151–163.
124 Bar, R., Sasson, Y. and Blum, J. (1984) *J. Mol. Catal.*, **26**, 327–332.
125 Ogo, S., Makihara, N. and Watanabe, Y. (1999) *Organometallics*, **18**, 5470–5474.
126 Ogo, S., Makihara, N., Kaneko, Y. and Watanabe, Y. (2001) *Organometallics*, **20**, 4903–4910.
127 Ogo, S., Abura, T. and Watanabe, Y. (2002) *Organometallics*, **21**, 2964–2969.
128 Ogo, S., Hayashi, H., Uehara, K. and Fukuzumi, S. (2005) *Appl. Organomet. Chem.*, **19**, 639–643.
129 Abura, T., Ogo, S., Watanabe, Y. and Fukuzumi, S. (2003) *J. Am. Chem. Soc.*, **125**, 4149–4154.
130 Kuo, L.Y., Weakley, T.J.R., Awana, K. and Hsia, C. (2001) *Organometallics*, **20**, 4969–4972.
131 Kuo, L.Y., Finigan, D.M. and Tadros, N.N. (2003) *Organometallics*, **22**, 2422–2425.
132 Canivet, J., Karmazin-Brelot, L. and Suss-Fink, G. (2005) *J. Organomet. Chem.*, **690**, 3202–3211.
133 Canivet, J., Labat, G., Stoeckli-Evans, H. and Suss-Fink, G. (2005) *Eur. J. Inorg. Chem.*, 4493–4500.
134 Canivet, J. and Suss-Fink, G. (2007) *Green Chem.*, **9**, 391–397.
135 Canivet, J., Suss-Fink, G. and Stepnicka, P. (2007) *Eur. J. Inorg. Chem.*, 4736–4742.
136 Govindaswamy, P., Canivet, J., Therrien, B., Suss-Fink, G., Stepnicka, P. and Ludvik, J. (2007) *J. Organomet. Chem.*, **692**, 3664–3675.
137 Therrien, B., Said-Mohamed, C. and Suss-Fink, G. (2008) *Inorg. Chim. Acta*, **361**, 2601–2608.
138 Diez, J., Gamasa, M. P., Lastra, E., Garcia-Fernandez, A. and Tarazona, M. P. (2006) *Eur. J. Inorg. Chem.*, 2855–2864.
139 Ikariya, T., Murata, K. and Noyori, R. (2006) *Org. Biomol. Chem.*, **4**, 393–406.
140 Wu, X.F., Liu, J.K., Li, X.H., Zanotti-Gerosa, A., Hancock, F., Vinci, D., Ruan, J.W. and Xiao, J.L. (2006) *Angew. Chem. Int. Ed.*, **45**, 6718–6722.
141 Sharma, A., Kumar, V. and Sinha, A.K. (2006) *Adv. Synth. Catal.*, **348**, 354–360.
142 Blaser, H.U., Malan, C., Pugin, B., Spindler, F., Steiner, H. and Studer, M. (2003) *Adv. Synth. Catal.*, **345**, 103–151.
143 Clapham, S.E., Hadzovic, A. and Morris, R.H. (2004) *Coord. Chem. Rev.*, **248**, 2201–2237.
144 Cornils, B., Herrmann, W.A. and Eckl, R.W. (1997) *J. Mol. Catal. A: Chemistry*, **116**, 27–33.
145 Everaere, K., Mortreux, A. and Carpentier, J.F. (2003) *Adv. Synth. Catal.*, **345**, 67–77.
146 Ikariya, T. and Blacker, A.J. (2007) *Acc. Chem. Res.*, **40**, 1300–1308.
147 Liu, J.K., Wu, X.F., Iggo, J.A. and Xiao, J.L. (2008) *Coord. Chem. Rev.*, **252**, 782–809.
148 Pinault, N. and Bruce, D.W. (2003) *Coord. Chem. Rev.*, **241**, 1–25.
149 Rautenstrauch, V., Hoang-Cong, X., Churlaud, R., Abdur-Rashid, K. and Morris, R.H. (2003) *Chem. Eur. J.*, **9**, 4954–4967.
150 Saluzzo, C. and Lemaire, M. (2002) *Adv. Synth. Catal.*, **344**, 915–928.

151 Samec, J.S.M., Backvall, J.E., Andersson, P.G. and Brandt, P. (2006) *Chem. Soc. Rev.*, **35**, 237–248.

152 Wu, X.F., Mo, J., Li, X.H., Hyder, Z. and Xiao, J.L. (2008) *Prog. Nat. Sci.*, **18**, 639–652.

153 Xiao, J.L., Wu, X.F., Zanotti-Gerosa, A. and Hancock, F. (2005) *Chim. Oggi Chem. Today*, **23**, 50–55.

154 Wu, X.F., Liu, J.K., Tommaso, D.D., Iggo, J.A., Catlow, C.R.A., Bacsa, J. and Xiao, J.L. (2008) *Chem. Eur. J.*, **14**, 7699–7715.

155 Wu, X.F., Li, X.H., McConville, M., Saidi, O. and Xiao, J.L. (2006) *J. Mol. Catal. A: Chemistry*, **247**, 153–158.

156 Wu, X.F., Vinci, D., Ikariya, T. and Xiao, J.L. (2005) *Chem. Commun.*, 4447–4449.

157 Wu, X.F., Li, X.G., King, F. and Xiao, J.L. (2005) *Angew. Chem. Int. Ed.*, **44**, 3407–3411.

158 Wu, X.F., Li, X.G., Hems, W., King, F. and Xiao, J.L. (2004) *Org. Biomol. Chem.*, **2**, 1818–1821.

159 Zeror, S., Collin, J., Fiaud, J.C. and Zouioueche, L.A. (2008) *Adv. Synth. Catal.*, **350**, 197–204.

160 Xu, Z., Mao, J.C., Zhang, Y.W., Guo, J. and Zhu, J.L. (2008) *Catal. Commun.*, **9**, 618–623.

161 Liu, J.T., Zhou, Y.G., Wu, Y.N., Li, X.S. and Chan, A.S.C. (2008) *Tetrahedron: Asymmetry*, **19**, 832–837.

162 Liu, J.T., Wu, Y.N., Li, X.S. and Chan, A.S.C. (2008) *J. Organomet. Chem.*, **693**, 2177–2180.

163 Cortez, N.A., Aguirre, G., Parra-Hake, M. and Somanathan, R. (2008) *Tetrahedron: Asymmetry*, **19**, 1304–1309.

164 Alza, E., Bastero, A., Jansat, S. and Pericas, M.A. (2008) *Tetrahedron: Asymmetry*, **19**, 374–378.

165 Akagawa, K., Akabane, H., Sakamoto, S. and Kudo, K. (2008) *Org. Lett.*, **10**, 2035–2037.

166 Ahlford, K., Lind, J., Maler, L. and Adolfsson, H. (2008) *Green Chem.*, **10**, 832–835.

167 Zhou, H.F., Fan, Q.H., Huang, Y.Y., Wu, L., He, Y.M., Tang, W.J., Gu, L.Q. and Chan, A.S.C. (2007) *J. Mol. Catal. A: Chemistry*, **275**, 47–53.

168 Wettergren, J., Zaitsev, A.B. and Adolfsson, H. (2007) *Adv. Synth. Catal.*, **349**, 2556–2562.

169 Li, L., Wu, J.S., Wang, F., Liao, J., Zhang, H., Lian, C.X., Zhu, J. and Deng, J.G. (2007) *Green Chem.*, **9**, 23–25.

170 Cortez, N.A., Aguirre, G., Parra-Hake, M. and Somanathan, R. (2007) *Tetrahedron Lett.*, **48**, 4335–4338.

171 Chen, G., Xing, Y., Zhang, H. and Gao, J.X. (2007) *J. Mol. Catal. A: Chemistry*, **273**, 284–288.

172 Zeror, S., Collin, J., Fiaud, J.C. and Zouioueche, L.A. (2006) *J. Mol. Catal. A: Chemistry*, **256**, 85–89.

173 Xing, Y., Chen, J.S., Dong, Z.R., Li, Y.Y. and Gao, J.X. (2006) *Tetrahedron Lett.*, **47**, 4501–4503.

174 Wu, J.S., Wang, F., Ma, Y.P., Cui, X.C., Cun, L.F., Zhu, J., Deng, J.G. and Yu, B.L. (2006) *Chem. Commun.*, 1766–1768.

175 Matharu, D.S., Morris, D.J., Clarkson, G.J. and Wills, M. (2006) *Chem. Commun.*, 3232–3234.

176 Li, X.H., Blacker, J., Houson, I., Wu, X.F. and Xiao, J.L. (2006) *Synlett*, 1155–1160.

177 Li, B.Z., Chen, J.S., Dong, Z.R., Li, Y.Y., Li, Q.B. and Gao, J.X. (2006) *J. Mol. Catal. A: Chemistry*, **258**, 113–117.

178 Jiang, L., Wu, T.F., Chen, Y.C., Zhu, J. and Deng, J.G. (2006) *Org. Biomol. Chem.*, **4**, 3319–3324.

179 Cortez, N.A., Rodriguez-Apodaca, R., Aguirre, G., Parra-Hake, M., Cole, T. and Somanathan, R. (2006) *Tetrahedron Lett.*, **47**, 8515–8518.

180 Arakawa, Y., Haraguchi, N. and Itsuno, S. (2006) *Tetrahedron Lett.*, **47**, 3239–3243.

181 Wang, F., Liu, H., Cun, L.F., Zhu, J., Deng, J.G. and Jiang, Y.Z. (2005) *J. Org. Chem.*, **70**, 9424–9429.

182 Mao, J.C., Wan, B.S., Wu, F. and Lu, S.W. (2005) *Tetrahedron Lett.*, **46**, 7341–7344.

183 Liu, P.N., Gu, P.M., Deng, J.G., Tu, Y.Q. and Ma, Y.P. (2005) *Eur. J. Org. Chem.*, 3221–3227.

184 Letondor, C., Humbert, N. and Ward, T.R. (2005) *Proc. Natl. Acad. Sci. USA*, **102**, 4683–4687.

185 Liu, P. N., Deng, J. G., Tu, Y. Q. and Wang, S. H. (2004) *Chem. Commun.*, 2070–2071.

186 Li, X.G., Wu, X.F., Chen, W.P., Hancock, F.E., King, F. and Xiao, J.L. (2004) *Org. Lett.*, **6**, 3321–3324.

187 Ajjou, A.N. and Pinet, J.L. (2004) *J. Mol. Catal. A: Chemistry*, **214**, 203–206.

188 Ma, Y.P., Liu, H., Chen, L., Cui, X., Zhu, J. and Deng, J.G. (2003) *Org. Lett.*, **5**, 2103–2106.

189 Himeda, Y., Onozawa-Komatsuzaki, N., Sugihara, H., Arakawa, H. and Kasuga, K. (2003) *J. Mol. Catal. A: Chemistry*, **195**, 95–100.

190 Rhyoo, H.Y., Park, H.J., Suh, W.H. and Chung, Y.K. (2002) *Tetrahedron Lett.*, **43**, 269–272.

191 Thorpe, T., Blacker, J., Brown, S.M., Bubert, C., Crosby, J., Fitzjohn, S., Muxworthy, J.P. and Williams, J.M.J. (2001) *Tetrahedron Lett.*, **42**, 4041–4043.

192 Bubert, C., Blacker, J., Brown, S.M., Crosby, J., Fitzjohn, S., Muxworthy, J.P., Thorpe, T. and Williams, J.M.J. (2001) *Tetrahedron Lett.*, **42**, 4037–4039.

193 Rhyoo, H. Y., Park, H. J. and Chung, Y. K. (2001) *Chem. Commun.*, 2064–2065.

194 Wu, X.F., Li, X.H., Zanotti-Gerosa, A., Pettman, A., Liu, J.K., Mills, A.J. and Xiao, J.L. (2008) *Chem. Eur. J.*, **14**, 2209–2222.

195 Venkatakrishnan, T.S., Mandal, S.K., Kannan, R., Krishnamurthy, S.S. and Nethaji, M. (2007) *J. Organomet. Chem.*, **692**, 1875–1891.

196 Palmer, M.J. and Wills, M. (1999) *Tetrahedron: Asymmetry*, **10**, 2045–2061.

197 Li, X.G., Chen, W.P., Hems, W., King, F. and Xiao, J.L. (2004) *Tetrahedron Lett.*, **45**, 951–953.

198 Li, X.G., Chen, W.P., Hems, W., King, F. and Xiao, J.L. (2003) *Org. Lett.*, **5**, 4559–4561.

199 Wu, X.F. (2007) PhD Thesis, University of Liverpool.

200 Micskei, K., Hajdu, C., Wessjohann, L.A., Mercs, L., Kiss-Szikszai, A. and Patonay, T. (2004) *Tetrahedron: Asymmetry*, **15**, 1735–1744.

201 Watts, C.C., Thoniyot, P., Cappuccio, F., Verhagen, J., Gallagher, B. and Singaram, B. (2006) *Tetrahedron: Asymmetry*, **17**, 1301–1307.

202 Gao, J.X., Ikariya, T. and Noyori, R. (1996) *Organometallics*, **15**, 1087–1089.

203 Faber, K. (2004) *Biotransformations in Organic Chemistry*, John Wiley & Sons Inc, New York.

204 Eder, U., Sauer, G. and Weichert, R., (1971) *Angew. Chem. Int. Ed. Engl.*, **10**, 496–497.

205 Faber, K. and Kroutil, W. (2005) *Curr. Opin. Chem. Biol.*, **9**, 181–187.

206 Hoffmann, S., Seayad, A.M. and List, B. (2005) *Angew. Chem. Int. Ed.*, **44**, 7424–7427.

207 Malkov, A.V., Liddon, A., Ramirez-Lopez, P., Bendova, L., Haigh, D. and Kocovsky, P. (2006) *Angew. Chem. Int. Ed.*, **45**, 1432–1435.

208 Edegger, K., Gruber, C.C., Poessl, T.M., Wallner, S.R., Lavandera, I., Faber, K., Niehaus, F., Eck, J., Oehrlein, R., Hafner, A. and Kroutil, W. (2006) *Chem. Commun.*, 2402–2404.

209 Zhu, D.M., Yang, Y. and Hua, L. (2006) *J. Org. Chem.*, **71**, 4202–4205.

210 Kaiser, E.T. and Lawrence, D.S. (1984) *Science*, **226**, 505–511.

211 Kramer, R. (2006) *Angew. Chem. Int. Ed.*, **45**, 858–860.

212 Mahammed, A. and Gross, Z. (2005) *J. Am. Chem. Soc.*, **127**, 2883–2887.

213 Panella, L., Broos, J., Jin, J.F., Fraaije, M.W., Janssen, D.B., Jeronimus-Stratingh, M., Feringa, B.L., Minnaard, A.J. and de Vries, J.G. (2005) *Chem. Commun.*, 5656–5658.

214 Roelfes, G., Boersma, A.J. and Feringa, B.L. (2006) *Chem. Commun.*, 635–637.

215 Roelfes, G. and Feringa, B.L. (2005) *Angew. Chem. Int. Ed.*, **44**, 3230–3232.

216 Wilson, M.E. and Whitesides, G.M. (1978) *J. Am. Chem. Soc.*, **100**, 306–307.

217 Letondor, C., Pordea, A., Humbert, N., Ivanova, A., Mazurek, S., Novic, M. and Ward, T.R. (2006) *J. Am. Chem. Soc.*, **128**, 8320–8328.
218 Thomas, C.M. and Ward, T.R. (2005) *Chem. Soc. Rev.*, **34**, 337–346.
219 Ward, T.R., (2005) *Chem. Eur. J.*, **11**, 3798–3804.
220 Liu, K., Haeussinger, D. and Woggon, W.D. (2007) *Synlett*, 2298–2300.
221 Schlatter, A., Kundu, M.K. and Woggon, W.D. (2004) *Angew. Chem. Int. Ed.*, **43**, 6731–6734.
222 Schlatter, A. and Woggon, W.D. (2008) *Adv. Synth. Catal.*, **350**, 995–1000.
223 Darensbourg, D.J., Stafford, N.W., Joo, F. and Reibenspies, J.H. (1995) *J. Organomet. Chem.*, **488**, 99–108.
224 Poth, T., Paulus, H., Elias, H., Ducker-Benfer, C. and van Eldik, R. (2001) *Eur. J. Inorg. Chem.*, 1361–1369.
225 Wang, F., Chen, H., Parsons, S., Oswald, I.D.H., Davidson, J.E. and Sadler, P.J., (2003) *Chem. Eur. J.*, **9**, 5810–5820.
226 Malmstrom, T., Wendt, O. F. and Andersson, C. (1999) *J. Chem. Soc., Dalton Trans.*, 2871–2875.
227 Frost, B.J. and Mebi, C.A. (2004) *Organometallics*, **23**, 5317–5323.
228 Kovacs, G., Schubert, G., Joo, F. and Papai, I. (2005) *Organometallics*, **24**, 3059–3065.
229 Yin, C., Xu, Z., Yang, S.Y., Ng, S.M., Wong, K.Y., Lin, Z. and Lau, C.P. (2001) *Organometallics*, **20**, 1216–1222.
230 Ohnishi, Y., Nakao, Y., Sato, H. and Sakaki, S. (2006) *Organometallics*, **25**, 3352–3363.
231 Handgraaf, J.W. and Meijer, E.J. (2007) *J. Am. Chem. Soc.*, **129**, 3099–3103.

# 6
# Coupling Reactions in Water
*Lucie Leseurre, Jean-Pierre Genêt, and Véronique Michelet*

## 6.1
## Introduction

The chemistry of coupling reactions, including Grignard-type reactions, has been extensively studied in organoaqueous conditions since the discovery by Breslow's group [1] in the 1980s of the benefits of water as a solvent or a cosolvent. Water meets many of the stringent criteria imposed upon contemporary organic synthesis [2, 3]. Several books, reviews, and accounts have been published in this field, which proved that water is a highly valuable partner or solvent for carbon–hydrogen, carbon–carbon, and carbon–heteroatom bond formations. This chapter does not intend to cover all coupling reactions in organoaqueous media, but rather summarizes recent developments in this field with a focus on representative and historical examples. The first part is dedicated to Grignard-type reactions, including 1,2-additions to carbonyl derivatives and pinacol couplings. The second part places the stress on selected reactivity of alkenes and alkynes, such as the hydroformylation reaction, Mizoroki–Heck [4] coupling and some cyclization reactions. The final part focuses on more classic coupling reactions [5], including Tsuji–Trost [6], Suzuki–Miyaura [7], Sonogashira [8], Stille [9], Hartwig–Buchwald [10], and Hiyama [11] reactions. Special attention is devoted to recyclable organometallic systems [12], asymmetric catalysis [13], and the synthetic applications of such reactions.

## 6.2
## Reaction of Carbonyl Compounds and Derivatives

### 6.2.1
### Grignard-type Reactions

Barbier–Grignard-type reactions are among the most versatile and useful examples of the importance of carbanions in organic synthesis. However, a major requirement

*Handbook of Green Chemistry, Volume 5: Reactions in Water.* Edited by Chao-Jun Li
Copyright © 2010 WILEY-VCH Verlag GmbH & Co. KGaA, Weinheim
ISBN: 978-3-527-31591-7

in these reactions is the strict exclusion of air and water, which, besides the commonly used methods, leads to severe drawbacks considering the Principles of Green Chemistry [14]. Anhydrous conditions require large volumes of anhydrous organic solvents, which can potentially increase volatile organic emissions, excess drying agents, and the protection of functional groups such as hydroxy groups and carboxylic acids. These reactions carry a large amount of metal waste since they require stoichiometric amounts of metal in order to proceed. Moreover, the necessity for an inert gas atmosphere implies that the catalyst is prone to deactivation. Finally, the general use of organic halides as the carbanion source generates a stoichiometric amount of halide waste, in addition to requiring multi-step synthesis of the halides [15]. Although the preparation of arylmercuric chlorides in aqueous media has been known since 1905 [16], and the production of tribenzylstannyl in water since the 1960s [16], the exploration of aqueous organometallic reactions did not start before the end of the 1970s. One major discovery in this field was made by Wolinsky and co-workers in 1977, performing the zinc-mediated allylation reaction of carbonyl compounds with allyl bromide in 95% ethanol and *tert*-butanol, demonstrating the feasibility of organometallic reactions in hydroxylic solvents, including water [17].

#### 6.2.1.1 Allylation Reaction

The allylation of carbonyl compounds and their derivatives is one of the most important C—C bond-forming reactions in organic and pharmaceutical chemistry because homoallylic alcohols and amines are useful intermediates for the synthesis of natural products and biologically active compounds [18]. Among all the nucleophilic addition reactions of carbonyl compounds, the allylation reaction is the most successful in aqueous media, partly due to the relatively high reactivity of allyl halides. Various metals have been found to be effective in mediating the allylation reaction of carbonyl compounds in aqueous media (Scheme 6.1). It is interesting that for the traditional Barbier–Grignard-type reactions in organic solvents, the metals used tend to be the more reactive ones such as magnesium and lithium, whereas for the aqueous reactions, the metals tend to be less metallic, or "softer" [2].

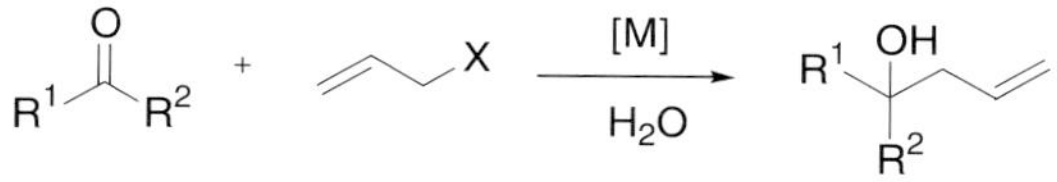

M: Sn, Zn, In, Ga, Mg, Co, Mn, Bi, Fe...
**Scheme 6.1**

Several metals [19] can be used for such a reaction, but investigations of these metals are rather limited. Zinc, tin, and indium appear to be the most efficient metals for allylation reactions in aqueous media on carbonyl compounds or derivatives [3a]. At the very start, tin required harsh conditions to mediate the reaction with a combination of allyl bromide, tin metal, a catalytic amount of hydrobromic acid, and metallic aluminum powder [20]. This reaction was extended to allyl chloride by using an alcohol–water–acetic acid mixture as the solvent [21]. Important improvements were made with the discovery that aluminum powder and hydrobromic acid could be replaced by higher temperatures or sonication [22]. The use of a phase-transfer catalyst was reported to enhance the tin-mediated reaction [23]. Sub-micron zinc powder and tin powder were discovered to mediate the allylation of aldehydes and ketones in high yields without any other assistance [24].

Applying this method, the zinc-mediated allylation reaction of carbonyl compounds with ethyl (2-bromomethyl)acrylate was reported for the synthesis of α-methylene-γ-butyrolactones (Scheme 6.2) [25]. The same reaction in anhydrous THF gives only low yields.

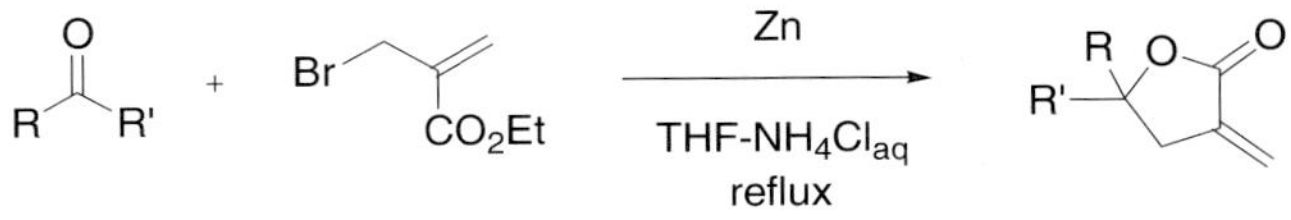

**Scheme 6.2**

Recently, an efficient method was developed to perform the indium-mediated allylation of aldehydes with fluorous-tagged acrylates and to purify the products directly by fluorous solid-phase extraction [26]. Further treatment of the esters with base generated α-methylene-γ-butyrolactones. The synthesis of allyl-substituted indenes was also recently described, employing a zinc-mediated allylation of indanones in THF–aqueous $NH_4Cl$ [27]. Chan and co-workers first described the use of metallic indium to mediate Barbier–Grignard-type reactions in water at room temperature without any promoter [28]. Compared with other metals, indium has proved to be attractive, due to its great stability towards air and water [29]. It is interesting that these processes can be applied to the synthesis of cyclopentanyl derivatives starting from 1,3-dicarbonyl compounds [30], and homoallylic amines starting from sulfonimines [31]. Moreover, water accelerates the indium-mediated allylation and benzylation of β, γ-unsaturated piperidinium ion, which was generated from β, γ-unsaturated α-methoxycarbonylpiperidine (Scheme 6.3) [32].

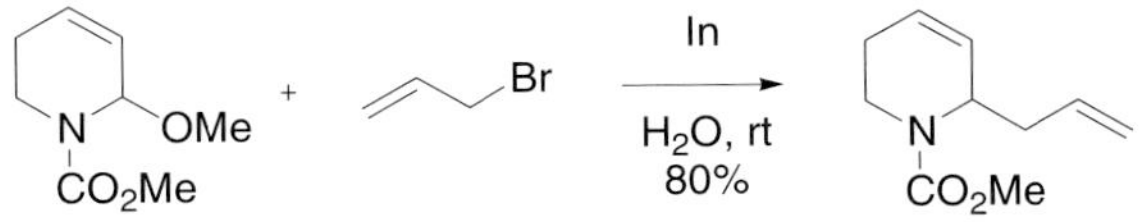

**Scheme 6.3**

The use of a zerovalent metal is not compulsory, as preformed organic tin reagents can smoothly mediate the reaction in the presence of a Lewis acid catalyst [33], such as scandium triflate [34]. The use of salts such as tin(II) chloride was also possible for mediating the allylation reaction [35]. However, $SnCl_2$, by itself, can hardly mediate the coupling between an allyl halide and a carbonyl compound in water, unless an Al(III), Pd(II), Cu(I), Ti(III), or Co(II) catalyst is introduced to the reaction [36]. Cyclopentadienylindium(I) has also been demonstrated to be an effective organometallic reagent that can react with aldehydes in aqueous media to give highly functionalized substituted cyclopentadienes [37].

An important issue in the metal-mediated allylation reaction is the synthesis of various allylic halides or allylmetal compounds. Allyltin intermediates were first

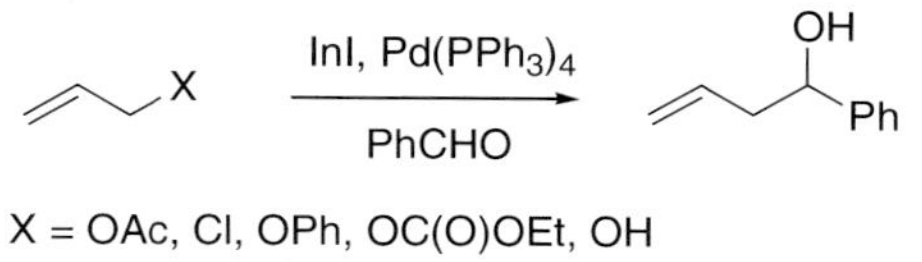

## 6 Coupling Reactions in Water

**Scheme 6.4**

X = OAc, Cl, OPh, OC(O)OEt, OH

prepared using a catalytic amount of Pd(II), starting from allylic alcohol [38], carboxylates, or methoxycarbonyloxyl groups [39]. The synthesis of various allylindium(III) reagents in the presence of a catalytic amount of Pd(0) was reported recently and they reacted smoothly with aldehydes to give the corresponding allylic alcohols (Scheme 6.4) [40].

In 2000, Chang and Cheng reported a highly regio- and stereoselective allylation of aldehydes in aqueous/organic media without any halide moiety [41]. Addition of allenes on aldehydes proceeds smoothly in the presence of tin(II) chloride, hydrochloric acid, and a catalytic amount of palladium(II).

Electrochemical methods are another approach to green chemistry because they can fundamentally eliminate the waste treatment and disposal of used redox agents. Based on the recycling of allyltin reported in 1984, the tin-mediated allylation reaction of aldehyde in water was described (Scheme 6.5) [42]. Tin(0) is generated *in situ* by electrolysis to mediate allylation and tin salts are regenerated.

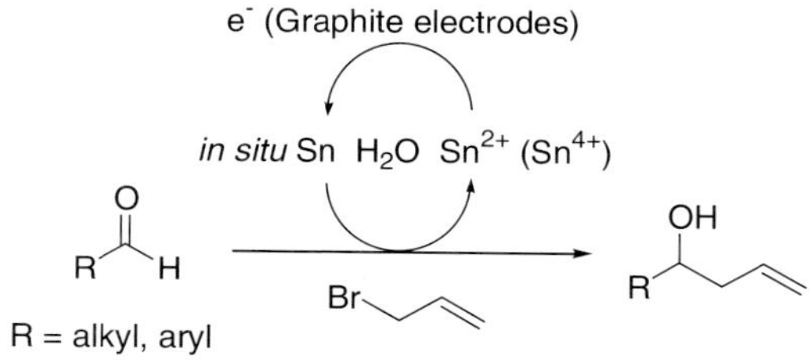

R = alkyl, aryl
**Scheme 6.5**

Recently, allylboron and allylsilicon reagents have been used as organometallic partners in allylation reactions in aqueous media. In the presence of a phase-transfer catalyst, which significantly accelerates the rate of the reaction, potassium allyl- and crotyltrifluoroborates can undergo allylation reaction on aldehydes in biphasic media providing homoallylic alcohols in high yields and with high diastereoselectivity (Scheme 6.6) [43].

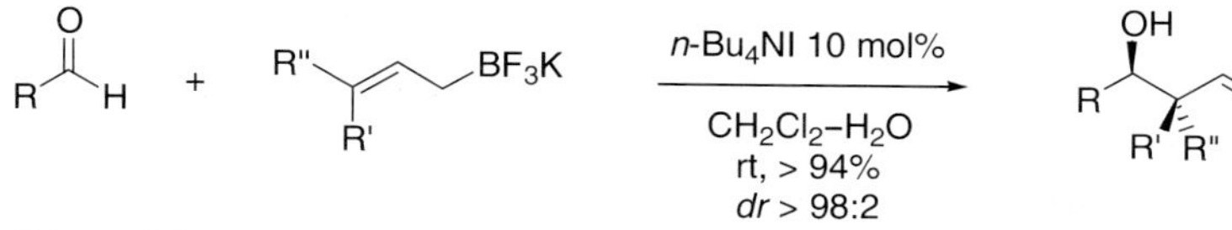

**Scheme 6.6**

In 1995, Yamamoto and co-workers first described the use of allylsilanes for the allylation reaction with aldehydes in the presence of tris(pentafluorophenyl)boron as Lewis acid catalyst [44]. This interest in allylsilanes is due to their low toxicity compared with allyltins and the simple removal of silicon by-products. The allylation reaction of carbonyl compounds using allyltrimethoxysilane in aqueous media in the presence of a catalytic amount of a CdF$_2$–terpyridine complex was reported [45].

The allylation reaction of carbonyl compounds with substituted allylic halides in aqueous media can generate two different regioisomers (Scheme 6.7). The regioisomer preferentially formed is the one where the substituent is α to the C–C bond to be formed: the γ-adduct [46]. Different factors are involved in directing the regioselectivity, the most important being the steric size of the substituent. Specific conditions were optimized to generate the α-adduct preferentially [47].

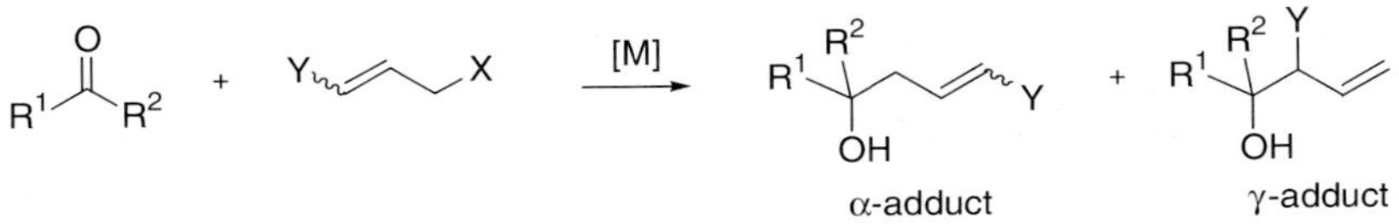

Scheme 6.7

The asymmetric allylation reaction has also been studied and represents one of the major recent developments in this field. Loh and Zhou described the first enantioselective indium-mediated allylation of aldehydes in the presence of pybox [(S,S)-2,6-bis(4-isopropyl-2-oxazolin-2-yl)pyridine] as a chiral ligand and Ce(OTf)$_4$ hydrate as a Lewis-acid promoter (Scheme 6.8) [48]. The reaction of allylindium with benzaldehyde in a mixture of water and ethanol affords the homoallylic alcohol in 90% yield and 92% ee.

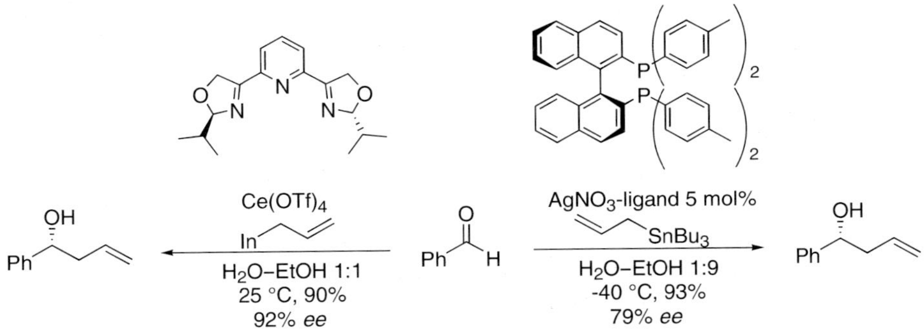

Scheme 6.8

A modified Yamamoto–Yanagisawa catalyst, (S)-TolBINAP/AgNO$_3$, was then successfully applied to the same transformation using allyltributyltin in the place of allylindium (Scheme 6.8) [49]. The use of a catalytic amount of BINOL–In(III) for the addition of allyltributyltin to aldehydes affords similar results [50]. Aryl ketones are allylated by a mixture of tetraallyltin and triallylbutyltin in the presence of monothiobinaphthol with high *ee* [51]. The presence of water enhances the enantiomeric excess. In order to introduce chirality, the use of chiral allylic compounds has been described. *N*-Trimethylsilylbenzaldimines can be allylated by *B*-allyldiisopinocampheylborane with good *ee* up to 92% [52]. Moreover, Yamamoto and co-workers described the use of a chiral π-allylpalladium complex as a catalyst for the allylation of diverse imines with allyltributyltin in good to high yields in the presence of 1 equiv. of water [53].

The asymmetric allylation of hydrazono esters in aqueous media with allylsilanes in the presence of a catalytic amount of zinc fluoride and a chiral diamine ligand generates (benzoyl)hydrazino-4-pentanoates with enantioselectivity up to 84% (Scheme 6.9) [54].

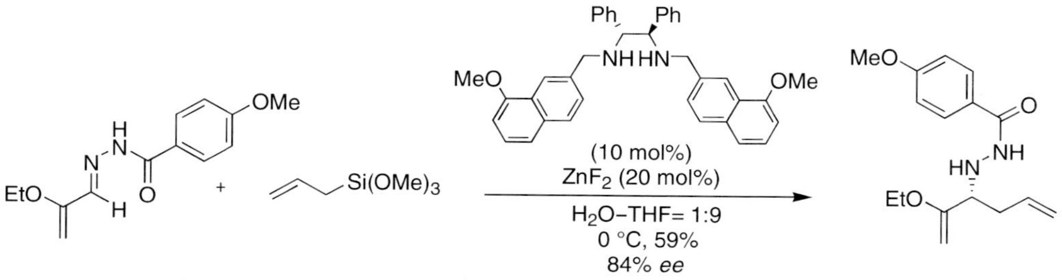

**Scheme 6.9**

### 6.2.1.2 Propargylation and Allenylation Reaction

The tin-mediated propargylation in aqueous media with aldehydes generates a mixture of propargylation and allenylation products with a rather low selectivity [55]. Instead of using zerovalent metals, these reactions can also be carried out in the presence of preformed organotin reagents [56]. Recently, a one-pot procedure for tin-mediated allylation and propargylation of aryl epoxides has been developed [57]. A combination of $SnCl_2$ and catalytic Pd(0) or Pd(II) promotes the reaction of organic halides and epoxides in a mixture of DMSO and water, leading to the regioselective formation of the corresponding homoallylic and homopropargylic alcohols in good yields. Zinc-mediated propargylation of various aldehydes was also studied in aqueous media [58]. A better selectivity in favor of propargylic alcohol is observed. When the reaction is carried out on a chiral aldehyde in the presence of water, homopropargylic alcohols are obtained with very high diastereoselectivity (Scheme 6.10) [59].

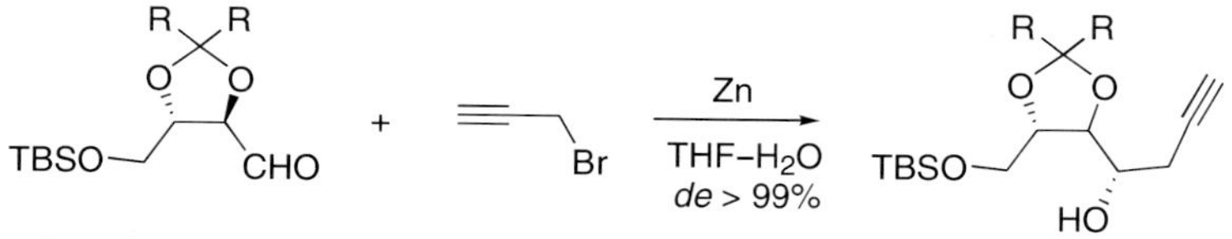

**Scheme 6.10**

Indium-mediated propargylation in aqueous media with prop-2-yn-1-yl bromide on both aliphatic and aromatic aldehydes affords homopropargylic alcohols as major products [60]. In contrast, γ-substituted propargyl bromides afford predominantly or exclusively allenylic alcohols [61] and this was applied to the synthesis of the antiviral, antitumor compound (+)-goniofurfurone [62]. Hammond and co-workers recently reported the indium- and zinc-mediated Barbier-type reaction of difluoropropargyl bromide with several aldehydes in aqueous media [63]. The reaction gives the corresponding β,β-difluorohomopropargylic alcohols with high regioselectivity. The indium-mediated coupling of propargyl bromide with a variety of imines, glyoxylic oxime ether, and imine oxides in aqueous environments affords homopropargylamine derivatives in high yields under mild conditions [64].

5-Methylisoxazoles can be obtained in high yields through a domino addition, C—O heterocyclization involving allenylindium bromide and benzonitrile oxide in aqueous media (Scheme 6.11) [65].

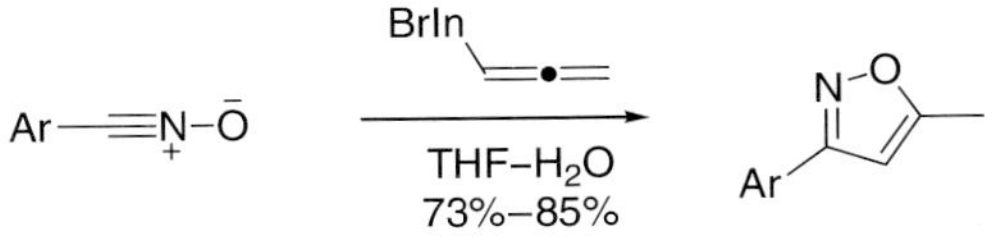

**Scheme 6.11**

### 6.2.1.3 Alkylation Reaction

Although the metal-mediated C—X bond activation of allyl and propargyl halides in aqueous media has been studied during the past decade, few efforts have been made to activate the C—X bond of alkyl halides by a metal in water because of the instability of alkyl organometallic reagents towards water. The carbon–nitrogen double bond of imine derivatives has emerged as a radical acceptor, and thus numerous, synthetically useful carbon–carbon bond-forming reactions are available [66]. In principle, reactions of strictly neutral species such as uncharged free radicals are not affected by the presence of water. Thus, alkyl radicals can be generated via sonication of alkyl iodide in the presence of Zn–CuI in water [67].

Zinc and indium appear to be effective mediators for such reactions on N-sulfonimines, oxime ethers, and hydrazones [68]. The indium-mediated alkylation reaction of the Oppolzer camphorsultam derivative of glyoxylic oxime ether proceeds with good diastereoselectivity in aqueous media, providing α-amino acids (Scheme 6.12) [69].

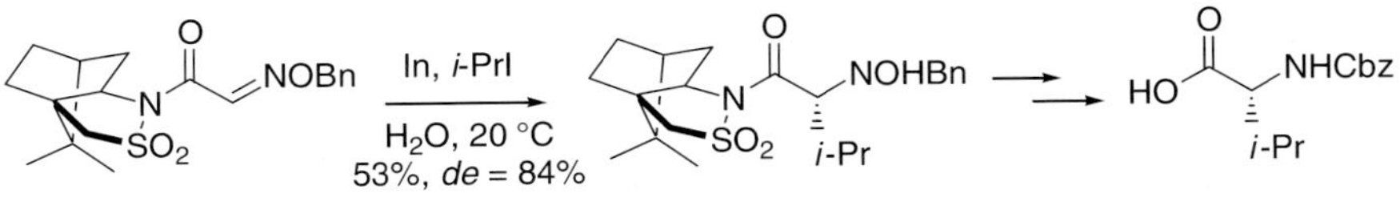

**Scheme 6.12**

The analogous metal-mediated alkylation reactions of simple imines in aqueous media have been more difficult and challenging since they are less reactive and easier to hydrolyze [70]. The synthesis of amines by alkylation reaction of simple imines in the presence of metallic zinc in aqueous ammonium chloride solution was described [71]. Recently, a radical-type alkylation reaction of imines by using Ti(III) and $PhN_2^+$ ($BF_4^-$ or $Cl^-$) in acidic aqueous media has been reported with limited substrates and moderate selectivities [72]. Efficient systems of In–CuI–$InCl_3$ and In or Zn–AgI–$InCl_3$ were developed for Barbier–Grignard-type alkylation reactions of simple imines, using a one-pot condensation of various aldehydes, amines (including aliphatic and chiral version), and primary or secondary alkyl iodides in aqueous media [73].

The direct addition of alkyl groups to aldehydes is the most challenging. In 1999, the indium-mediated alkylation of carbonyl compounds with α-sulfur-stabilized halides was reported [74]. However, the direct addition of simple alkyl groups is more difficult. In fact, unlike imine addition, aryl groups do not stabilize the intermediates [68, 71]. In 2003, a zinc–copper iodide-mediated Barbier–Grignard-type carbonyl alkylation with non-activated halides was reported in water (Scheme 6.13) [75].

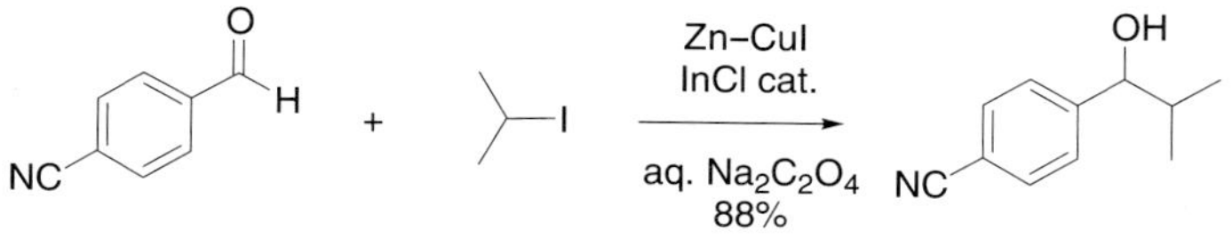

**Scheme 6.13**

Water permits the alkylation of aldehydes with trialkylboranes under nickel catalysis (Scheme 6.14) [76]. This reaction proceeds by $\eta^2$-coordination of the nickel complexes with aldehydes. It is worth noting that the addition of cyclodextrins increases the yield in case of difficult reactions.

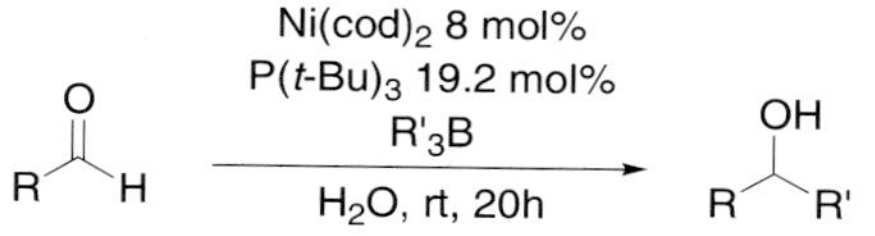

**Scheme 6.14**

#### 6.2.1.4 Arylation and Vinylation Reactions

In 1998, Miyaura and co-workers reported the rhodium-catalyzed addition of aryl- or alkenylboronic acids to aldehydes giving secondary alcohols in good to excellent yields (Scheme 6.15) [77]. A variety of arylboronic acids and aldehydes could be used. Optimal conditions were determined to be reaction of aldehyde with 2 equiv. of boronic acid in the presence of a catalyst generated *in situ* from [Rh(acac)(CO)$_2$] and dppf in a DME–$H_2O$ mixture at 80 °C (Table 6.1, Entry 1). The effect of added ligands was also examined in the addition of phenylboronic acid to aldehydes [78].

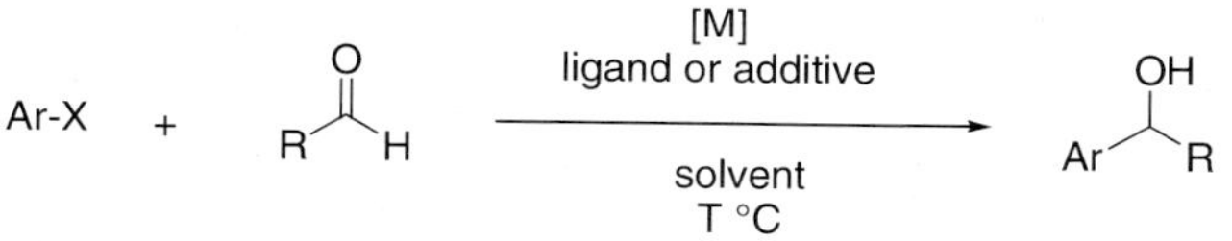

**Scheme 6.15**

Table 6.1 Metal-catalyzed 1,2-addition of boron and stannyl derivatives to aldehydes.

| Entry | Vinyl or aryl source | Catalyst | Ligand or additive | Solvent | Temperature (°C)[a] |
|---|---|---|---|---|---|
| 1 | $ArB(OH)_2$ | $[Rh(acac)(CO)_2]$ | dppf | $DME–H_2O$ | 80 |
| 2 | $ArB(OH)_2$ | $[Rh(acac)(COE)_2]$ | $(t\text{-}Bu)_3P$ | $DME–H_2O$ | r.t. |
| 3 | $ArB(OH)_2$ | $RhCl_3 \cdot 3H_2O$ | NHC | $DME–H_2O$ | 80 |
| 4 | $ArB(OH)_2$ | Ir(cod)X NHC complex | — | $DME–H_2O$ | 80 |
| 5 | $ArB(OH)_2$ | $[Rh_2(OAc)_4]$ | NHC | $DME–H_2O$ | 40–90 |
| 6 | $ArB(OH)_2$ | Rh⁻ K⁺ quinonoid (cod) | — | $H_2O$ | 75 |
| 7 | $ArB(OH)_2$ | $RhCl_3 \cdot 3H_2O$ | t-Bu₂P-CH₂-NMe₃Cl (t-Bu-AMPHOS) | $CH_3CN–H_2O$ | 80 |
| 8 | $ArBF_3K$ | $[Rh(C_2H_4)]$ | $(t\text{-}Bu)_3P$ | Toluene–$H_2O$ | r.t. |
| 9 | R–B(pinacol), R = aryl or styryl | $Ni(cod)_2$ | R—≡—R | Dioxane–$H_2O$ | 80 |
| 10 | $ArSnR_3$ | $Rh(cod)_2BF_4$ | — | $H_2O$ | 110 |

[a] rt, room temperature.

Tri(*tert*-butyl)phosphane was reported to have a more significant effect in accelerating the reaction than dppf even at room temperature (Table 6.1, Entry 2). *N*-Heterocyclic carbine (NHC) ligands, which are known for their strong σ-donor and weak π-acceptor properties [79], are superior surrogates for air-sensitive and rather expensive monophosphanes such as P(*t*-Bu)₃. Fürstner and Krause successfully applied these ligands, in organic solvents, to the arylation and vinylation of aldehydes with boronic acids in the presence of $RhCl_3 \cdot 3H_2O$ as a catalyst [80]. Further developments were realized in aqueous media (Table 6.1, Entry 3) [81]. Moreover, Ir–NHC complexes can also be used for such reactions; they appear to be less active yet more selective catalysts (Table 6.1, Entry 4) [82]. Polymer-supported NHC and hexadentate imidazolium salts were also recently developed [83]. More recently, a performant catalytic system was described for this reaction in the presence of dirhodium(II) complexes, which are readily available and highly versatile (Table 6.1, Entry 5) [84]. An anionic rhodium η⁴-quinonoid complex was reported to be a good catalyst for the coupling of arylboronic acids and aldehydes in water (Table 6.1, Entry 6) [85]. Recently, Shaughnessy *et al.* described a combination of $RhCl_3 \cdot 3H_2O$ and *t*-Bu-AMPHOS as a recyclable catalyst for this reaction (Table 6.1, Entry 7) [86]. Many organoboranes are not stable under atmospheric conditions [87]. One solution emerged with the discovery of potassium organotrifluoroborates. Batey *et al.* were the first to describe the feasibility of the addition of potassium aryltrifluoroborates to aryl aldehydes in aqueous media [88]. Indeed, in the presence of catalytic amounts of Rh(acac)(CO)$_2$ and a bidentate ligand (dppf), good yields of carbinols were obtained. However, the reaction was generally limited to electron-deficient arylaldehydes. More efficient conditions involved tri-*tert*-butylphosphane [78] as ligand (Table 6.1, Entry 8) [89].

Highly hindered diarylmethanols could be formed under these conditions, and aliphatic aldehydes were also reactive. Moreover, the reaction could be run at room temperature for many substrates. Nickel-catalyzed addition of organoboronates (vinyl- or arylboronates) was reported using alkynes as activators in the presence of one equivalent of water (Table 6.1, Entry 9) [90]. Alkynes do not react with organoboron compounds or aldehydes but activate them.

In 2000, Li and Meng found that cationic rhodium complexes could be used to catalyze the addition of arylstannanes to aldehydes in neat water at 110 °C with good yields (Table 6.1, Entry 10) [91]. Arylboron and arylbismuth reagents reacted equally well to generate the aldehyde addition product. A strong electronic effect was also reported [92]. No reaction was observed with arylmetal halides under neutral conditions, but the replacement of halide by alkyl, alkoxyl, hydroxyl, or aryl groups provided the product efficiently.

The arylation of imines by phenyltrimethyltin or phenyltrimethyllead in the presence of a rhodium catalyst in air and water under ultrasonic irradiation at 35 °C to give the corresponding diarylmethylamines in good yields was described [93]. Recently, Bolshan and Batey reported the addition of organoboronic acids and potassium aryltrifluoroborates to chiral sulfinimines [94]. The reaction proceeded under mild conditions at room temperature in the presence of water as a cosolvent. It is noteworthy that the absence of water leads to very low conversion. The sulfinamide adducts were formed with high diastereoselectivities, providing a convenient route to the synthesis of enantiomerically enriched chiral benzylic amines.

Petasis and Zavialov reported an efficient addition of vinylboronic acid to iminium salts [95]. Whereas no reaction was observed when acetonitrile was used as a solvent, the reaction went smoothly in water to give allylamines. When the iminium ions are generated from glyoxylic acid and amines, the reaction of the boron reagent afforded novel α-amino acids (Scheme 6.16).

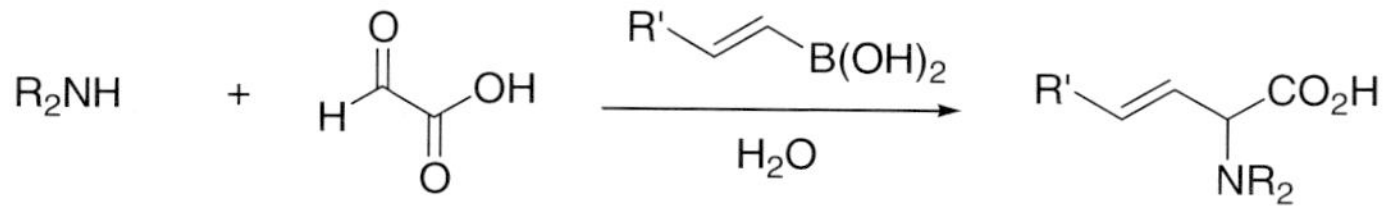

$$R_2NH \ + \ \underset{H}{\overset{O}{\|}}{\overset{}{\text{C}}}{\text{--}}{\overset{O}{\|}}{\text{C}}{\text{--}}OH \ \xrightarrow[H_2O]{R' \diagdown B(OH)_2} \ R' \diagdown \overset{NR_2}{\overset{|}{\text{C}}} CO_2H$$

**Scheme 6.16**

Examples of asymmetric arylation of carbonyl compounds (or derivatives) with arylboronic acid are fairly limited. Miyaura and co-workers initially reported the enantioselective Rh-catalyzed addition of phenylboronic acid to naphthaldehyde by using the (S)-MeO-MOP ligand, giving naphthylphenylmethanol in 41% *ee* [77]. This result had long been the highest enantioselectivity for this reaction until recently,

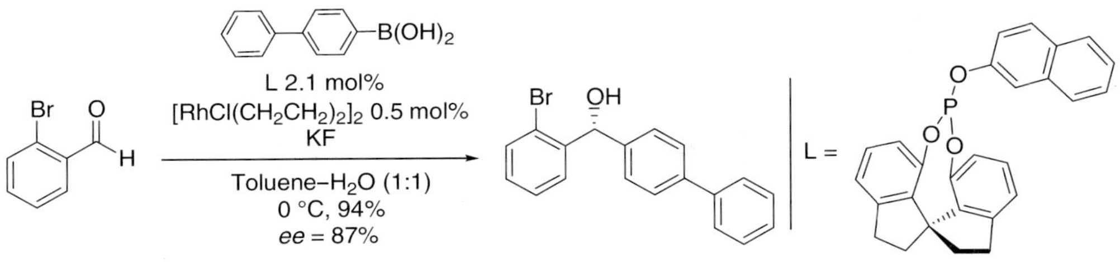

**Scheme 6.17**

when Zhou and co-workers described a highly efficient rhodium-catalyzed asymmetric addition of arylboronic acids to aldehydes by using chiral spiro monophosphite ligands, affording diarylmethanols with good ees of up to 87% (Scheme 6.17) [96].

### 6.2.1.5 Alkynylation Reaction

The alkynylation reaction of carbonyl compounds and derivatives in aqueous media is of great interest because no halide is involved. This reaction proceeds via C—H activation to give Grignard-type nucleophilic addition products in water [5]. Basing their work on the efficiency of ruthenium to catalyze C—H activation [97], Li's group reported the first alkynylation reaction of aldehydes by phenylacetylene under aqueous conditions in the presence of a bimetallic Ru–In catalytic system (Scheme 6.18) [98]. The same authors described an effective Ru–Cu-catalyzed addition of terminal alkynes to arylimines in water or without solvent [99].

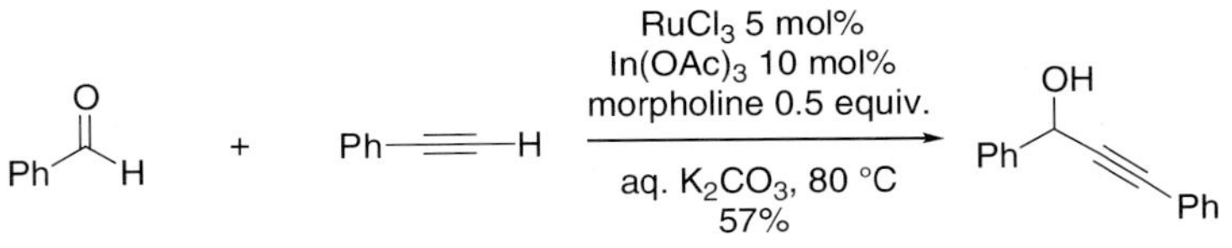

**Scheme 6.18**

A copper-mediated coupling of alkynes with N-acylimines and N-acyliminium ions (generated in situ from amines containing good leaving groups) in water to generate propargyl amide derivatives was also described [100]. Nevertheless, an excess amount of copper bromide is required. Tu and co-workers reported the role of microwave irradiation in accelerating copper-catalyzed three-component coupling of aldehyde, amine, and alkyne ($A^3$) via C—H activation in water [101].

Li and co-workers reported the efficiency of AgCl, AgBr, and AgI as a catalyst for this three-component coupling in aqueous conditions (Scheme 6.19a) [102]. Silver catalysts are especially effective for reactions involving aliphatic aldehydes. The addition of a phosphane ligand changes the product of this reaction from exclusive $A^3$ coupling product to exclusive aldehyde–alkyne coupling product (Scheme 6.19b) [103].

Three-component $A^3$ coupling via C—H activation in water is also highly efficient with a gold catalyst [104]. Che's group reported a three-component coupling of aldehydes, amines, and alkynes in water using a gold(III)–salen complex as precursor [105]. When chiral prolinol derivatives were used as the amine component,

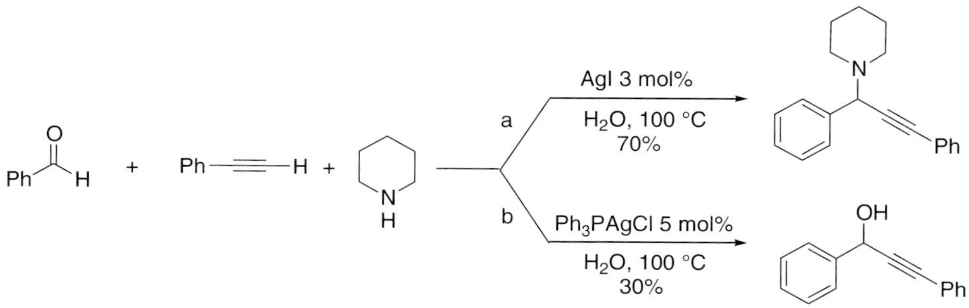

**Scheme 6.19**

diastereoselectivities of up to 99:1 were attained. Yao and Li reported a water-triggered gold-catalyzed cascade alkynylation–cyclization on o-alkynylarylaldehydes leading to 1-alkynyl-1H-isochromenes [106]. Interestingly, no alkynylation product was observed using the same catalyst in the presence of phenylacetylene and benzaldehyde. Recently, Liu and co-workers described a three-component gold-catalyzed addition–cycloisomerization reaction of pyridine-2-carboxaldehyde, amine, and alkyne under neat conditions or in water to provide substituted aminoindolizines [107].

A copper-catalyzed enantioselective direct alkynimine (generated *in situ* from aldehyde and amine) addition was developed in water in the presence of a tridentate bis(oxazolinyl)pyridine (pybox) [108]. The process provides a diverse range of propargylic amines in good yield and with *ee* up to 91% (Scheme 6.20).

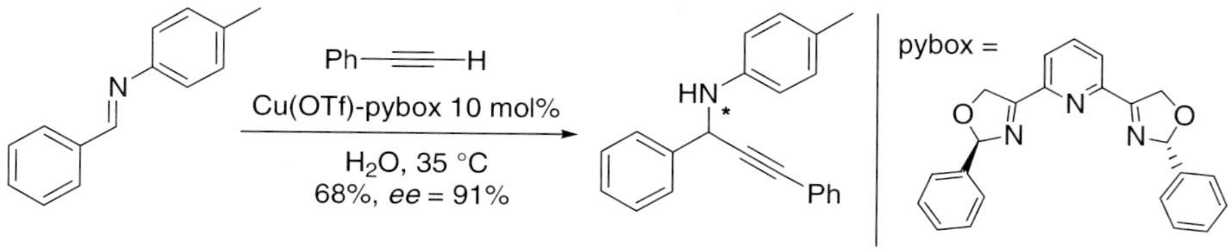

**Scheme 6.20**

## 6.2.2
## Pinacol Coupling

The pinacol coupling, known as the coupling of carbonyl compounds to give 1,2-diols was discovered in the nineteenth century by Fittig [109]. Important examples of the use of this reaction in total synthesis are the syntheses of taxol by the groups of Mukaiyama [110] and Nicolaou [110]. Pinacol coupling in aqueous media was first discovered using titanium as a mediator [111]. Later, pinacol coupling was developed in aqueous media using other metals. such as Zn–Cu, Mn, Mg, In, Al, and Sm (Scheme 6.21) [112].

Arylimines (generated from aromatic aldehydes and aromatic amines) can also perform reductive coupling in aqueous media. In aqueous ethanol, indium-mediated reductive coupling affords diamines without a trace of side product due to unimolecular

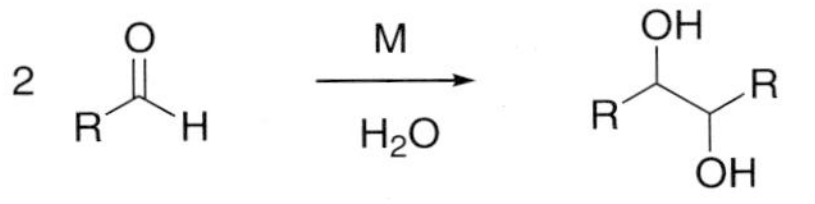

Scheme 6.21

M = Ti, Zn-Cu, Mn, Mg, In, Al, Sm...

reduction [113]. Zinc and samarium can also be used as mediator for reductive coupling of aldimines [114].

## 6.3
## Reaction of Alkenes and Alkynes

### 6.3.1
### Reaction of Unconjugated Alkenes and Alkynes

#### 6.3.1.1 Hydroformylation Reaction

The hydroformylation reaction, discovered by Roelen in 1938, is one of the most important reactions and was industrialized by Rührchemie/Rhône-Poulenc using the ligand TPPTS [tris(m-sulfonatophenyl)phosphane trisodium salt] [115], and is nowadays well known as the oxo process (>600 000 t per year). Alkenes, such as propene initially, react with carbon monoxide and hydrogen in the presence of a transition metal catalyst to give a mixture of linear and branched aldehyde. A Rh–TPPTS complex was used efficiently for this process and starting from propene led to n-butyraldehyde in high yield (95%) and isomer ratio (up to 95 : 5 for linear/branched aldehydes) (Scheme 6.22). The homogeneous two-phase catalysis offers the advantages of an easy separation of aldehydes, after the reaction, and the recycling of the water-soluble catalyst. The selectivity of the reaction and the application of this particularly clean and economic reaction to a large variety of alkenes have been studied in the presence of several water-soluble ligands [116].

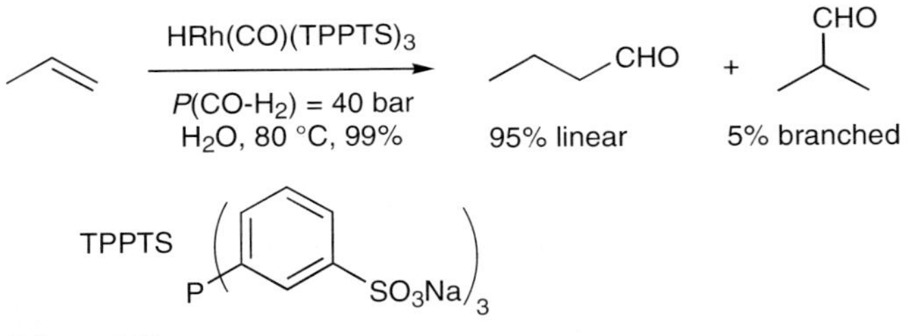

Scheme 6.22

Several concepts have been suggested for increasing the rates in aqueous-phase catalytic conversion of higher substrates such as addition of phase-transfer agents, promoter ligands, or cosolvents [117]. The hydroformylation of various long-chain olefins (1-hexene to 1-octene) has been conducted in the presence of anionic, non-ionic, and cationic surfactants, metal ions, and in microemulsions [118]. Functionalized cyclodextrins also appear as inverse phase-transfer agents leading to similar effects (enhancement of yield and moderate linear/branched ratio) than in the presence of surfactants or microemulsions [119]. Another alternative for achieving higher efficiency and easy catalyst recovery lies in the heterogenization of the system, which has been elegantly illustrated in supported aqueous phase catalysis (SAPC) and supported homogeneous film catalysis (SHFC) [120]. The catalytic materials consist of a thin film that resides on a high surface area support such as controlled-pore glass, silica, or recently apatitic tricalcium phosphate supports [121]. The combination of the above-mentioned techniques and the search for alternative transition metal catalysts have led to the emergence of efficient systems with platinum, ruthenium, palladium, and cobalt [122]. Methyl formate has been used as the source of carbon monoxide: hydrogen was generated by the water gas shift reaction and cycloalkanemethanols were thus prepared [123]. Considering the widespread improvements of the hydroformylation reaction, its application to substituted and functionalized olefins recently attracted considerable attention. Many functionalized aldehydes are useful building blocks and are involved in the synthesis of flavors, fragrances, and pharmaceuticals [124]. In the presence of an amine, a hydroaminomethylation reaction may occur and was described in the presence of rhodium and iridium complexes in aqueous medium [125]. The asymmetric hydroformylation was investigated but gave moderate $ees$ [126].

### 6.3.1.2 Hydroxycarbonylation

The hydroxycarbonylation process, which promotes the synthesis of carboxylic acids in aqueous–organic two-phase systems, may be seen as the addition of formic acid to an activated alkene [127]. Alkyl- or arylalkenes are indeed highly reactive in acidic medium in the presence of carbon monoxide, water, and a water-soluble palladium catalyst. The best results are obtained on vinyl aromatics and small chain $\alpha$-olefins using $Pd^{II}$–TPPTS, [Pd(Pyca)(TPPTS)](OTs) (Pyca = pyridine-2-carboxylato), systems in the presence of HCl, APTS, or other Brønsted acids in either water or water–toluene media (Scheme 6.23) [128].

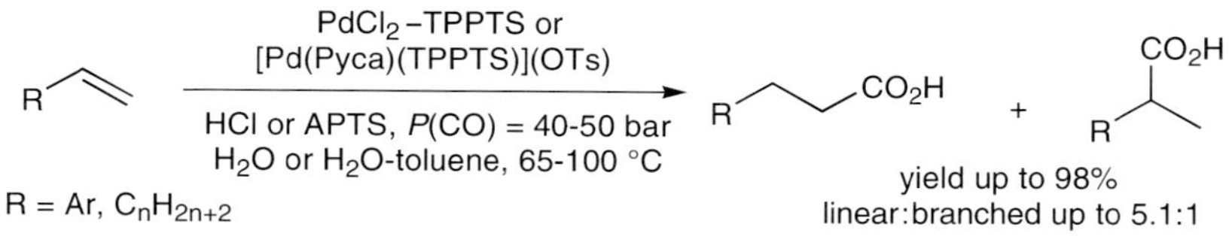

Scheme 6.23

PdCl$_2$–TPPTS or [Pd(Pyca)(TPPTS)](OTs)
HCl or APTS, $P(CO)$ = 40–50 bar
H$_2$O or H$_2$O-toluene, 65–100 °C

R = Ar, C$_n$H$_{2n+2}$

yield up to 98%
linear:branched up to 5.1:1

In the same manner as for the hydroformylation, limitations are encountered for low water-soluble higher alkenes. Several mass transfer promoters have been added including alkali metal salts, protective-colloid agents, and modified cyclodextrins, the latter being the most efficient [129]. The key intermediates of the hydroxycarbonylation of olefins in the aqueous phase have recently been identified [130]. It should be noted that the synthesis of carboxylic acid under CO pressure may also be performed starting from benzyl or allylic halides [131].

### 6.3.1.3 Metathesis, Polymerization Reactions, and Carbene Reactivity

In 1965, Rhinehart and Smith described the first example of norbornene polymerization in the presence of an iridium catalyst in organoaqueous medium, but the isolated yields were particularly low [132]. Polymerization of norbornene and functionalized norbornenes bearing alcohols and sugar residues were then described with either palladium [133] or ruthenium [134] in aqueous media with or without sodium dodecyl sulfate (SDS), leading to valuable latexes in high isolated yields. The reactivity of ethylene was also observed under CO pressure and led to the synthesis of polyketones [135]. The polymerization reactions were also extended to unsaturated compounds, including styrene and oct-1-ene, in the presence of ruthenium carbene catalyst [135]. One should be emphasized that carbenes can be generated in a variety of ways in aqueous conditions, and many of these complexes are stable in water [136]. Whereas ROMP is an important method for making polymers, ring-closing metathesis (RCM) has been used extensively in synthesis in aqueous conditions (Scheme 6.24) [137].

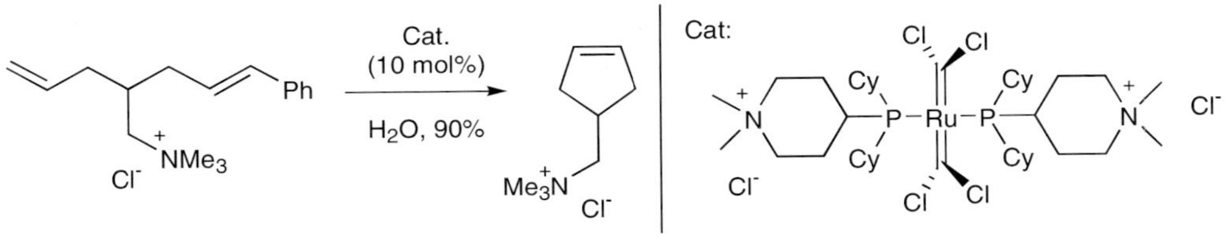

Scheme 6.24

Another key application involves the cyclopropanation of styrene in the presence of diazoacetate in aqueous media using Rh(II) carboxylates, Ru(II) pybox, or Co(II) salen complexes (Scheme 6.25) [138].

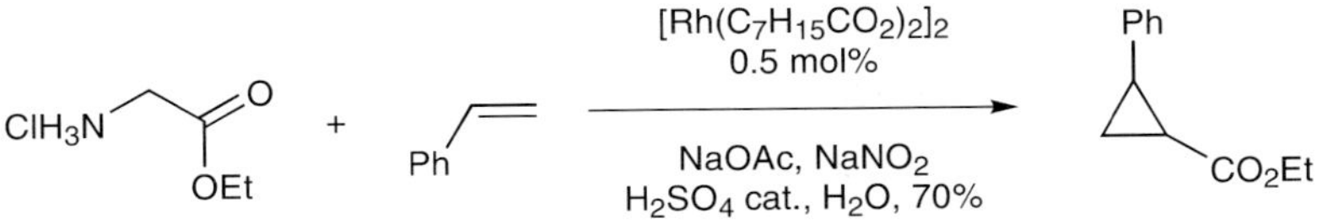

Scheme 6.25

### 6.3.1.4 Isomerization of Alkenes

The preparation of ketones and aldehydes is also possible via classic isomerization of allylic alcohols catalyzed by water-soluble rhodium and ruthenium complexes (Scheme 6.26) [139b]. The reaction was tested by high-throughput screening in microreactors for a limited number of substrates, however [139]. Nickel-catalyzed isomerization of olefins in a two-phase system with a Brønsted acid was also described [140]. It is noteworthy that homoallylic alcohols undergo structural reorganization in which both the hydroxyl group and the olefin have been reshuffled in water in the presence of $RuCl_2(PPh_3)_3$ [141].

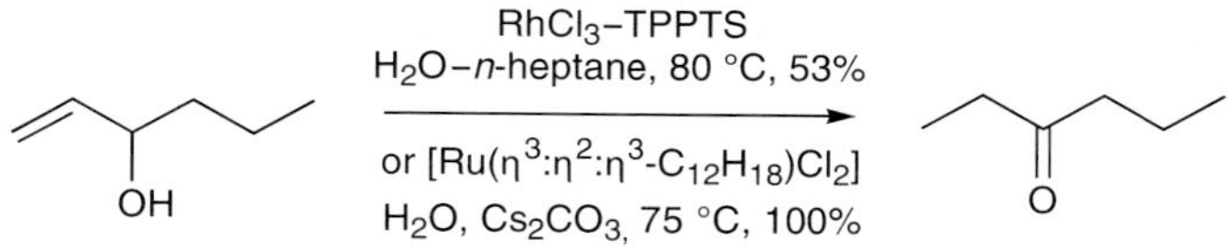

Scheme 6.26

### 6.3.1.5 Coupling of Alkynes

The dimerization of alkynes, known as the Glaser, Eglinton and Cadiot–Chodkiewicz coupling reaction, is one of the oldest organometallic coupling reactions. In 1882, Baeyer synthesized indigo using potassium ferricyanide as the oxidant in water (Scheme 6.27) [142]. An elegant synthesis of water-soluble conjugated rotaxane has also been reported using the Glaser homo-coupling reaction with a copper salt in water [143].

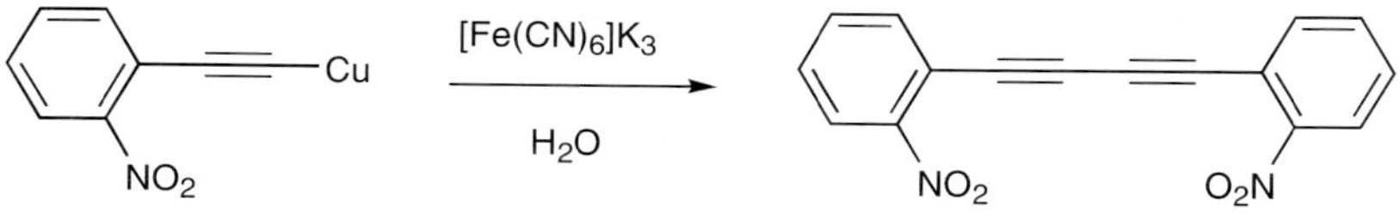

Scheme 6.27

In aqueous media, the addition of unactivated alkynes to unactivated alkenes to form Alder-ene products has been realized by using a ruthenium catalyst (Scheme 6.28) [144]. A polar medium (DMF–H$_2$O) favors the reaction and benefits the selectivity. Remarkably, no homocoupling products from either alkyne or alkene partner are observed.

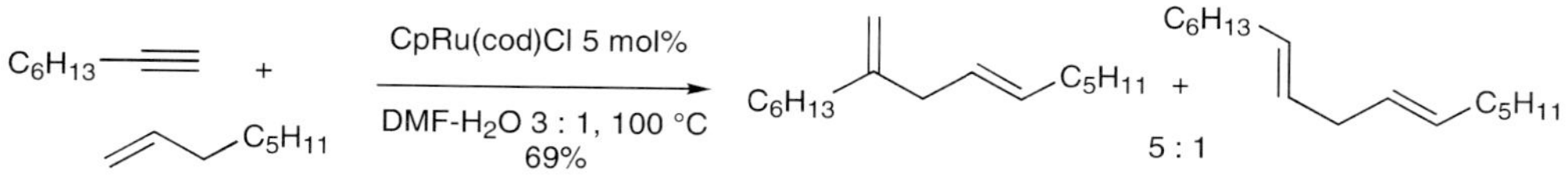

Scheme 6.28

When the unactivated alkene was replaced by an allylic alcohol, the addition to alkynes still occurred and afforded γ,δ-unsaturated ketones and aldehydes [145]. A sequential alkyne–alkene coupling and a Lewis-acid-catalyzed Prins-type reaction in a DMF–H$_2$O mixture was described that gave rise to an entry to 1,5-oxygen-bridged medium-sized carbocycles [146].

### 6.3.1.6 Mizoroki–Heck Reaction and Related Hydroarylation Reactions

One of the most popular coupling reactions is the coupling of vinyl or aryl halides with alkenes in the presence of palladium catalysts, namely the Mizoroki–Heck reaction [147]. It was originally performed on activated alkenes, but was very rapidly applied to a large variety of alkenyl compounds. It represents a powerful tool for building up new carbon–carbon bonds. Such reaction was initially performed under ligand-free conditions in water [148], but then a myriad of water-tolerant and

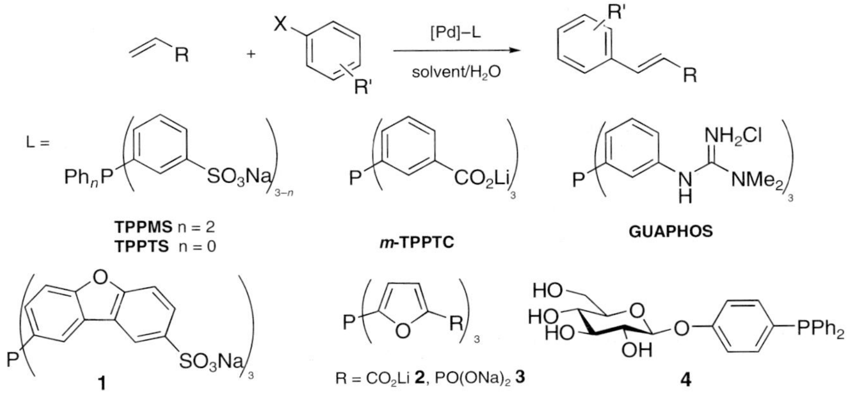

## 6.3 Reaction of Alkenes and Alkynes

**Scheme 6.29**

water-accelerating conditions were described. The use of water-soluble ligands such as TPPMS, TPPTS, *m*-TPPTC, GUAPHOS, or ligands **1–4** allowed short reaction times and an easy catalyst–product separation (Scheme 6.29) [149].

Several modifications have been introduced to increase the product–catalyst separation and catalyst recycling such as the use of a biphasic toluene/EtOH–$H_2O$ system, or supercritical $CO_2$ [150]. The use of supported liquid phase catalysis (SLPC), a similar concept to SAPC, has appeared in the literature with different metal complexes, liquid films, and anchored materials [151].

A Mizoroki–Heck-type reaction was also conducted in the presence of a rhodium catalyst to give styrenyl derivatives. The use of water-soluble ligands such as TPPDS and *m*-TPPTC gave rise to the addition of various boronic acids to styrenyl derivatives in water (Scheme 6.30) [152].

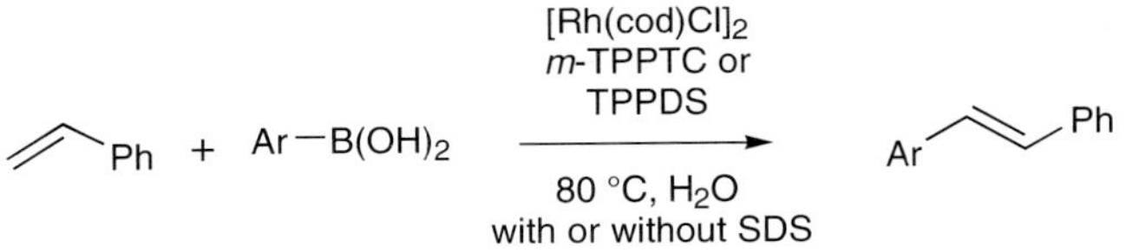

**Scheme 6.30**

Some Rh–dppf and Rh–water-soluble ligands were recently described for the hydroarylation of alkynes leading to a wide array of substituted alkenes (Scheme 6.31) [153]. The use of the sulfonated ligand **5** was limited to the addition to 2-pyridyl-substituted alkynes, whereas Rh–*m*-TPPTC gave rise to regio- and stereoselective reactions. Recent developments have broadened the scope of the hydroarylation reaction by incorporating an electrophilic functionality such as an aldehyde or a ketone [154].

### 6.3.1.7 Cyclization and Cyclotrimerization of Polyfunctional Unsaturated Derivatives

The development of routes for the synthesis of functionalized five- and six-membered rings has also attracted attention via organoaqueous metal-catalyzed reactions.

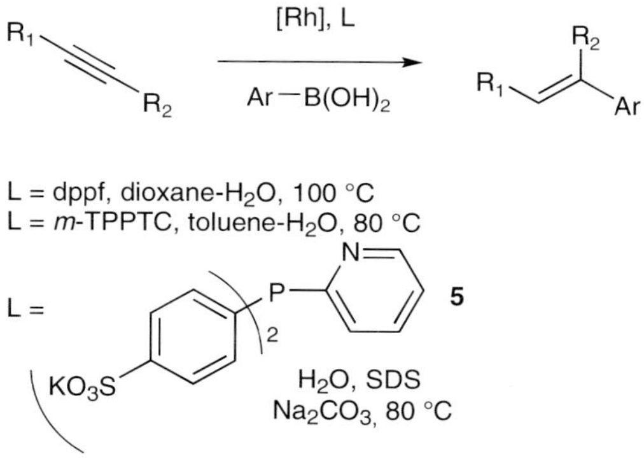

**Scheme 6.31**

Efficient metallo-ene reactions have been reported using water-soluble palladium, rhodium, and nickel catalysts [155]. The last is particularly attractive as the cyclization occurs at room temperature and for a large variety of substrates (Scheme 6.32).

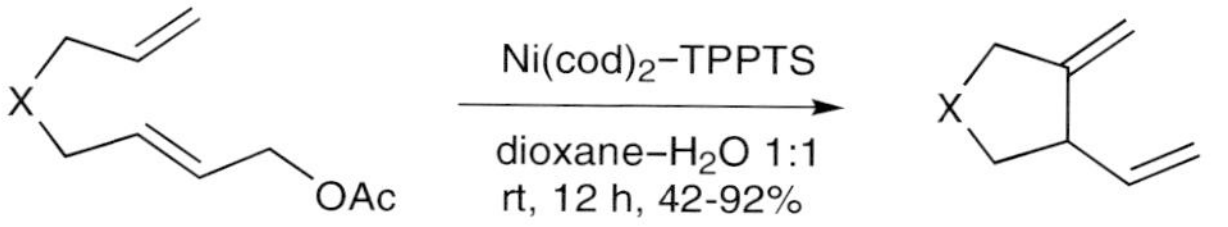

$X = C(CO_2Me)_2, C(SO_2Ph)_2, C(CN)_2, NTs$

**Scheme 6.32**

A rhodium catalyst associated with a new water-soluble bidentate phosphane ligand 6 was shown to be highly effective for [5 + 2] enyne cyclization in water, giving a seven-membered ring product (Scheme 6.33). It is noteworthy that the water-soluble catalyst was reused eight times without any significant decrease in catalytic activity [156].

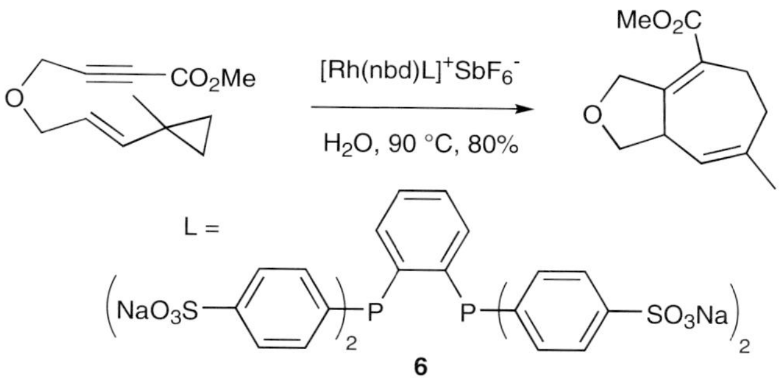

**Scheme 6.33**

The cyclization of functionalized enyne substrates gives bicyclic ketones under aqueous Pauson–Khand-type conditions. The first aqueous Pauson–Khand reaction was reported in 1997, involving aqueous ammonium hydroxide as a reaction

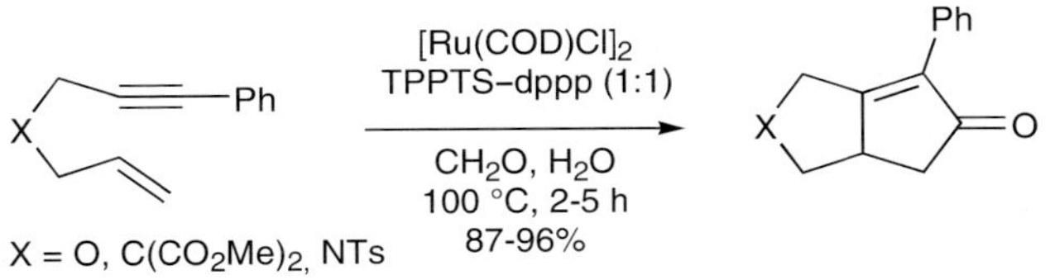

**Scheme 6.34**

X = O, C(CO₂Me)₂, NTs

Reagents: [Ru(COD)Cl]₂, TPPTS–dppp (1:1), CH₂O, H₂O, 100 °C, 2–5 h, 87–96%

medium [157]a. Aqueous colloidal cobalt nanoparticles were also excellent catalysts [157b]. A very elegant system involved a rhodium–water-soluble ligand TPPTS in water, using formaldehyde as CO source, and led to the ketone in 96% yield (Scheme 6.34) [158].

A new atom-economy reaction was discovered when propargylic enynes were subjected to PdCl$_2$–TPPTS and PtCl$_2$–TPPTS systems in a homogeneous medium. The clean cyclization is accompanied by diastereoselective hydroxyl functionalization (Scheme 6.35) [159]. This reaction may be applied to several enynes in the presence of gold catalysts [160]. On the basis of the same findings, a palladium-catalyzed oxidative carbohydroxylation of allene-substituted conjugated dienes in a water–THF medium was recently described [161].

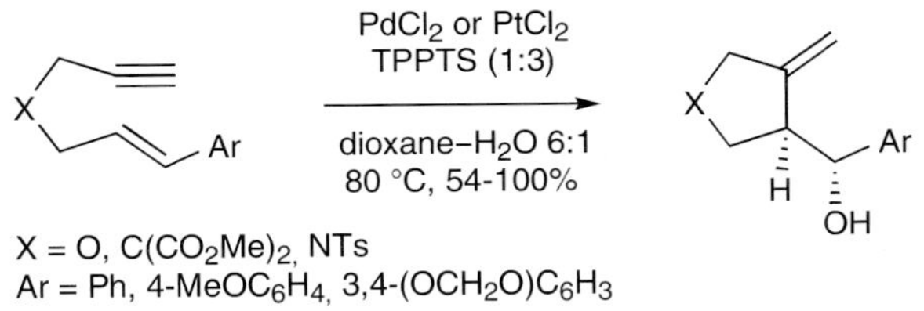

**Scheme 6.35**

Reagents: PdCl₂ or PtCl₂, TPPTS (1:3), dioxane–H₂O 6:1, 80 °C, 54–100%

X = O, C(CO₂Me)₂, NTs
Ar = Ph, 4-MeOC₆H₄, 3,4-(OCH₂O)C₆H₃

The addition of water may also occur on a ruthenium intermediate, which comes from the addition of CpRu(CH$_3$CN)$_2$PF$_6$ to an enyne bearing a ketone moiety [162]. The formation of 1,5-diketone was therefore observed in aqueous acetone with camphorsulfonic acid as a cocatalyst (Scheme 6.36). Other enyne rearrangements [163] involving the addition of water have been reported recently, such as a Claisen–heterocyclization cascade in the presence of a gold catalyst to prepare 3,6-*syn*-substituted pyrans [164].

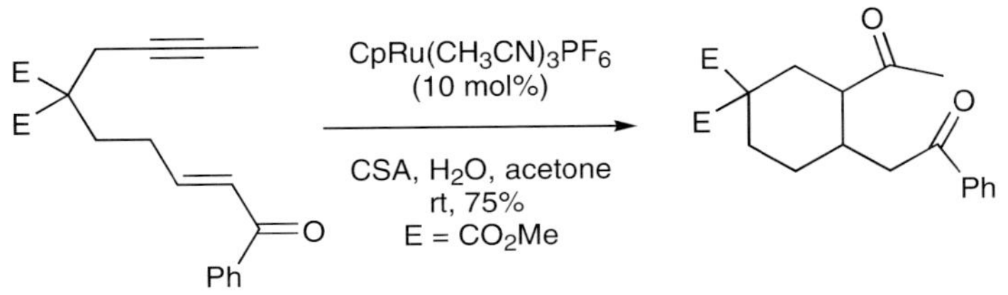

**Scheme 6.36**

Reagents: CpRu(CH₃CN)₃PF₆ (10 mol%), CSA, H₂O, acetone, rt, 75%, E = CO₂Me

Alkyne cyclotrimerization reactions are important synthetic processes, since aromatic rings are key units for many pharmaceutical, biological, and polymer molecules. Some cyclotrimerization reactions have been reported in organoaqueous media using cobalt or rhodium catalysts. The water-soluble CpCo-$\eta^4$-cyclooctadiene catalyst promoted alkyne [2 + 2 + 2] cycloaddition in water, protection of functional groups such as hydroxyl, amine, and carboxylic acids therefore not being necessary [165]. Other cobalt and rhodium catalysts were also described for performing similar cyclizations in an organoaqueous medium and water [166]. An original example involved a nitrile with two alkynes for the synthesis of highly functionalized pyridines (Scheme 6.37) [167].

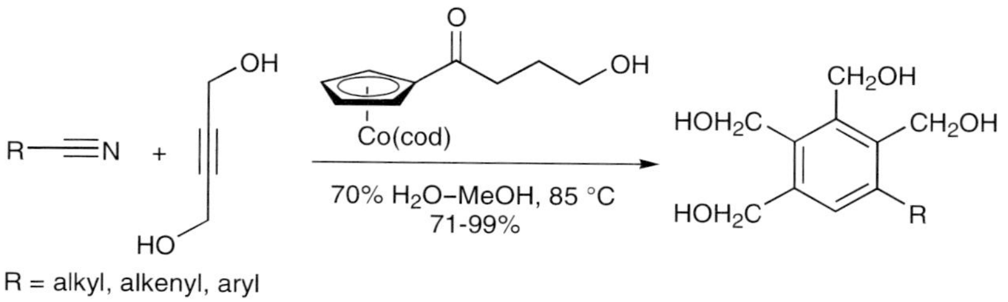

R = alkyl, alkenyl, aryl
**Scheme 6.37**

A rhodium-catalyzed [2 + 2 + 2] cyclotrimerization of alkynes was reported using [Rh(cod)Cl]$_2$–TPPTS catalyst in biphasic ether–water solvents at room temperature (Scheme 6.38) [168]. Adding HCl and NaCl could broaden the reaction to the annelation between a 1,6-diyne and propargylic alcohols, although the influence of such additives is not yet clear.

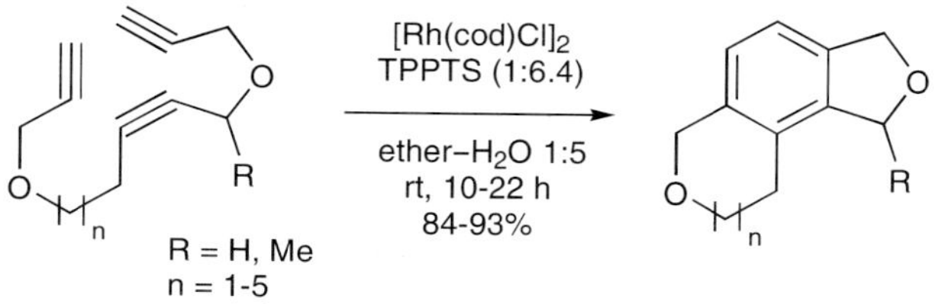

R = H, Me
n = 1-5
**Scheme 6.38**

## 6.3.2
## Reaction of Conjugated Alkenes

### 6.3.2.1 Telomerization of Dienes
Under organoaqueous conditions, conjugated alkenes undergo addition of various nucleophiles. The telomerization of 1,3-dienes with carbonucleophiles was used industrially for vitamin E precursor synthesis with an Rh–TPPTS system (Scheme 6.39) [169].

The telomerization of 1,3-dienes with other nucleophiles such as polyfunctional alcohols has also been described using the trisulfonated ligand. The monotelomers

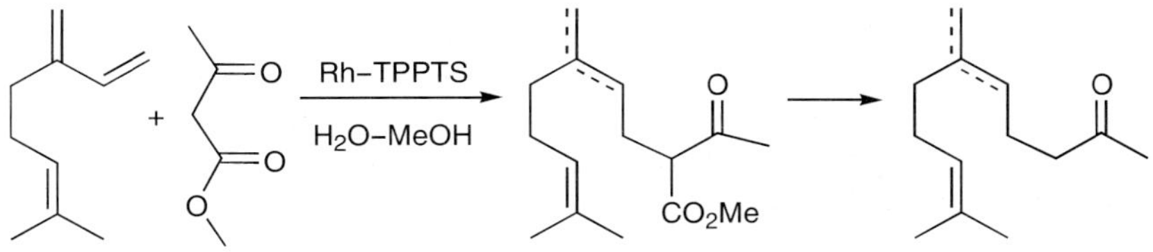

**Scheme 6.39**

obtained from ethylene glycol and butadiene, used as plasticizers for polymers such as PVC, were isolated in 80% yield with more than 95% selectivity using a Pd(acac)$_2$–TPPTS system [170]. Sucrose and starch were also used as nucleophiles in the Pd-catalyzed telomerization of butadiene. In the case of sucrose, numerous conditions were tested and the use of a 1 M NaOH–2-propanol mixture afforded mono- and dioctadienyl ethers selectively [171].

### 6.3.2.2 1,4-Addition to α,β-Unsaturated Derivatives

Other coupling reactions generally involve unsaturated compounds bearing electron-withdrawing group. In 1997, Miyaura et al. reported a novel Michael-type addition of boronic acids to enones, which opened up new opportunities in catalysis (Scheme 6.40) [172]. This work was followed by several seminal works for the addition of boron reagents to other substrates (ketones, esters, amides, phosphonates, nitroalkenes) and for asymmetric 1,4-addition of vinyl and aryl groups. Since then, several organometallic partners have been involved in such coupling, such as potassium organotrifluoroborates [87], organotin, organoindium, organobismuth, organolead, and organosilicon compounds [5]. In all these systems, water was used as a proton source.

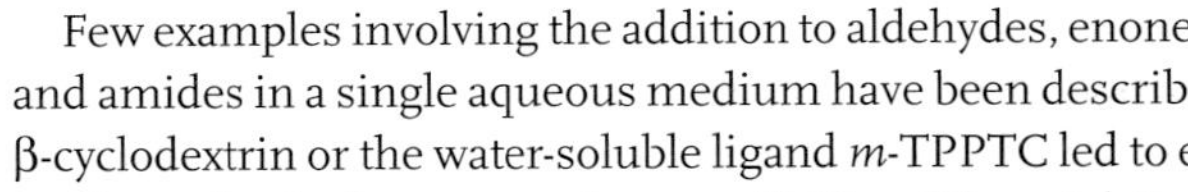

M = B, In, Sn, Bi, Pb, Si...
**Scheme 6.40**

Few examples involving the addition to aldehydes, enones, α,β-unsaturated esters, and amides in a single aqueous medium have been described [173]. The use of either β-cyclodextrin or the water-soluble ligand m-TPPTC led to excellent yields in arylated products. In the latter case, the recyclibility of the catalytic system was proven [173]. It should be emphasized that asymmetric inductions have rarely been observed in water. The use of modified BINAP ligands **7** and **8** afforded moderate to excellent enantiomeric excesses (Scheme 6.41) [174]. It is noteworthy that the supported ligand **7** gave rise to a recyclable system whereas the diguanidinium ligand **8** associated with rhodium offered the possibility of decreasing the catalyst loading [turnover number (TON) = 13 200].

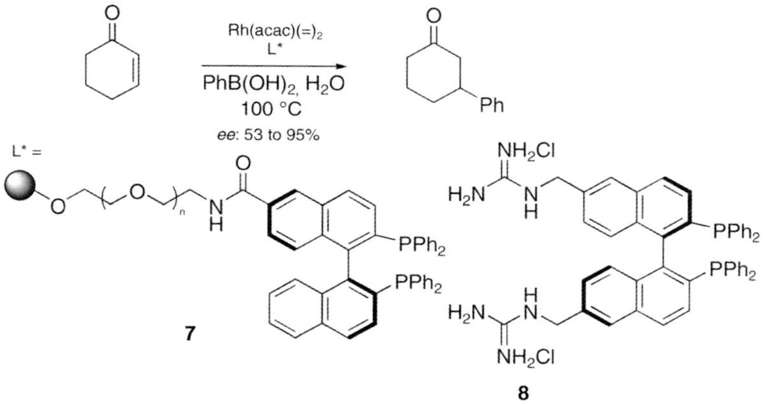

# 172 | 6 Coupling Reactions in Water

**Scheme 6.41**

## 6.4
## Reaction of Organic Halides and Derivatives

Many classical reactions in organic synthesis involve the reactivity of organic halides in reductive coupling, transition metal reactions, and nucleophile substitution reactions. All these transformations can be performed in aqueous media.

### 6.4.1
### Homo- and Heterocoupling Reactions

In aqueous media, homo- and cross-coupling of alkyl halides can be mediated by manganese–cupric chloride to give the coupling products in good yields [175]. Allylgallium reagents were effective for radical allylation of α-iodo- or α-bromocarbonyl compounds. The addition of water as cosolvent improved the yields of coupled products (Scheme 6.42) [176].

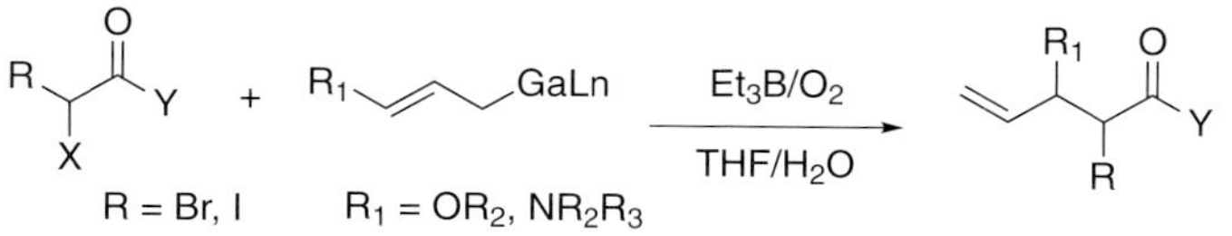

**Scheme 6.42**

Li and co-workers reported a novel palladium-catalyzed Ullman-type reductive coupling of aryl halides under an air atmosphere and in aqueous acetone using Pd/C. The addition of a catalytic amount of crown ether in a single aqueous medium provided better yields (Scheme 6.43) [177].

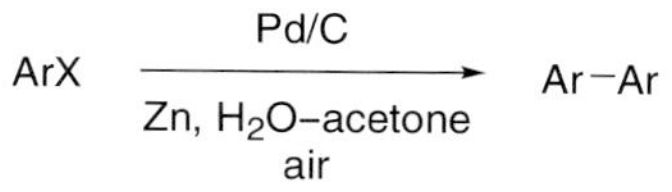

**Scheme 6.43**

Sasson and co-workers further developed this reaction by using PEG as additive; in this case, aryl chlorides also worked effectively [178]. Ullman coupling of aryl halides in the presence of carbon dioxide, using Pd/C, was found to be very effective with various aromatic halides, including aryl chlorides [179]. Lemaire and co-workers reported homocoupling of aryl/heteroaryl bromides and iodides in the presence of Pd(OAc)$_2$ under reducing conditions using DMF–water [180]. Oxidative homocoupling of arylboronic acids in aqueous solvent using Pd(OAc)$_2$ as catalyst afforded biphenyl compounds in good yield (Scheme 6.44) [181].

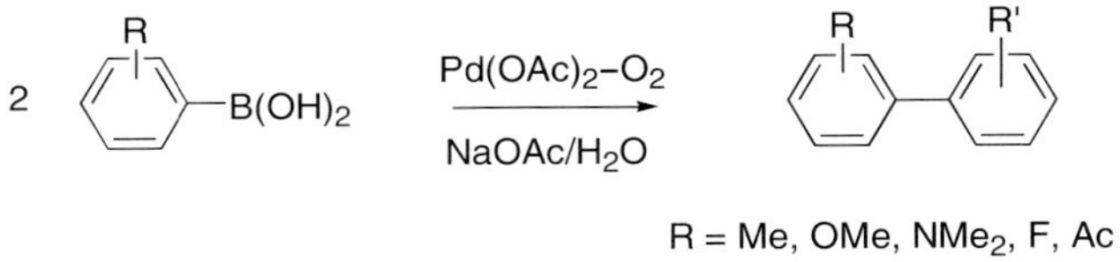

$$2 \ \text{R-C}_6\text{H}_4\text{-B(OH)}_2 \xrightarrow[\text{NaOAc/H}_2\text{O}]{\text{Pd(OAc)}_2-\text{O}_2} \text{R-C}_6\text{H}_4\text{-C}_6\text{H}_4\text{-R'}$$

R = Me, OMe, NMe$_2$, F, Ac

**Scheme 6.44**

### 6.4.2
### Suzuki–Miyaura (S–M) Reaction

#### 6.4.2.1 Palladium-catalyzed Reactions (Aryl and Vinyl Iodides, Triflates, Bromides, and Diazoniums salts)

Among metal-mediated cross-coupling reactions, the Suzuki–Miyaura (S–M) reaction between different types of organoboron compounds and various electrophiles such as halides and triflates in the presence of base provides powerful and general methodology for the formation of C–C bonds [7]. Therefore, a standard technique of performing the S–M reaction is the use of organic/aqueous solvents to increase the solubility of the inorganic salts (Scheme 6.45). These coupling conditions have been applied to the preparation of dienes and some pharmaceutical compounds such as losartan [182], an angiotensin II receptor antagonist.

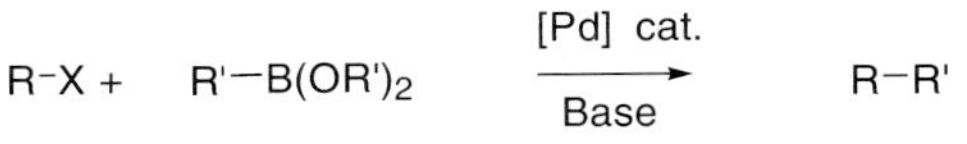

$$\text{R-X} + \text{R'-B(OR')}_2 \xrightarrow[\text{Base}]{[\text{Pd}] \ \text{cat.}} \text{R-R'}$$

**Scheme 6.45**

The S–M reaction requires the presence of bases such as Na$_2$CO$_3$, KOH, and NaOH. The TlOH-promoted S–M reaction offers several advantages [183], such as a broad tolerance of a wide range of functional groups; remarkably, in Kishi's palitoxin synthesis, the reaction was carried out at room temperature [184]. This procedure has been applied to many other interesting biologically active compounds, such as kijanolide [185], rutamycin [186], (12$R$)-hydroxyeicosatetraenoic acid (HETE) [187], and pharmaceutical intermediates [188].

The method in which palladium complexes with hydrophilic phosphanes are used in a biphasic system of water–organic solvent can be considered complementary to

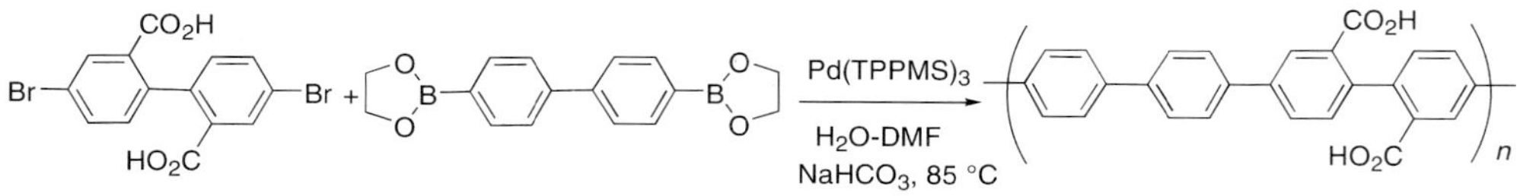

**Scheme 6.46**

the standard protocol. Casalnuovo and Calabrese reported the first reaction using preformed Pd(TPPMS)$_3$ catalyst in acetonitrile or neat water [189]. The reaction required prolonged heating at 80 °C, thus having no specific advantages over the standard protocol. However, this procedure has been successfully applied to the preparation of water-soluble polyphenylenes containing carboxylic groups (Scheme 6.46) [190].

Genêt and co-workers have demonstrated through a series of kinetic and $^{31}$P NMR experiments that a mixture of Pd(OAc)$_2$ and TPPTS affords spontaneously a zerovalent palladium complex [191]. This system offered a new and mild selective method suitable both for common needs [192] and for complex reactions with fragile substrates. The reaction can be run in aqueous acetonitrile at room temperature using Pd(OAc)$_2$–TPPTS and applied to the cross-coupling of vinyl iodides (without the need for toxic TlOH) with an amine as a base (Scheme 6.47) [193].

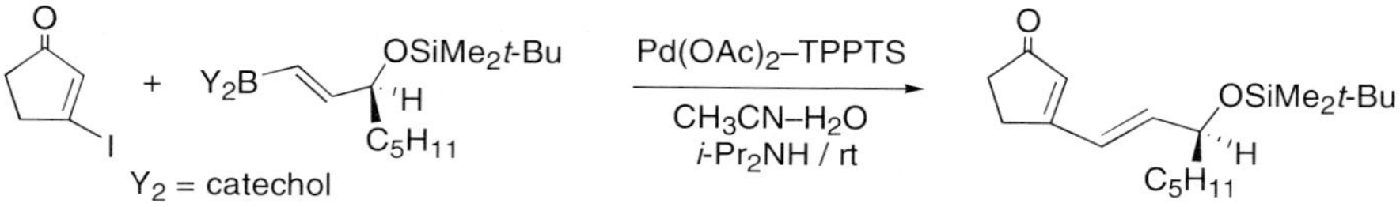

**Scheme 6.47**

Mapp and Heathcock used this procedure in the last step of the synthesis of myxalamide A, a polyene antibiotic (Scheme 6.48) [194]. This Pd(OAc)$_2$–TPPTS

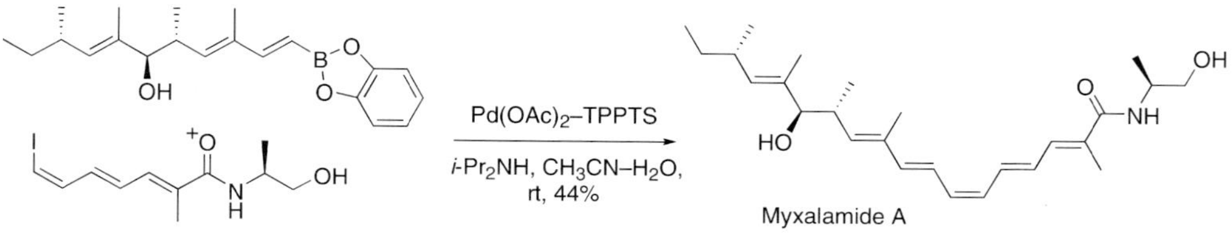

**Scheme 6.48**

system has also been extended to aryl bromides; good turnovers were observed and the catalyst can be recycled without loss of activity [195].

Shaughnessy and co-workers [196] reported an efficient synthesis of unprotected halonucleosides using Pd(OAc)2–TPPTS or Pd(OAc)$_2$–TXTPS catalysts (see formulae below): good TONs are observed (Scheme 6.49).

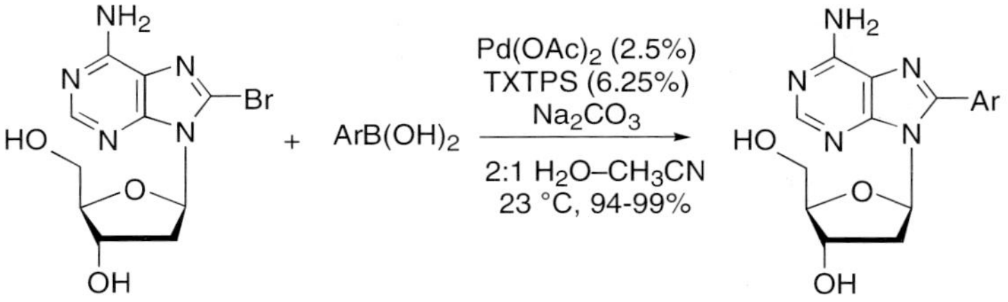

**Scheme 6.49**

Increasing attention has been devoted to the design of new hydrophilic catalysts. For example, the combination of tri(8-sulfonatodibenzofuran-2-yl)phosphane **1** with Pd(OAc)$_2$ gave a modestly active catalyst in S–M coupling [197]. The catalytic activity is somewhat lower than with TPPTS. Trialkylphosphane with quaternary ammonium substituents were found to give active catalysts, *t*-Bu-Pip-phos at room temperature for S–M coupling of aryl bromides. A catalyst derived from *t*-Bu-Amphos–Na$_2$PdCl$_4$ gave good yields in S–M coupling with hydrophobic and hydrophilic aryl bromides and was used in the synthesis of anti-inflammatory drugs [198]. Beller and Miyaura reported the preparation of a series of carbohydrate-modified ligands **4** and GLCAphos; these ligands gave active catalysts for the S–M reaction of aryl bromides and activated aryl chlorides [149g–h, 199]. Complete conversion was achieved with a Pd catalyst derivatized with GLCAphos using a low catalyst loading (TON = 95 000) in coupling of 4-bromoacetophenone at 80 °C in neat water [199].

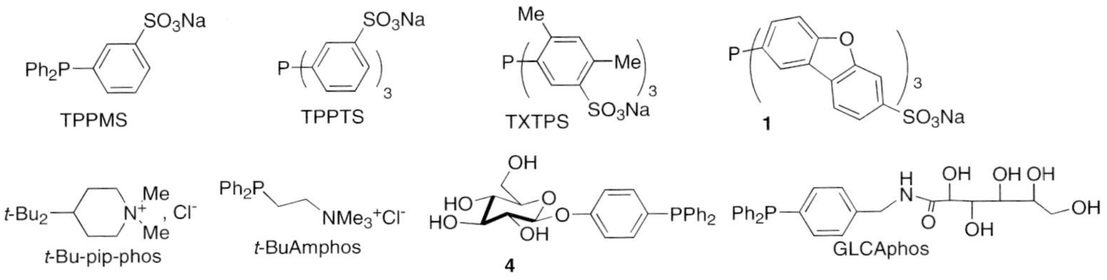

The presence of water is an important factor, which has led to the development of aqueous phosphane-free and biphasic phosphane-assisted techniques. The S–M reaction catalyzed by Pd(OAc)$_2$ phosphane-free in water proceeds in most cases with high catalytic efficiently (250 000 catalytic cycles) [200]. An interesting S–M coupling of benzylic halides and arylboronic acids using PdCl$_2$ in the absence of ligand has been reported [201]. This ligandless system has been employed in an elegant synthesis of conjugated rotaxanes and polyrotaxanes [202]. The addition of an additive such as tetrabutylammonium bromide (TBAB) significantly increases the efficiency of ligand-free S–M coupling for the coupling of hydrophilic substrates [203]. Suzuki coupling of activated aryl bromides catalyzed by PdCl$_2$–1,1′-*N*-disubstituted ferrocenyl ligand in aqueous medium gave good yields of biaryl products [204]. Another interesting phosphane-free S–M reaction was reported by Hirao and coworkers using Pd/C in aqueous $K_2CO_3$ solution; the catalyst can be recovered and reused several times [205]. Suzuki-type reactions in air and water have also been studied, with the first example by Li and co-workers using Pd/C as catalyst and more recently using a combination of $Pd(OAc)_2$ and tetramethylguanidine [206].

Several Pd-supported systems have been developed for Suzuki reactions in an aqueous medium (Scheme 6.50), such as a polystyrene N-heterocyclic Pd catalyst **9** [207], an N-anchored 2-aza-1,3-bis(diphenylphosphino)propane catalyst **10** [208], a silica-supported 1,2-bis(ethylthio)ethanepalladium **11** [209] and a mercaptopropyl-modified mesoporous silica SBA-SH-15 palladium **12** [210]; these catalysts can be recycled (Scheme 6.50).

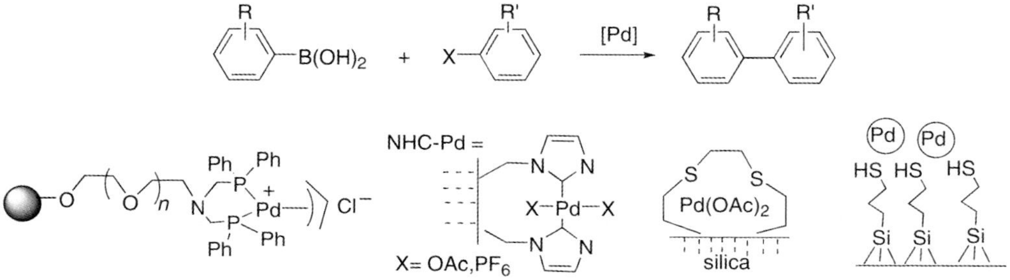

**Scheme 6.50**

Pd nanoparticles stabilized by poly(N-vinyl)-2 pyrrolidone, dodecanethiolate, and cross-linked resin capture nanoparticulate palladium are efficient catalysts for aryl iodides and bromides [211]. Reversed-phase glass beads have also been employed in Suzuki reactions to provide, in aqueous media, a route to diverse polar compounds in good yields and with low levels of palladium leaching [212]. Pd(II)-exchanged NaY zeolite showed high activity in the Suzuki cross-coupling reactions of aryl bromides and iodides without added ligands [213]. Sepiolite, a natural clay, has been used for S–M coupling of 4-bromophenol with phenylboronic acid in water in air at room temperature, giving high yields of product [214]. Other supports for palladium catalysts, such as mesoporus silica [215], aminophosphane–$Al_2O_3$ [216], Fibrecat-1000 [217], and metal–organic framework (MOF) [218] have also been used for Suzuki reactions in water. These materials are preserved throughout catalysis and are reusable.

A drawback of S–M coupling is the phase transfer arising from water-insoluble substrates. To circumvent this problem, the use of surfactants or mass transfer promoters has been explored [219]. Cyclodextrins or calixarenes possessing extended hydrophobic host cavities and surface-active properties were found to be very efficient as mass-transfer promoters for the palladium-mediated Suzuki cross-coupling reaction of 1-iodo-4-phenylbenzene and phenylboronic acid in an aqueous medium. The cross-coupling rates were up to 92 times higher than those obtained without addition of any compound [220].

Microwave conditions, which have been used extensively in homogeneous catalysis [221], have been employed in ligand-free palladium-catalyzed S–M reactions [222]. Leadbeater and co-workers reported that using either conventional or microwave heating, the S–M coupling could be promoted by palladium with very low loading (50–2.5 ppb of palladium), which could be supplied by the impurities in the base used ($Na_2CO_3$): this "no catalyst added" procedure has been optimized and offers new opportunities for S–M reactions [222]. This was originally thought to be a transition metal-free process [223].

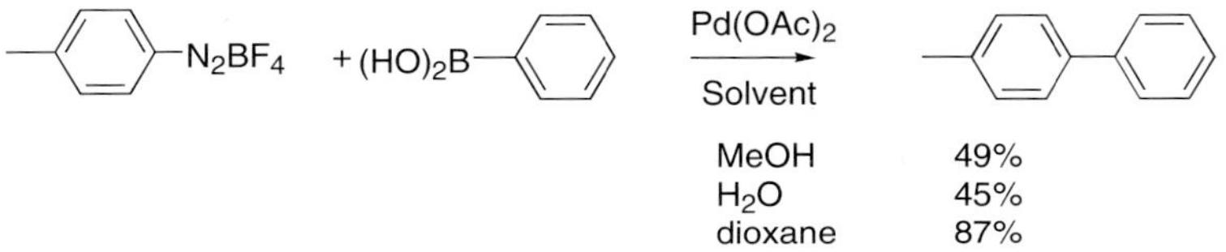

**Scheme 6.51**

In addition to aryl halides, vinyl halides, or triflates, it has been shown that the highly stable and non-explosive arene diazonium tetrafluoroborates are good partners in the S–M reaction. The reaction takes place under mild conditions at 20 °C in dioxane using $Pd(OAc)_2$ or Pd/C without addition of phosphane [224]. In protic solvents such as methanol or water, the yields are moderate (Scheme 6.51).

This S–M coupling was applied to functionalized electrophiles, the diazonium group being far more reactive than bromo or triflate functionalities [225]. This methodology has recently been applied for a practical synthesis of unsymmetrical terphenyls (Scheme 6.52) [226].

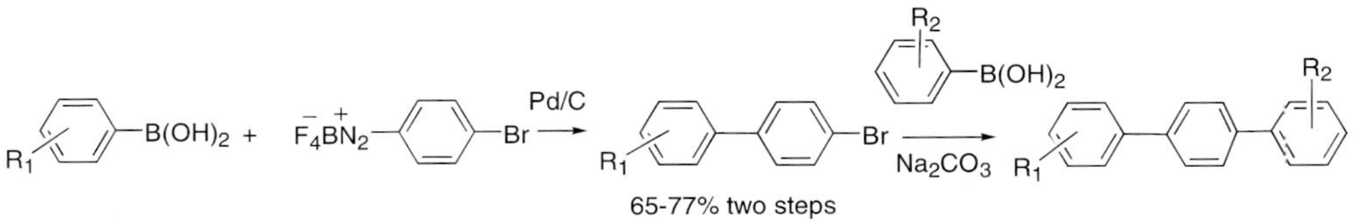

**Scheme 6.52**

It was also shown that this phosphane-free reaction of diazonium salts could be applied to boronate esters in aqueous medium, although in this case this procedure gave no distinct advantages over anhydrous organic solvents [227]. It was reported by Darses et al that highly stable potassium organoborates [87] readily react with aryl diazonium salts [228]. This palladium phosphane-free cross-coupling reaction is very efficient in methanol and without addition of base as used in the original

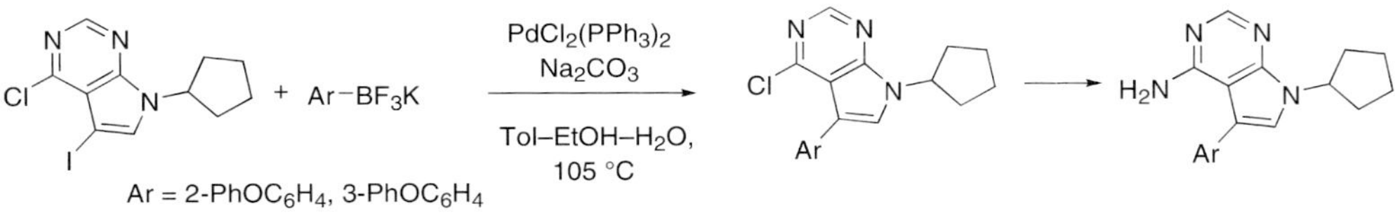

**Scheme 6.53**

Ar = 2-PhOC$_6$H$_4$, 3-PhOC$_6$H$_4$

protocol of the S–M reaction. In 1998, a patent by Knoll AG Chemische Fabriken (BASF) reported the cross-coupling reaction of an aryl halide with potassium aryltrifluoroborates in the presence of a base [229]. Indeed, the cross-coupling reaction of 5-iodopyrrolo[2,3-d]pyrimidine with potassium aryltrifluoroborates was efficiently catalyzed by PdCl$_2$(PPh$_3$)$_2$ in the presence of Na$_2$CO$_3$ as a base (Scheme 6.53). The resulting pyrrolo[2,3-d]pyrimidines were reported to be tyrosine kinase inhibitors and useful in treating proliferative diseases and disorders of the immune systems in mammals.

In 2000, a patent by Hofmann-La Roche described the introduction of a vinyl substituent on pyrimidine derivatives using potassium trifluoro(vinyl)borate [230]. The reaction conditions were optimized and it was found that the combination of PdCl$_2$(dppf) and an aliphatic amine in an alcoholic solvent showed the highest activity and the highest conversions. The conditions established proved to be general for the introduction of an alkenyl moiety on aryl halides or pseudo-halides. Indeed, Molander's group showed that under these conditions [PdCl$_2$(dppf), 2-PrOH–H$_2$O and t-BuNH$_2$], potassium alken-1-yltrifluoroborates effected the cross-coupling reaction with aryl iodides, bromides, and triflates, allowing efficient access to alken-1-yl-substituted aromatic compounds. The groups of Batey, Molander, and Buchwald used similar catalytic systems in the Pd-catalyzed cross-coupling of a wide range of potassium organotrifluoroborates with aryl halides and pseudo-halides in an alcoholic solvent or DME–water [231]. Recently, Molander and Jean-Gerard reported that alkyl β-aminoethyltrifluoroborates are efficient partners in S–M coupling reactions with aryl iodides, bromides, and triflates under basic conditions in toluene–water. This procedure allows the easy preparation of protected phenethylamines [232].

For a long time, an important limitation of the Suzuki–Miyaura reaction was the inefficiency of aryl chlorides. However, in view of the increased availability and decreased expense of aryl chlorides relative to aryl bromides and iodides, an efficient procedure for cross-coupling aryl chlorides was highly desirable. The groups of Fu and Buchwald reported that the use of Pd$_2$dba$_3$ in the presence of electron-rich phosphanes afforded the corresponding biaryl compounds in excellent yields [233]. The groups of Najera and Corma reported the use of hydrophilic oxime-derived palladacycles **13** and **14** for S–M coupling of aryl chlorides in refluxing water using low catalyst loadings (Scheme 6.54) [234]. Anderson and Buchwald have shown that water-soluble aryl chlorides and heterocyclic halides react in organic aqueous medium or neat water in the presence of a Pd(OAc)$_2$–water-soluble electron-rich phosphane **15** system [235].

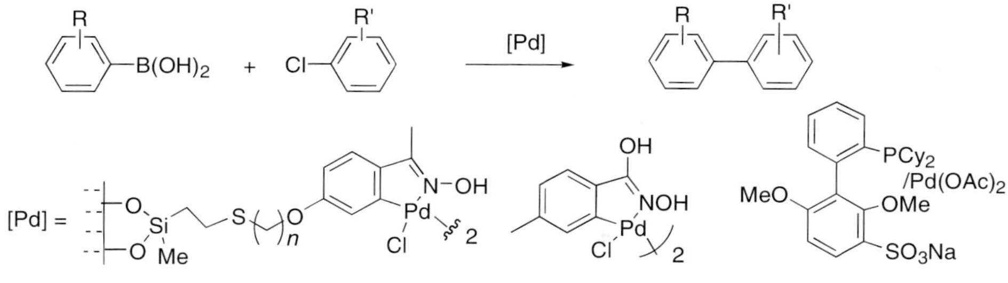

*6.4 Reaction of Organic Halides and Derivatives* | **179**

**Scheme 6.54**

Recently, several other palladium catalysts have been developed for cross-coupling of aryl chlorides in the S–M reaction, such as dendritic Buchwald-type catalysts [236], Pd-disulfonated 9-(3-phenylpropyl)-9′-PCyc$_2$-fluorene **16** [237], Pd-sulfonated water-soluble imidazolium **17** [238], *in situ*-generated Pd nanoparticles in PEG-400 [239], Pd-heterogeneous(*tert*-butylarylphosphino)polystyrene **18** [240], ferrocenyltetraphosphane **19**, Pd–ferrocenylamine **20**, palladacycles **21** [241], and pincer-type Pd catalysts bearing aminophosphane **22** [242].

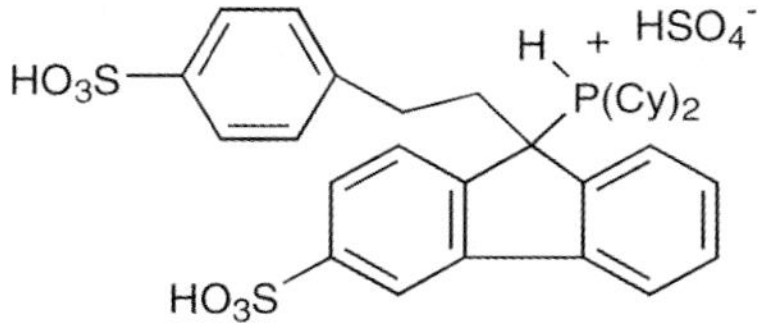

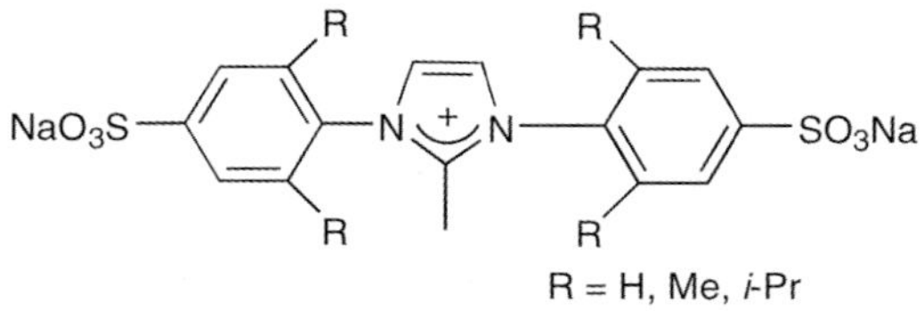

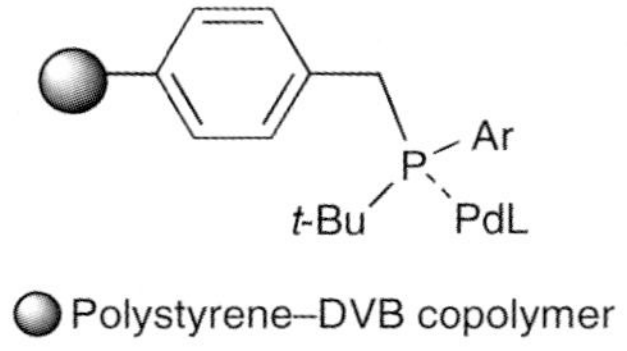

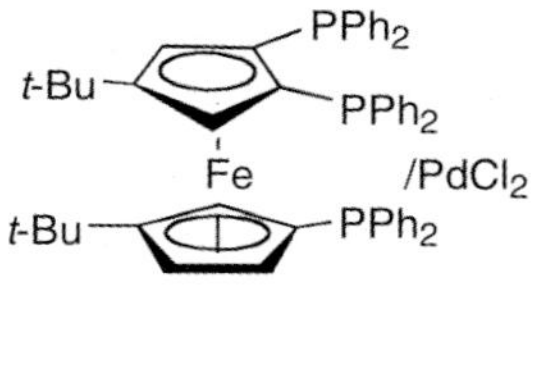

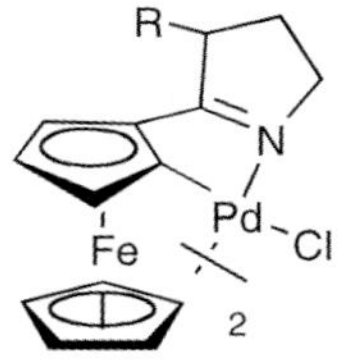

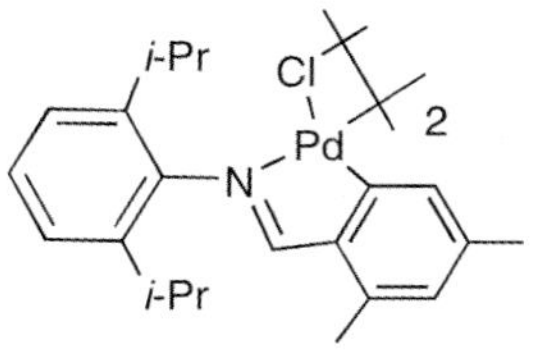

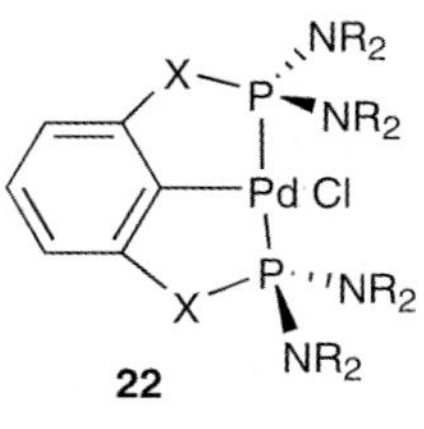

Lee *et al.* have developed a supramolecular reactor from self-assembly of rod–coil molecules in aqueous solution. This system is efficient for the S–M coupling of aryl halides, including activated aryl chlorides, with phenylboronic acids in the absence of organic solvents at room temperature [243]. Hor and co-workers recently reported a benzene imidazole-functionalized imidazolinium-based N-heterocyclic complex of Pd(II) active toward S–M coupling of aryl bromides and activated chlorides and boronic acids, giving a TON up to 11 750 at room temperature [244].

#### 6.4.2.2 Nickel- and ruthenium-catalyzed Reactions

The synthesis of biarylic compounds based on the S–M reaction is often based on palladium-catalyzed coupling between aryl-electrophiles (diazonium salts, triflates, aryl halides) and boronic acids. Again, the use of aryl chlorides remains challenging since they are generally less reactive towards oxidative addition of zerovalent palladium catalysts. Interestingly, Ni(0)-catalyzed cross-couplings of aryl chlorides and arylboronic acids have been reported using hydrophobic phosphane–nickel (DPPF) catalysis [245] and hydrophilic phosphane (TPPTS) biphasic assisted technology (Scheme 6.55) [246]. Unsymmetrical biaryls with both electron-withdrawing and electron-donating groups were obtained in good yields. This method allows easier and more cost-efficient synthesis of biaryl-functionalized compounds.

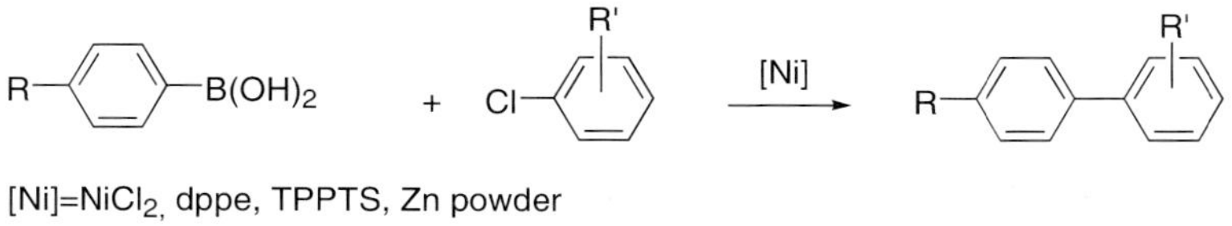

R—⟨⟩—B(OH)$_2$ + Cl—⟨⟩-R'  →[Ni]  R—⟨⟩—⟨⟩-R'

[Ni]=NiCl$_2$, dppe, TPPTS, Zn powder
**Scheme 6.55**

Saito and Fu established that commercially available 1,2-diamines may serve as an effective ligand for the Ni-catalyzed cross-coupling of unactivated alkyl bromides with alkylboranes at room temperature under basic conditions in t-BuOH–dioxane solvent [247].

RuCl$_2$(p-cymene)$_2$ immobilized on Al$_2$O$_3$ (1 mol% catalyst) has also been shown to be active for the S–M reaction of aryl iodides in DME–water (1:1) at 60–90 °C, furnishing products with high chemical yields and purity. It is assumed that the catalytically active species are ruthenium colloids with zero oxidation state. The ruthenium catalyst was recovered and reused in up to several runs with consistent efficiency [248].

### 6.4.3
### Stille Reaction

The organostannanes have also emerged as the reagents of choice in various transition metal-catalyzed reactions [9]. The basic technique for carrying out the Stille reaction required palladium complexes with phosphane ligands in anhydrous organic solvents, often at elevated temperatures. However, this reaction has been less investigated in alternative solvents [249]. Zhang and Daves found that the addition of small amounts of water to the system helps to increase the selectivity and yields in Stille coupling (Scheme 6.56) [250].

The main drawback in the Stille reaction is the utilization of only one of the four organic radicals, which leads to the formation of highly toxic waste R$_3$SnX. The monosubstituted RSnX$_3$, which are less volatile and less toxic, were known to be

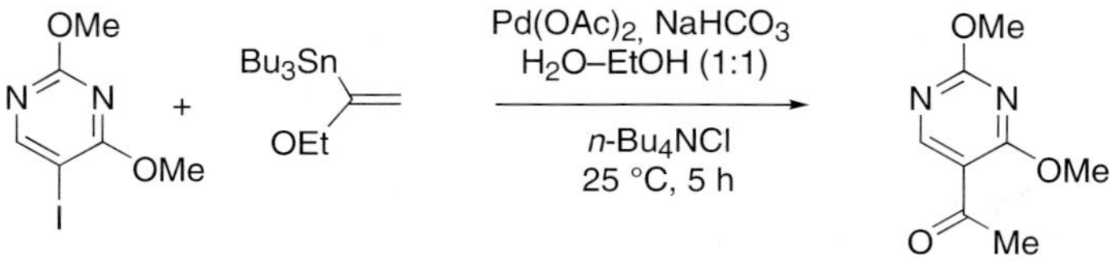

## Scheme 6.56

less reactive: the groups of Beletskaya and Collum have found that the cross-coupling of trichloroorganotin may be conducted with water-soluble aryl and vinyl halides in water [251]. The RSnCl$_3$ compounds are dissolved in basic aqueous conditions to give [RSn(OH)$_{3+n}$]$^{n-}$ stannate salts, which are fairly reactive in cross-coupling with a wide range of substrates. Coupling with p-iodobenzoic acid could be conducted either with or without water-soluble phosphanes (Scheme 6.57).

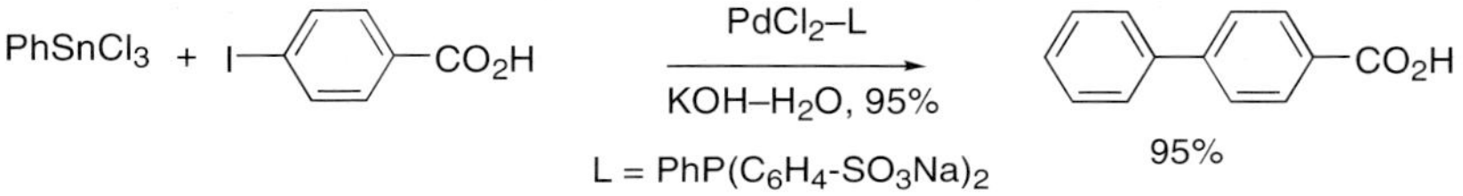

## Scheme 6.57

Aryldiazonium and aryliodinium salts are also cross-coupled in aqueous acetonitrile using Pd(OAc)$_2$ as catalyst to give unsymmetrical arenes in high yields [252]. Gallagher and co-workers reported a one-pot tandem Pd-catalyzed hydrostannylation–Stille coupling protocol for the stereoselective generation of vinyltin and their subsequent union with various electrophiles [253]. An efficient Stille cross-coupling reaction using a variety of aryl chlorides and bromides in neat water has been developed, using a palladium–phosphinous catalyst, [(t-Bu)$_2$POH]$_2$PdCl$_2$. This system allows the formation of biaryls from aryl halides in good yields: the recycling of this air-stable catalyst has also been demonstrated (Scheme 6.58) [254].

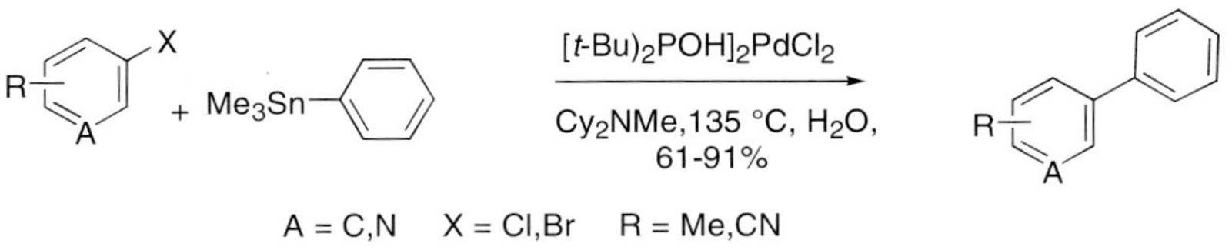

## Scheme 6.58

Li and co-workers have also reported a Stille reaction in water under an atmosphere of air [255].

### 6.4.4
**Sonogashira Reaction, Alkyne Oxidative Dimerization**

The coupling of terminal alkynes with organic halides or triflates, known as the Sonogashira reaction, is widely used in organic synthesis [8]. This Pd cross-coupling is commonly performed in an anhydrous solvent in the presence of tertiary amines with copper salts as cocatalysts. Casalnuovo and Calabrese reported that a combination of a water-soluble catalyst, [Pd(TPDMS)$_3$], and CuI promoted the reaction under very mild conditions in CH$_3$CN–water [189]. A demonstration of the synthetic utility of the palladium cross-coupling in an aqueous medium was illustrated in an alternative synthesis of T-505, a chain-terminating nucleotide reagents used in DNA sequencing and labeling. In the commercial syntheses of these derivatives, the acetylene coupling is carried out prior to the hydrophilic phosphate introduction. The convergent route developed by DuPont can be conducted in the final step using the water-soluble Pd(TPPMS)$_3$ catalyst (Scheme 6.59) [189].

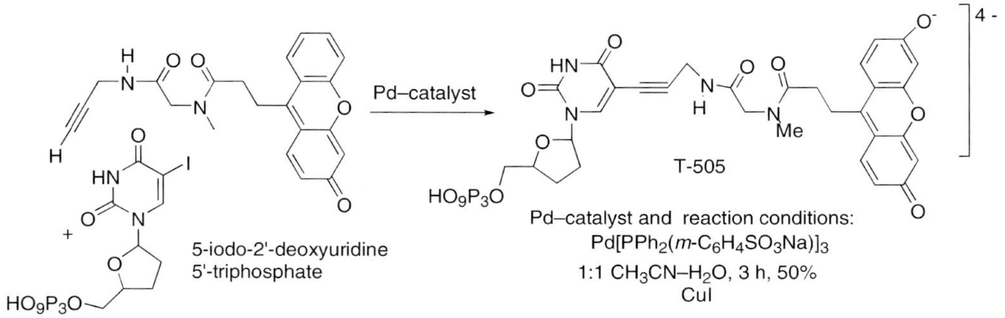

**Scheme 6.59**

Genêt and co-workers carried out more detailed studies and found that the Pd(OAc)$_2$–TPPTS system was also very efficient in homogeneous aqueous media. The reactions are performed at room temperature without copper salt [191]. The water-soluble Pd–TPPTS system has also proved its efficiency in diyne synthesis (Scheme 6.60) [192].

$$RC\equiv Cl \; + \; HC\equiv CR' \xrightarrow[\text{MeCN–H}_2\text{O, rt}]{\text{Pd(OAc)}_2, \text{ TPPTS, Et}_3\text{N}} RC\equiv C-C\equiv CR$$

R = n-Bu, R' = CMe$_2$OH ; R = Me$_3$Si, R' = CH(OH)CH$_2$n-Bu ;
R = Me$_3$Si, R' = CMe$_2$OH; R = CEt$_2$NH$_2$, R' = CH(OH)CH$_2$n-Bu

**Scheme 6.60**

The Pd–TPPTS copper-free system has also been employed for the coupling of 2-iodophenols and 2-iodoanilines with terminal alkynes, giving the corresponding benzofurans or indoles in good yields [191]. More recently, this transformation has been accomplished using Pd/C in the presence of copper salts–PPh$_3$ and prolinol in water at 80 °C [256]. A wide range of 2-aryl/alkyl-substituted benzo[b]furans/nitrobenzo[b]furans were obtained in yields ranging from 68 to 88% (Scheme 6.61).

This methodology has been applied as a key step in the synthesis of eutypine, an antibacterial substance isolated from culture medium of *Eutypa lata* [191]. Recently, Michelet, Genêt and co-workers reported the use of the carboxylated ligand m-TPPTC in Sonogashira coupling [149d, 153d]. Interestingly, in contrast to the Pd–TPPTS system, the use of Pd(OAc)$_2$ in combination with m-TPPTC affords linear

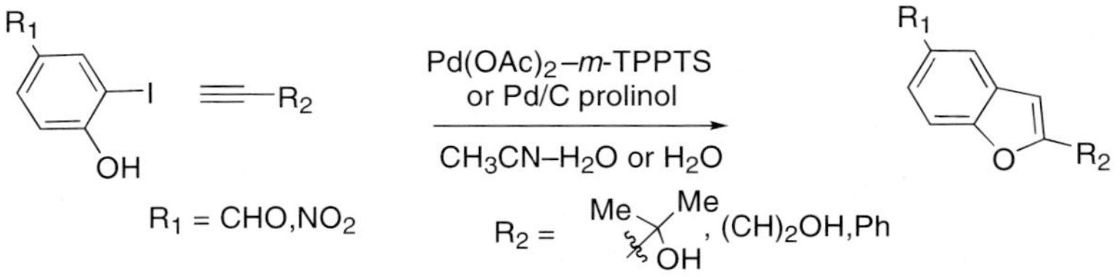

## Scheme 6.61

products (Scheme 6.62). This behavior could be explained by the higher basicity of m-TPPTC compared with its sulfonated analog. This methodology has been applied for the preparation of novel fluorophores [257].

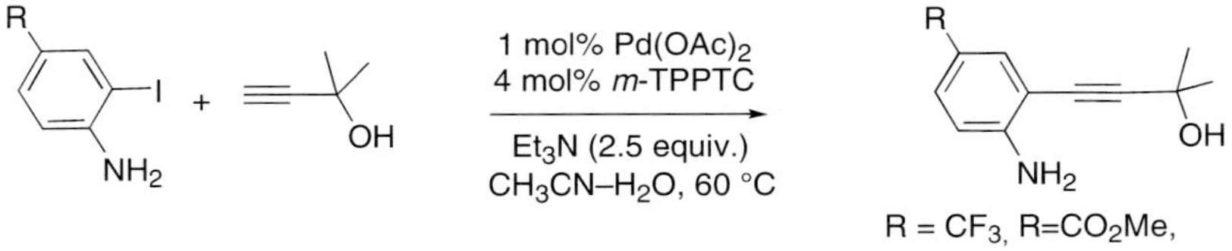

## Scheme 6.62

Beletskaya and co-workers have shown that the coupling of water-soluble aromatic iodides and propargyl alcohols can be performed in water in the presence of both $Ph_3P$ and TPPTS ligands with CuI as additive (Scheme 6.63) [258]. They also established that the participation of hydrophilic phosphane is not necessary for the coupling of terminal alkynes with iodoarenes, as the reaction can be performed by the water-insoluble palladium complex $PdCl_2(PPh_3)_2$ in the presence of $K_2CO_3$ and $(n\text{-Bu})_3N$ in water [259]. It was also shown that diaryliodinium salts are more reactive than aryl iodides under the same conditions as the reaction takes place within 10 min at ambient temperature.

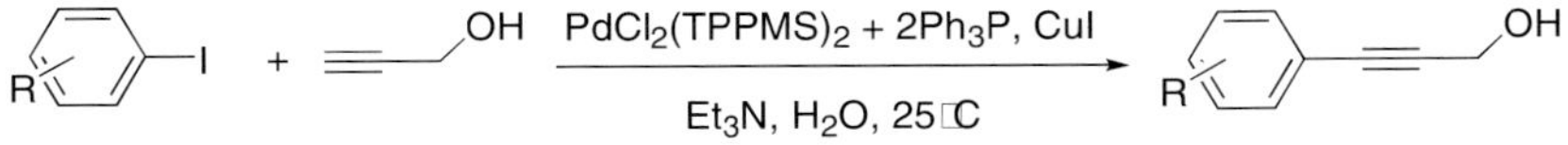

R = p-COOH, m-COOH, p-$NO_2$, p-MeCO

## Scheme 6.63

Li and co-workers reported a copolymerization of diiodobenzoic acid with acetylene using $Pd(OAc)_2$–TPPTS in aqueous acetonitrile [260]. This system is also efficient for bis-coupling of acetylene gas with aryl diiodides under basic conditions, providing a high molecular weight (~60 000) zig-zag areneethynylene polymer (Scheme 6.64) [261].

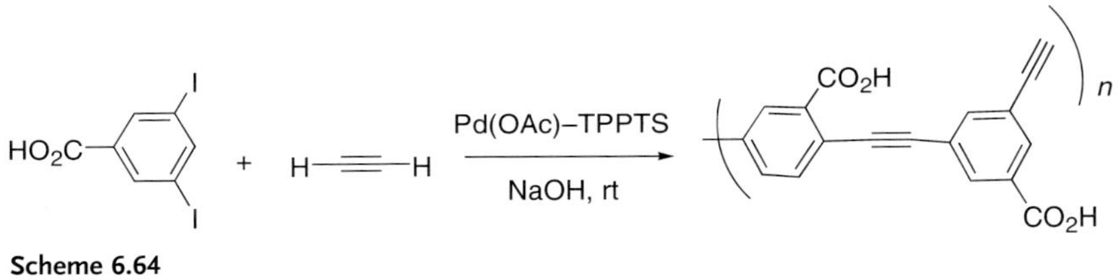

## Scheme 6.64

Cationic phosphanes with guanidinium groups such as TPPDG in combination with palladium were also developed by Stelzer and co-workers for cross-coupling reactions of water-soluble acetylenes with aryl iodides [262]. Other ligands such as m-TPPTC and ligands **15** and **23** were also efficient.

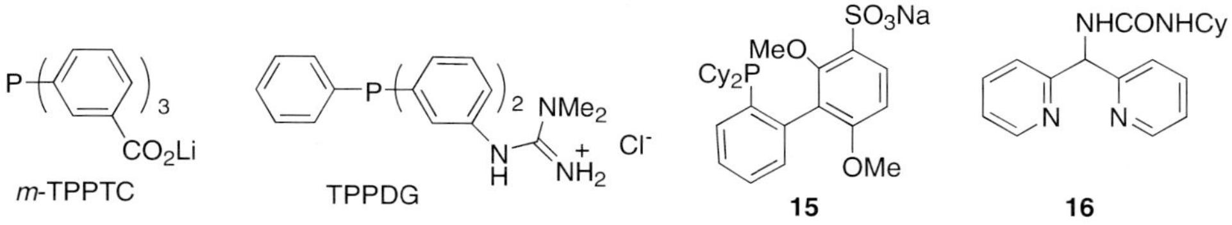

m-TPPTC

TPPDG

**15**

**16**

These catalytic systems using the Sonogashira reaction have found interesting applications in bioconjugation reactions. For example, Dibowski and Schmidtchen further demonstrated the coupling of propargyl glycine to an 11-residue peptide containing iodophenylalanine (Scheme 6.65) [263].

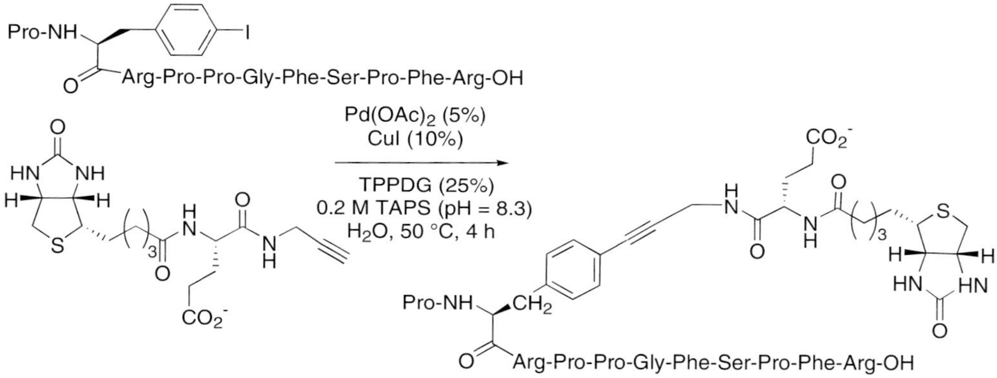

**Scheme 6.65**

The authors advocated the use of guanidinylated phosphanes over sulfonated ligands, suggesting that the net negative charge borne by many proteins would inhibit the approach of the negatively charged palladium complex, limiting the coupling efficiency. However, Bong and Ghadiri [264] have demonstrated the utility of $Pd(OAc)_2$–TPPTS system [191, 192] to construct protein in aqueous medium. A tricoupling of highly charged peptides of considerable length, from 17 to 33 residues, with a trialkyne by using a water-soluble sulfonated Pd(0) catalyst in water, giving large structures (MW 12 000) very efficiently (Scheme 6.66) [264].

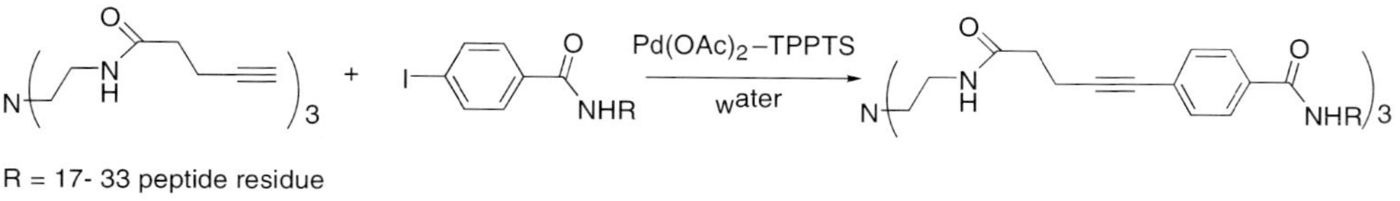

R = 17- 33 peptide residue

**Scheme 6.66**

Palladium/charcoal with copper additives could also serve as catalysts in aqueous media for Sonogashira coupling [265]. Granja and co-workers used Pd/C in combination with 4-triphenylphosphinobenzoic acid (4-DPPBA) ligand to couple peptides containing alkynes with heteroaryl electrophiles in aqueous DMF [266].

Sinou and co-workers reported a Pd-catalyzed copper-free coupling of terminal alkynes with aryl halides under mild conditions in the presence of quaternary ammonium salt in aqueous acetonitrile [267]. Following this report, the combination of Pd(PPh$_3$)$_4$ or PdCl$_2$(PPh$_3$)$_2$ with copper iodide was reported to be effective for Sonogashira coupling without the need for any additives [268]. The Pd/Cu system was also found to be efficient for the preparation of enyne derivatives [269]. Interestingly, hypervalent iodine compounds have also been coupled with terminal alkynes using a phosphane-free catalyst to give diynes and enynes. Remarkable chemoselectivity was found in the Pd-assisted coupling of an alkenyl iodo salt bearing an additional triflate moiety at the β-position; the triflate group remained intact in the coupling (Scheme 6.67) [270].

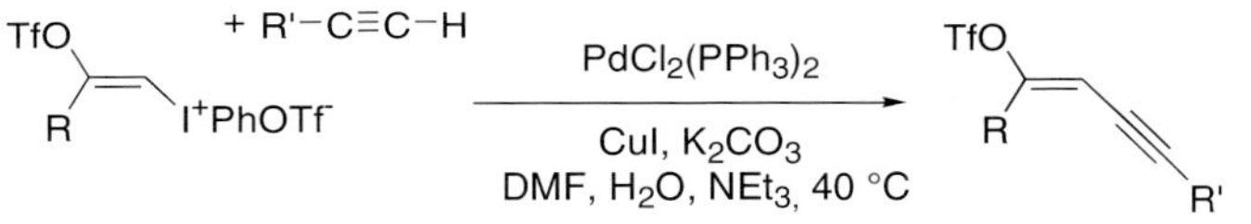

R = H, Ph, nBu, R' = SiMe$_3$, Ph, CH$_2$Cl$_2$, etc.

**Scheme 6.67**

Over the last several years, a number of reactions of terminal alkynes in copper-free Sonogashira coupling in aqueous medium have been reported. Anderson and Buchwald reported a Sonogashira coupling with aryl chlorides using the combination of Pd(II) salts and a basic sulfonated ligand **15** [235]. The reaction takes place in H$_2$O–CH$_3$CN (1:1) at 60–80 °C, giving the corresponding alkynes in good yields. This new system represents an interesting alternative for Sonogashira coupling of fragile substrates such as electron-deficient propiolate acids, which have a propensity to polymerize in the presence of palladium catalysts.

A palladium catalyst with a nitrogen ligand, (dipyridin-2-ylmethyl)amine **23**, has been developed in water with pyrrolidine as base and tetra-n-butylammonium as additive with high TON for the coupling of terminal alkynes with aryl iodides and bromides [271]. A mild protocol using PdCl$_2$ with pyrrolidine under aerobic conditions has been developed for coupling aryl iodides and terminal acetylenes [272]. An interesting application of the Sonogashira reaction is the coupling of cationic porphyrins using Pd(OAc)$_2$–TPPTS in water (Scheme 6.68) [273].

Beletskaya et al. reported an inexpensive alternative to the palladium-catalyzed reaction using an Ni(PPh$_3$)$_2$Cl$_2$ complex and CuI in a dioxane–water mixture [274]. The catalyst system gave high yields of the coupled products between functionalized arylacetylenes and aryl/heteroaryl iodides (Scheme 6.69).

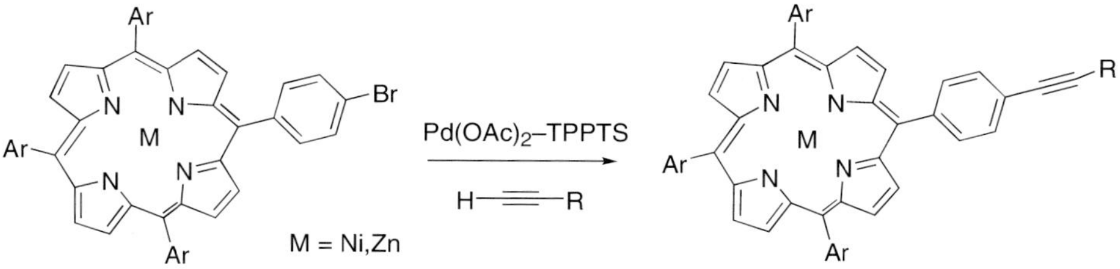

**Scheme 6.68**

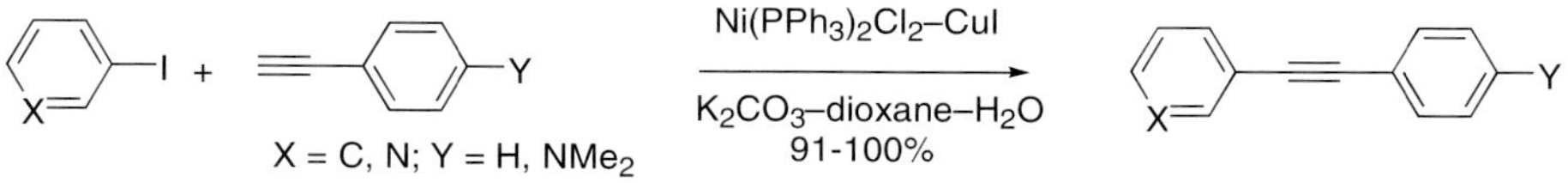

**Scheme 6.69**

### 6.4.5
### Tsuji–Trost Reaction

Allylic substitutions, well known as the Tsuji–Trost reaction, is also one of the most widely used palladium-catalyzed reactions in organic syntheses [6]. The first examples were reported by Safi and Sinou [275] and Genêt et al. [192] using Pd(0) catalysts generated *in situ* from Pd(OAc)$_2$ or Pd$_2$(dba)$_3$ and TPPTS ligand. The design of recyclable systems was possible using water-insoluble nitriles (butyronitrile or benzonitrile) in allylic substitutions [276]. Very easy separation of the products from the catalyst is also efficient when the reaction is performed under biphasic conditions [5]. The water-soluble catalysts are efficient for substitution of allylic substrates with compounds bearing active hydrogen (α-nitro esters, malonates, acetylacetone, etc.) and heteroatom nucleophiles (primary and secondary amines, hydroxylamine, sodium tolylsulfinates, azides, uracils) under mild conditions (Scheme 6.70). The substitution of an allylic alcohol is possible by carbon nucleophiles in water using [PdCl(η$^3$-C$_3$H$_5$)]$_2$–TPPTS catalyst without the help of any activating reagents. The role of water in the reaction was suggested to be to promote the hydration of the hydroxyl group for generation of the η$^3$-allylpalladium species [277]. Palladium-catalyzed allylic substitutions of allylic alcohols have also been realized in water using tetrakis(triphenylphosphane)–palladium catalysts in the presence of carboxylic acids [278].

Moreno-Manas and co-workers reported a remarkable selectivity in the Pd-allylation of uracil and thiouracil [279]. In DMSO, the allylic substitution of cinnamyl acetate in the presence of Pd(OAc)$_2$–PPh$_3$ gave a mixture of substituted products at N-1 and

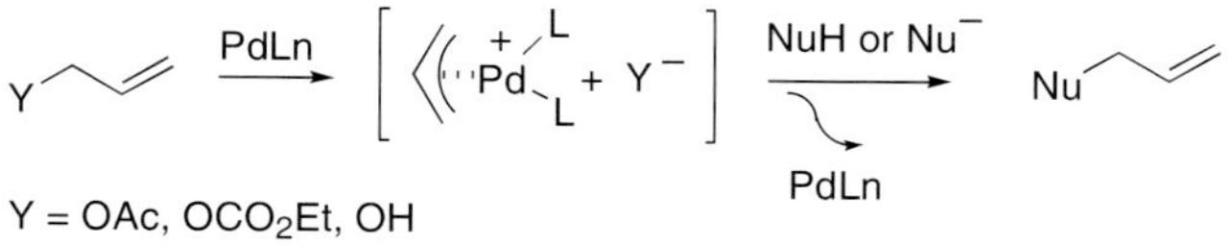

Y = OAc, OCO$_2$Et, OH
**Scheme 6.70**

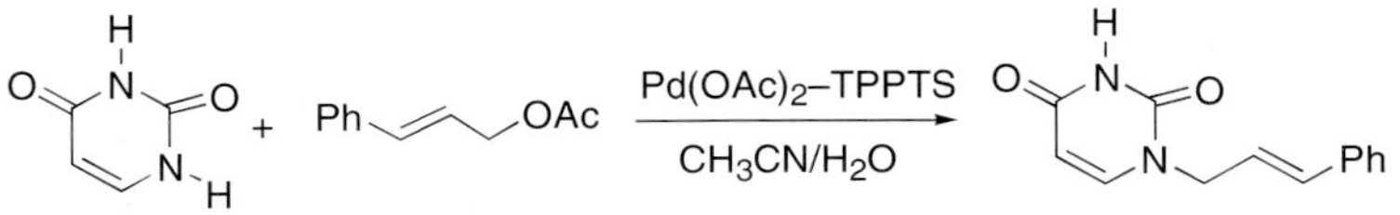

## Scheme 6.71

N-3 whereas the reaction in aqueous acetonitrile catalyzed by a Pd–TPPTS complex gave selectively the $N^1$-cinnamyluracil compound in 90% yield (Scheme 6.71).

An interesting application of palladium-catalyzed alkylation in homogeneous (MeCN–H$_2$O) or biphasic media (n-PrCN–H$_2$O) is the removal of the allyl- and alkyloxycarbonyl group (Alloc) from allylic esters and carbamates in the presence of Et$_2$NH as an allyl scavenger [279]. The Pd(0) catalyst can be recycled 10 times in the butyronitrile–water biphasic system [280]. Fast and chemoselective deprotections of primary and secondary alcohols and amines occur smoothly, as shown in (Scheme 6.72) [281].

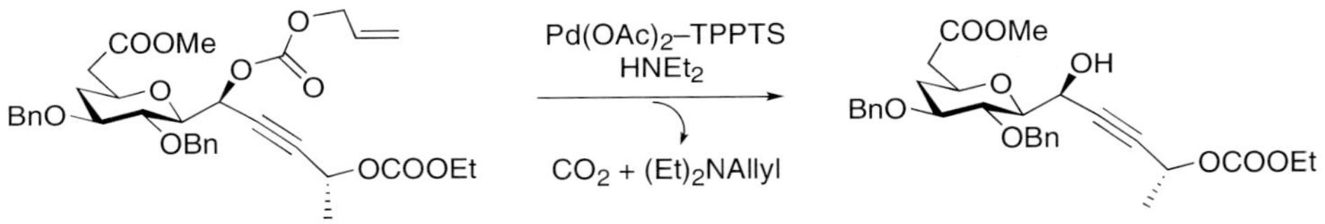

## Scheme 6.72

A very interesting chemoselective deprotection has been found using Pd–TPPTS catalyst and diethylamine as an allyl scavenger (Scheme 6.73) [282]. This methodology has been applied using biphasic conditions to the synthesis of tetrapeptides [283]. Modified cyclodextrins have also been reported to influence selectivity in the cleavage of allylic carbonates in aqueous medium [284].

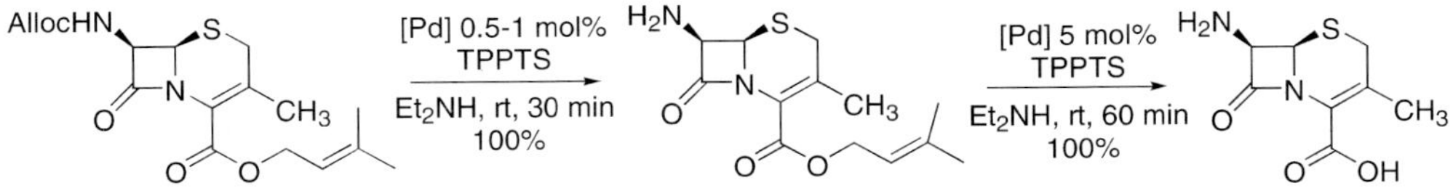

## Scheme 6.73

An efficient synthesis of medium- and large-sized lactones in an ethyl acetate–aqueous organic biphasic system using Pd(TPPTS)$_n$ catalyst has been reported by Oshima and co-workers [285] (Scheme 6.74). In pure water the desired lactone is obtained in only a 1% yield.

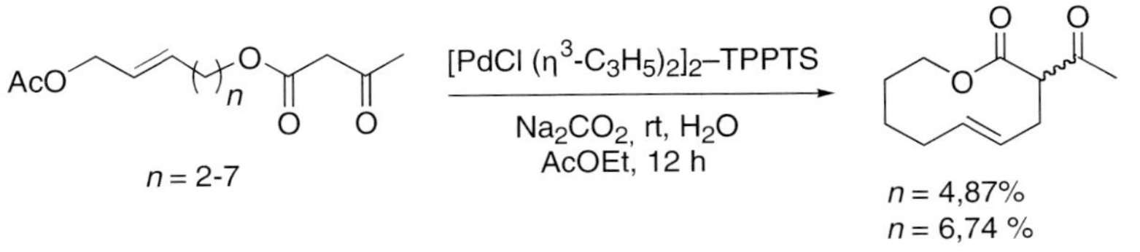

## Scheme 6.74

Other catalytic systems have been tested in aqueous organic solvents and neat water, such as Pd/C [286], Pd(OAc)$_2$–TPPTS supported on silica [287], glass beads [288], and colloidal dispersions [289]. These catalysts are very robust and usable in Pd-catalyzed processes. Water-soluble polymer-bound Pd(0)–phosphane catalyst **24** can be used and recycled several times in pure water or an aqueous organic solvent in nucleophilic substitutions [290]. Hayashi and co-workers have developed efficient amphiphilic palladium–phosphane complexes bound to poly(ethylene glycol) **25**; good yields were achieved with a Pd loading of 2 mol% [291]. A catalyst derived from the tedicyp–($\eta^3$-allyl)PdCl$_2$ system has been found to be highly active for the Pd-allylic amination of allylic acetate in neat water. TONs up to 980 000 have been achieved [292].

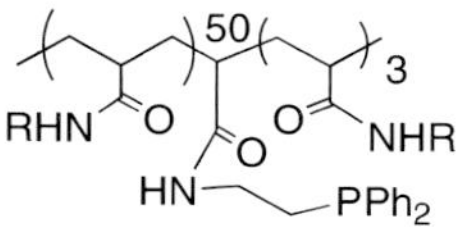

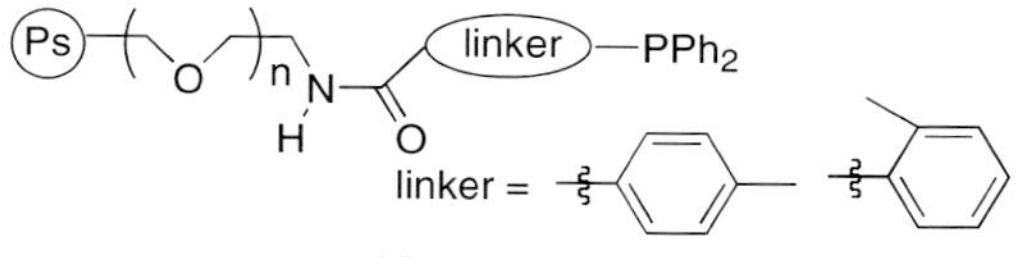

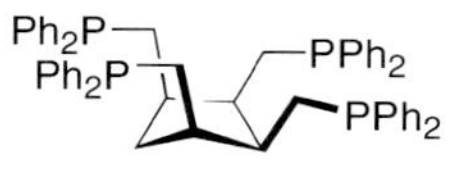

Asymmetric palladium-catalyzed alkylation is now a well-established methodology in organic synthesis [6]. Sinou and co-workers have reported an efficient alkylation of 1,3-diphenyl-3-acetoxyprop-1-ene with dimethyl malonate in water in the presence of surfactant, K$_2$CO$_3$, and BINAP as chiral ligand with enantioselectivity up to 93% [293]. Hayashi and co-workers designed an amphiphilic resin-supported MOP ligand, which was also an effective catalyst for allylic substitution with carbonucleophiles under aqueous K$_2$CO$_3$ conditions [294]. An interesting example of Pd-allylic amination of a cyclic allylic carbonate has been reported using a chiral polystyrene–poly(ethylene glycol)-supported catalyst **26** (Scheme 6.75) [295]. Cycloalkenyl substrates were aminated with dibenzylamine in water to give the corresponding amines with *ee* up to 98%. The catalyst was recycled several times without loss of enantioselectivity.

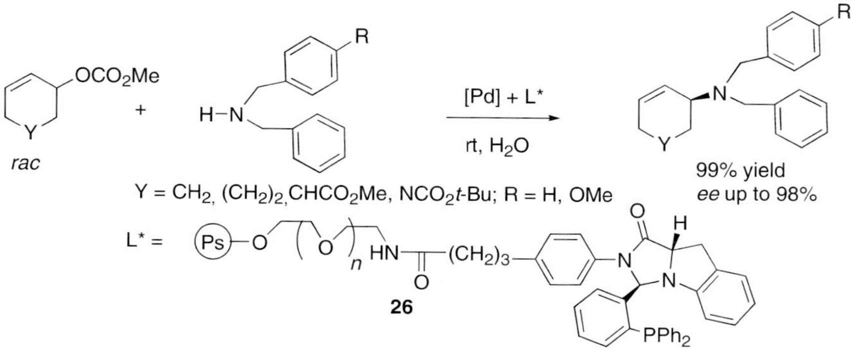

Scheme 6.75

### 6.4.6
### Hartwig–Buchwald Coupling

The use of the Hartwig–Buchwald reaction for palladium-catalyzed N–C coupling between amides/amines and aryl halides is becoming important in both academic research and industrial applications [10, 296]. It has been observed that the addition of water is beneficial. This has been reported in Pd-amination of aryl chlorides, bromides [297], and sulfonates (Scheme 6.76) [298].

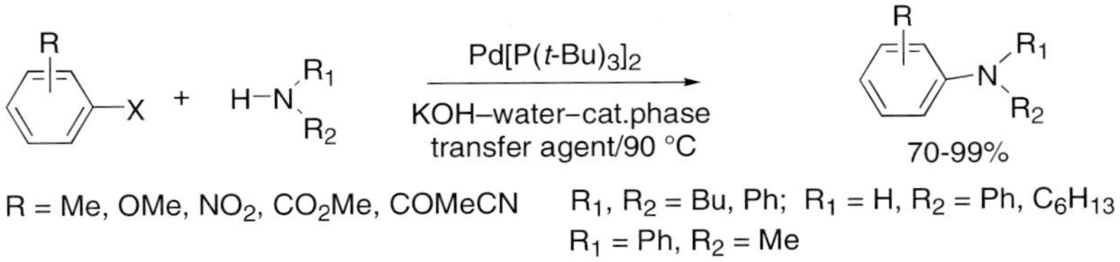

$$Ar-X + H-NR_1R_2 \xrightarrow[\text{transfer agent/90 °C}]{\text{Pd[P(}t\text{-Bu)}_3]_2 \\ \text{KOH–water–cat.phase}} Ar-NR_1R_2$$

R = Me, OMe, $NO_2$, $CO_2Me$, COMeCN

$R_1$, $R_2$ = Bu, Ph; $R_1$ = H, $R_2$ = Ph, $C_6H_{13}$
$R_1$ = Ph, $R_2$ = Me

70-99%

**Scheme 6.76**

Buchwald and co-workers reported a copper-catalyzed coupling of primary amides with aryl iodides using cesium carbonate, which could be efficient in the presence of water [299]. Yin *et al.* also found that N-arylation of aminothiazoles could be improved by addition of 1 equiv. of water [300].

Dallas and Gothelf reported a more detailed study of coupling reactions between amide and aryl derivatives in aqueous organic solvents using $Pd_2(dba)_3$–xantphos catalyst in the presence of $Cs_2CO_3$ as base [301] (Scheme 6.77). This system is very tolerant and beneficial to the addition of water, giving high yields of the corresponding amides.

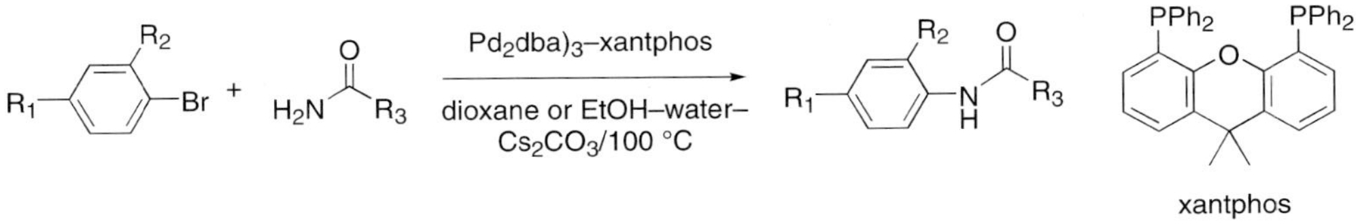

$$ArBr + H_2N-C(O)R_3 \xrightarrow[\text{dioxane or EtOH–water–}Cs_2CO_3/100\,°C]{Pd_2(dba)_3\text{–xantphos}} Ar-NH-C(O)R_3$$

xantphos

**Scheme 6.77**

### 6.4.7
### Hiyama Reaction

Organometallic compounds containing a main group metal such as Si, Cd, and Hg are also good partners in transition metal coupling reactions with organic halides. Over the years, Hiyama and others have explored the possibility of silicon-based cross-coupling employing organosilicon compounds for synthetic purposes [11]. Biaryls were obtained in good yields by reacting diphenyldifluorosilane with aryl halides in aqueous DMF at 120 °C in the presence of a catalytic amount of $PdCl_2$ (Scheme 6.78) [302]. Phosphane ligands are not necessary for the reaction; $Pd(OAc)_2$ and Pd/C are effective as catalysts in the open air and in water [303].

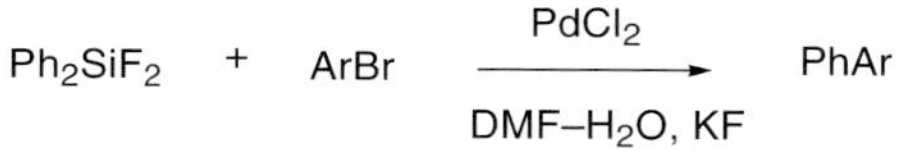

Ph$_2$SiF$_2$ + ArBr $\xrightarrow[\text{DMF–H}_2\text{O, KF}]{\text{PdCl}_2}$ PhAr

**Scheme 6.78**

Hiyama and co-workers reported cross-coupling of triallyl(aryl)silanes with aryl bromides and chlorides in water using ($\eta^3$-C$_3$H$_5$)PdCl$_2$ catalyst in the presence of electron-rich monophosphanes (Scheme 6.79) [304]. The high efficiency of the reaction is attributed to the spontaneous cleavage of the allyl groups on silicon upon treatment with TBAF. Water may provide an appropriate silicate species that eases the transmetalation step.

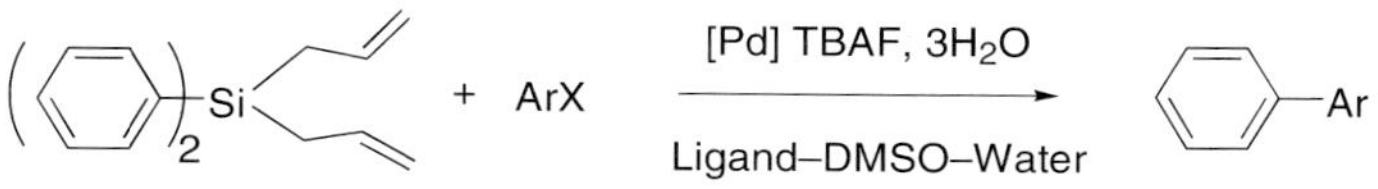

(Ph-Si≤)$_2$ + ArX $\xrightarrow[\text{Ligand–DMSO–Water}]{\text{[Pd] TBAF, 3H}_2\text{O}}$ Ph-Ar

**Scheme 6.79**

Wolf and Lerebours developed a palladium–phosphinous acid-catalyzed NaOH-promoted cross-coupling of arylsiloxanes with aryl chlorides and bromides in water [305]. This Pd-catalyzed reaction is compatible with various functional groups and affords biaryls in up to 99% yields. The coupling does not require additives such as surfactants or organic cosolvents and proceeds under air. Najera and co-workers also described the Hiyama coupling reaction of arylsiloxanes and vinylalkoxysilanes with aryl chlorides under solvent- and fluoride-free conditions using oxime-derived palladacycles **27** and **28** under heating (120 °C) or microwave irradiation (Scheme 6.80) [306].

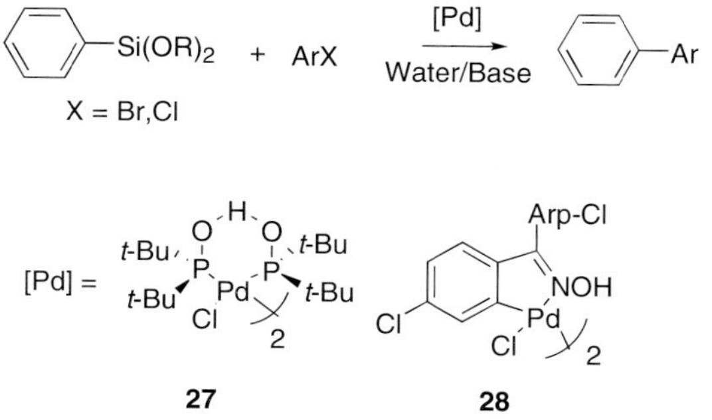

Ph-Si(OR)$_2$ + ArX $\xrightarrow[\text{Water/Base}]{\text{[Pd]}}$ Ph-Ar

X = Br,Cl

[Pd] = 

**27**: [t-Bu(t-Bu)P(O)-H···O-P(t-Bu)(t-Bu)]Pd-Cl dimer

**28**: oxime-palladacycle with Ar$_p$-Cl, NOH, Pd, Cl (dimer)

**Scheme 6.80**

Recently, symmetric and unsymmetric 1,2-diarylethenes were synthesized from aryl bromides by consecutive one-pot Hiyama–Heck reactions in water and under air using Pd(OAc)$_2$ with NaOH as additive and poly(ethylene glycol) (Scheme 6.81) [307].

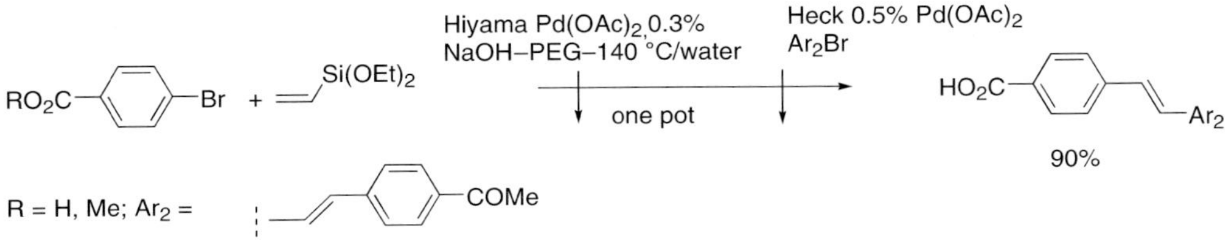

**Scheme 6.81**

Reagents shown: RO$_2$C–C$_6$H$_4$–Br + Si(OEt)$_2$ with R = H, Me; Ar$_2$ = CH=CH–C$_6$H$_4$–COMe. Conditions: Hiyama Pd(OAc)$_2$ 0.3%, NaOH–PEG–140 °C/water; then Heck 0.5% Pd(OAc)$_2$, Ar$_2$Br, one pot → HO$_2$C–C$_6$H$_4$–CH=CH–Ar$_2$, 90%.

## 6.5 Conclusion

This chapter demonstrates how it is being recognized that water as a medium can promote various reactions for carbon–carbon and carbon–heteroatom bond formation. During the past two decades, the field of aqueous catalyzed reactions has been growing very rapidly. The types of reactions that can be carried out in water are as diverse as those in non-aqueous conditions. New reactivities have been discovered by using water as a solvent. The opportunities or pursuing unconventional chemical reactivities provide the driving force for future innovation for academic and industrial applications in synthetic organic chemistry.

One of the methods to solve the main problem of homogeneous catalysis, namely the separation and recycling of the catalyst, is to partition the catalyst and the product into two separate and immiscible phases. Significant progress has been made in the development of aqueous-phase catalysts by varying the central atom of the catalyst, the ligand (hydrophilic or supported), and the phase in which it is used. Thus, the catalyst can be separated from organic products and sometimes recycled several times without loss of activity and low leaching. However, continued development and design of novel hydrophilic and supported catalysts are needed to provide improved catalytic activity. In this respect, catalyzed aqueous syntheses will provide more efficient synthetic routes, increasing product selectivity and reducing volatile organic consumption, and will lead to more sophisticated fields of chemistry under the most environmentally friendly conditions.

## References

1 (a) Breslow, R. (1991) *Acc. Chem. Res.*, **24**, 159; (b) Breslow, R. (2004) *Acc. Chem. Res.*, **37**, 471.

2 For recent books, see: (a) Li, C.J. and Chan, T.H. (1997) *Organic Reactions in Aqueous Media*, Wiley, New York; (b) Grieco, P.A. (ed), (1998) *Organic Synthesis in Water*, Thomson Science, Glasgow; (c) Cornils, B. and Hermann, W.A. (eds.),(1998) *Aqueous-Phase Organometallic Catalysis*, Wiley-VCH Verlag GmbH, Weinheim; (d) Lubineau, A. and Augé, J. (1999) *Modern Solvent in Organic Synthesis. Topics in Current Chemistry*, Vol. 206 (ed. P. Knochel), Springer, Berlin, p. 1; (e) Sinou, D. (1999) *Modern Solvent in Organic Synthesis. Topics in Current Chemistry*, Vol. 206 (ed. P. Knochel), Springer-Verlag, Berlin, p. 41.

3. For selected reviews, see: (a) Li, C.J. (2005) *Chem. Rev.*, **105**, 3095; (b) Pinault, N. and Bruce, D.W. (2003) *Coord. Chem. Rev.*, **241**, 1; (c) Lindström, U.M. (2002) *Chem. Rev.*, **102**, 2751; (d) Li, C.J. (1993) *Chem. Rev.*, **93**, 2023; (e) Li, C.-J. and Chan, T.H. (1999) *Tetrahedron*, **55**, 11149 (f) Lubineau, A., Augé, J. and Queneau, Y. (1994) *Synthesis*, 741; (g) Herrmann, W.A. and Kohlpaintner, C.W. (1993) *Angew. Chem. Int. Ed. Engl.*, **32**, 1524.

4. For recent reviews and monographs, see: (a) Beletskaya, I.P. and Cheprakov, A.V. (2000) *Chem. Rev.*, **100**, 3009 (b) Withcombe, N., Hii (Mimi), K.K. and Gibson, S. (2001) *Tetrahedron*, **57**, 7449; (c) de Vries, J.G. (2001) *Can. J. Chem.*, **79**, 1086; (d) Larhed, M. and Hällberg, A. (2002) *Handbook of Organopalladium Chemistry for Organic Synthesis* (eds E. Negishi, and A. de Meijere), John Wiley & Sons, Ltd, Chichester, p 1133; (e) Diederich, F. and de Meijere, A. (eds) (2004) *Metal-Catalyzed Cross-Coupling Reactions*, 2nd edn, Wiley-VCH Verlag GmbH, Weinheim; (f) Beller, M., and Bolm, C. (eds) (2004) *Transition Metals for Organic Synthesis, Building Blocks and Fine Chemicals*, 2nd edn, Wiley-VCH Verlag GmbH, Weinheim; (g) Tsuji, J. (2004) *Palladium Reagents and Catalysis*, John Wiley & Sons Ltd, Chichester; (h) Alonso, F., Beletskaya, I.P. and Yus, M. (2005) *Tetrahedron*, **61**, 11711; (i) Polshettiwar, V. and Molnár, A. (2007) *Tetrahedron*, **63**, 6949.

5. (a) Herrerias, C.I., Yao, X., Li, Z. and Li, C.J. (2007) *Chem. Rev.*, **107**, 3095; (b) Shaughnessy, K.H. (2006) *Eur. J. Org. Chem.*, 1827; (c) Shaughnessy, K.H. and DeVasher, R.B. (2005) *Curr. Org. Chem.*, **9**, 585; (d) Li, C.J. (2002) *Acc. Chem. Res.*, **35**, 533; (e) Chen, L. and Li, C.J. (2006) *Adv. Synth. Catal.*, **348**, 1459; (f) Genêt, J.-P. and Savignac, M. (1999) *J. Organomet. Chem.*, **576**, 305; (g) Genêt, J.-P., Savignac, M. and Lemaire-Audoire, S. (1999) Monographs, in *Transition Metal Catalyzed Reactions, A Chemistry for the 21st Century* (eds S-.I. Murahashi and S.G. Davies), Blackwell Science, New York, p. 55; (h) Michelet, V., Savignac, M. and Genêt, J.-P., (2004) *Electronic Encyclopedia of Reagents for Organic Synthesis* (eds L. Paquette, P. Fuchs, D. Crich and P. Wipf), John Wiley & Sons, Ltd, Chichester.

6. (a) Tsuji, J. (2002) *Handbook of Organopalladium Chemistry for Organic Synthesis*, Vol. 2 (ed. E. Negishi), John Wiley & Sons, Inc, Hoboken, NJ, p. 1669; (b) Trost, B.M. and Lee, C. (2000) *Catalytic Asymmetric Synthesis*, 2nd edn (ed. I. Ojima), Wiley-VCH Verlag GmbH, Weinheim, p. 593; (c) Pfaltz, A. and Lautens, M. Comprehensive Asymmetric Catalysis Vol II, (eds E.N. Jacobsen, A. Pfaltz and H. Yamamoto), Springer, Berlin, 834.

7. (a) Miyaura, N. and Suzuki, A. (1995) *Chem. Rev.*, **95**, 2457; (b) Alonso, F., Beletskaya, I.P. and Yus, M. (2008) *Tetrahedron*, **64**, 3047.

8. (a) Sonogashira, K. (2002) *Handbook of Organopalladium Chemistry for Organic Synthesis* (ed. E. Negishi), John Wiley & Sons, Inc, Hoboken, NJ, p. 493; (b) Chinchilla, R. and Najera, C. (2007) *Chem. Rev.*, **107**, 874; (c) Doucet, H. and Hierso, J.-C. (2007) *Angew. Chem. Int. Ed.*, **46**, 834.

9. Stille, J.K. (1986) *Angew. Chem. Int. Ed. Engl.*, **25**, 508.

10. (a) Muci, A.R. and Buchwald, S.L. (2002) *Top. Curr. Chem.*, **219**, 131; (b) Hartwig, J.F. (1998) *Angew. Chem. Int. Ed.*, **37**, 2046; (c) Hartwig, J.F. (2002) *Handbook of Organopalladium Chemistry for Organic Synthesis* (eds E. Negishi and A. de Meijere), Wiley-Interscience, New York, p. 1051; (d) Hartwig, J.F. (1999) *Pure Appl. Chem.*, **71**, 1417.

11. (a) Hiyama, T. and Shirakawa, E. (2002) *Top. Curr. Chem.*, **219**, 61; (b) Hiyama, T. (2002) *J. Organomet. Chem.*, **653**, 58; (c) Hiyama, T. and Hatanaka, Y. (1994) *Pure Appl. Chem.*, **66**, 1471; (d) Hatanaka, Y. and Hiyama, T. (1991) *Synlett*, 845; (e) Denmark, S.E. and Sweis, R.F. (2004)

*Metal-Catalyzed Cross-Coupling Reactions* (eds A. de Meijere and F. Diederich), Wiley-VCH Verlag GmbH, Weinheim, p. 163.

12 Fan, Q.-H., Li, Y.-M. and Chan, A.S.C. (2002) *Chem. Rev.*, **102**, 3385.
13 Sinou, D. (2002) *Adv. Synth. Catal.*, **344**, 221.
14 Anastas, P.T. and Wagner, J.C. (1998) *Green Chemistry: Theory and Practice*, Oxford University Press, Oxford.
15 Li, C.-J. (2002) *Green Chem.*, **4**, 1.
16 (a) Peters, W. (1905) *Chem. Ber.*, **38**, 2567; (b) Sisido, K., Takeda, Y. and Kinugawa, Z. (1961) *J. Am. Chem. Soc.*, **83**, 538; (c) Sisido, K., Kozima, S. and Hanada, T. (1967) *J. Organomet. Chem.*, **9**, 99; (d) Sisido, K. and Kozima, S. (1968) *J. Organomet. Chem.*, **11**, 503.
17 Killinger, T.A., Boughton, N.A., Runge, T.A. and Wolinsky, J. (1977) *J. Organomet. Chem.*, **124**, 131.
18 (a) Roush, W.R. (1991) *Comprehensive Organic Synthesis*, Vol. 2 (eds B.M. Trost, I. Fleming and C.H. Heathcock Pergamon Press, Oxford, p. 1; (b) Gao, Y. and Sato, F. (1995) *J. Org. Chem.*, **60**, 8136; (c) Yamamoto, Y. and Asao, N. (1993) *Chem. Rev.*, **93**, 2207; (d) Hoppe, D. (1996) *Houben-Weyl: Methods of Organic Chemistry*, Vol. 21, (eds G. Helmchen, R.W. Hoffmann, J. Mulzer, and E. Schaumann), Georg Thieme, Stuttgart, p. 1357; (e) Bloch, R. (1998) *Chem. Rev.*, **98**, 1407; (f) Kobayashi, S. and Ishitani, H. (1999) *Chem. Rev.*, **99**, 1069; (g) Rottlander, M., Boymond, L., Berillon, L., Lepretre, A., Varchi, G., Avolio, S., Laaziri, H., Queguiner, G., Ricci, A., Cahiez, G. and Knochel, P. (2000) *Chem. Eur. J.*, **6**, 76; (h) Neipp, C.E., Humpherey, J.M. and Martin, S.F. (2001) *J. Org. Chem.*, **66**, 531; (i) Nicolaou, K.C., Kim, D.W. and Baati, R. (2002) *Angew. Chem. Int. Ed.*, **41**, 3701; (j) Denmark, S.E. and Fu, J. (2003) *Chem. Commun.*, 163; (k) Zhang, Z.-J., Wan, B.-S. and Chen, H.-L. (2003) *Chin. J. Org. Chem.*, **23**, 636; (l) Kennedy, J.W.J. and Hall, D.G. (2003) *Angew. Chem. Int. Ed.*, **42**, 4732.

19 **Mg**: (a) Wada, M., Fukuma, T., Morioka, M., Takahashi, T. and Miyoshi, N. (1997) *Tetrahedron Lett.*, **38**, 8045; (b) Li, C.-J. and Zhang, W.-C. (1998) *J. Am. Chem. Soc.*, **120**, 9102; (c) Zhang, W.-C. and Li, C.-J. (1999) *J. Org. Chem.*, **64**, 3230; **Mn**: Li, C.-J., Meng, Y., Yi, X.-H., Ma, J. and Chan, T.-H., (1998) *J. Org. Chem.*, **63**, 7498; **Sm**: Laskar, D., Prajapati, D. and Sandhu, J.S., (2001) *Tetrahedron Lett.*, **42**, 7883; **Sb**: (a) Wang, W., Shi, L. and Huang, Y., (1990) *Tetrahedron*, **46**, 3315; (b) Li, L.-H. and Chan, T.-H. (2000) *Tetrahedron Lett.*, **41**, 5009; (c) Li, L.-H. and Chan, T.-H. (2001) *Can. J. Chem.*, **79**, 1536; (d) Fukuma, T., Lock, S., Miyoshi, N. and Wada, M. (2002) *Chem. Lett.*, 376. (a) **Bi**: Wada, M., Okhi, H. and Akiba, K.Y., (1990) *Bull. Chem. Soc. Jpn.*, **63**, 1738; (b) Miyamoto, H., Daikawa, N. and Tanaka, K. (2003) *Tetrahedron Lett.*, **44**, 6963; (c) Katrinsky, A.R., Shobana, N. and Harris, P.A. (1992) *Organometallics*, **11**, 1381; (d) Xu, X., Zha, Z., Miao, Q. and Wang, Z. (2004) *Synlett*, 1171. **Pb**: Zhou, J.-Y., Jia, Y., Sun, G.-F. and Wu, S.-H., (1997) *Synth. Commun.*, **27**, 1899. **Ga**: Wang, Z., Yuan, S. and Li, C.-J., (2002) *Tetrahedron Lett.*, **43**, 5097. **Fe**: Chan, T.C., Lau, C.P. and Chan, T.-H., (2004) *Tetrahedron Lett.*, **45**, 4189. **Co**: Khan, R.H. and Prasada Rao, T.S.R., (1998) *J. Chem. Res.(S)*, 202.
20 Nokami, J., Otera, J., Suto, T. and Okawara, R. (1983) *Organometallics*, **2**, 191.
21 Uneyama, K., Kamaki, N., Moriya, A. and Torii, S.J. (1985) *Org. Chem.*, **50**, 5396.
22 (a) Wu, S.H., Huang, B.Z., Zhu, T.M., Yiao, D.Z. and Chu, Y.L. (1990) *Acta Chim. Sin.*, **48**, 372; (b) Petrier, C. and Luche, J.L. (1985) *J. Org. Chem.*, **50**, 910; (c) Petrier, C., Einhorn, J. and Luche, J.L. (1989) *Tetrahedron Lett.*, **26**, 1449.
23 Zha, Z., Wang, Y., Yang, G., Zhang, L. and Wang, Z. (2002) *Green Chem.*, **4**, 578.
24 (a) Durand, A., Delplancke, J.L., Winand, R. and Reisse, J. (1995) *Tetrahedron Lett.*, **36**, 4257; (b) Wang, Z., Zha, Z. and Zhou, C. (2002) *Org. Lett.*, **4**, 1683.

25 (a) Mattes, H. and Benezra, C. (1985) *Tetrahedron Lett.*, **26**, 5697; (b) Zhou, J.Y., Lu, G.D. and Wu, S.H. (1992) *Synth. Commun.*, **22**, 481.

26 Reid, C.S., Zhang, Y. and Li, C.-J. (2007) *Org. Biomol. Chem.*, **5**, 3589.

27 Silver, S., Leppänen, A.-S., Sjöholm, R., Penninkangas, A. and Leino, R. (2005) *Eur. J. Org. Chem.*, 1058.

28 (a) Li, C.-J. and Chan, T.-H. (1991) *Tetrahedron Lett.*, **32**, 7071; (b) Chan, T.-H. and Yang, Y. (1999) *J. Am. Chem. Soc.*, **121**, 3228.

29 Li, C.-J. and Chan, T.-H. (1999) *Tetrahedron*, **55**, 11149.

30 (a) Li, C.-J. and Lu, Y.-Q. (1995) *Tetrahedron Lett.*, **36**, 2721; (b) Li, C.-J. and Lu, Y.-Q. (1996) *Tetrahedron Lett.*, **37**, 471.

31 (a) Lu, W. and Chan, T.-H. (2000) *J. Org.Chem.*, **65**, 8589; (b) Hirashita, T., Hayashi, Y., Mitsui, K. and Araki, S. (2003) *J. Org. Chem.*, **68**, 1309; (c) Estevam, I.H.S. and Bieber, L.W. (2003) *Tetrahedron Lett.*, **44**, 667.

32 Matsumura, Y., Onomura, O., Suzuki, H., Furukubo, S., Maki, T. and Li, C.-J. (2003) *Tetrahedron Lett.*, **44**, 5519.

33 (a) Boaretto, A., Marton, D., Tagliavini, G. and Gambaro, A. (1985) *J. Organomet. Chem.*, **286**, 9; (b) Boaretto, A., Marton, D. and Tagliavini, G. (1985) *J. Organomet. Chem.*, **297**, 149; (c) Furlani, D., Marton, D., Tagliavini, G. and Zordan, M. (1988) *J. Organomet. Chem.*, **341**, 345; (d) Yanagisawa, A., Inoue, H., Morodome, M. and Yamamoto, H. (1993) *J. Am. Chem. Soc.*, **115**, 10356.

34 (a) Hachiya, I. and Kobayashi, S. (1993) *J. Org. Chem.*, **58**, 6958; (b) Kobayashi, S., Wakabayashi, T. and Oyamada, H. (1997) *Chem. Lett.*, 831; (c) McCluskey, A. (1999) *Green Chem.*, **1**, 167; (d) Kobayashi, S., Hamada, T. and Manabe, K. (2001) *Synlett*, 1140.

35 (a) Mukaiyama, T., Harada, T. and Shoda, S. (1980) *Chem. Lett.*, **96**, 1507; (b) Talaga, P., Schaeffer, M., Benezra, C. and Stampf, J.-L. (1990) *Synthesis*, 530.

36 (a) Tan, X.-H., Hou, Y.-Q., Huang, C., Liu, L. and Guo, Q.-X. (2004) *Tetrahedron*, **60**, 6129; (b) Uneyama, K., Ueda, K. and Torii, S. (1986) *Tetrahedron Lett.*, **27**, 2395; (c) Okano, T., Kiji, J. and Doi, T. (1998) *Chem. Lett.*, 5; (d) Kundu, A., Prabahkar, S., Vairamani, M. and Roy, S. (1997) *Organometallics*, **16**, 4796; (e) Tan, X.-H., Shen, B., Liu, L. and Guo, Q.-X. (2002) *Tetrahedron Lett.*, **43**, 9373; (f) Tan, X.-H., Deng, W., Zhao, H., Liu, L. and Guo, Q.-X. (2003) *Org. Lett.*, **5**, 1833; (g) Chaudhuri, M.K., Dehury, S.K. and Hussian, S. (2005) *Tetrahedron Lett.*, **46**, 6247; (h) Zhou, C., Zhou, Y., Jiang, J., Xie, Z., Wang, Z., Zhang, J., Wu, J. and Yin, H. (2004) *Tetrahedron Lett.*, **45**, 5537.

37 Yang, Y. and Chan, T.H. (2000) *J. Am. Chem. Soc.*, **122**, 402.

38 (a) Masuyama, Y., Takahara, J.P. and Kurusu, Y. (1988) *J. Am. Chem. Soc.*, **110**, 4473; (b) Masuyama, Y., Takahara, J.P. and Kurusu, Y. (1989) *Tetrahedron Lett.*, **30**, 3437.

39 (a) Masuyama, Y., Nimura, Y. and Kurusu, Y. (1991) *Tetrahedron Lett.*, **32**, 225; (b) Sati, M. and Sinou, D. (1991) *Tetrahedron Lett.*, **32**, 2025; (c) Fontana, G., Lubineau, A. and Scherrmann, M.-C. (2005) *Org. Biomol. Chem.*, **3**, 1375.

40 Araki, S., Kamei, T., Hirashita, T., Yamamura, H. and Kawai, M. (2000) *Org. Lett.*, **2**, 847.

41 Chang, H.-M. and Cheng, C.-H. (2000) *Org. Lett.*, **2**, 3439.

42 (a) Uneyama, K., Matsuda, H. and Torii, S. (1984) *Tetrahedron Lett.*, **25**, 6017; (b) Zha, Z., Hui, A., Zhou, Y., Miao, Q., Wang, Z. and Zhang, H. (2005) *Org. Lett.*, **7**, 1903.

43 Thadani, A.N. and Batey, R. A. (2002) *Org. Lett.*, **4**, 3827.

44 Ishihara, K., Hanaki, N., Funahashi, M., Miyata, M. and Yamamoto, H. (1995) *Bull. Chem. Soc. Jpn*, **68**, 1721.

45 Aoyama, N., Hamada, T., Manabe, K. and Kobayashi, S. (2003) *Chem. Commun.*, 676.

46 (a) Chan, T.-H. and Li, C.-J. (1992) *Can. J. Chem.*, **70**, 2726; (b) Isaac, M.B. and

Chan, T.-H. (1995) *Tetrahedron Lett.*, **36**, 8957.

47 (a) Matsuyama, Y., Kishida, M. and Kurusu, Y. (1995) *J. Chem. Soc., Chem. Commun.*, 1405; (b) Ito, A., Kishita, M., Kurusu, Y. and Matsuyama, Y. (2000) *J. Org. Chem.*, **65**, 494; (c) Tan, K.-T., Chng, S.-S., Cheng, H.-S. and Loh, T.-P. (2003) *J. Am. Chem. Soc.*, **125**, 2958; (d) Isaac, M.B. and Chan, T.-H. (1995) *Tetrahedron Lett.*, **36**, 8957.

48 Loh, T.-P. and Zhou, J.-R. (1999) *Tetrahedron Lett.*, **40**, 9115.

49 Loh, T.-P. and Zhou, J.-R. (2000) *Tetrahedron Lett.*, **41**, 5261.

50 Teo, Y.-C., Goh, E.-L. and Loh, T.-P. (2005) *Tetrahedron Lett.*, **46**, 6209.

51 Cunningham, A. and Woodward, S. (2002) *Synlett*, 43.

52 Chen, G.-M., Ramachandran, P.V. and Brown, H.C. (1999) *Angew. Chem. Int. Ed.*, **38**, 825.

53 Fernandes, R.A., Stimac, A. and Yamamoto, Y. (2003) *J. Am. Chem. Soc.*, **125**, 14133.

54 Hamada, T., Manabe, K. and Kobayashi, S. (2003) *Angew. Chem. Int. Ed.*, **42**, 3927.

55 Wang, S.H. and Huang, B.Z. (1990) *Synth. Commun.*, **20**, 1279.

56 (a) Boaretto, A., Marton, D., Tagliavini, G. and Gambaro, A. (1985) *J. Organomet. Chem.*, **286**, 9; (b) Boaretto, A., Marton, D. and Tagliavini, G. (1985) *J. Organomet. Chem.*, **297**, 149; (c) Furlani, D., Marton, D., Tagliavini, G. and Zordan, M. (1988) *J. Organomet. Chem.*, **341**, 345; (d) Hachiya, I. and Kobayashi, S. (1993) *J. Org. Chem.*, **58**, 6958.

57 Roy, U.K. and Roy, S. (2006) *Tetrahedron*, **62**, 678.

58 (a) Chattopadhyay, A. (1996) *J. Org. Chem.*, **61**, 6104; (b) Keltjens, R., Vadivel, S.K., deGelder, R., Bieber, L.W., da Silva, M.F., da Costa, R.C. and Silva, L.O.S. (1998) *Tetrahedron Lett.*, **39**, 3655; (c) Klunder, A.J.H. and Zwanenburg, B. (2003) *Eur. J. Org. Chem.*, 1749.

59 Chattopadhyay, A. and Dhotare, B. (1998) *Tetrahedron: Asymmetry*, **9**, 2715.

60 Isaac, M.B. and Chan, T.-H. (1995) *J. Chem. Soc., Chem. Commun.*, 1003.

61 (a) Alcaide, B., Almendros, P., Aragoncillo, C. and Rodriguez-Acebes, R. (2001) *J. Org. Chem.*, **66**, 5208; (b) Alcaide, B., Almendros, P., Aragoncillo, C. and Rodriguez-Acebes, R. (2003) *Synthesis*, 1163; (c) Alcaide, B., Almendros, P. and Rodriguez-Acebes, R. (2005) *J. Org. Chem.*, **70**, 3198; (d) Alcaide, B., Almendros, P. and Rodriguez-Acebes, R. (2005) *Chem. Eur. J.*, **11**, 5708.

62 Yi, X.H., Meng, Y., Hua, X.G. and Li, C.J. (1998) *J. Org. Chem.*, **63**, 7472.

63 (a) Arimitsu, S. and Hammond, G.B. (2006) *J. Org. Chem.*, **71**, 8665; (b) Arimitsu, S., Jacobsen, J.M. and Hammond, G.B. (2007) *Tetrahedron Lett.*, **48**, 1625.

64 (a) Prajapati, D., Laskar, D.D., Gogoi, B.J. and Devi, G. (2003) *Tetrahedron Lett.*, **44**, 6755; (b) Miyabe, H., Yamaoka, Y., Naito, T. and Takemoto, Y. (2004) *J. Org. Chem.*, **69**, 1415.

65 Kumar, H.M.S., Singh, P.P., Shafi, S., Reddy, P.B., Shravankumar, K. and Reddy, D.M. (2007) *Tetrahedron Lett.*, **48**, 887.

66 (a) Fallis, A.G. and Brinza, I.M. (1997) *Tetrahedron*, **53**, 17543; (b) Naito, T. (1999) *Heterocycles*, **50**, 505.

67 (a) Petrier, C., Dupuy, C. and Luche, J.-L. (1986) *Tetrahedron Lett.*, **27**, 3149; (b) Giese, B., Damm, W., Roth, M. and Zehnder, M. (1992) *Synlett*, 441; (b) Erdmann, P., Schäfer, J., Springer, R., Zeitz, H.-G. and Giese, B. (1992) *Helv. Chim. Acta*, **75**, 638.

68 (a) Miyabe, H., Ueda, M., Nishimura, A. and Naito, T. (2002) *Org. Lett.*, **4**, 131; (b) Ueda, M., Miyabe, H., Nishimura, A., Sugino, H. and Naito, T. (2003) *Tetrahedron: Asymmetry*, **14**, 2857; (c) Miyabe, H., Ueda, M., Nishimura, A. and Naito, T. (2004) *Tetrahedron*, **60**, 4227; (d) Miyabe, H., Ueda, M. and Naito, T. (2000) *Chem. Commun.*, 2059.

69 (a) Miyabe, H., Nishimura, A., Ueda, M. and Naito, T. (2002) *Chem. Commun.*,

1454; (b) Ueda, M., Miyabe, H., Sugino, H. and Naito, T. (2005) *Org. Biomol. Chem.*, **3**, 1124.
70 Kalyanam, N. and Rao, G.V. (1993) *Tetrahedron Lett.*, **34**, 1647.
71 Huang, T., Keh, C.C.K. and Li, C.-J. (2002) *Chem. Commun.*, 2440.
72 Cannella, R., Clerici, A., Pastori, N., Ragolini, E. and Porta, O. (2005) *Org. Lett.*, **7**, 645.
73 (a) Shen, Z.-L. and Loh, T.-P. (2007) *Org. Lett.*, **9**, 5413; (b) Shen, Z.-L., Cheong, H.-L. and Loh, T.-P., (2008) *Chem. Eur. J.*, **14**, 1875.
74 (a) Engstrom, G., Morelli, M., Palomo, C. and Mitzel, T.M. (1999) *Tetrahedron Lett.*, **40**, 5967; (b) Mitzel, T.M., Palomo, C. and Jendza, K. (2002) *J. Org. Chem.*, **67**, 136.
75 Keh, C.C.K., Wei, C. and Li, C.-J. (2003) *J. Am. Chem. Soc.*, **125**, 4062.
76 Hirano, K., Yorimitsu, H. and Oshima, K. (2006) *Adv. Synth. Catal.*, **348**, 1543.
77 Sakai, M., Ueda, M. and Miyaura, N. (1998) *Angew. Chem. Int. Ed.*, **37**, 3279.
78 Ueda, M. and Miyaura, N. (2000) *J. Org. Chem.*, **65**, 4450.
79 (a) Bourissou, D., Guerret, O., Gabbai, F.P. and Bertrand, G. (2000) *Chem. Rev.*, **100**, 39; (b) Herrmann, W.A. and Köcher, C. (1997) *Angew. Chem. Int. Ed.*, **36**, 2162.
80 Fürstner, A. and Krause, H. (2001) *Adv. Synth. Catal.*, **343**, 343.
81 Moreau, C., Hague, C., Weller, A.S. and Frost, C.G. (2001) *Tetrahedron Lett.*, **42**, 6957.
82 Imlinger, N., Mayr, M., Wang, D., Wurst, K. and Buchmeiser, M.R. (2004) *Adv. Synth. Catal.*, **346**, 1836.
83 (a) Chen, J., Zhang, X., Feng, Q. and Luo, M. (2006) *J. Organomet. Chem.*, **691**, 470; (b) Yan, C., Zeng, X., Zhang, W. and Luo, M. (2006) *J. Organomet. Chem.*, **691**, 3391.
84 Gois, P.M.P., Trindade, A.F., Veiros, L.F., André, V., Duarte, M.T., Afonso, C.A.M., Caddick, S. and Cloke, F.G.N. (2007) *Angew. Chem. Int. Ed.*, **46**, 5750.
85 Son, S.U., Kim, S.B., Reingold, J.A., Carpenter, G.B. and Sweigart, D.A. (2005) *J. Am. Chem. Soc.*, **127**, 12238.
86 Huang, R. and Shaughnessy, H. (2005) *Chem. Commun.*, 4484.
87 Darses, S. and Genêt, J.-P. (2008) *Chem. Rev.*, **108**, 288.
88 Batey, R.A., Thadani, A.N. and Smil, D.V. (1999) *Org. Lett.*, **1**, 1683.
89 Pucheault, M., Darses, S. and Genêt, J.-P. (2005) *Chem. Commun.*, 4714.
90 Takahashi, G., Shirakawa, E., Tsuchimoto, T. and Kawakami, Y. (2005) *Chem. Commun.*, 1459.
91 Li, C.-J. and Meng, Y. (2000) *J. Am. Chem. Soc.*, **122**, 9538.
92 Huang, T., Meng, Y., Venkatraman, S., Weng, D. and Li, C.-J. (2001) *J. Am. Chem. Soc.*, **123**, 7451.
93 Ding, R., Zhao, C.-H., Chen, Y.-J., Liu, L., Wang, D. and Li, C.-J. (2004) *Tetrahedron Lett.*, **45**, 2995.
94 Bolshan, Y. and Batey, R.A. (2005) *Org. Lett.*, **7**, 1481.
95 Petasis, N.A. and Zavialov, I.A. (1997) *J. Am. Chem. Soc.*, **119**, 445.
96 Duan, H.-F., Xie, J.-H., Shi, W.-J., Zhang, Q. and Zhou, Q.-L. (2006) *Org. Lett.*, **8**, 1479.
97 Trost, B.M., Toste, F.D. and Pinkerton, A.B. (2001) *Chem. Rev.*, **101**, 2067.
98 (a) Wei, C. and Li, C.-J. (2002) *Green Chem.*, **4**, 39; (b) Wang, M. and Li, C.-J. (2004) *Topics Organomet. Chem.*, **11**, 321.
99 Li, C.-J. and Wei, C. (2002) *Chem. Commun.*, 268.
100 Zhang, J., Wei, C. and Li, C.-J. (2002) *Tetrahedron Lett.*, **43**, 5731.
101 Shi, L., Tu, Y.-Q., Wang, M., Zhang, F.-M. and Fan, C.-A. (2004) *Org. Lett.*, **6**, 1001.
102 Wei, C., Li, Z. and Li, C.-J. (2003) *Org. Lett.*, **5**, 4473.
103 Yao, X. and Li, C.-J. (2005) *Org. Lett.*, **7**, 4395.
104 Wei, C. and Li, C.-J. (2003) *J. Am. Chem. Soc.*, **125**, 9584.
105 Lo, V.K.-Y., Liu, Y., Wong, M.-K. and Che, C.-M. (2006) *Org. Lett.*, **8**, 1529.
106 Yao, X. and Li, C.-J. (2006) *Org. Lett.*, **8**, 1953.

107 Yan, B. and Liu, Y. (2007) *Org. Lett.*, **9**, 4323.
108 Wei, C. and Li, C.-J. (2002) *J. Am. Chem. Soc.*, **124**, 5638.
109 Fittig, R. (1859) *Justus Liebigs Ann Chem.*, **110**, 23.
110 (a) Mukaiyama, T., Shiina, I., Iwadare, H., Saitoh, M., Nishimura, T., Ohkawa, N., Sakoh, H., Nishimura, K., Tani, Y.-I., Hasegawa, M., Yamada, K. and Saitoh, K. (1999) *Chem. Eur. J.*, **5**, 121; (b) Nicolaou, K.C., Yang, Z., Liu, J.J., Ueno, H., Nantermet, P.G., Guy, R.K., Claiborne, C.F., Renaud, J., Couladouros, E.A., Paulvannan, K. and Sorensen, E.J. (1994) *Nature*, **367**, 630.
111 (a) Clerici, A. and Porta, O. (1982) *J. Org. Chem.*, **47**, 2852; (b) Clerici, A. and Porta, O. (1989) *J. Org. Chem.*, **54**, 3872; (c) Barden, M.C. and Schwartz, J. (1996) *J. Am. Chem. Soc.*, **118**, 5484.
112 (a) Delair, P. and Luche, J.-L. (1989) *J. Chem. Soc., Chem. Commun.*, 398; (b) Tanaka, K., Kishigmi, S. and Toda, F. (1990) *J. Org. Chem.*, **55**, 2981; (c) Li, C.-J., Meng, Y., Yi, X.-H., Ma, J. and Chan, T.-H. (1997) *J. Org. Chem.*, **62**, 8632; (d) Li, C.-J., Meng, Y., Yi, X.-H., Ma, J. and Chan, T.-H. (1998) *J. Org. Chem.*, **63**, 7498; (e) Zhang, W.C. and Li, C.-J. (1998) *J. Chem. Soc., Perkin Trans. 1*, 3131; (f) Zhang, W.C. and Li, C.-J. (1999) *J. Org. Chem.*, **64**, 3230; (g) Wang, L. and Zhang, Y. (1998) *Tetrahedron*, **54**, 11129; (h) Matsukawa, S. and Hinakubo, Y. (2003) *Org. Lett.*, **5**, 1221; (i) Lim, H.J., Keum, G. and Kim, Y. (1998) *Tetraedron Lett.*, **39**, 4367; (j) Nair, V., Ros, S., Jayan, C.N. and Rath, N.P. (2002) *Tetrahedron Lett.*, **43**, 8967; (k) Wang, C., Pan, Y. and Wu, A. (2007) *Tetrahedron*, **63**, 429; (l) Meciarova, M. and Toma, S. (1999) *Green Chem.*, **1**, 257; (m) Li, L.-H. and Chan, T.-H. (2000) *Org. Lett.*, **2**, 1129; (n) Yuan, S.-Z., Wang, Z.-Y. and Li, Z. (2006) *Chin. J. Chem.*, **24**, 141; (o) Buchammagari, H., Toda, Y., Hirano, M., Hosono, H., Takeuchi, D. and Osakada, K. (2007) *Org. Lett.*, **9**, 4287.
113 Kalyanam, N. and Venkateswara, R.G. (1993) *Tetrahedron Lett.*, **34**, 1647.
114 (a) Dutta, M.P., Baruah, B., Boruah, A., Prajapati, D. and Sandhu, J.S. (1998) *Synlett*, 857; (b) Tsukinoki, T., Nagashima, S., Mitoma, Y. and Tashiro, M. (2000) *Green Chem.*, **2**, 117; (c) Liu, X., Liu, Y. and Zhang, Y. (2002) *Tetrahedron Lett.*, **43**, 6787; (d) Kim, M., Knettle, B.W., Dahlen, A., Hilmersson, G. and Flowers, R.A. (2003) *Tetrahedron*, **59**, 10397.
115 (a) Kohlpaintner, C.W., Fischer, R.W. and Cornils, B. (2001) *Appl. Catal. A*, **221**, 219; (b) Cornils, B. (1998) *Org. Process Res. Dev.*, **2**, 121; (c) Cornils, B. and Kuntz, E.G. (1995) *J. Organomet. Chem.*, **502**, 177; (d) Kuntz, E.G. (1987) *Chemtech*, 570.
116 (a) Herrmann, W.A., Kohlpainter, C.W., Bahrmann, H. and Konkol, W. (1992) *J. Mol. Catal.*, **73**, 191; (b) Herrmann, W.A., Bach, H., Frohning, D.C., Kleiner, J.H., Lappe, P., Peters, D., Regnat, D. and Bahrmann, H. (1997) *J. Mol. Catal. A: Chemistry*, **116**, 49; (c) Buhling, A., Kamer, J.C.P. and van Leeuwen, M.N.W.P. (1995) *J. Mol. Catal. A: Chemistry*, **98**, 69; (d) Buhling, A., Kamer, J.C.P. and van Leeuwen, M.N.W.P. (1997) *Organometallics*, **16**, 3027; (e) Tilloy, S., Genin, E., Hapiot, F., Landy, D., Fourmentin, S., Genêt, J.-P., Michelet, V. and Monflier, E. (2006) *Adv. Synth. Catal.*, **348**, 1547.
117 (a) Kalck, P., Dessoudeix, M. and Schwarz, S. (1999) *J. Mol. Catal. A: Chemistry*, **143**, 41; (b) Chaudhari, R.V., Bhanage, B.M., Deshpande, R.M. and Delmas, H. (1995) *Nature*, **373**, 501; (c) Chen, H., Li, Y., Chen, J., Cheng, P. and Li, X. (2002) *Catal. Today*, **74**, 131; (d) Purwanto, P. and Delmas, H. (1995) *Catal. Today*, **24**, 135.
118 (a) Wang, L., Chen, H., He, Y., Li, Y., Li, M. and Li, X. (2003) *J. Mol. Catal. A: Chemistry*, **242**, 85; (b) Riisager, A. and Hanson, B.E. (2002) *J. Mol. Catal. A: Chemistry*, **189**, 195; (c) Ding, H. and Hanson, B.E. (1995) *J. Mol. Catal. A: Chemistry*, **99**, 131; (d) Haumann, M., Yildiz, H., Koch, H. and Schomäcker, R. (2002) *Appl. Catal. A*, **236**, 173; (e) Tic, W.,

Miesiac, I. and Szymanowski, J. (2001) *J. Colloid Interface Sci.*, **244**, 423.

119 (a) Dessoudeix, M., Urrutigoïty, M. and Kalck, P. (2001) *Eur. J. Inorg. Chem.*, 1797; (b) Monflier, E., Frémy, G., Castanet, Y. and Mortreux, A. (1995) *Angew. Chem. Int. Ed. Engl.*, **34**, 2269; (c) Monflier, E., Blouet, E., Barbaux, Y. and Mortreux, A. (1994) *Angew. Chem. Int. Ed. Engl.*, **33**, 2100; (d) Mathivet, T., Méliet, C., Castanet, Y., Mortreux, A., Caron, L., Tilloy, S. and Monflier, E. (2001) *J. Mol. Catal. A: Chemistry*, **176**, 105.

120 (a) Arhancet, J.P., Davis, M.E., Merola, J.S. and Hanson, B.E. (1989) *Nature*, **339**, 454; (b) Naughton, M.J. and Drago, R.S. (1995) *J. Catal.*, **155**, 383.

121 (a) Dessoudeix, M., Jauregui-Haza, U.J., Heughebaert, M., Wilhelm, A.M., Delmas, H., Lebugle, A. and Kalck, P. (2002) *Adv. Synth. Catal.*, **344**, 406; (b) Anson, M.S., Leese, M.P., Tonks, L. and Williams, J.M.J. (1998) *J. Chem. Soc., Dalton Trans.*, 3529; (c) Zhu, H., Ding, Y., Yin, H., Yan, L., Xiong, J., Lu, Y., Luo, H. and Lin, L. (2003) *Appl. Catal. A*, **245**, 111; (d) Jauregui-Haza, U.J., Pardillo-Fontvila, E., Kalck, P., Wilhelm, A.M. and Delmas, H. (2003) *Catal. Today*, **79–80**, 409; (e) Li, Z., Peng, Q. and Yuan, Y. (2003) *Appl. Catal. A*, **239**, 79.

122 (a) Guo, I., Hanson, B.E., Toth, I. and Davis, M.E. (1991) *J. Organomet. Chem.*, **403**, 221; (b) Khna, M.M.T., Halligudi, S.B. and Abdi, S.H.R. (1988) *J. Mol. Catal.*, **48**, 313; (c) Haumann, M., Koch, H. and Schomäcker, R. (2003) *Catal. Today*, **79–80**, 43–49. (d) Beller, M. and Krauter, J.G.E. (1999) *J. Mol. Catal. A: Chemistry*, **143**, 31.

123 Jenner, G. (1991) *Tetrahedron Lett.*, **32**, 505.

124 (a) Frémy, G., Monflier, E., Carpentier, J.-F., Castanet, Y. and Mortreux, A. (1998) *J. Mol. Catal. A: Chemistry*, **129**, 35; (b) Frémy, G., Monflier, E., Carpentier, J.-F., Castanet, Y. and Mortreux, A. (1995) *Angew. Chem. Int. Ed. Engl.*, **34**, 1474; (c) Verspui, G., Elbertse, G., Papadogianakis, G. and Sheldon, R.A. (2001) *J. Organomet. Chem.*, **621**, 337; (d) Verspui, G., Elbertse, G., Papadogianakis, G. and Sheldon, R.A. (2000) *Chem. Commun.*, 1363; (e) Bahrmann, H., Haubs, M., Müller, T., Schöpper, N. and Cornils, B. (1997) *J. Organomet. Chem.*, **545–546**, 139; (f) Botteghi, C., Corrias, T., Marchetti, M., Paganelli, S. and Piccolo, O. (2002) *Org. Process Res. Dev.*, **6**, 379.

125 (a) Zimmermann, B., Herwig, J. and Beller, M. (1999) *Angew. Chem. Int. Ed.*, **38**, 2372; (b) Wang, Y.Y., Luo, M.M., Li, Y.Z., Chen, H. and Li, X.J. (2004) *Appl. Catal. A: General*, **272**, 151.

126 (a) Miquel-Serrano, D.M., Aghmiz, A., Dieguez, M., Masdeu-Bulto, M.A., Claver, C. and Sinou, D. (1999) *Tetrahedron: Asymmetry*, **10**, 4463; (b) Herrmann, W.A., Eckl, W.R. and Priermeier, T. (1997) *J. Organomet. Chem.*, **532**, 243; (c) Miquel-Serrano, D.M., Masdeu-Bulto, M.A., Claver, C. and Sinou, D. (1999) *J. Mol. Catal. A: Chemistry*, **143**, 49; (d) Köckritz, A., Bischoff, S., Kant, M. and Siefken, R. (2001) *J. Mol. Catal. A: Chemistry*, **174**, 119.

127 Bertoux, F., Monflier, E., Castanet, Y. and Mortreux, A. (1999) *J. Mol. Catal. A: Chemistry*, **143**, 11.

128 (a) Verspui, G., Feiken, J., Papadogianakis, G. and Sheldon, R.A. (1999) *J. Mol. Catal. A: Chemistry*, **146**, 299; (b) Bertoux, F., Tilloy, S., Monflier, E., Castanet, Y. and Mortreux, A. (1999) *J. Mol. Catal. A: Chemistry*, **138**, 53; (c) Papadogianakis, G., Verspui, G., Maat, L. and Sheldon, R.A. (1997) *Catal. Lett.*, **47**, 43; (d) Jayasree, S., Seayad, A. and Chaudhari, R.V. (2000) *Chem. Commun.*, 1239.

129 (a) Bertoux, F., Monflier, E., Castanet, Y. and Mortreux, A. (1999) *J. Mol. Catal. A: Chemistry*, **143**, 23; (b) Tilloy, S., Monflier, E., Bertoux, F., Castanet, Y. and Mortreux, A. (1997) *New J. Chem.*, **21**, 857.

130 Verspui, G., Moiseev, I.I. and Sheldon, R.A. (1999) *J. Organomet. Chem.*, **586**, 196.

131 (a) Alper, H. and Abbayes, H.D. (1977) *J. Organomet. Chem.*, **134**, C11; (b) Cassar, L. and Foa, M. (1977) *J. Organomet. Chem.*, **134**, C15; (c) Joó, F. and Alper, H. (1985) *Organometallics*, **4**, 1775; (d) Kohlpaintner, C.W. and Beller, M. (1997) *J. Mol. Catal. A: Chemistry*, **116**, 259; (e) Kiji, J., Okano, T., Nishiumi, W. and Konishi, H. (1988) *Chem. Lett.*, 957; (f) Kiji, J., Okano, T. and Okabe, N. (1992) *Bull. Chem. Soc. Jpn.*, **65**, 2589.

132 (a) Rhinehart, R.E. and Smith, H.P. (1965) *Polym. Lett.*, **3**, 1049; (b) recent review: Mecking, S., Held, A. and Bauers, F.M., (2002) *Angew. Chem. Int. Ed.*, **41**, 544.

133 Puech, L., Perez, E., Rico-Lattes, I., Bon, M. and Lattes, A. (2000) *Colloids Surf. A*, **167**, 123.

134 (a) Claverie, J.P., Viala, S., Maurel, V. and Novat, C. (2001) *Macromolecules*, **34**, 382; (b) Lynn, M.D., Mohr, B. and Grubbs, H.R. (1998) *J. Am. Chem. Soc.*, **120**, 1627; (c) Lynn, M.D., Mohr, B., Grubbs, H.R., Henling, M.L. and Day, W.M. (2000) *J. Am. Chem. Soc.*, **122**, 6601.

135 (a) Mul, P.W., Dirkzwager, H., Broekhuis, A.A., Heeres, J.H., van der Linden, J.A. and Orpen, G.A. (2002) *Inorg. Chim. Acta*, **327**, 147; (b) Jiang, Z. and Sen, A. (1994) *Macromolecules*, **27**, 7215; (c) Verspui, G., Schanssema, F. and Sheldon, A.R. (2000) *Angew. Chem. Int. Ed.*, **39**, 804; (d) Opstal, T. and Verpoort, F. (2003) *New J. Chem.*, **27**, 257.

136 (a) Grubbs, R.H. (ed.), (2003) *Handbook of Metathesis*, John Wiley & Sons, Inc, Hoboken, NJ. (b) Bernasconi, C.F. and Ruddat, V. (2002) *J. Am. Chem. Soc.*, **124**, 14968.

137 (a) Kirkland, T.A., Lynn, D.M. and Grubbs, R.H. (1998) *J. Org. Chem.*, **63**, 9904; (b) Rölle, T. and Grubbs, R.H. (2002) *Chem. Commun.*, 1070; (c) Mendez-Andino, J. and Paquette, L.A. (2002) *Adv. Synth. Catal.*, **344**, 303; (d) Audouard, C., Fawcett, J., Grif.ths, G.A., Percy, J.M., Pintat, S. and Smith, C.A. (2004) *Org. Biomol. Chem.*, **2**, 528; (e) Connon, S.J. and Blechert, S. (2002) *Bioorg. Med. Chem. Lett.*, **12**, 1873.

138 (a) Iwasa, S., Takezawa, F., Tuchiya, Y. and Nishiyama, H. (2001) *Chem. Commun.*, 59; (b) Wurz, R.P. and Charette, A.B. (2002) *Org. Lett.*, **4**, 4531; (c) Dehmlow, E.V. (1971) *Tetrahedron*, **27**, 4071.

139 (a) McGrath, D.V. and Grubbs, R.H. (1994) *Organometallics*, **13**, 224; (b) De Bellefon, C., Tanchoux, N., Caravieilhes, S., Grenouillet, P. and Hessel, V. (2000) *Angew. Chem. Int. Ed.*, **39**, 3442; (c) Cadierno, V., Garcia-Garrido, S.E. and Gimeno, J. (2004) *Chem. Commun.*, 232.

140 Bricout, H., Mortreux, A. and Monflier, E. (1998) *J. Organomet. Chem.*, **553**, 469.

141 Li, C.-J., Wang, D. and Chen, D.L. (1995) *J. Am. Chem. Soc.*, **117**, 12867.

142 (a) Baeyer, A. (1882) *Ber. Dtsch. Chem. Ges.*, **15**, 50. (b) review: Cadiot, P. and Chodkiewicz, W. (1969) Chemistry of Acetylenes, (ed. H.G. Viehe), M. Dekker, New York, p. 597.

143 Anderson, S. and Anderson, H.L. (1996) *Angew. Chem. Int. Ed. Engl.*, **35**, 1956.

144 (a) Trost, B.M. and Indolese, A. (1993) *J. Am. Chem. Soc.*, **115**, 4361; (b) Trost, B.M., Indolese, A.F., Müller, T.J.J. and Treptow, B. (1995) *J. Am. Chem. Soc.*, **117**, 615.

145 (a) Trost, B.M., Martinez, J.A., Kulawiec, R.J. and Indolese, A.F. (1993) *J. Am. Chem. Soc.*, **115**, 10402; (b) Dérien, S., Jan, D. and Dixneuf, P.H. (1996) *Tetrahedron*, **52**, 5511.

146 Lopez, F., Castedo, L. and Mascarenas, J.L. (2002) *J. Am. Chem. Soc.*, **124**, 4218.

147 (a) Mizoroki, T., Mori, K. and Ozaki, A. (1971) *Bull. Chem. Soc. Jpn.*, **44**, 581; (b) Heck, R.F. and Nolley, J.P. (1972) *J. Org. Chem.*, **37**, 2320.

148 (a) Bumagin, N.A., Andryuchova, N.P. and Beletskaya, I.P. (1988) *Izv. Akad. Nauk SSSR*, **6**, 1449; (b) Bumagin, N.A., Bykov, V.V., Sukhomlinova, L.I., Tolstaya, T.P. and Beletskaya, I.P. (1995) *J. Organomet. Chem.*, **486**, 259.

149 (a) Casalnuovo, A.L. and Calabrese, J.C. (1990) *J. Am. Chem. Soc.*, **112**, 4324;

(b) Genêt, J.-P., Blart, E. and Savignac, M. (1992) *Synlett*, 715; (c) Hessler, A., Stelzer, O., Dibowski, H., Worm, K. and Schmidtchen, F.P. (1997) *J. Org. Chem.*, **62**, 2362; (d) Amengual, R., Genin, E., Michelet, V., Savignac, M. and Genêt, J.-P. (2002) *Adv. Synth. Catal.*, **344**, 393; (e) Gelpke, A.E.S., Veerman, J.N.J., Goedheijt, M.S., Kamer, P.C.J., van Leuwen, P.W.N.M. and Hiemstra, H. (1999) *Tetrahedron*, **55**, 6657; (f) Eymery, F., Burattin, P., Mathey, F. and Savignac, P. (2000) *Eur. J. Org. Chem.*, **54**, 2425; (g) Zapf, A., Krauter, J.G.E. and Beller, M. (1997) *Angew. Chem. Int. Ed. Engl.*, **36**, 772; (h) Zapf, A., Krauter, J.G.E., Beller, M. and Bogdanovic, S. (1999) *Catal. Today*, **48**, 279.

150 (a) Bhanage, B.M., Ikushima, Y., Shirai, M. and Arai, M. (1999) *Tetrahedron Lett.*, **40**, 6427; (b) Beller, M., Krauter, J.G.E. and Zapf, A. (1997) *Angew. Chem. Int. Ed. Engl.*, **36**, 772.

151 (a) Wan, K.T. and Davis, M.E. (1994) *Nature*, **370**, 449; (b) Bhanage, B.M., Shirai, M. and Arai, M. (1999) *J. Mol. Catal. A: Chemistry*, **145**, 69; (c) Fujita, S.I., Yoshida, T., Bhanage, B.M. and Arai, M. (2002) *J. Mol. Catal. A: Chemistry*, **188**, 37; (d) Bhanage, B.M., Fujita, S.I., Yoshida, T., Sano, Y. and Arai, M. (2003) *Tetrahedron Lett.*, **44**, 3505; (e) Tonks, L., Anson, M.S., Hellgardt, K., Mirza, A.R., Thompson, D.F. and Williams, J.M.J. (1997) *Tetrahedron Lett.*, **38**, 4319.

152 (a) Lautens, M., Roy, A., Fukuoka, K., Fagnou, K. and Martin-Matute, B. (2001) *J. Am. Chem. Soc.*, **123**, 5358; (b) Amengual, R., Michelet, V. and Genêt, J.-P. (2002) *Tetrahedron Lett.*, **43**, 5905.

153 (a) Hayashi, T., Inoue, K., Tanigushi, N. and Ogasawara, M. (2001) *J. Am. Chem. Soc.*, **123**, 9918; (b) Lautens, M. and Yoshida, M. (2002) *Org. Lett.*, **4**, 123; (c) Genin, E., Michelet, V. and Genêt, J.-P. (2004) *Tetrahedron Lett.*, **45**, 4157; (d) Genin, E., Michelet, V. and Genêt, J.-P. (2004) *J. Organomet. Chem.*, **689**, 3820; (d) Genin, E., Amengual, R., Michelet, V., Savignac, M., Jutand, A., Neuville, L. and Genêt, J.-P. (2004) *Adv. Synth. Catal.*, **346**, 1733.

154 (a) Hayashi, T., Inoue, K., Tanigushi, N. and Ogasawara, M. (2001) *J. Am. Chem. Soc.*, **123**, 9918; (b) Miura, T., Shimada, M. and Murakami, M. (2007) *Tetrahedron*, **63**, 6131.

155 Michelet, V., Galland, J.-C., Charruault, L., Savignac, M. and Genêt, J.-P. (2001) *Org. Lett.*, **3**, 2065.

156 Wender, P.A., Love, J.A. and Williams, T.J. (2003) *Synlett*, 1295.

157 (a) Sugihara, T., Yamada, M., Ban, H., Yamaguchi, M. and Kaneko, C. (1997) *Angew. Chem. Int. Ed. Engl.*, **36**, 2801; (b) Son, S.U., Lee, S.I., Chung, Y.K., Kim, S.-W. and Hyeon, T. (2002) *Org. Lett.*, **4**, 277.

158 Fuji, K., Morimoto, T., Tsutsumi, K. and Kabiuchi, K. (2003) *Angew. Chem. Int. Ed.*, **42**, 2409.

159 (a) Nevado, C., Charruault, L., Michelet, V., Nieto-Oberhuber, C., Munoz, M.P., Mendez, M., Rager, M.-N., Genêt, J.-P. and Echavarren, A.M. (2003) *Eur. J. Org. Chem.*, 706; (b) Galland, J.-C., Savignac, M. and Genêt, J.-P. (1997) *Tetrahedron Lett.*, **38**, 8695; (c) Charruault, L., Michelet, V. and Genêt, J.-P. (2002) *Tetrahedron Lett.*, **43**, 4757; (d) Charruault, L., Michelet, V., Taras, R., Gladiali, S. and Genêt, J.-P. (2004) *Chem. Commun.*, 850.

160 (a) Genin, E., Leseurre, L., Toullec, P.-Y., Genêt, J.-P. and Michelet, V. (2007) *Synlett*, 1780; (b) Cabello, N., Rodriguez, C. and Echavarren, A.M. (2007) *Synlett*, 1753; (c) Buzas, A.K., Istrate, F.M. and Gagosz, F. (2007) *Angew. Chem. Int. Ed.*, **46**, 1141; (d) Méndez, M., Muñoz, M.P. and Echavarren, A.M. (2000) *J. Am. Chem. Soc.*, **122**, 11549; (e) Nieto-Oberhuber, C., Muñoz, M.P., Nevado, C., Cárdenas, D.J. and Echavarren, A.M. (2004) *Angew. Chem. Int. Ed.*, **43**, 2402; (f) Méndez, M., Muñoz, M.P., Nevado, C., Cárdenas, D.J. and Echavarren, A.M. (2001) *J. Am. Chem. Soc.*, **123**, 10511.

**161** Piera, J., Persson, A., Caldentey, X. and Bäckvall, J.-E. (2007) *J. Am. Chem. Soc.*, **129**, 14120.

**162** Trost, B.M., Brown, R.E. and Toste, F.D. (2000) *J. Am. Chem. Soc.*, **122**, 5877.

**163** Michelet, V., Toullec, P.Y. and Genêt, J.-P., (2008) *Angew. Chem. Int. Ed.*, **47**, 4268.

**164** Sherry, B.D. and Toste, F.D. (2004) *J. Am. Chem. Soc.*, **126**, 15978.

**165** Sigman, M.S., Fatland, A. and Eaton, B.E. (1998) *J. Am. Chem. Soc.*, **120**, 5130.

**166** (a) Yong, L. and Butenschçn, H. (2002) *Chem. Commun.*, 2852; (b) Uozumi, Y. and Nakazono, M. (2002) *Adv. Synth. Catal.*, **344**, 274.

**167** Fatland, A.W. and Eaton, B.E. (2000) *Org. Lett.*, **2**, 3132.

**168** Kinoshita, H., Shinokubo, H. and Oshima, K. (2003) *J. Am. Chem. Soc.*, **125**, 7784.

**169** (a) Bortollo, M.H., Lavenot, L., Larpent, C., Roucoux, A. and Patin, H. (1997) *Appl. Catal. A*, **156**, 347; (b) Mercier, C. and Chabardes, P. (1994) *Pure Appl. Chem.*, **66**, 1509; (c) Morel, D., Mignani, G. and Colleuille, Y. (1986) *Tetrahedron Lett.*, **27**, 2591.

**170** Behr, A. and Urschey, M. (2003) *J. Mol. Catal. A: Chemistry*, **197**, 101.

**171** (a) Pennequin, I., Meyer, J., Suisse, I. and Mortreux, A. (1997) *J. Mol. Catal. A: Chemistry*, **120**, 139; (b) Donze, C., Pinel, C., Gallezot, P. and Taylor, P.L. (2002) *Adv. Synth. Catal.*, **344**, 906.

**172** Miyaura, N., Hayashi, H. and Sakai, M. (1997) *Organometallics*, **16**, 4229. For recent reviews, see. (a) Hayashi, T. (2001) *Synlett*, 879; (b) Hayashi, T. and Yamasaki, K. (2003) *Chem. Rev.*, **103**, 2829; (c) Hayashi, T. (2004) *Bull. Chem. Soc. Jpn.*, **77**, 13; (d) Hayashi, T. (2004) *Pure Appl. Chem.*, **76**, 465.

**173** (a) Huang, T.-S., Meng, Y., Venkatraman, S., Wang, D. and Li, C.-J. (2001) *J. Am. Chem. Soc.*, **123**, 7451; (b) Miyaura, N., Iguchi, Y. and Itooka, R. (2001) *Chem. Lett.*, 722; (c) Amengual, R., Michelet, V. and Genêt, J.-P. (2002) *Tetrahedron Lett.*, **43**, 5905.

**174** (a) Otomaru, Y., Senda, T. and Hayashi, T. (2004) *Org. Lett.*, **6**, 3357; (b) Amengual, R., Michelet, V. and Genêt, J.-P. (2002) *Synlett*, 1791.

**175** Ma, J. and Chan, T.-H. (1998) *Tetrahedron Lett.*, **39**, 2499.

**176** Usugi, S., Yorimitsu, H. and Oshima, K. (2001) *Tetrahedron Lett.*, **42**, 4535.

**177** (a) Venkatraman, S. and Li, C.-J. (1999) *Org. Lett.*, **1**, 1133; (b) Venkatraman, S. and Li, C.-J. (2000) *Tetrahedron Lett.*, **41**, 4831; (c) Venkatraman, S., Huang, T. and Li, C.-J. (2002) *Adv. Synth. Catal.*, **344**, 399.

**178** (a) Mukhopadhyay, S., Yaghmur, A., Baidossi, M., Kundu, B. and Sasson, Y. (2003) *Org. Process Res. Dev.*, **7**, 641; (b) Mukhopadhyay, S., Rothenberg, G. and Sasson, Y. (2001) *Adv. Synth. Catal.*, **343**, 274; (c) Mukhopadhyay, S., Joshi, A.V., Peleg, L. and Sasson, Y. (2003) *Org. Process Res. Dev.*, **7**, 44.

**179** (a) Li, J.-H., Xie, Y.-X. and Yin, D.-L. (2003) *J. Org. Chem.*, **68**, 9867; (b) Li, J.-H. and Xie, Y.-X. (2004) *Chin. J. Chem.*, **22**, 966; (c) Li, J.-H., Xie, Y.-X., Jiang, H. and Chen, M. (2002) *Green Chem.*, **4**, 424.

**180** Hassan, J., Penalva, V., Lavenot, L., Cozzi, C. and Lemaire, M. (1998) *Tetrahedron*, **54**, 13793.

**181** Panish, J.P., Jung, Y.C., Floyd, R.J. and Jung, K.W. (2002) *Tetrahedron Lett.*, **43**, 7899.

**182** Larsen, D., King, A.O., Chen, C.Y., Corley, E.G., Foster, B.S., Roberts, F.E., Yang, C., Lieberman, D.R., Reamer, R.A., Tschaen, D.M., Verhoeven, T.R., Reider, P.J., Lo, Y.S., Rossano, L.T., Brookes, A.S., Meloni, D., Moore, J.R. and Arnett, J.F. (1994) *J. Org. Chem.*, **59**, 6391.

**183** Miyaura, N., Ishiyama, T., Sasaki, H., Ishikawa, M., Satoh, M. and Suzuki, A. (1989) *J. Am. Chem. Soc.*, **111**, 314.

**184** Armstrong, R.W., Beau, J.M., Cheon, S.H., Christ, W.J., Fujioka, H., Ham, W.H., Hawkins, L.D., Jin, H., Kang, S.H., Kishi, Y., Martinelli, M.J., McWhorter, W.W., Mizuno, M., Nakata, M., Stutz, A.E., Talamas, F.X., Taniguchi, M., Tino, J.A., Ueda, K., Uenishi, J.I., White, J.B.

and Yonaga, M. (1989) *J. Am. Chem. Soc.*, **111**, 7525.
185 Roush, W.R. and Brown, B.B. (1993) *J. Org. Chem.*, **58**, 2162.
186 Evans, D.A., Ng, H.P. and Rieger, D.L. (1993) *J. Am. Chem. Soc.*, **115**, 11446.
187 Nicolaou, K.C., Ramphal, J.Y., Palazon, J.M. and Spanevello, R.A. (1989) *Angew. Chem. Int. Ed. Engl.*, **28**, 587.
188 Ennis, D.S., McManus, J., Wood-Kaszmar, W., Richarson, J., Smith, G.E. and Carstairs, A. (1999) *Org. Process Res. Dev.*, **3**, 248.
189 Casalnuovo, A.L. and Calabrese, J.C. (1990) *J. Am. Chem. Soc.*, **112**, 4324.
190 Wallow, T.I. and Novak, B. (1991) *J. Am. Chem. Soc.*, 7411.
191 (a) Amatore, C., Blart, E., Genêt, J.P., Jutand, A., Lemaire-Audoire, S. and Savignac, M. (1995) *J. Org. Chem.*, **60**, 6829; (b) Papadogiarnakis, G., Peters, J.A., Maat, L. and Sheldon, R.A. (1995) *J. Chem. Soc., Chem. Commun.*, 1105.
192 Genêt, J.P., Blart, E. and Savignac, M. (1992) *Synlett*, 715.
193 Blart, E., Mouries, V., Linquist, A., Savignac, M., Genêt, J.P. and Vaultier, M. (1995) *Tetrahedron Lett.*, **36**, 1443.
194 Mapp, A.K. and Heathcock, C.H. (1999) *J. Org. Chem.*, **64**, 23.
195 Dupuis, C., Charruault, L., Savignac, M., Michelet, V., Adiey, K. and Genêt, J.P. (2001) *Tetrahedron Lett.*, **42**, 6523.
196 (a) Western, E.C., Datf, J.R., Johnson, E.M., II, Gannett, P.M. and Shaughnessy, K.H. (2003) *J. Org. Chem.*, **68**, 6767; (b) Shaughnessy, K.H. and Booth, R.S. (2001) *Org. Lett.*, **3**, 2757; (c) Moore, L.R. and Shaughnessy, K.H. (2004) *Org. Lett.*, **6**, 225.
197 Gelke, A.E.S., Veerman, J.J.N., Goedheijt, M.S., Kamer, P.C., van Leeuwen, P.W.N.N. and Hiemstra, H. (1999) *Tetrahedron*, **55**, 6657.
198 (a) Hannah, J., Ruyle, W.V., Jones, H., Matzuk, K.W., Kelly, K.W., Witzel, B.E., Holtz, W.J., Houser, R.A., Shen, T.Y., Sarett, L.H., Lotti, V.J., Risley, E.A., Van Arman, C.G. and Winter, C.A. (1978) *J. Med. Chem.*, **21**, 1093; (b) Adamski-Werner, S.L., Palaninathan, S.K., Sachettini, J.C. and Kelly, J.W. (2004) *J. Med. Chem.*, **47**, 355.
199 Nishimura, M., Veda, M. and Miyaura, N. (2002) *Tetrahedron*, **58**, 5779.
200 Bumagin, N.A. and Bykov, V.V. (1997) *Tetrahedron*, **53**, 14437.
201 Bandgar, B.P., Bettigeri, S.V. and Phopase, J. (2004) *Tetrahedron Lett.*, **45**, 6959.
202 (a) Stanier, C.A., O'Connell, M.H., Anderson, H.L. and Clegg, W. (2001) *Chem. Commun.*, 493; (b) Taylor, P.N., O'Connell, M.H., McNeill, L.A., Hall, M.J., Aplin, R.T. and Anderson, H.L. (2000) *Angew. Chem. Int. Ed.*, **39**, 3456.
203 Badone, D., Baroni, M., Cardamone, R., Ielmini, A. and Guzzi, U. (1997) *J. Org. Chem.*, **62**, 7170.
204 Weng, Z., Koh, L.L. and Hor, T.S.A. (2004) *J. Organomet. Chem.*, **689**, 18.
205 (a) Sakurai, H., Tsukuda, T. and Hirao, T. (2002) *J. Org. Chem.*, **67**, 2721; (b) Kitamura, Y., Sakurai, A., Udzu, T., Maegawa, T., Monguchi, Y. and Sajikiri, H. (2007) *Tetrahedron*, **63**, 1056.
206 (a) Venkatraman, S. and Li, C.J. (1999) *Org. Lett.*, **1**, 1133; (b) Li, S., Lin, Y., Cao, J. and Zhang, S. (2007) *J. Org. Chem.*, **72**, 4067.
207 (a) Kang, T., Feng, Q. and Luo, M. (2005) *Synlett*, **15**, 2305; (b) Kim, J.H., Kim, J.-W., Shokouhimehr, M. and Lee, Y.S. (2005) *J. Org. Chem.*, **70**, 6714.
208 (a) Uozumi, Y. and Nakai, Y. (2002) *Org. Lett.*, **4**, 2997; (b) Yamada, Y.M.A., Takeda, K., Takahashi, H. and Ikegami, S. (2003) *J. Org. Chem.*, **68**, 7733.
209 Al-Hashimi, M., Oazi, A., Sullivan, A.C. and Wilson, J.R.H. (2007) *J. Mol. Catal. A: Chemical*, **278**, 160.
210 (a) Crudden, C.M., Sateesh, M. and Lewis, R. (2005) *J. Am. Chem. Soc.*, **127**, 10045; (b) Richardson, J.M. and Jones, C.W. (2007) *J. Catal.*, **251**, 80.
211 (a) Li, Y., Hong, X.M., Collard, D.M. and El-Sayed, M.A. (2000) *Org. Lett.*, **2**, 2385; (b) Lu, F., Ruiz, J. and Astruc, D. (2004)

*Tetrahedron Lett.*, **45**, 9443; (c) Cho, J.K., Najman, R., Dean, T.W., Ichihara, O., Muller, C. and Bradley, M. (2006) *J. Am. Chem. Soc.*, **128**, 6276; (d) Brudoni, G., Lorma, A., Garcia, H. and Primo, A. (2007) *J. Catal.*, **251**, 345; (e) Zheng, P. and Zhang, W. (2007) *J. Catal.*, **250**, 324.

212 Lawason Daku, K.M., Newton, R.F., Pearce, S.P., Vile, J. and Williams, J.M.J. (2003) *Tetrahedron Lett.*, **44**, 5095.

213 (a) Bulut, H., Artok, L. and Yilmazu, S. (2003) *Tetrahedron Lett.*, **44**, 289; (b) Artok, L. and Bulut, H. (2004) *Tetrahedron Lett.*, **45**, 3881.

214 Shimizu, K.I., Maruyama, R., Komai, S.-I., Kodama, T. and Kitayama, Y. (2004) *J. Catal.*, **227**, 202.

215 Yan, P., Ma, S., Li, J., Xiao, F. and Xiong, H. (2006) *Chem. Commun.*, 2495.

216 Lin, B., Liu, Z., Liu, M., Pan, C., Ding, J., Wu, H. and Cheng, J. (2007) *Catal. Commun.*, **8**, 2150.

217 Colacot, T.J., Gore, E.S. and Kuber, A. (2002) *Organometallics*, **21**, 3301.

218 Llabresi Xamena, F.X., Abad, A., Corma, A. and Garcia, H. (2007) *J. Catal.*, **256**, 294.

219 Arcadi, A., Cerichelli, G., Chiarini, M., Correa, M. and Zorzan, D. (2003) *Eur. J. Org. Chem.*, 4080.

220 Hapiot, F., Lyskawa, J., Bricout, H., Tilloy, S. and Monflier, E. (2004) *Adv. Synth. Catal.*, **346**, 83.

221 Larhed, M., Moberg, C. and Hallberg, A. (2002) *Acc. Chem. Res.*, **35**, 717.

222 (a) Leadbeater, N.E. (2005) *Chem. Commun.*, 2881; (b) Arvela, R.K., Leadbeater, N.E., Mack, T.L. and Kromos, C.M. (2006) *Tetrahedron Lett.*, **47**, 217.

223 (a) Leadbeater, N.E. and Marco, M. (2003) *Angew. Chem. Int. Ed.*, **42**, 1407; (b) Leadbeater, N.E. and Marco, M. (2002) *Org. Lett.*, **4**, 2973; (c) Leadbeater, N.E. and Marco, M. (2003) *J. Org. Chem.*, **68**, 888; (d) Bai, L., Wang, J.-X. and Zhang, Y. (2003) *Green Chem.*, **5**, 615; (e) Leadbeater, N.E. and Marco, M. (2003) *J. Org. Chem.*, **68**, 5660; (f) Li, C.-J. (2003) *Angew. Chem. Int. Ed.*, **42**, 4856.

224 Darses, S., Jeffery, T., Brayer, J.L., Demoute, J.P. and Genêt, J.P. (1996) *Tetrahedron Lett.*, **37**, 3857.

225 Darses, S., Jeffery, T., Brayer, J.L., Demoute, J.P. and Genêt, J.P. (1996) *Bull. Soc. Chim. Fr.*, **133**, 1095.

226 Taylor, R.H. and Felpin, F.X. (2007) *Org. Lett.*, **9**, 2911.

227 Willis, D.M. and Strongin, R.N. (2000) *Tetrahedron Lett.*, **41**, 6271.

228 (a) Darses, S., Michaud, G. and Genêt, J.P. (1999) *Eur. J. Org. Chem.*, 1875; (b) Darses, S., Michaud, G. and Genêt, J.P. (1998) *Tetrahedron Lett.*, **39**, 5045.

229 Calderwood, D.J., Johnston, D.N., Rafferty, P., Twigger, H.L., Munschauer, R. and Arnold, L. (1998) (Knoll AG Chemische Fabriken) Patent WO 9841525.

230 Puenteuer, K. and Scalone, M. (2000) (Hoffmann-La Roche AG) European Patent EP 1057831A.

231 (a) Batey, R.A. and Qauch, T.D. (2002) *Tetrahedron Lett.*, **42**, 7601; (b) Molander, G.A. and Bernardi, C.R. (2002) *J. Org. Chem.*, **67**, 8424; (c) Molander, G.A. and Biolatto, B. (2002) *Org. Lett.*, **4**, 1867; (d) Barder, T.E. and Buchwald, S.L. (2004) *Org. Lett.*, **6**, 2649.

232 Molander, G.A. and Jean-Gerard, L. (2007) *J. Org. Chem.*, **72**, 8422.

233 (a) Littke, A. and Fu, G. (1998) *Angew. Chem. Int. Ed.*, **37**, 3387; (b) Old, D.W., Wolfe, J.P. and Buchwald, S.L. (1998) *J. Am. Chem. Soc.*, **120**, 9722.

234 (a) Najera, C., Gil-Molto, J. and Karlström, S. (2004) *Adv. Synth. Catal.*, **346**, 1798; (b) Botella, L. and Najera, C. (2002) *Angew. Chem. Int. Ed.*, **41**, 179; (c) Botella, L. and Najera, C. (2002) *J. Organomet. Chem.*, **663**, 46; (d) Botella, L. and Najera, C. (2004) *Tetrahedron Lett.*, **45**, 1833; (e) Baleizao, C., Corma, A., Garcia, H. and Leyva, A. (2004) *J. Org. Chem.*, **69**, 439.

235 Anderson, K.W. and Buchwald, S.L. (2005) *Angew. Chem. Int. Ed.*, **44**, 6173.

236 Limo, J., Henri, K. and Astruc, D. (2007) *Chem. Commun.*, 4351.

237 (a) Fleckenstein, C.A. and Plenio, H. (2007) *Green Chem.*, **9**, 1287; (b) Fleckenstein, C.A. and Plenio, H. (2007) *Chem. Eur. J.*, **13**, 2701.

238 Fleckenstein, C.A., Roy, S., Leuthauber, S. and Plenio, H. (2007) *Chem. Commun.*, 2870.

239 Han, W. and Jin, Z.-L. (2007) *Org. Lett.*, **9**, 4005.

240 Schweizer, S., Becht, J.M. and Le Drian, C. (2007) *Org. Lett.*, **9**, 3777.

241 (a) Hierso, J.C., Fihri, A., Armadeil, P., Meunier, P., Doucet, H., Santelli, M. and Monnadieu, B. (2003) *Organometallics*, **22**, 4490; (b) review: Hierso, J.C., Beauperin, M. and Meunier, P. (2007) *Eur. J. Inorg. Chem.*, 3767.

242 Bollinger, J.L., Blacque, O. and Frech, C.M. (2007) *Angew. Chem. Int. Ed.*, **46**, 6514.

243 Lee, M., Jang, C.-J. and Ryu, J.-H. (2004) *J. Am. Chem. Soc.*, **126**, 8082.

244 Li, F., Bai, S. and Hor, T.S. (2008) *Organometallics*, **27**, 672.

245 (a) Saito, S., Sakai, M. and Miyaura, N. (1996) *Tetrahedron Lett.*, **37**, 2993; (b) Indolese, A.F. (1997) *Tetrahedron Lett.*, **38**, 3513.

246 Galland, J.C., Savignac, M. and Genêt, J.P. (1999) *Tetrahedron Lett.*, **40**, 2323.

247 Saito, B. and Fu, G.C. (2007) *J. Am. Chem. Soc.*, **129**, 9602.

248 Na, Y., Park, S., Han, S.B., Han, H., Ko, S. and Chang, S. (2004) *J. Am. Chem. Soc*, **126**, 250.

249 Tueting, D.R., Echavarren, A.M. and Stille, J.K. (1989) *Tetrahedron*, **45**, 979.

250 Zhang, H.C. and Daves, G.D. Jr.* (1993) *Organometallics*, **12**, 1499.

251 (a) Roshchin, A.I., Bumagin, N.A. and Beletskaya, I.P. (1995) *Tetrahedron Lett.*, **36**, 125; (b) Rai, R., Aubrecht, K.B. and Collum, D.B. (1995) *Tetrahedron Lett.*, **36**, 3111.

252 Bumagin, N.A., Sukhomlinova, L.I., Tolstaya, T.P. and Beletskaya, I.P. (1994) *Russ. J. Org. Chem.*, **30**, 1605.

253 (a) Maleczka, R.E., Jr, Gallagher, W.P. and Terstiege, I. (2000) *J. Am. Chem. Soc.*, **122**, 384; (b) Gallagher, W.P., Terstiege, I. and Maleczka, R.E. Jr.* (2001) *J. Am. Chem. Soc.*, **123**, 3194.

254 Wolf, C. and Lerebours, R. (2003) *J. Org. Chem.*, **68**, 7551.

255 (a) Venkatraman, S. and Li, C.J. (1999) *Org. Lett.*, **1**, 1133; (b) Venkatraman, S., Huang, T.S. and Li, C.J. (2002) *Adv. Synth. Catal.*, **344**, 399.

256 Pal, M., Subramanian, V. and Yeleswarapu, K.R. (2003) *Tetrahedron Lett.*, **44**, 8221.

257 Métivier, R., Amengual, R., Leray, I., Michelet, V. and Genêt, J.P. (2004) *Org. Lett.*, **6**, 739.

258 (a) Bumagin, N.A., Bykov, V.V. and Beletskaya, I.P. (1995) *Zh. Org. Khim.*, **31**, 385; (b) Luzikova, E.V., Bumagin, N.A. and Beletskaya, I.P. (1993) *Bull. Acad. Sci. Russ.*, **42**, 585; (c) Davydov, D.V. and Beletskaya, I.P. (1995) *Russ. Chem. Bull.*, **44**, 965.

259 (a) Bumagin, N.A., Sukhomlinova, L.I., Luzikova, E.V., Tolstaya, T.P. and Beletskaya, I.P. (1996) *Tetrahedron Lett.*, **37**, 897; (b) Kang, S.-K., Lee, H.-W., Jang, S.-U. and Ho, P.-S. (1996) *Chem. Commun.*, 835.

260 (a) Li, C.J., Slaven, W.T., IV, Chen, Y.P., John, V.T. and Rachakonda, S.H. (1998) *Chem. Commun.*, 1351; (b) Li, C.J., Chen, D.L. and Costello, C.W. (1997) *Org. Res. Process Dev.*, **1**, 315.

261 Li, C.J., Slaven, W.T., John, V.T. and Banerjee, S. (1997) *J. Chem. Soc., Chem. Commun.*, 1569.

262 (a) Dibowski, H. and Schmidtchen, F.P. (1995) *Tetrahedron*, **51**, 2325; (b) Hessler, A., Stelzer, O., Dibowski, H., Worm, K. and Schmidtchen, F.P. (1997) *J. Org. Chem.*, **62**, 2362; (c) Machnitzki, P., Tepper, M., Wenz, K., Stelzer, O. and Herdtweck, E. (2000) *J. Organomet. Chem.*, **602**, 158.

263 Dibowski, H. and Schmidtchen, F.P. (1998) *Angew. Chem. Int. Ed.*, **37**, 476.

264 Bong, D.T. and Ghadiri, M.R. (2001) *Org. Lett.*, **3**, 2509.

265 (a) Bleicher, L.L. and Cosford, D.P. (1995) *Synlett*, 1115; (b) Batchu, V.R., Subramanian, V., Parasuraman, K., Swany, N.K., Kumar, S. and Pal, M. (2005) *Tetrahedron*, **61**, 9869; (c) review: Felpin, F.-X., Ayad, T. and Mitra, S. (2006) *Eur. J. Org. Chem.*, 2679.

266 Lopez-Deber, M.P., Castedo, L. and Granja, J.R. (2001) *Org. Lett.*, **3**, 2823.

267 Nguefack, J.F., Bolitt, V. and Sinou, D. (1996) *Tetrahedron Lett.*, **37**, 5525.

268 (a) Bhattacharya, S. and Sengupta, S. (2004) *Tetrahedron Lett.*, **45**, 8733; (b) Ahmed, M.S.M. and Mori, A. (2004) *Tetrahedron*, **60**, 9977.

269 Hoshi, M., Nikayababu, H. and Shirakawa, K. (2005) *Synthesis*, 1991.

270 (a) Pirguliyev, N.Sh., Brel, V.K., Zefirov, N.S. and Stang, P.J. (1999) *Tetrahedron*, **55**, 12377; (b) Pirguliyev, N.Sh., Brel, V.K., Zefirov, N.S. and Stang, P.J. (2001) *Tetrahedron Lett.*, **42**, 5759.

271 (a) Gil-Molto, J. and Najera, C. (2005) *Eur. J. Org. Chem.*, 4073; (b) Najera, C., Gil-Molto, J., Kalstrom, S. and Falvello, L.R. (2003) *Org. Lett.*, **5**, 1451.

272 Yang, Z. (2005) *J. Org. Chem.*, **70**, 391.

273 Tremblay-Morin, J.P., Ali, H. and Van Lier, E. (2005) *Tetrahedron Lett.*, **46**, 6999.

274 Beletskaya, I.P., Latyshev, G.V., Tsvetkov, A.V. and Lukashev, N.V. (2003) *Tetrahedron Lett.*, **44**, 5011.

275 Safi, M. and Sinou, D. (1991) *Tetrahedron Lett.*, **32**, 2025.

276 Blart, E., Genêt, J.-P., Safi, M., Savignac, M. and Sinou, D. (1994) *Tetrahedron*, **50**, 505.

277 Kinoshita, H., Shinokubo, H. and Oshima, K. (2004) *Org. Lett.*, **6**, 4085.

278 (a) Manabe, K. and Kobayashi, S. (2003) *Org. Lett.*, **5**, 3241; (b) Patil, N.T. and Yamamoto, Y. (2004) *Tetrahedron Lett.*, **45**, 3101; (c) Yang, S.C., Hsu, Y.C. and Gan, K.H. (2006) *Tetrahedron*, **62**, 3949.

279 Sigismondi, S., Sinou, D., Perez, M., Moreno-Manas, M., Pleixats, R. and Villarroya, M. (1994) *Tetrahedron Lett.*, **35**, 7085.

280 (a) Genêt, J.P., Blart, E., Savignac, M., Lemeune, S. and Paris, J.M. (1993) *Tetrahedron Lett.*, **34**, 4189; (b) Genêt, J.P., Blart, E., Savignac, M., Lemaire-Audoire, S., Paris, J.M. and Bernard, J.M. (1994) *Tetrahedron*, **50**, 497.

281 (a) Michelet, V., Adiey, K., Bulic, B., Genêt, J.-P., Dujardin, G., Rossignol, S., Brown, E. and Toupet, L. (1999) *Eur. J. Org. Chem.*, 2885; (b) Michelet, V., Adiey, K., Tanier, S., Dujardin, G. and Genêt, J.-P. (2003) *Eur. J. Org. Chem.*, 2947.

282 (a) Lemaire-Audoire, S., Genêt, J.P. and Savignac, M. (1994) *Tetrahedron Lett.*, **35**, 8783; (b) Lemaire-Audoire, S., Genêt, J.P., Savignac, M., Pourcelot, G. and Bernard, J.M. (1997) *J. Mol. Catal. A: Chemical*, **116**, 247.

283 Lemaire-Audoire, S., Genêt, J.P. and Savignac, M. (1997) *Tetrahedron Lett.*, **38**, 2955.

284 Torque, C., Bricout, H., Hapiot, F. and Monflier, E. (2004) *Tetrahedron*, **60**, 6487.

285 Kinoshita, H., Shinokubo, H. and Oshima, K. (2005) *Angew. Chem. Int. Ed.*, **44**, 2397.

286 Felpin, F.C. and Landais, Y. (2005) *J. Org. Chem.*, **70**, 6441.

287 (a) Schneider, P., Quignard, F., Choplin, A. and Sinou, D. (1996) *New. J. Chem.*, **20**, 545; (b) Dos Santos, S., Tong, Y., Quignard, F., Choplin, A., Sinou, D. and Dutasta, J.P. (1998) *Organometallics*, **17**, 78.

288 Tonks, L., Anson, M.S., Hellgardt, K., Mirza, A.R., Thompson, D.F. and Williams, M.J. (1997) *Tetrahedron Lett.*, **38**, 4319.

289 Kobayashi, S., Lam, W.W. and Manabe, K. (2000) *Tetrahedron. Lett.*, **41**, 6115.

290 Bergbreiter, D.E. and Liu, Y.-C. (1997) *Tetrahedron Lett.*, **38**, 7843.

291 (a) Uozumi, Y., Danjo, H. and Hayashi, T. (1997) *Tetrahedron Lett.*, **38**, 3557; (b) Danjo, H., Tanaka, D., Hayashi, T. and Uozumi, Y. (1999) *Tetrahedron*, **55**, 14341.

292 Feuerstein, M., Laurenti, D., Doucet, H. and Santelli, M. (2001) *Tetrahedron Lett.*, **42**, 2313.

293 (a) Sinou, D., Rabeyrin, C. and Nguefack, J.F. (2003) *Adv. Synth. Catal.*, **345**, 357; (b) Rabeyrin, C. and Sinou, D. (2003) *Tetrahedron: Asymmetry*, **14**, 3891; (c) Rabeyrin, C., Nguefack, C. and Sinou, D. (2000) *Tetrahedron Lett.*, **41**, 7461.

294 Uozumi, Y., Danjo, H. and Hayashi, T. (1998) *Tetrahedron Lett.*, **39**, 8303.

295 (a) Uozumi, Y., Tanaka, H. and Shibatomi, K. (2003) *Org. Lett.*, **5**, 281; (b) Uozumi, Y. and Shibatomi, K. (2001) *J. Am. Chem. Soc.*, **123**, 2919.

296 (a) Yang, B.H. and Buchwald, S.L. (1999) *J. Organomet. Chem.*, **576**, 125; (b) Prim, D., Campagne, J.-M., Joseph, D. and Andrioletti, B. (2002) *Tetrahedron*, **58**, 2041; (c) Wolfe, J.P. and Buchwald, S.L. (2000) *J. Org. Chem.*, **65**, 1144; (d) Schlummer, B. and Scholz, U. (2004) *Adv. Synth. Catal.*, **346**, 1599.

297 Kuwano, R., Utsunomiya, M. and Hartwig, J.F. (2002) *J. Org. Chem.*, **67**, 6479.

298 Huang, X.H., Anderson, K.W., Zim, D., Jiang, L., Klapars, A. and Buchwald, S.L. (2003) *J. Am. Chem. Soc.*, **125**, 6653.

299 Klapars, A., Huang, X. and Buchwald, S.L. (2002) *J. Am. Chem. Soc.*, **124**, 7421.

300 Yin, J., Zhao, K.M., Huffman, M.A. and McNamara, J.M. (2002) *Org. Lett.*, **4**, 3481.

301 Dallas, A.S. and Gothelf, K.V. (2005) *J. Org. Chem.*, **70**, 3321.

302 Roshchin, A.I., Bumagin, N.A. and Beletskaya, I.P. (1994) *Dokl. Chem.*, **334**, 47.

303 Huang, T.S. and Li, C.J. (2002) *Tetrahedron Lett.*, **43**, 403.

304 Sahoo, A.K., Oda, T., Nakao, Y. and Hiyama, T. (2004) *Adv. Synth. Catal.*, **346**, 1715.

305 Wolf, C. and Lerebours, R. (2004) *Org. Lett.*, **6**, 1147.

306 (a) Alacid, E. and Najera, C. (2006) *Adv. Synth. Catal.*, **348**, 2085; (b) Alacid, E. and Najera, C. (2006) *Adv. Synth. Catal.*, **348**, 945.

307 Gordillo, A., de Jesus, E. and Lopez-Mardomingo, C. (2007) *Chem. Commun.*, 4056.

# 7
## "On Water" for Green Chemistry

Li Liu and Dong Wang

## 7.1
### Introduction

Water is a unique green solvent for chemical syntheses in view of its low cost, safety, and environmentally benign properties [1]. However, for more than 100 years, most organic reactions have been carried out in organic solvents [2]. There are two main drawbacks that limit the utility of water to act as a reaction medium: poor solubility of most organic compounds in water and the hydrolysis or decomposition of some substrates and organometallic complexes. Many efforts have been made to improve sluggish organic reactions in aqueous media and enhance the selectivities by increasing the water-solubility, such as by introducing some polar functional groups on the substrates and catalysts, and also the use of surfactants.

In 1980, Rideout and Breslow reported that the Diels–Alder reaction could be accelerated 700-fold in dilute aqueous solution compared with an organic solvent owing to the hydrophobic effect and interaction through hydrogen bonding [3]. This discovery attracted much attention and led to research on unusual properties and advantages of water as solvent. Since that time, many modifications, such as increasing water solubility of the reagents, addition of a surfactant or an organic cosolvent, and so on, have been made for highly efficient reactions in water. In these cases, homogeneity of the reaction system is still desired. However, in a dilute aqueous solution, it is not possible to perform preparative-scale reactions. Moreover, in general, the work-up procedure after an aqueous reaction needs an organic solvent to extract the products. There is a wide gap between the use of water as a reaction medium and practical green synthesis.

In 2005, Sharpless and co-workers showed that several mono- and bimolecular reactions, including cycloadditions, ene reactions, Claisen rearrangements, and nucleophilic ring opening of an epoxide, are dramatically accelerated when water-insoluble reactants are stirred in aqueous suspensions, denoted "on water" conditions [4], which has been commonly used to refer to water-based reactions of water-insoluble organic substrates. Various reaction media, such as non-protic and protic

*Handbook of Green Chemistry, Volume 5: Reactions in Water.* Edited by Chao-Jun Li
Copyright © 2010 WILEY-VCH Verlag GmbH & Co. KGaA, Weinheim
ISBN: 978-3-527-31591-7

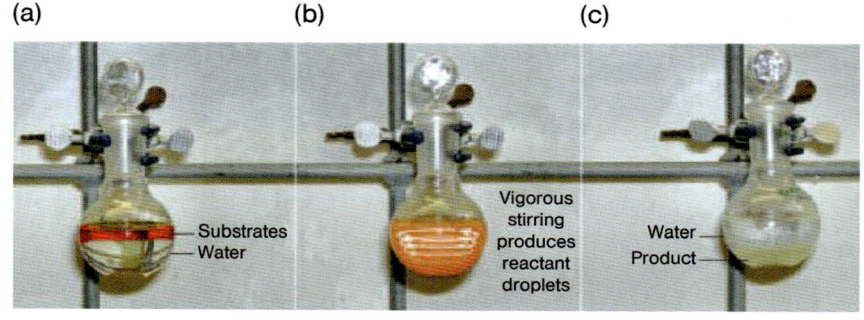

**Figure 7.1** The reaction of quadricyclane with dimethyl azodicarboxylate under "on water" conditions. (a) Reactants float on the surface of water; (b) the reaction suspension is stirred; (c) the reaction product is formed. Reproduced with permission from [5].

organic solvents, plain water, and solvent-free, were examined. In all cases, the use of plain water as a reaction medium is an optimal reaction condition, and the efficiency of "on water" reactions is remarkable. Furthermore, even in aqueous media, heterogeneity is crucial for observing large rate accelerations. The reactant loading can be scaled up regardless of their water solubility. Mixings of liquid–liquid, liquid–solid and even solid–solid reactants are possible under "on water" conditions. On the other hand, the amount of water used in "on water" reactions is not strict provided that enough water is present to disperse all reactants and make a clear phase separation. The use of water as a reaction medium has been applied not only for safety reasons by controlling the reaction thermal effect and the mixing of reagents, but also to accelerate the reaction rate.

A simple procedure for the reaction of quadricyclane with dimethyl azodicarboxylate "on water" is shown in Figure 7.1 [5]. Initially, the insoluble reactants float on top of the water (a). Then the mixture was stirred vigorously, and an aqueous suspension was observed (b). Finally, an insoluble product was formed, which could be easily isolated by phase separation or filtration (c). "On water" reactions are not only suitable for syntheses on a large scale, but are also favorable for the isolation of the products without the use of any organic extractors.

Although it has been only for a short time since the "on water" effect was pointed out, many examples have now been reported. Several organic reactions in aqueous suspensions are summarized in this chapter to help understand further the "on water" effect and to apply such an effect to practical green synthesis.

## 7.2
### Pericyclic Reactions

The first reaction exhibiting the "on water" effect, reported by Sharpless, was the $2\sigma + 2\sigma + 2\pi$ cycloaddition reaction of quadricyclane with azodicarboxylates [4].

**Table 7.1** Reaction of quadricyclane with dimethyl azodicarboxylate in various solvents.

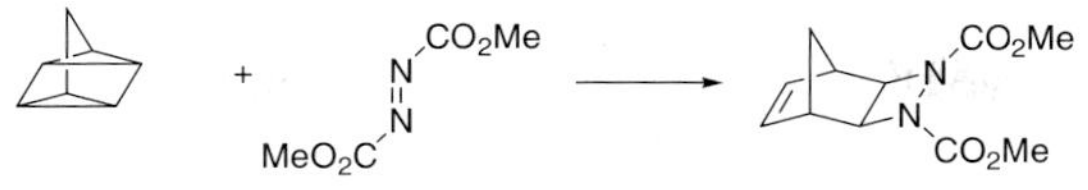

| Solvent | Concentration (M)[a] | Time |
|---|---|---|
| Toluene | 2 | >120 h |
| EtOAc | 2 | >120 h |
| $CH_3CN$ | 2 | 84 h |
| $CH_2Cl_2$ | 2 | 72 h |
| DMSO | 2 | 36 h |
| MeOH | 2 | 18 h |
| Neat | 4.53 | 48 h |
| "On water" | 4.53 | 10 min |
| MeOH–$H_2O$ (3:1, homogeneous) | 2 | 4 h |
| MeOH–$H_2O$ (1:1, heterogeneous) | 4.53 | 10 min |
| MeOH–$H_2O$ (1:3, heterogeneous) | 4.53 | 10 min |

[a] Concentrations of the neat and heterogeneous reactions are calculated from the measured density of a 1:1 mixture of two reactants.

In organic solvents the reaction took a long time to reach completion, from 18 h to several days, whereas when the two water-insoluble reactant liquids were vigorously stirred in water, the reaction could be completed in 10 min to afford the products, 1,2-diazetidines, in 82% yield (Table 7.1). A concentration effect on the rate acceleration was excluded since the corresponding solvent-free conditions were shown to be unfavorable. The impressive rate acceleration by water was also observed in heterogeneous reaction systems. For example, homogeneous reaction in $H_2O$–MeOH (1:3) is noticeably slower than heterogeneous reaction in $H_2O$–MeOH (1:1 or 3:1), 4 h versus 10 min (Table 7.1).

Azodicarboxylates were also employed in the ene reaction with cyclohexene on water for 8 h to provide the allylamine product in 91% yield, whereas the reactions in toluene or under neat conditions required at least 24 h, affording the product in 60–70% yields (Scheme 7.1) [4].

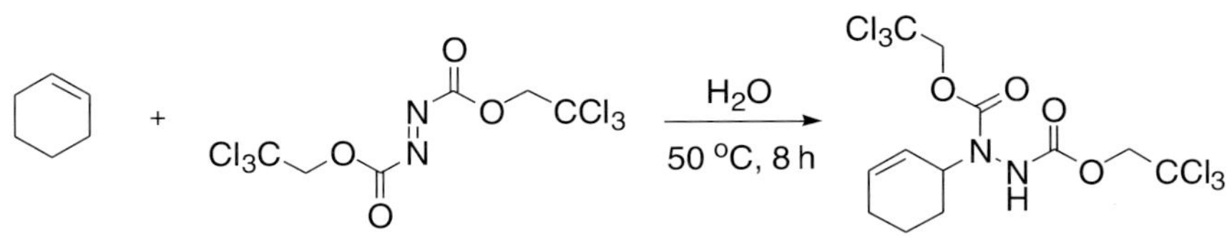

**Scheme 7.1**

The Diels–Alder reaction is one of the most important reactions for C–C bond formation. The rate acceleration in the Diels–Alder reaction was also observed

**Table 7.2** Comparison of water with organic solvents for a typical Diels–Alder reaction

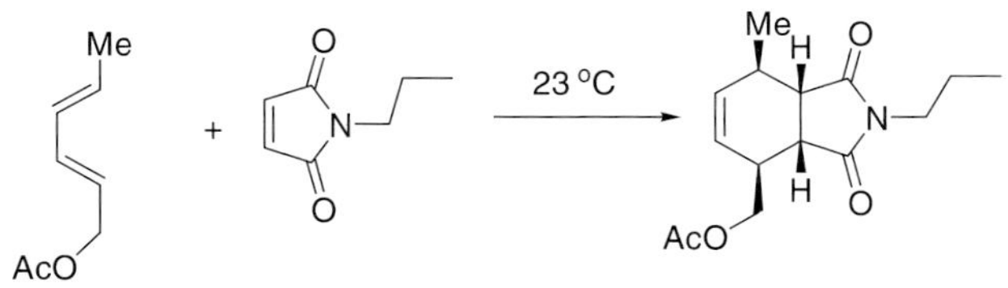

| Solvent | Concentration (M)[a] | Time (h) | Yield (%) |
|---|---|---|---|
| Toluene | 1 | 144 | 79 |
| $CH_3CN$ | 1 | >144 | 43 |
| MeOH | 1 | 48 | 82 |
| Neat | 3.69 | 10 | 82 |
| $H_2O$ | 3.69 | 8 | 81 |

[a]Calculated from the measured density of a 1:1 mixture of two reactants/.

in an aqueous suspension system. The addition reaction of water-insoluble *trans,trans*-2,4-hexadienyl acetate and *N*-propylmaleimide proceeded on water in 8 h and provided the corresponding product in 81% yield, which was similar to that under solvent-free conditions, whereas in an organic solvent the reaction took at least 48 h for completion (Table 7.2) [4].

Based on their experimental results, Sharpless and co-workers indicated that nonpolar liquids could be ideal candidates for "on water" reactions. The reactants needed be mixed adequately and vigorous stirring promoted the "on water" reaction noticeably.

de Armas and co-workers reported the first example of regioselective and organocatalyzed 1,3-dipolar cycloaddition reactions between conjugated alkynoates and nitrones "on water" (Scheme 7.2) [6]. In the presence of $PR_3$ catalyst or a tertiary amine, reagents suspended in water ("on water" conditions) and reacted with *in situ*-generated β-phosphonium (or ammonium) allenolates provided 2,3-dihydroisoxazole with complete regioselectivity. It could be found that the amount of water was not important provided that enough water was present to fuse the solid nitrone and bring all of the reactants in close contact. Vigorous stirring was necessary, but no reaction was observed under the same conditions in organic solvents.

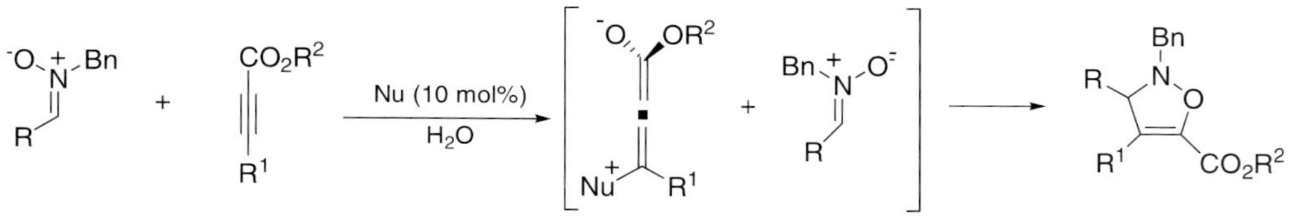

**Scheme 7.2**

## 7.3
## Addition of Heteronucleophiles to Unsaturated Carbonyl Compounds

Various conjugate additions have been successfully carried out in aqueous media. Soft heteroatom-based nucleophiles such as amines, thiols, and phosphines can be compatible with soft electron-deficient alkenes and alkynes in the addition of heteronucleophiles to unsaturated carbonyl compounds. Significant rate acceleration was observed in the conjugate addition of thiols to $\alpha,\beta$-unsaturated carbonyl compounds in water in comparison with organic solvent or solvent-free conditions [7]. In the aza-Michael addition of various primary and secondary amines (except aniline) to $\alpha,\beta$-unsaturated carboxylic esters, ketones, nitriles, and amides, the transformations in a heterogeneous aqueous system proceeded smoothly and provided the best yields (85–90%). In contrast, the reactions in common organic solvents or under solvent-free conditions were very slow, with relatively low yields (30–70%) (Scheme 7.3) [8]. However, aromatic amines and $\alpha,\beta$-unsaturated aldehydes could not be used in this reaction.

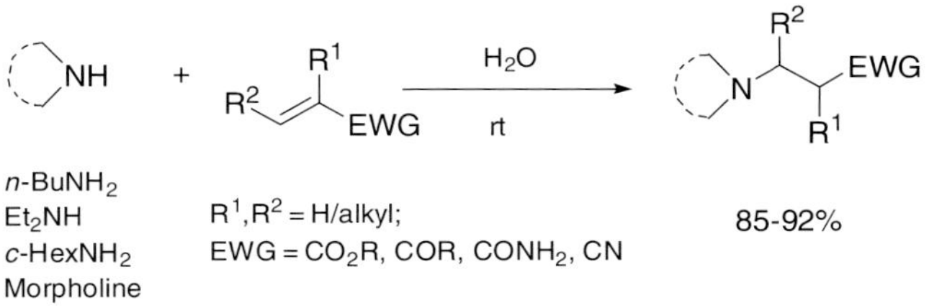

Scheme 7.3

The dual activation by water of both C-electrophiles and N-nucleophiles is proposed in Scheme 7.4. The promotion by water of the aza-Michael addition reaction may be through hydrogen bond formation of water with both the carbonyl oxygen atom of the $\alpha,\beta$-unsaturated carbonyl compound and the nitrogen atom of the amine.

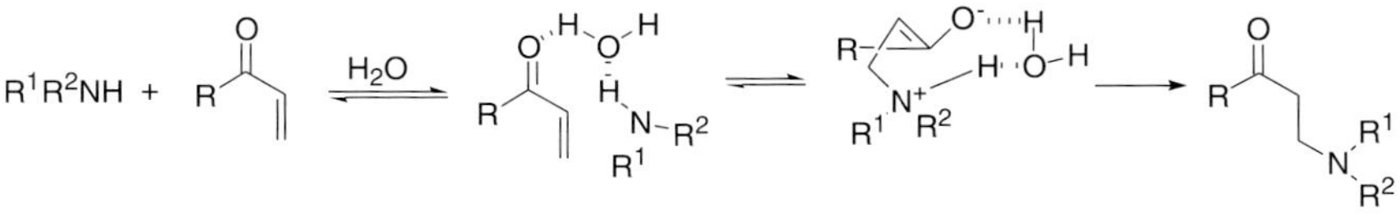

Scheme 7.4

In a one-pot, three-component condensation reaction of an aldehyde, 2-aminobenzothiazole, and 2-naphthol or 6-hydroxyquinoline in aqueous suspension at 90 °C, 2′-aminobenzothiazolomethylnaphthols or 5-(2′-aminobenzothiazolomethyl)-6-hydroxyquinolines were obtained in high yields [9]. The reaction did not proceed in organic solvents and ionic liquids even after 24 h. Under solvent-free conditions, the reaction was very slow and the yields of products were only 10–35% (Table 7.3).

The proposed mechanism might be through the *in situ* formation of *ortho*-quinone methides, followed by aza-Michael addition of 2-aminobenzothiazole to *ortho*-quinone

**Table 7.3** Solvent effects on the reaction of 2-naphthol and p-methylbenzaldehyde with 2-aminobenzothiazole.

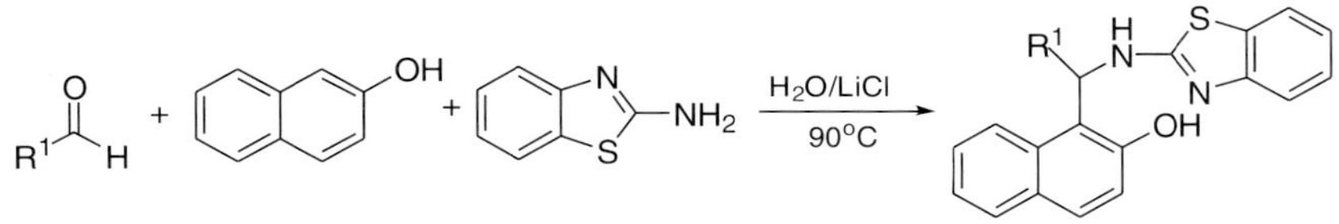

| Entry | Solvent | Time (h) | Temperature (°C) | Yield (%) |
|---|---|---|---|---|
| 1 | $CH_2Cl_2$ | 24 | Reflux | 0 |
| 2 | $CHCl_3$ | 24 | Reflux | 0 |
| 3 | $CH_3CO_2Et$ | 24 | Reflux | 0 |
| 4 | $CH_3CN$ | 24 | Reflux | 0 |
| 5 | $CH_3OH$ | 24 | Reflux | 0 |
| 6 | [Bmim]Br | 24 | 90 | 0 |
| 7 | [Bmim]$PF_6$ | 24 | 90 | 0 |
| 8 | Water | 24 | 90 | 81 |
| 9 | Water | 6 | 90 | 80 |
| 10 | Water | 3 | 90 | 52 |
| 11 | – | 24 | 90 | 10 |

methides (Scheme 7.5). If LiCl was added to the aqueous reaction system, the yields increased noticeably, which provided evidence for the existence of a hydrophobic effect in the reaction system. The hydrophobic effect of water may play an important role in the rate acceleration and is very efficient for multi-component reactions in which the entropy of reaction is decreased in the transition state.

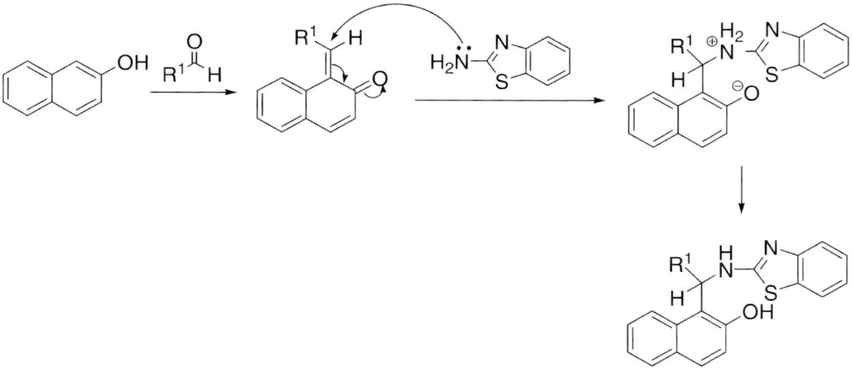

**Scheme 7.5**

Highly efficient condensation reactions of 2-aminothiophenol with aldehydes were carried out "on water" in the absence of any acid–base catalyst (Scheme 7.6) [10].

The aromatic aldehydes afforded benzothiazoles, whereas aliphatic aldehydes gave benzothiazolines in good yields. No Michael addition occurred for cinnamaldehyde and no competitive dithioacetal formation was observed. The reactions can be carried out smoothly in tap water, so that the consumption of energy in preparing distilled water was avoided. The condensation reaction is faster on water than under neat conditions or in organic solvent. The consumption of 2-aminothiophenol substrate required 20, 60, 90, 120, 120, 180, 40 and 50 min "on water "and in DMSO, EtOH, THF, PhMe, dioxane, and NMP, and under solvent-free conditions, respectively. "On water" catalysis, in which water was used alone as the reaction medium and reagents were insoluble in water, provided the best ratio of thiazole to thiazoline (100:0) after 3 h, in comparison with other reaction conditions.

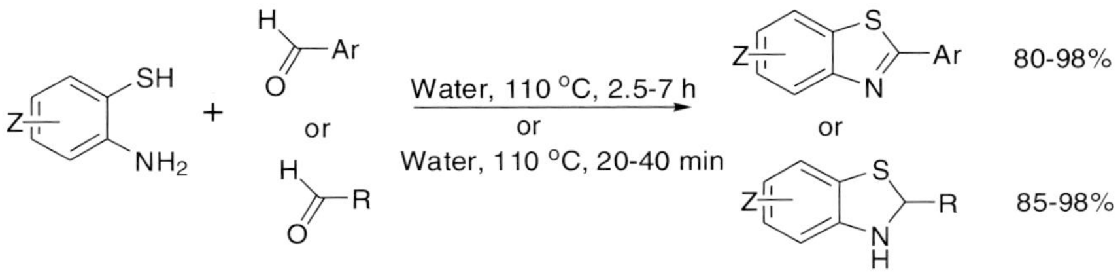

**Scheme 7.6**

Water may exhibit a dual activation effect by hydrogen bond formation with the aldehyde carbonyl oxygen and the SH hydrogen of 2-aminothiophenol, facilitating the formation of thiazoline. Dehydrogenation of thiazoline by dissolved oxygen in water provided thiazoles for aromatic aldehydes, in which a large conjugated system is formed with lower product energy (Scheme 7.7).

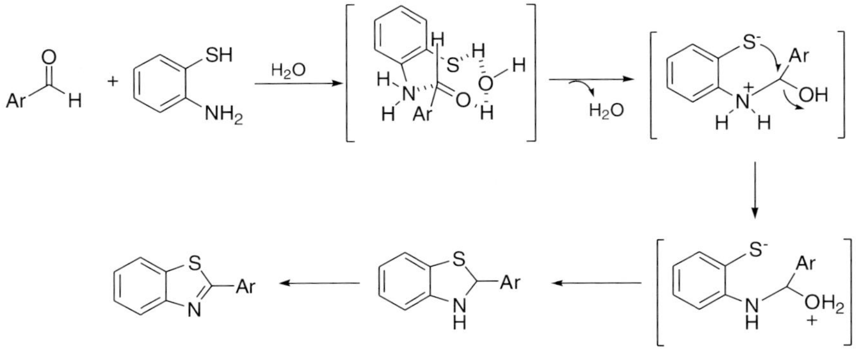

**Scheme 7.7**

## 7.4 Enantioselective Direct Aldol Reactions

Although there has been great progress in developing enantioselective organic reactions catalyzed by organic catalysts, satisfactory results of organocatalytic enantioselective transformations in aqueous media are still limited. For successful processes, either surfactants were required or some of the reactions were carried out in mixed aqueous–organic solvents [11]. The development of chiral organic catalysts suitable for stereoselective reactions in pure water is the true challenge. Benaglia and co-workers reported that 1,1'-binaphthyl-2,2'-diamine-based (S)-prolinamides combined with an acid additive can catalyze the direct condensation of cyclohexanone or other ketones with various aldehydes "on water" in good yields and with high diastereoselectivity, and also high enantioselectivity (Scheme 7.8) [12]. On the other hand, in the same catalytic system, only by using 2-butanone as a reaction medium did the reaction provide the product in high yield, but with lower enantioselectivity. Clearly, water was beneficial for the enantioselective control in the condensation process, probably due to the hydrophobic effect by keeping the reactants and the catalyst in close contact.

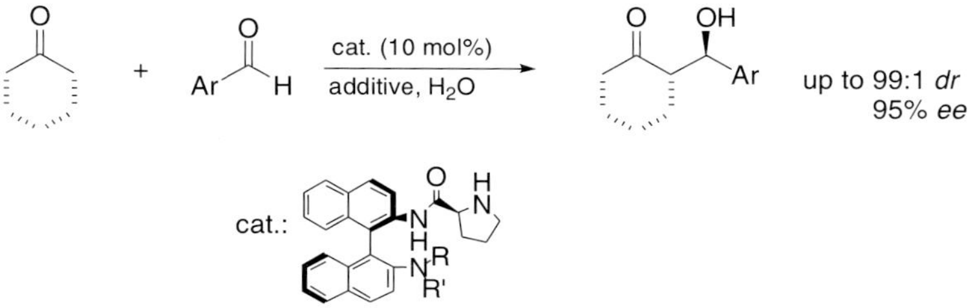

**Scheme 7.8**

In most organocatalytic direct aldol reactions, a large excess of ketone is employed. Recently, Armstrong and co-workers reported highly efficient asymmetric direct stoichiometric aldol reactions "on water" (Scheme 7.9) [13]. The derivative of proline exhibited greatly improved catalytic activity in the "on water" reaction. When cyclohexanone was used as both substrate and solvent, the reaction slowed substantially and the yield of the adduct decreased, with relatively low diastereoselectivity and enantioselectivity (Table 7.4, entry 1). When the reaction was carried out "on water", the enantio- and diastereoselectivities increased greatly (Table 7.4, entry 2), indicating that water participated in the reaction transition states through hydrogen bonding and enhanced the stereoselectivity of the reaction. Further addition of water did not increase the yield or stereoselectivity (Table 7.4).

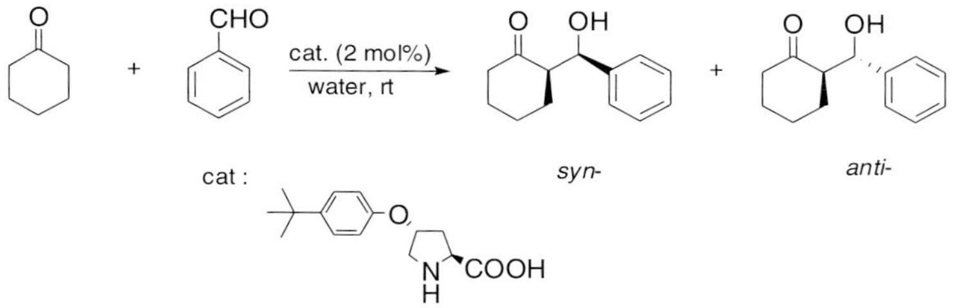

**Scheme 7.9**

Table 7.4 Effect of water on the reaction of cyclohexanone with benzaldehyde catalyzed by *tert*-butylphenoxyproline.

| Entry | H$_2$O (mol%) | Yield (%) | anti:syn | ee (%) |
| --- | --- | --- | --- | --- |
| 1 | 0 | 85 | 69:31 | 61 |
| 2 | 50 | 84 | 89:11 | 94 |
| 3 | 100 | 84 | 90:10 | 91 |
| 4 | 200 | 84 | 90:10 | 94 |
| 5 | 300 | 83 | 89:11 | 93 |

## 7.5
## Coupling Reactions

### 7.5.1
### Transition Metal-catalyzed Cross-coupling Reactions

The development of transition metal-catalyzed coupling reactions in aqueous media has broadened the scope of "on water" chemistry and contributed to green chemistry. The Suzuki–Miyaura coupling reaction, which is usually catalyzed by palladium reagents, has been a powerful tool in organic synthesis. However, such coupling reactions with nitrogen-containing heterocycles are difficult to perform. Recently, Suzuki–Miyaura coupling of N-heteroaryl halides with N-heteroarylboronic acids was found to be catalyzed efficiently by a palladium complex in pure water at 100 °C (Scheme 7.10) [14]. The Pd catalyst loading can be as low as 0.02–0.05 mol%. Although the reaction was carried out homogeneously in aqueous solution, the gram-scale coupling reaction resulted in the water-insoluble product being deposited as an oily liquid on the water surface, which could be separated easily from the aqueous phase. The purity of the crude product is excellent (>98%), and further purification is not needed. The advantages derived from the use of aqueous media, instead of organic solvents, are attributed to the fact that in the presence of water, the nitrogen atoms in the substrate molecule prefer to form hydrogen bonds with water rather than to coordinate to the soft Pd, leading to much higher catalytic activities.

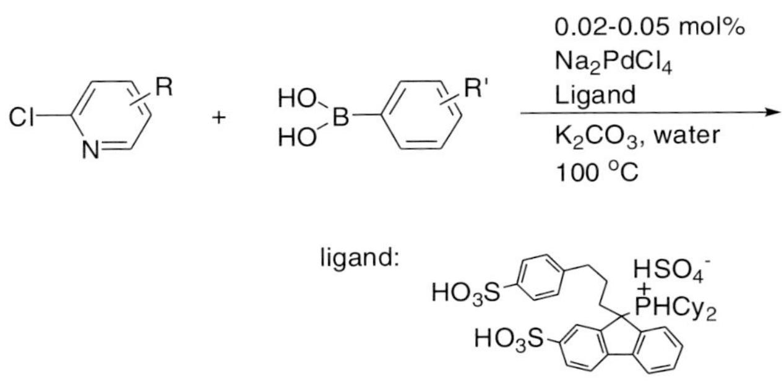

Scheme 7.10

In contrast to cross-coupling reactions involving stoichiometric organometallic compounds, such as arylboronic acids and arylstannes, direct intermolecular reactions of aromatic compounds with formally inactivated C–H bonds as a nucleophilic coupling partner have high atom economy. A direct arylation of thiazoles "on water" under mild conditions was developed by Greaney and coworkers [15]. Under catalysis by [Pd(dppf)Cl$_2$]·CH$_2$Cl$_2$–PPh$_3$ with Ag$_2$CO$_3$ as a base, the coupling of 2-phenylthiazole and aryl iodides provided higher yields in water (Table 7.5).

Although in most cases the reaction could be carried out in MeCN, water was found to be the best solvent, with higher yields and shorter reaction times (Table 7.5). For some substrates, such as 2-iodo-1,3,5-trimethylbenzene and 2-iodopyrazine, the reactions could not proceed in MeCN, but achieved quantitative yields in aqueous conditions (Table 7.5, entries 6 and 7). In water–MeCN mixed solvent, small amount of MeCN were compatible, but the conversion decreased with increase in the proportion of MeCN. This underlined the requirement for heterogeneity for "on-water" chemistry. The insoluble reactants, reagents, and products in water confer heterogeneous properties on the reaction system. Due to similar results obtained under solvent-free conditions, the rate acceleration in the "on water" coupling reaction was suggested to be a concentration effect. The direct arylation "on water" methodology provides a mild and efficient route to the synthesis of heterocycles for medicinal chemistry screening and bioactive research.

In a Cu-catalyzed intermolecular O-arylation of deoxybenzoin derivatives, water was found to be an efficient solvent [16]. Under catalysis by a Cu–TMEDA complex, various deoxybenzoin derivatives could be converted in water into the corresponding benzofurans in good to excellent yields (Scheme 7.11). Water-soluble inorganic bases or organic bases, such as K$_3$PO$_4$ and Et$_3$N, were not effective in this reaction. On the other hand, some 1,2-diamine derivatives could improve the reaction remarkably, probably due to the double behavior as both a ligand and a base.

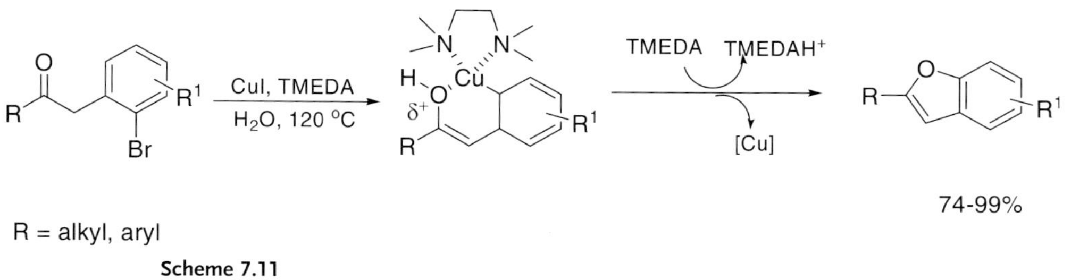

R = alkyl, aryl

Scheme 7.11

## 7.5.2
### Dehydrogenative Coupling Reactions

Dehydrogenative coupling reactions represent green methodology to construct C–C bonds by directly using two unfunctionalized C–H bonds. Substituted 1,4-benzoqui-

**Table 7.5** Reaction of 2-phenylthiazole with aryl iodides[a].

Ar-I + (2-phenylthiazole) → [Pd(dppf)Cl$_2$]·CH$_2$Cl$_2$, PPh$_3$, Ag$_2$CO$_3$, 60 °C, MeCN 72 h or H$_2$O 24 h → Ar-(thiazole)-Ph

| Entry | ArI | Product | Yield in MeCN (%) | Yield in water (%) |
|---|---|---|---|---|
| 1 | Cl-C$_6$H$_4$-I | Cl-C$_6$H$_4$-thiazole-Ph | 81 | 95 |
| 2 | OMe-C$_6$H$_4$-I | OMe-C$_6$H$_4$-thiazole-Ph | 78 | >99 |
| 3 | 2,4,6-Me$_3$C$_6$H$_2$-I | 2,4,6-Me$_3$C$_6$H$_2$-thiazole-Ph | 0 | >99 |
| 4 | 2-pyridyl-I | 2-pyridyl-thiazole-Ph | 57 | 71 |
| 6 | 3-pyridyl-I | 3-pyridyl-thiazole-Ph | 36 | >99 |
| 7 | 2-pyrazinyl-I | 2-pyrazinyl-thiazole-Ph | 0 | >99 |

[a]The data selected from [15].

none derivatives are distributed widely in Nature and exhibit various important biological activities. Recently, we reported the water-promoted cross-dehydrogenative coupling reaction of aromatic compounds with 1,4-benzoquinones [17]. A highly efficient direct coupling of indole compounds with 1,4-benzoquinones can be carried out "on water" to provide 3-indolylquinones in excellent yields without using any catalyst (Scheme 7.12). The yields of the coupled product were very poor in organic solvents (either polar, non-polar, or homogeneous aqueous–organic) and under solvent-free conditions (Table 7.6).

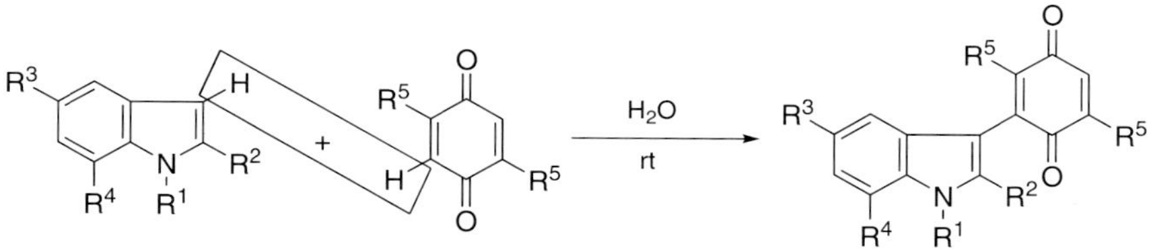

**Scheme 7.12**

The reactants were insoluble in water, forming an aqueous suspension on vigorous stirring. No molten starting materials were observed during the reaction when both reactants were solid. Increasing the reaction temperature to 50 °C decreased the yield (Table 7.7, entry 10). The use of an aqueous LiCl (2.5 M) or a solution of glucose (1 M *in* water) as reaction media neither changed the yield of the product in an aqueous reaction nor accelerated the reaction.

Moreover, water benefited the formation of bis(indolyl)quinone when excess indoles were used. Thus, cross-bis(indolyl)quinone could be synthesized by stepwise condensation of quinone first with an active indole and then another less active indole via the "on water" reaction (Scheme 7.13).

When other aromatic compounds such as anisole or aniline derivatives were used as nucleophiles, a Lewis acid was needed in order to obtain good yields, and In(OTf)$_3$ was found to be the best catalyst (Scheme 7.14) [18]. Dehydrogenative

Table 7.6 Dehydrative coupling reaction of 2-methylindole with 1,4-benzoquinone in various solvents.[a]

| Entry | Solvent | Yield (%) |
|---|---|---|
| 1 | CH$_2$Cl$_2$ | 13 |
| 2 | CH$_3$CN | –[b] |
| 3 | Et$_2$O | –[b] |
| 4 | THF | –[b] |
| 5 | Toluene | Trace |
| 6 | C$_2$H$_5$OH | 38 |
| 7 | THF–H$_2$O (1:2) | 55 |
| 8 | C$_2$H$_5$OH–H$_2$O (10:1) | 51 |
| 9 | H$_2$O | 82 |
| 10 | H$_2$O[c] | 39 |
| 11 | None | 20 |

[a] Room temperature for 10 h.
[b] No product was detected.
[c] At 50 °C.

**Table 7.7** Oxidative coupling reactions of phenol derivatives.

| Substrate | Product | Yield (%) | | | |
|---|---|---|---|---|---|
| | | Water | CH$_3$CN | Toluene | Neat |
| 2-naphthol | BINOL | 100 | 18 | 100 | 100 |
| OHC-C$_6$H$_3$(OMe)(OH) | bisaryl OHC/OMe/OH | 57 | 3 | 0 | 0 |
| 2,6-dimethylphenol | biphenyl + diquinone | 87 | 16 | 15 | 0 |
| 2,4-dimethylphenol | tetramethyl biphenol | 75 | 0 | 8 | 69 |
| NaO$_3$S-naphthol | NaO$_3$S-BINOL | 100 | 0 | 0 | 0 |

coupling reactions of substituted benzene with 1,4-quinone catalyzed by 5 mol% In(OTf)$_3$ in water could provide mono- and bisaryl-substituted 1,4-quinones. High regioselectivities were observed for bisaryl-substituted 1,4-quinones which depended on the reaction conditions and substrates.

The mechanism of cross-dehydrogenative coupling reaction of aromatic compounds with 1,4-benzoquinones may proceed by Friedel–Crafts-type conjugate addition followed by oxidation (Scheme 7.15). In organic solvents, no bis-substituted

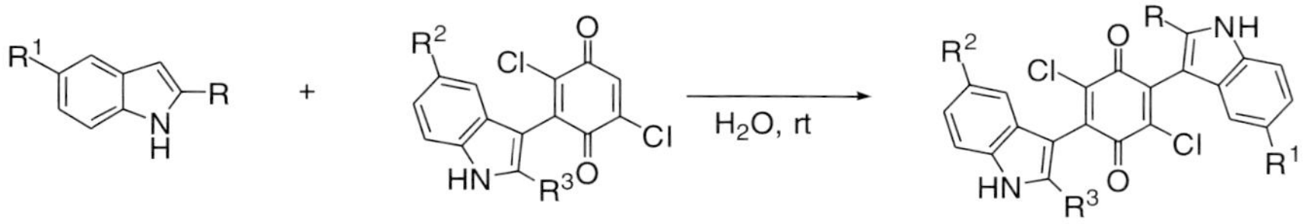

R = CH$_3$, R$^1$ = H, R$^2$ = OMe, R$^3$ = H             76%
R = CH$_3$, R$^1$ = CH$_3$, R$^2$ = OMe, R$^3$ = H         85%
R = CH$_3$, R$^1$ = H, R$^2$ = H, R$^3$ = (CH$_3$)$_2$CH   92%
R = CH$_3$, R$^1$ = CH$_3$, R$^2$ = H, R$^3$ = (CH$_3$)$_2$CH  90%
R = CH$_3$, R$^1$ = H, R$^2$ = H, R$^3$ = (CH$_3$)$_3$C    72%
R = CH$_3$, R$^1$ = OMe, R$^2$ = H, R$^3$ = (CH$_3$)$_3$C  80%

**Scheme 7.13**

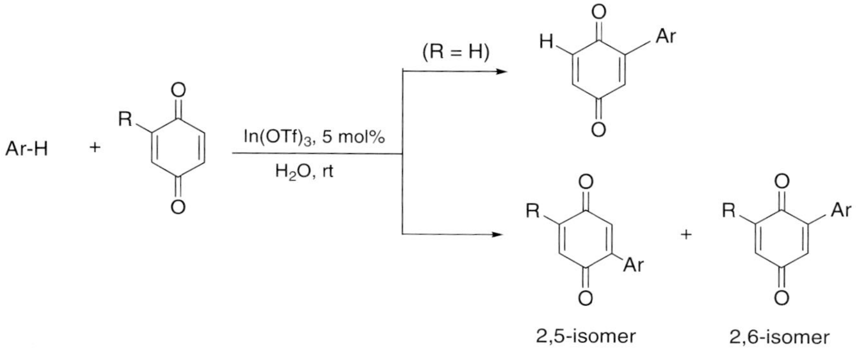

**Scheme 7.14**

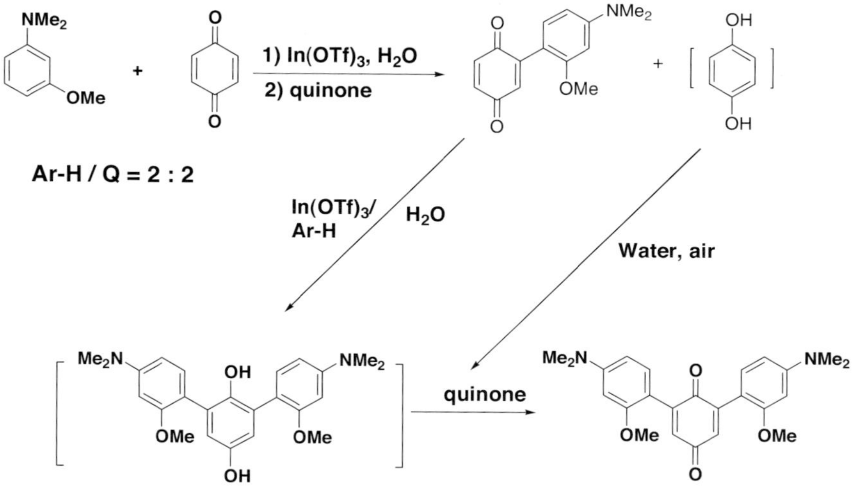

Ar-H / Q = 2 : 2

**Scheme 7.15**

products were detected. Water plays an essential role in the formation of bisaryl-substituted 1,4-quinone compounds.

In the $FeCl_3$-catalyzed oxidative coupling of phenol derivatives "on water" [19], biphasic reactions of poorly water-soluble substrates in contact with aqueous solutions of $FeCl_3$ could provide various biaryl coupling products for various phenols, whereas many homogeneous reactions in organic solvents, and also under solvent-free conditions, afford little or no coupling product (Scheme 7.16, Table 7.7).

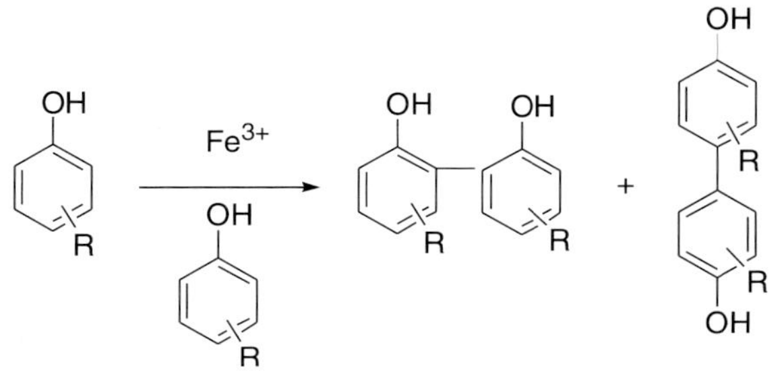

**Scheme 7.16**

## 7.6
## Oxidation

Direct oxidation with oxygen or air in water is very attractive in green chemistry. Direct aerobic oxidation of enol silyl ethers to α-hydroxy ketones in water without a catalyst can be performed (Table 7.8) [20]. Enol silyl ethers are stable in organic solvents and under an air atmosphere, but are hydrolyzed in a homogeneous water–ethanol mixture or in the presence of a Lewis acid catalyst. Although the solubility of oxygen in water is not as high as in organic solvents, a clean oxidation of aromatic silyl enol ethers in water by air without the use of any catalyst generates α-hydroxy ketones in good yields after 4 days. A possible mechanism was proposed for the oxidation reaction of enolate, activated by counterion complexation in a six-membered transition state.

Most recently, Shapiro and Vigalok reported the "on water" oxidation of aldehydes by using air or molecular oxygen as the oxidant [21]. It was found that the oxidation of aliphatic and aromatic aldehydes proceeded very smoothly upon simply stirring their aqueous emulsion in air to give the corresponding carboxylic acids in high yields (Table 7.9). Under the same reaction conditions, no oxidation reactions were observed when performed in methanol and dichloromethane instead of water. Moreover, the addition of a small amount of THF (5%) also completely inhibited the reaction. As both the starting materials and the products are insoluble in water, they can be easily isolated from the aqueous phase.

The "on water" oxidation process can be extended to the multi-component Passerini reaction "on water" or in water (Scheme 7.17) [21,22] In this three-component

**Table 7.8** Oxidation of 1-phenyl-1-trimethylsiloxypropene under various conditions.

| Solvent | Time (days) | Yield (%) 2 | Yield (%) 3 |
|---|---|---|---|
| $C_2H_5OH$ | 4 | – | – |
| $C_2H_5OH–H_2O(1:1)$ | 4 | 25 | 35 |
| $C_2H_5OH–H_2O(1:9)$ | 4 | 74 | 16 |
| $H_2O$ | 4 | 86 | 9 |
| Acetone | 4 | – | – |
| THF | 4 | – | – |
| $C_6H_6$ | 4 | – | – |
| None | 4 | – | – |
| $H_2O–Ga(OTf)_3$ (20 mol%) | 2 | 81 | 8 |
| $H_2O–Cu(OTf)_2$ (20 mol%) | 2 | 53 | 40 |

**Table 7.9** Aldehyde oxidation "on water"

| Entry | Aldehyde | Oxidative conditions[a] | Yield (%) |
|---|---|---|---|
| 1 | $c$-$C_6H_{11}CHO$ | A | 87 |
| 2 |  | B | 87 |
| 3 |  | C | 66 |
| 4 | $n$-$C_7H_{15}CHO$ | A | 50 |
| 5 |  | B | 60 |
| 6 | 2-Ethylhexanal | A | 86 |
| 7 |  | B | 88 |
| 8 |  | C | 75 |
| 9 | Benzaldehyde | A | 83 |
| 10 |  | B | 81 |
| 11 |  | C | 53 |
| 12 | $n$-$C_4H_9CHO$ | A | 11 |

[a]Condition: A, a suspension in pure water was stirred in the presence of air; B, a suspension in pure water was stirred in pure molecular oxygen at 1 atm; C, a suspension in pure water was stirred in the presence of air on large scale of aldehyde (5 ml).

reaction, only two substrates were added to the reaction mixture, the third component being generated *in situ*. The reactions of hydrophobic aldehydes with isocyanides were carried out "on water" very smoothly to give the Passerini reaction products. However, the use of more water-soluble aldehydes, which was closer to the reaction in water, gave a mixture of the Passerini reaction products with another product, α-hydroxyamide.

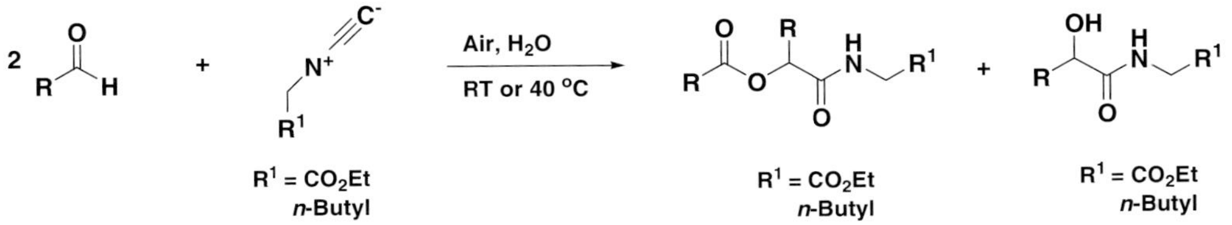

$$2\ RCHO + R^1CH_2N^+\equiv C^- \xrightarrow[\text{RT or 40 °C}]{\text{Air, H}_2\text{O}} R\underset{O}{\overset{O}{\|}}C-O-\underset{R}{\overset{H}{C}}-C(O)NHR^1 + R-\underset{OH}{\overset{H}{C}}-C(O)NHR^1$$

R¹ = CO₂Et, n-Butyl

**Scheme 7.17**

## 7.7
### Bromination Reactions

It is known that reactions through free radical intermediates can be carried out in water, and water does not interfere with the radical chain process due to strong OH bonds in water. Free radical benzylic bromination (the well-known Wohl–Ziegler reaction) of various hydrophobic 4-substituted toluenes was effectively conducted using N-bromosuccinimide (NBS) in pure water. The radical chain reaction is activated by visible light (40 W incandescent light bulb) (Scheme 7.18), hence additional initiator and heating are saved [23]. The hydrophobic substrates formed a layer "on water" during the reaction, but they became denser and sank to the bottom as soon as the bromination occurred. After the aqueous bromination reaction has finished, the by-product, water-soluble succinimide, can be easily separated from the products. The aqueous bromination reactions of the substrates with a ketone

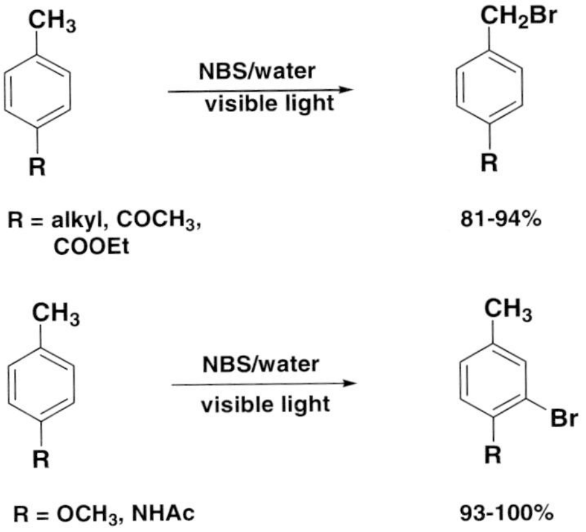

4-R-C₆H₄-CH₃ →(NBS/water, visible light)→ 4-R-C₆H₄-CH₂Br

R = alkyl, COCH₃, COOEt    81–94%

4-R-C₆H₄-CH₃ →(NBS/water, visible light)→ 3-Br-4-R-C₆H₃-CH₃

R = OCH₃, NHAc    93–100%

**Scheme 7.18**

functionality on the aromatic ring give benzylic bromide products selectively. Moreover, for aromatics with OMe and NHAc activating groups, the NBS–$H_2O$ system can be applied for electrophilic bromination solely on the aromatic ring.

Oxidative bromination reactions of various 1,3-diketones, keto esters, and ketones with an aqueous $H_2O_2$–HBr system "on water" without the use of a catalyst were effective, giving α-brominated products in 69–97% yields (Scheme 7.19) [24]. The reactions were also greener, because the reactions proceeded under mild, organic solvent-free and organic waste-free conditions and with higher atom economy. The reaction passed through *in situ* oxidation of HBr into bromine by $H_2O_2$. It was proposed that water would activate the reaction by promoting the formation of the enol form of the ketones in the course of the reaction. Since no organic waste is produced, an environmentally friendly work-up procedure, simple filtration or separation, can be used to isolate the products.

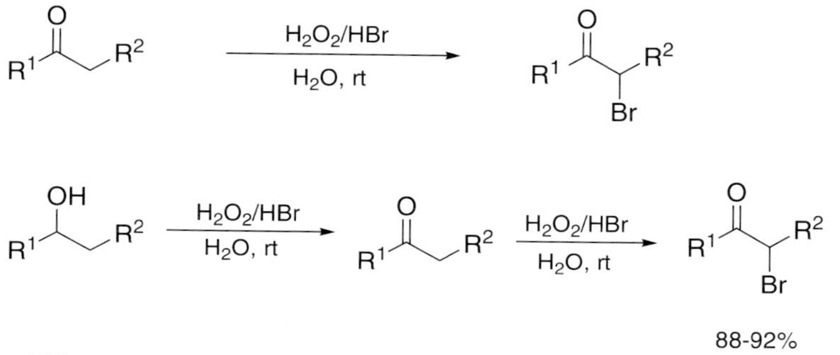

Scheme 7.19

Furthermore, the $H_2O_2$–HBr system could be applied to tandem oxidation–bromination directly from alcohols for the synthesis of α-bromo ketones (Scheme 7.19). When a higher amount of the reagents ($H_2O_2$:$H_2O$ = 4:1.7) was used, the α-brominated products were obtained in excellent yields of 88–92% based on the corresponding alcohols.

## 7.8
## Miscellaneous Reactions

### 7.8.1
### Nucleophilic Substitution

Nucleophilic substitution of ferrocenyl alcohols is effectively promoted "on water" without any catalyst. The common Brønsted acids were not effective catalysts for this reaction and the presence of a Lewis acid was not necessary. Chiral ferrocenyl alcohols could react with indoles, pyrroles, and thiols in aqueous suspension at 80 °C to provide the corresponding optically active products in excellent enantioselectivities with the same configuration (Scheme 7.20) [25]. It is worth noting that the ferrocenyl cations generated are stable enough to participate in the subsequent aqueous reaction with indoles and keep the configuration of the chiral ferrocenyl alcohols unchanged. It was suggested that in slightly basic or neutral conditions the intermediates of the $S_N1$ reaction could be trapped with electron-rich π-systems rather than with water.

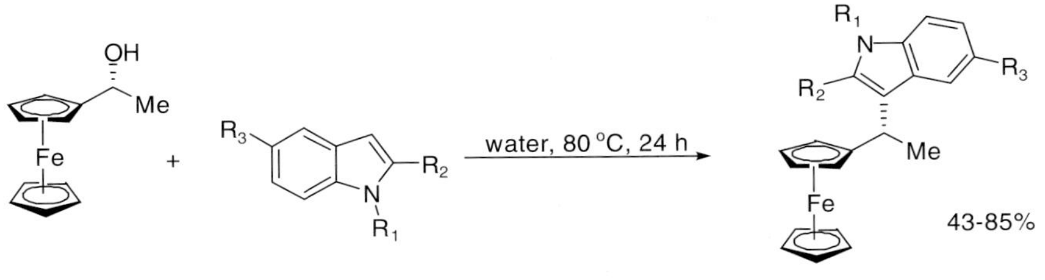

Scheme 7.20

### 7.8.2
### Functionalization of SWNTs

Single-walled carbon nanotubes (SWNTs) are attracting much attention in material science. Functionalized SWNTs can efficiently increase interactions between the SWNTs and the host materials. Non-surfactant-wrapped SWNTs could be functionalized by an "on water" process [26]. In the presence of a substituted aniline and isoamyl nitrite as an oxidative species, the reaction of SWNTs in aqueous suspension produced a higher degree of functionalization and better reproducibility (Scheme 7.21). The diazonium species formed "on water" through the oxidation of the aniline is more effective than the diazonium salts prepared separately. The biphasic system of water and organic solvents provided a low degree of functionalization and the use of brine instead of pure water did not give an improvement in functionalization. Homogenization during the reaction did not improve the degree of functionalization either. The "on water" procedure represents a "green", or environmentally friendly, process.

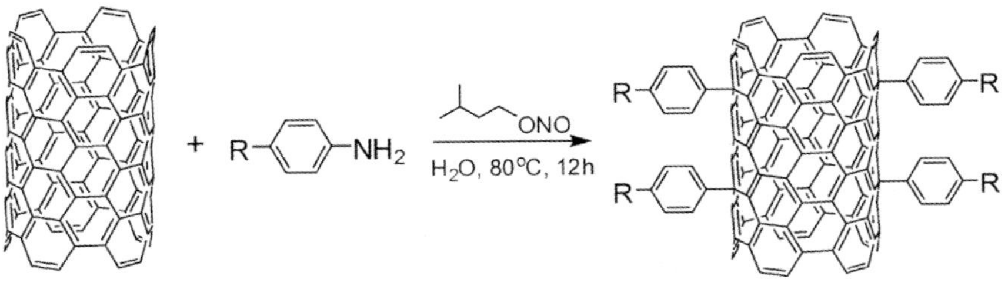

Scheme 7.21

## 7.9
## Theoretical Studies

Since water is used as a solvent in organic reactions, rate acceleration by water has been observed in many reactions. The effect of water on the enhancement of rate or selectivities has been studied with various theories such as hydrophobic aggregation, hydrogen bonding effect, cohesive energy density, and ground-state destabilization. For non-polar reactants, the reaction with negative activation volumes, as indicated by its pressure dependence, could be accelerated in aqueous media over an organic solvent [27]. However, the dramatic rate acceleration for "on water" reactions, which require heterogeneity, is remarkable compared with homogeneous reactions. A crucial aspect of "on water" chemistry is related to the unique properties of water and reactants at an oil–water phase boundary.

Recently, Jung and Marcus reported theoretical studies on the nature of "on water" rate enhancment [28]. The large acceleration of the "on water" [$2\sigma + 2\sigma + 2\pi$] cycloaddition reaction of quadricyclane (4) with dimethyl azodicarboxylate (DMAD) was chosen as the model reaction (Scheme 7.22). Since both homogeneous and heterogeneous cycloaddition reactions are faster than the neat reaction, hydrophobic aggregation of reactants is a less important factor for the rate acceleration.

The special structure of water at the oil–water interface plays a key role in catalyzing reactions via its free ("dangling") OH group that protrudes into the organic phase by the formation of hydrogen bonds (Figure 7.2).

Three rate constants, $k_N$, $k_S$ and $k_H$ in the neat, surface, and aqueous homogeneous reactions, respectively, were obtained by experiment and calculation using an approximate transition state (TS) theory. The rate constant of the surface reaction $k_S$ is higher than that of the neat reaction $k_N$ by approximately five orders of magnitude and higher than that of the homogeneous reaction $k_H$ by a factor of 600. DFT calculations show that the surface reaction has a lower activation energy by about 7 kcal mol$^{-1}$ than the neat reaction due to H-bond formation. Although the H-bond-mediated barrier lowering applies to both homogeneous and heterogeneous aqueous reactions, the surface reaction is still more efficient than the homogeneous reaction. The structural arrangement of water molecules in homogeneous solution and in an emulsion is different (Figure 7.2). In aqueous solution, the twist H-bonding net of water molecules forms a cage around the hydrocarbon solutes, preventing the loss of their usual H-bonding at the expense of entropy. In contrast, in a heterogeneous aqueous medium, free and dangling OH groups of water at the extended hydrophobic surface are available to protrude into the oil phase and ready to catalyze

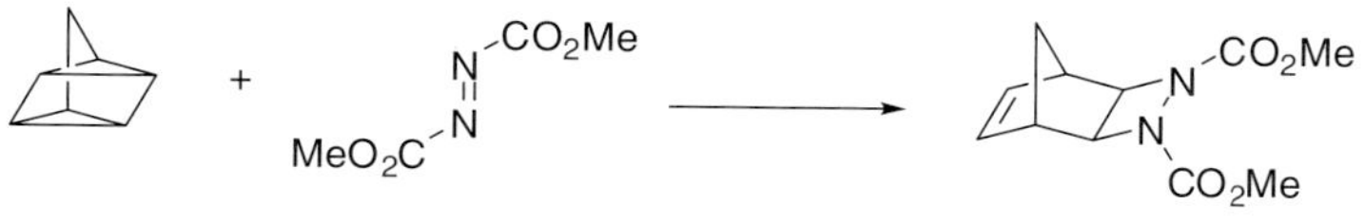

Scheme 7.22

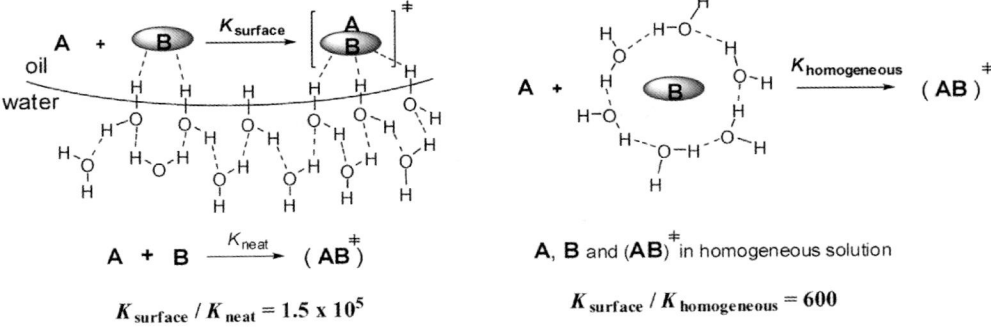

**Figure 7.2** "On water" catalysis in comparison with the neat and aqueous homogeneous reactions.

$$A + B \xrightarrow{K_{surface}} [AB]^{\ddagger}$$

$$A + B \xrightarrow{K_{neat}} (AB)^{\ddagger}$$

$$K_{surface} / K_{neat} = 1.5 \times 10^5$$

A, B and $(AB)^{\ddagger}$ in homogeneous solution

$$A + B \xrightarrow{K_{homogeneous}} (AB)^{\ddagger}$$

$$K_{surface} / K_{homogeneous} = 600$$

reactions. Compared with an "on water" reaction, breaking an existing hydrogen bond network in homogeneous solution is needed in order to obtain free OH to catalyze reactions. Therefore, the reaction in homogeneous aqueous solution is intrinsically slower than the surface reaction.

## 7.10
## Conclusion

Water as a green solvent shows some particular properties, such as high hydrogen bonding ability and hydrophobic effect, to benefit some organic reactions. In practical terms, "on water" reactions are more reproducible and convenient than those reactions under solvent-free conditions and in homogeneous aqueous media. With the increasing interest in green chemistry research, more and more organic reactions under "on water" conditions are expected to emerge.

## References

1. (a) Grieco, P.A. (1998) *Organic Synthesis in Water*, Blackie Academic & Professional, London; (b) Li, C.-J. and Chan, T.-H. (2007) *Comprehensive Organic Reactions in Aqueous Media*, John Wiley & Sons, Inc, Hoboken, NJ.
2. Reichardt, C. (2002) *Solvents and Solvent Effects in Organic Chemistry*, 3rd edn, Wiley-VCH Verlag GmbH, Weinheim.
3. Rideout, D.C. and Breslow, R. (1980) *J. Am. Chem. Soc.*, **102**, 7816.
4. Narayan, S., Muldoon, J., Finn, M.G., Fokin, V.V., Kolbe, H.C. and Sharpless, K.B. (2005) *Angew. Chem., Int. Ed.*, **44**, 3275.
5. Klijn, J.E. and Engberts, J.B.F.N. (2005) *Nature*, **435**, 746.
6. (a) González-Cruz, D., Tejedor, D., de Armas, P. and García-Tellado, F. (2007) *Chem. Eur. J.*, **13**, 4823; (b) González-Cruz, D., Tejedor, D., de Armas, P., Moralesa, E.Q. and García-Tellado, F. (2006) *Chem. Commun.*, 2798.

7 Khatik, G.L., Kumar, R. and Chakraborti, A.K. (2006) *Org. Lett.*, **8**, 2433.
8 Ranu, B.C. and Banerjee, S. (2007) *Tetrahedron Lett.*, **48**, 141.
9 Shaabani, A., Rahmati, A. and Farhangi, E. (2007) *Tetrahedron Lett.*, **48**, 7291.
10 Chakraborti, A.K., Rudrawar, S., Jadhav, K.B., Kaur, G. and Chankeshwara, S.V. (2007) *Green Chem.*, **9**, 1335.
11 (a) Brogan, A.P., Dickerson, T.J. and Janda, K.D. (2006) *Angew. Chem. Int. Ed.*, **45**, 8100; (b) Hayashi, Y. (2006) *Angew. Chem. Int. Ed.*, **45**, 8103; (c) Blackmond, D.G., Armstrong, A., Coombe, V. and Wells, A. (2007) *Angew. Chem. Int. Ed.*, **46**, 3798.
12 Guizzetti, S., Benaglia, M., Raimondi, L. and Celentano, G. (2007) *Org. Lett.*, **9**, 1247.
13 Huang, J., Zhang, X. and Armstrong, D.W. (2007) *Angew. Chem. Int. Ed.*, **46**, 9073.
14 Fleckenstein, C.A. and Plenio, H. (2007) *Green Chem.*, **9**, 1287.
15 Turner, G.L., Morris, J.A. and Greaney, M.F. (2007) *Angew. Chem. Int. Ed.*, **46**, 7996.
16 Carril, M., SanMartin, R., Tellitu, I. and Domínguez, E. (2006) *Org. Lett.*, **8**, 1467.
17 Zhang, H.-B., Liu, L., Chen, Y.-J., Wang, D. and Li, C.-J. (2006) *Eur. J. Org. Chem.*, 869.
18 Zhang, H.-B., Liu, L., Chen, Y.-J., Wang, D. and Li, C.-J. (2006) *Adv. Synth. Catal.*, **348**, 229–235.
19 Wallis, P.J., Booth, K.J., Pattiab, A.F. and Scott, J.L. (2006) *Green Chem.*, **8**, 333.
20 Li, H.-J., Zhao, J.-L., Chen, Y.-J., Liu, L., Wang, D. and Li, C.-J. (2005) *Green. Chem.*, **7**, 61.
21 Shapiro, N. and Vigalok, A. (2008) *Angew. Chem. Int. Ed.*, **47**, 2849.
22 Pirrung, M.C. and Das Sarma, K. (2004) *J. Am. Chem. Soc.*, **126**, 444.
23 Podgoršek, A., Stavber, S., Zupanb, M. and Iskra, J. (2006) *Tetrahedron Lett.*, **47**, 1097.
24 Podgoršek, A., Stavber, S., Zupanab, M. and Iskra, J. (2007) *Green Chem.*, **9**, 1212.
25 Cozzi, P.G. and Zoli, L. (2007) *Green Chem.*, **9**, 1292.
26 Price, B.K. and Tour, J.M. (2006) *J. Am. Chem. Soc.*, **128**, 12899.
27 Pirrung, M.C. (2006) *Chem. Eur. J.*, **12**, 1312.
28 Jung, Y. and Marcus, R.A. (2007) *J. Am. Chem. Soc.*, **129**, 5492.

# 8
# Pericyclic Reactions in Water. Towards Green Chemistry
*Jaap E. Klijn and Jan B.F.N. Engberts*

> *There are not enough lips to pronounce your transient names,*
> *O water...*  Wislawa Szymborska

## 8.1
## Introduction

### 8.1.1
### Pericyclic Reactions

Pericyclic reactions comprise a broad array of chemical transformations, characterized by an activated complex with a cyclic organization of atoms and a corresponding positioning of interacting orbitals in a process of reorganization of σ and π bonds. The majority of these reactions are concerted (i.e. do not involve intermediates) rather than stepwise and possess negative volumes of activation in accord with a bimolecular process [1]. Bond breaking and bond making may be synchronous, but not necessarily so. Usually high regio- and stereoselectivities are observed. For a general discussion of the interesting mechanistic and stereochemical aspects of pericyclic reactions, including the importance of conservation of orbital symmetry, the reader is referred to recent textbooks and reviews [2, 3].

Pericyclic reactions are among the most important carbon–carbon bond-forming reactions in organic chemistry. Examples include the famous Diels–Alder reactions, for which the simplest thermal [4 + 2] cycloaddition is shown (Scheme 8.1).

The diene **1** (four π electrons) reacts in a single reaction step with the dienophile **2** (two π electrons) via the cyclic activated complex (AC) **3** to form the product **4** with stereochemical control, which means that the stereochemistry of substituents in the diene and dienophile is retained in the product. The binding interactions [3] in **3** involve the highest occupied molecular orbital (HOMO) of **1** and the lowest unoccupied molecular orbital (LUMO) of **2**. These cycloadditions were termed normal electron-demand Diels–Alder reactions. As a consequence, reaction rates for normal

---

*Handbook of Green Chemistry, Volume 5: Reactions in Water.* Edited by Chao-Jun Li
Copyright © 2010 WILEY-VCH Verlag GmbH & Co. KGaA, Weinheim
ISBN: 978-3-527-31591-7

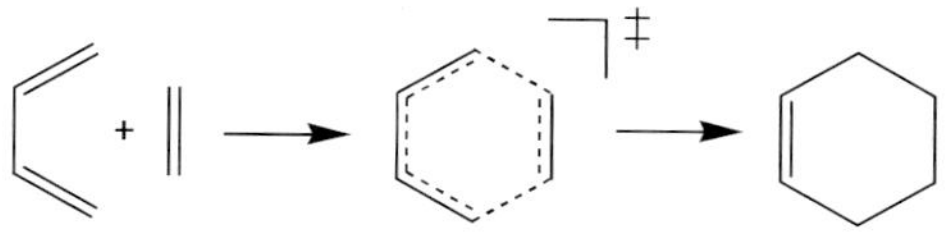

**Scheme 8.1** The Diels–Alder reaction.

electron-demand Diels–Alder reactions are increased by electron-donating substituents in **1** and electron-attracting substituents in **2**. The reverse situation applies for inverse electron-demand Diels–Alder reactions.

Photochemical [$2\pi + 2\pi$] cycloadditions are less important than their thermal counterparts. The stereochemistry of the photoadduct provides clues regarding the question of whether or not the process is concerted. Lack of concerted behavior is indicative of diionic or diradical intermediates [3].

Other synthetically important classes of pericyclic reactions are 1,3-dipolar cycloadditions (Scheme 8.2) and sigmatropic rearrangements (Scheme 8.3). In a 1,3-dipolar cycloaddition, a 1,3-dipolar substrate **5** reacts with a 1,3-dipolarophile **6** to give the cycloadduct **8**. In the activated complex **7**, the dipolar features of substrate **5** are partly lost.

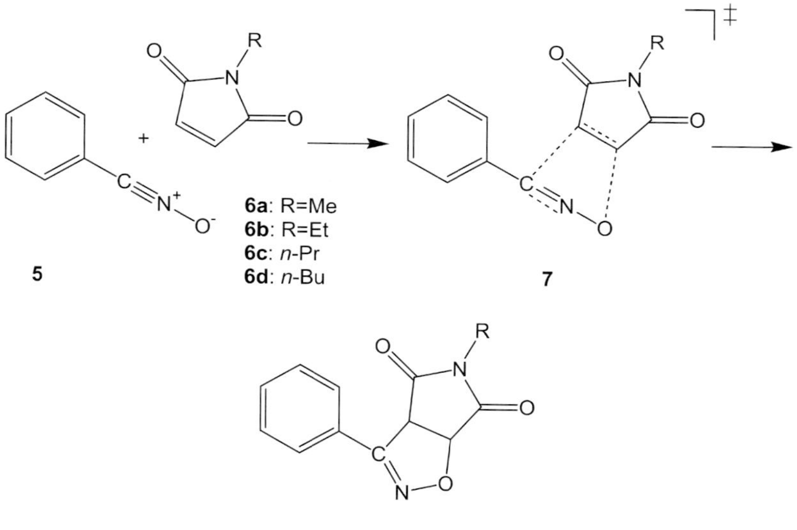

6a: R=Me
6b: R=Et
6c: n-Pr
6d: n-Bu

**Scheme 8.2** An example of a 1,3-dipolar cycloaddition.

The synthetically useful Claisen rearrangement is a monomolecular sigmatropic reaction that proceeds, for the given example (**9**), via the cyclic activated complex **10** to yield the rearranged product **11**.

Solvent effects on pericyclic reactions have a highly interesting history. Already at an early stage it was recognized that reaction rates varied only very modestly over a

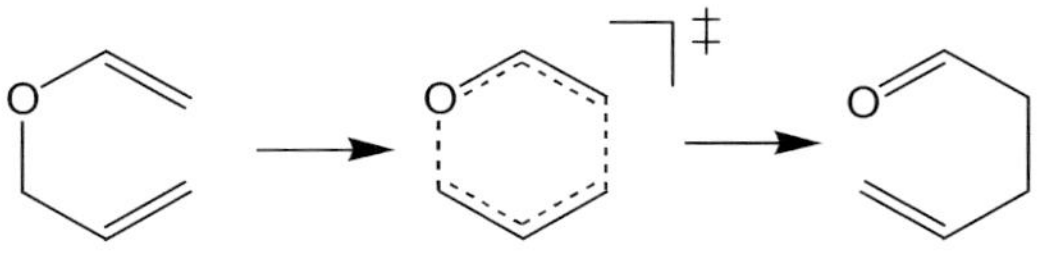

**Scheme 8.3** Claisen rearrangement of allyl vinyl ether **9**.

wide range of organic solvents, from apolar hydrocarbons to relatively polar alcohols. An extreme example is the dimerization of cyclopentadiene, for which the second-order rate constant varies by only a factor of four for a large variety of organic solvents. These observations were in accord with the mechanism of these reactions. Going from the reactant(s) to the activated complex, only minor changes in charge are involved and the Gibbs energy of activation responds accordingly to changes in the reaction medium. Most quantitative work was carried out on Diels–Alder reactions. Although Diels–Alder reactions had been carried out occasionally in water, organic solvents were clearly the media of choice, primarily because of the usually highly limited aqueous solubilities of diene and dienophile. For many years, this led to a "hydrophobic attitude" of the chemists who performed cycloaddition reactions, despite the obvious environmental disadvantages.

In 1980, there was a breakthrough. Rideout and Breslow [4] reported kinetic evidence for substantial rate accelerations of several Diels–Alder reactions performed in water. These reactions could be easily monitored at low concentrations. The rate enhancements were clearly dependent on the nature of the diene and dienophile, but they were all outside the magnitudes of the solvent effects found in organic solvents. It was immediately evident that the specific properties of water had to be involved to rationalize these fascinating results. The authors ascribed the rate accelerations exclusively to hydrophobic interactions between the apolar substrates, whereas Grieco and co-workers suggested (pre)micellar catalysis [5, 6]. This interpretation was later corrected in more in-depth mechanistic and computational studies, which revealed the often dominating importance of hydrogen bond interactions with the (small) water molecules in the activated complex. Shortly after Rideout and Breslow's paper appeared, it was found that aqueous rate accelerations were also possible for many other types of bimolecular and also intramolecular organic reactions. However, the problem of substrate solubilities remained unsolved and continued to make synthetic applications troublesome. Apart from using binary aqueous solvent mixtures, many more ingenious attempts were also made to cope with these problems. An example is the attachment of a hydrophilic group to a substrate, which can be easily removed after reaction in water [7].

The second breakthrough came in 2005, when Sharpless and co-workers [8, 9] obtained compelling evidence that the highly limited aqueous solubilities can be turned into a *beneficial* effect on organic transformations, including pericyclic reactions, in water – a totally unexpected and even counterintuitive result! It was shown that reactions in two-layer mixtures of water with two liquid (or one liquid and one solid) substrates can induce, upon efficient stirring, high reaction rates, product yields, and reaction specificities. The first detailed study of the mechanism of these "on water" reactions has now been reported by Jung and Marcus [10].

Synthetic aqueous organic synthesis was born, also because efficient work-up procedures are possible for these "on-water" reactions. For a general review, the reader is referred to Chapter 7.

### 8.1.2
### Water, the Ultimate Green Solvent

The rapid growth of the world population, coupled with an increasing advance of large-scale chemical technologies to produce huge quantities of chemical products, had immense consequences for the health and quality of the environment. Since the 1960s, attempts to find remedies for these problems initially led to end-of-pipeline control, but this was later replaced by a prime emphasis on pollution *prevention*. The history of these developments is interesting and has been laid down in reviews and books [11, 12]. "Benign by design" became a popular issue, primarily focused on environmental care in combination with economic and political considerations. Combined efforts by representatives of industry, academia, and government resulted in large and comprehensive programs aimed at protection of the environment and achieving economic prosperity. The United States had a leading role from the beginning, but many other industrialized countries followed in its footsteps. It was hoped that these activities would also result in a better image of the chemical industry.

In the second half of the 1990s, it was recognized that "benign by design" was too strongly restricted to environmental care, and Anastas and Williamson [13] even concluded, "one obvious but important point: nothing is benign." Therefore, the term was replaced by the softer "green chemistry," which cannot be quantitatively defined but clearly expresses the intention to achieve a combination of environmental and economic prosperity. "Green chemistry" was first used in 1990 by Cathcart [14] and later employed by Anastas in his many influential publications. The holistic concept of green chemistry became accepted world-wide. Although attempts have been made to quantify the degree of greenness, particularly by Sheldon [15], it has not been found possible to include the many complex factors that determine "greenness" in generally applicable parameters. This obviously gave "space" in the use of the term green chemistry. An interesting historic overview of the origins of green chemistry has been reported by Linthorst [16].

A recent thematic issue of *Chemical Reviews* [17] presents 21 articles on different aspects of sustainable chemistry.

In a handbook published in 1998 [18], 12 "principles" of green chemistry, perhaps better viewed as "guidelines," were summarized. One of the key guidelines reads: "The use of auxiliary substances (e.g. solvents, separation agents, etc.) should be made unnecessary wherever possible and innocuous when used."

The present chapter makes an attempt to show that many synthetically important pericyclic reactions can be carried out in or on *water*, probably the best solvent from the viewpoint of green chemistry. Moreover, it will be shown that the use of aqueous media for these reactions possesses many other advantages, which include the often high reaction rates, high product yields, good product stereochemistry, and simple and efficient work-up procedures.

Traditionally, water has been considered to be an unfavorable solvent for organic synthesis. Of course, there are a few exceptions, such as a variety of Mannich-type condensation reactions, but in many cases the major problem was the limited solubility of apolar organic substrates in water. Life processes occur in aqueous media and, as beautifully summarized by Ball in a recent review [19], take advantage of water for a good performance. There are strong reasons for viewing water as part of the biomolecules. However, the reaction conditions in the living cell are greatly different from those employed for straightforward organic synthesis in the laboratory or in industrial processes. By contrast, mechanistic studies of organic reactions have usually been performed in water for a number of good reasons.

The exceptional solvent properties of water, as viewed from the standpoint of a physical-organic chemist interested in reactivity, have recently been summarized [20]. The isolated water molecule is very small (hard-sphere diameter 2.75 Å) and has two hydrogen bond donor sites and two hydrogen bond acceptor sites. As a consequence, liquid water contains a large, highly dynamic, three-dimensional hydrogen bond network, with the Gibbs energy determined by complex enthalpic and entropic factors. Hydration of organic solutes responds to this structural property of water. Because of the rather strong intermolecular water–water interactions, the creation of a cavity in water, as required for the dissolution of an organic molecule, is a costly process [21], and for most apolar solutes insufficiently favorable Gibbs energy is gained from putting the solute into the cavity and turning on solute–water interactions. This explains the hydrophobicity of apolar solutes. Water tends to keep as many water–water hydrogen bonds intact and this results in a preference for tangential orientation of water O—H bonds relative to a non-polar surface of a relatively small solute. This contributes to the unfavorable entropic consequences of dissolution of apolar solutes. The Gibbs energy of hydration is entropy controlled and, as a result, at a characteristic concentration of the solute, the solute molecules tend to stick together, thereby reducing the solute's surface area in contact with water. This is a very important process in water, called *hydrophobic interactions* [22, 23]. The exact mechanism is complicated and still not all issues are fully understood. At ambient temperature and for not too large molecules, these interactions are entropy driven and play an important role in pericyclic reactions in water. Of course, upon further increasing the solute concentration, the unavoidable next step will be phase separation.

Water is a polar solvent [$\varepsilon_r = 78.30$, $E_T(30) = 63.1$]. The molecules in the bulk possess a substantial dipole moment (2.8 D), but a low polarizability and a low internal pressure. The cohesive energy density, which measures the total molecular cohesion per unit volume, is very high (2302 versus 703 J cm$^{-3}$ for ethanol and 225 J cm$^{-3}$ for *n*-hexane).

From the viewpoint of green chemistry, there is little doubt that water is the best choice for chemical reactions. Non-toxic, non-flammable, no smell, not mutagenic, no carcinogenic properties, cheap, not explosive, easy temperature control, often good potential for product separation, often no protection–deprotection protocols necessary for reactive functional groups – these are major reasons for this statement. There are also a few disadvantages: the high heat of vaporization, solubility limitations for apolar solutes, relatively high costs of recycling and purification of contaminated aqueous media, and the instability of highly reactive and hydrolyzable substrates and intermediates in water.

Sharpless and co-workers [24] suggested that water, apart from being environmentally benign, is the most desirable solvent for "click chemistry". Moreover, large-scale green industrial processes have been developed with an emphasis on biocatalysis and on aqueous biphasic techniques [25].

In the following sections, we will briefly review different types of pericyclic reactions that have been performed in water. We will put the main emphasis on reactivity as determined by the aqueous medium and the mechanistic details of the reactions. A number of pericyclic reactions in water will be reviewed and particular attention will be paid to water as a highly favorable solvent medium for green chemistry.

## 8.2
## Pericyclic Reactions in Aqueous Media

### 8.2.1
### Introduction

Pericyclic reactions in water provide excellent possibilities for designing environmentally friendly chemical transformations. First, the bimolecular varieties have a 100% atom economy: two reactants combine to a single product of reaction. Second, water is a favorable solvent for these reactions because often reaction rates, yields, and stereochemical issues are better than in organic solvents.

An important aspect, discussed in the following sections, involves the question of how these aqueous pericyclic reactions depend on reactant structure and on mechanistic details. We will focus our attention on these considerations rather than giving a comprehensive summary of the large number of pericyclic reactions that have been performed in water.

### 8.2.2
### Normal and Inverse Electron-demand Diels–Alder Reactions in Water

As noted above, Rideout and Breslow [4] carried out some simple Diels–Alder reactions in water. For example, cyclopentadiene (**12**) reacted with butenone (**13**) at 20 °C to give **14** with a more than 700-fold higher second-order rate constant than that in isooctane as the solvent (Scheme 8.4). Compared with methanol, the rate constant was larger by a factor of about 60. Anthacene-9-carbinol (**15**) as the diene reacted with N-ethylmaleimide (**6b**) about 28 times faster in water than in isooctane (Scheme 8.5). It was concluded that solvent polarity was not the dominant influence but that hydrophobic effects had to be invoked to explain these (at that time) unexpected solvent effects. Consistent with this explanation, it was also found that for Diels–Alder reactions that gave rise to *endo/exo* isomers, there occurred an increased preference for the *endo* product. This will be, in part, a consequence of the high solvent polarity of water [3]. However, another contribution to the preferred formation of the *endo* isomer in water is due to the fact that the activated complex for the formation of the *endo* isomer has the smallest water-accessible surface area and is, therefore, most easily formed in water. Experimental support is found in the more favorable volume of activation for the *endo*-cycloaddition reaction.

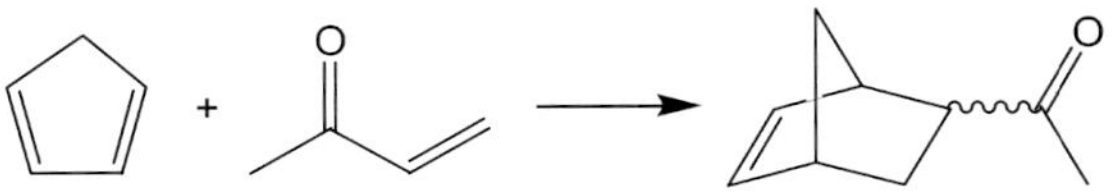

**Scheme 8.4** Diels–Alder reaction of **12** with **13**.

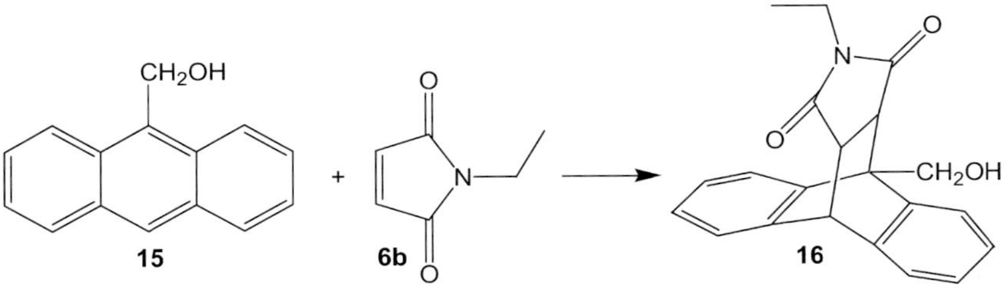

**Scheme 8.5** Diels–Alder reaction of **15** with **6b**.

This work set the stage for extensive and detailed studies of organic reactions in water, including different types of pericyclic reactions. Aqueous Diels–Alder reactions were examined by many groups around the world and this work was focused on both the mechanism of the aqueous rate enhancements and on synthetic applications, including natural product synthesis [26–31]. The synthetic studies were also carried out in heterogeneous aqueous systems, and it was recognized that there was a great potential for green organic chemistry.

Following in the footsteps of Breslow and co-workers, the mechanistic studies were focused on hydrophobic interactions as the origin of the rate accelerations and, indeed, a number of observations supported this interpretation. For example, rates of a number of Diels–Alder reactions in organic solvents and in water could be correlated with the solvophobic parameter $Sp$ [32]. But was this the whole story? In 1991, we showed [33, 34] that the Gibbs energy of activation for the Diels–Alder reaction of cyclopentadiene (**12**) with 5-methoxy-1,4-naphthoquinone (**17**) (Scheme 8.6) decreased with increasing value of $E_T(30)$, the most popular microscopic parameter for solvent polarity (Figure 8.1). In fact, for 19 different solvents, there were *two* linear plots of $\Delta^\ddagger G^\circ$ vs $E_T(30)$, one for the aprotic solvents (smallest slope) and one for the protic solvents (largest slope).

**Scheme 8.6** Diels–Alder reaction of **12** with **17**.

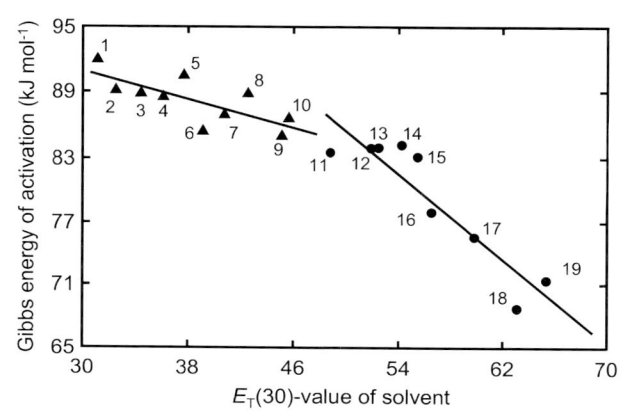

**Figure 8.1** Gibbs energy of activation for the Diels–Alder reaction of **12** with **17** in (1) n-hexane, (2) carbon tetrachloride, (3) benzene, (4) 1,4-dioxane, (5) THF, (6) chloroform, (7) dichloromethane, (8) acetone, (9) DMSO, (10) acetonitrile, (11) 2-propanol, (12) ethanol, (13) N-methylacetamide, (14) N-methylformamide, (15) methanol, (16) glycol, (17) TFE, (18) water, and (19) HFP as a function of the $E_T(30)$ value of the aprotic (▲) and protic (•) solvents. Reproduced with permission from J. Org. Chem. 1994, 59, 5372–5376.

This was strong evidence that the Diels–Alder reaction responded to solvent polarity and particularly to hydrogen bonding interactions. However, the fact that the rate in water was the largest and even larger than that in the more acidic trifluoroethanol (TFE) indicated that water provided another rate accelerating effect. Hydrophobic association of diene and dienophile was often considered, but this could be excluded on the basis of vapor pressure measurements and was unrealistic for the low substrate concentrations employed in the kinetic measurements [35]. It was clear that more in-depth studies were required, that could distinguish between solvent effects on the reactants and on the activated complex. The results of such studies, published in 1991 and 1992, were revealing [33, 35]. Solvent effects on the thermodynamic activation parameters and on the thermodynamic transfer parameters of diene and dienophile (and *endo* reaction product) were determined for the Diels–Alder reaction of cyclopentadiene (**12**) with methyl vinyl ketone (MVK, **19a**) and ethyl vinyl ketone (EVK, **19b**) in mixtures of 1-propanol and water over the whole mole fraction range of water ($X_{H_2O}$) (Scheme 8.7) [33]. Results are shown in Figures 8.2 and 8.3.

**Scheme 8.7** Diels–Alder reaction of **12** with MVK and EVK.

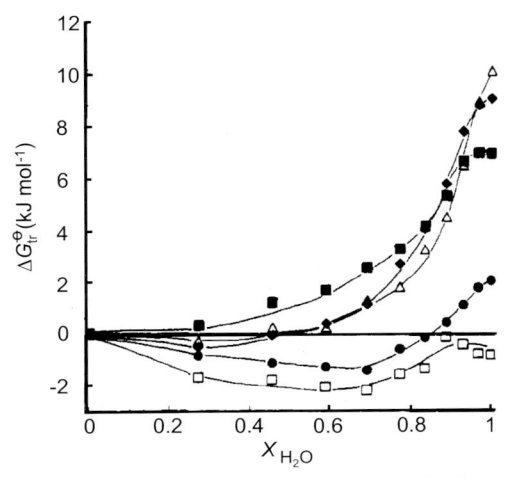

**Figure 8.2** Relative standard chemical potentials of cyclopentadiene (**12**; ■), methyl vinyl ketone (MVK, **19a**; •), initial state (MVK + CPD) (◆), activated complex (□), and *endo* product (△) (kJ mol$^{-1}$, standard state = 1 mol dm$^{-3}$) for the reaction of **12** with MVK in aqueous mixtures of 1-propanol as a function of the mole fraction of water at 298 K. Reproduced with permission from *J. Am. Chem. Soc.* 1992, *114*, 5440–5442.

Gibbs energies of transfer of the activated complex from 1-propanol–water to water are not experimentally accessible but were calculated using the equation

$$\Delta G_{tr}^{\sigma}(AC) = \Delta G_{tr}^{\sigma}(\text{diene}) + \Delta G_{tr}^{\sigma}(\text{dienophile}) - \delta \Delta^{\ddagger} G^{\sigma} \quad (8.1)$$

It was clear that the intriguing aqueous rate accelerations were primarily caused by the large destabilization of diene and dienophile in the water mole fraction range 0.7–1.0, as, in fact, expected for non-polar reactants. Particularly surprising was the relative *insensitivity* of the activated complex to changes in the water mole fraction over the complete mole fraction range. Its behavior in the water-rich region showed that it has almost completely lost its hydrophobic character. This shows that the activated complex can adapt itself remarkably well to substantial changes in the reaction medium.

The following picture emerged. For a bimolecular Diels–Alder reaction to take place, the relatively hydrophobic diene and dienophile should approach each other

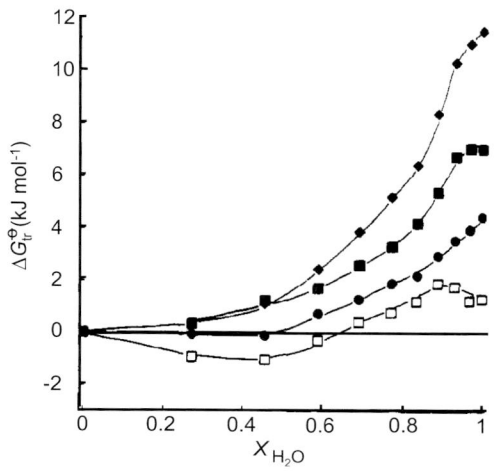

**Figure 8.3** Relative standard chemical potentials of cyclopentadiene (**12**; ■), ethyl vinyl ketone (EVK, **19b**; ●), initial state (EVK + CPD) (♦), and activated complex (□) (kJ mole$^{-1}$, standard state = 1 mol dm$^{-3}$) for the reaction of **12** with EVK in aqueous mixtures of 1-propanol as a function of the mole fraction of water at 298 K. Reproduced with permission from *J. Am. Chem. Soc.* 1992, *114*, 5440–5442.

along a reaction coordinate that allows maximum, orbital-symmetry-controlled, bond formation through HOMO–LUMO overlap. For this to occur, a rather close approach between the two reactants is required and formation of the expected complex (AC) necessitates substantial desolvation of the reactants and formation of a newly developed hydration shell of the AC. These solvation effects constitute a crucial contribution to the Gibbs energy of activation that determines the reaction rate. For a conventional $S_N$-2 reaction, Jorgensen [36] has beautifully shown the enormous impact of solvation by comparing the reaction in the gas phase, in an apolar organic solvent, and in water. During the activation process of the Diels–Alder reaction, the water-accessible surface area of the reactants is decreased (more for the *endo* than for the *exo* product formation), with its magnitude dependent on the nature of both reaction components. In water this is an *advantage* compared with an organic solvent. We have introduced the term "*enforced hydrophobic interaction*," because the favorable hydrophobic effect is just part of the activation process. No initial association of diene and dienophile has to be invoked to explain the aqueous rate enhancements, consistent with the finding that *intramolecular* Diels–Alder reactions are also speeded in water. The data in Figures 8.2 and 8.3 reveal that the aqueous rate acceleration originates almost completely from an initial state hydration effect because the AC is only slightly sensitive to the transfer from 1-propanol to water. For the two Diels–Alder reactions considered here, there is an enhanced polarization of the carbonyl group in the AC and the accompanying hydrogen bond stabilization of the AC apparently offsets the unfavorable hydrophobic contribution. This interpretation has now been widely accepted and has been strongly supported by advanced Monte Carlo computer simulations by Furiani and Gao [37] and Jorgensen and co-workers [38, 39]. *Ab initio* MO calculations have also been performed, supporting the conclusions based on the simulations [40]. In these studies, a remarkable correspondence was obtained between the experimental and computed solvent effects. This work has recently been extended [41] using QM/MM calculations, again with a most remarkable correspondence between calculated and experimental rate enhancements on going from aprotic organic solvents to water.

Jorgensen and co-workers had already shown previously that hydrogen bonding is sensitive to small charge shifts [39]. This effect is enhanced for the small water molecules, which can closely approach the polarized carbonyl group in the AC. A significant increase in hydrogen bond stabilization was found even for a minor change in charge of 0.03e (e is the electronic charge). The possibility has been considered that the charge separation in the AC in protic, polar solvents, including water, is larger than that in apolar solvents, and an analysis of substituent effects on an inverse electron-demand hetero-Diels–Alder reaction indicated that this is borne out in practice [42].

For the reaction of cyclopentadiene (**12**) with **19a** (Scheme 8.7), the *endo* product has a significantly stronger hydrophobic nature than the AC, although its molecular volume will be smaller. This can be explained by assuming that the more polarizable AC is a stronger hydrogen bond acceptor.

At this stage it is of interest to note that water is also a better solvent than room temperature ionic liquids (RTILs) as shown by Tiwari and Kumar [43] for the Diels–Alder reactions of cyclopentadiene (**12**) with methyl acrylate (**21a**), ethyl acrylate (**21b**), and *n*-butyl acrylate (**21c**) to yield the products **22a–c** (Scheme 8.8). Second-order rate constants are 3–10 times higher in water than in the RTILs (**23a–c**, Scheme 8.9) because of the hydrophobic interactions and stronger hydrogen bonding.

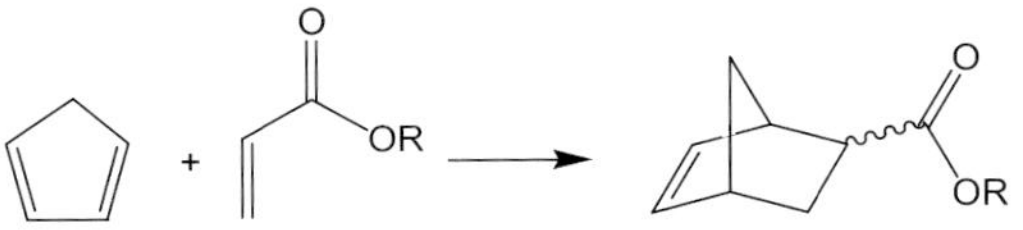

**12**

**21a**: R=CH$_3$
**21b**: R=C$_2$H$_5$
**21c**: R=n-C$_4$H$_9$

**22a**: R=CH$_3$
**22b**: R=C$_2$H$_5$
**22c**: R=n-C$_4$H$_9$

**Scheme 8.8** Diels–Alder reaction of **12** with alkyl acrylates (**21a–c**).

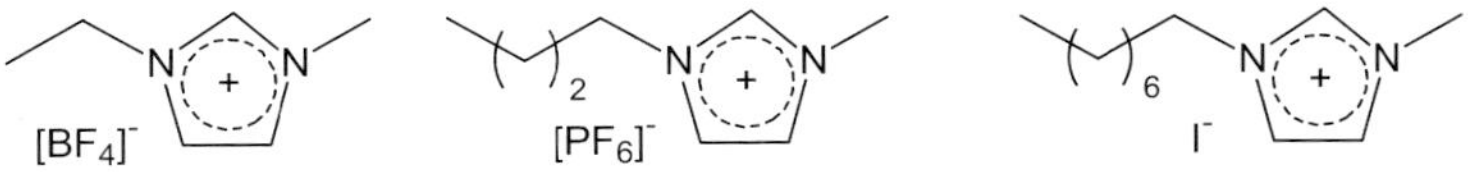

**23a** [BF$_4$]$^-$

**23b** [PF$_6$]$^-$

**23c** I$^-$

**Scheme 8.9** Room temperature ionic liquids used by Tiwari and Kumar [43].

It is, of course, highly valuable to know how variations in hydrophobicity of the diene and dienophile affect the magnitude of the rate enhancements in water. This question was examined in a study published in 1998 [44]. A series of *N*-alkyl-substituted maleimides (**6a–d**) of increasing hydrophobicity were employed in Diels–Alder reactions with three different dienes: cyclopentadiene (**12**), 2,3-dimethyl-1,3-butadiene (**25**), and 1,3-cyclohexadiene (**27**) (Scheme 8.10). In all reactions the *endo* adduct was formed predominantly, and in the following discussion we concentrate our attention on the *endo*-AC.

Second-order rate constants were measured in water and in a series of organic solvents. In all cases, there is a substantially increased reactivity in water. Here we give values for the rate increase (RI) in water relative to 1-propanol. A series of important observations were made.

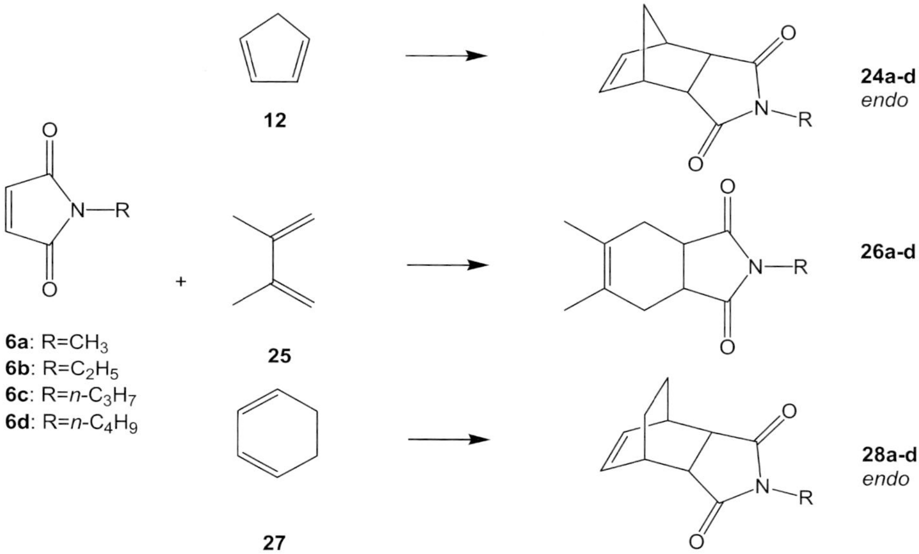

**6a**: R=CH$_3$
**6b**: R=C$_2$H$_5$
**6c**: R=n-C$_3$H$_7$
**6d**: R=n-C$_4$H$_9$

**Scheme 8.10** Diels–Alder reactions of dienes and dienophiles of variable hydrophobicities.

First, the RI for **25** and **27** is larger than that for **12**. For example, for dienophile **6d**, the RIs are as follows: 55 (**12**), 213 (**25**), and 300 (**27**). This is consistent with the fact that in case of **25** and **27** there is an enhanced hydrophobicity near the reaction center compared with **12**.

Second, for the reaction of **25** with **6a–d**, the second-order rate constant is increased approximately linearly with the length of the N-alkyl substituent in the dienophile. A similar effect was *not* found for the dienes **12** and **27**. A closer look at the AC for the reactions of **25** with **6a–d** reveals the possibility of hydrophobic interactions between the methyl substituents in **25** and the N-alkyl group, leading to a further increase in enforced hydrophobic interactions for larger N-alkyl groups. Such an effect is clearly not possible for **12** and **27**.

A third important observation was made when the Gibbs energies were determined for transfer of **12**, **27**, and **6a** and **b** from 1-propanol to water [44]. Adding the contributions of diene and dienophile provides the Gibbs energies of transfer for the reactants (initial state, IS) of the Diels–Alder reactions. As expected, the destabilization of the IS on going from 1-propanol to water becomes more pronounced for an increase in the hydrophobicity of both the diene and the dienophile (Table 8.1).

Table 8.1 Gibbs energies of transfer for **12**, **27** and **6a** and **b** from 1-propanol to water at 25 °C [44].

| Compound | Gibbs energy (kJ mol$^{-1}$) |
| --- | --- |
| 12 | 8.82 |
| 27 | 11.5 |
| 6a | 0.74 |
| 6b | 3.26 |

Combination of these Gibbs energies of transfer with the relevant Gibbs energies of activation for the Diels–Alder reactions gives the Gibbs energies of transfer for the ACs (Equation 8.1). Most significant is that an increase in the hydrophobicity of **6a** and **b** destabilizes the ACs, but this does not occur for the diene. We conclude that hydrophobic groups *near the reaction center* lose their hydrophobic character completely in the AC. Apparently hydrophobic hydration of these groups is hampered by the hydrogen bonding interactions that stabilize the polarizable AC. Non-polar groups more distant from the reaction center are able to retain their hydrophobicity throughout the reaction pathway.

A careful comparison of the thermodynamic activation parameters of Diels–Alder reactions in water and in organic solvents provides further support for the rationalization of the aqueous rate effects. It has been found that for the Diels–Alder reaction of cyclopentadiene (**12**) with methyl vinyl ketone (**19a**) in water and in methanol (hydrogen bond donating, no hydrophobic effects), the higher rate in water is primarily due to a more favorable entropy of activation [45], in accord with a dominant contribution of enforced hydrophobic interactions. It has often been observed that chemical processes facilitated by hydrophobic interactions are primarily entropy stimulated. By contrast, the rate acceleration on going from 1,4-dioxane (not hydrogen bond donating, no hydrophobic effects) to water is dominated by a more favorable enthalpy of activation [46]. These findings suggest that the hydrogen bonding part of the aqueous rate effects is, at least largely, due to enthalpic effects. Indeed, Diels–Alder reactions in water benefit from *both* favorable enthalpic and entropic effects, in accord with the two mechanisms responsible for the aqueous rate enhancements.

It will be clear that the relative contributions from enforced hydrophobic interactions and hydrogen bonding will be determined by the specific hydration properties of the diene and dienophile. Particularly the work of Jorgensen's group [38–41] has suggested that the hydrogen bonding contribution will often be dominant. The absence of "hydrophobic packing" (using Rideout and Breslow's nomenclature) between diene and dienophile, at least at the concentrations employed in the kinetic studies, has an advantage. In case there were to be association, the binding process could well be rate retarding and counterproductive because the orientation of both components in the complex might not be optimal or even strongly unfavorable for the formation of the AC for the cycloaddition reaction.

Several things should be noted here. First, on the basis of the above theory, aqueous rate accelerations are expected for both normal and inverse electron-demand Diels–Alder reactions. This has been tested and found to be correct [42]. Second, for

Diels–Alder reactions that possess no polarizable group(s) in the dienophile, the reaction rate will be enhanced to a smaller extent because the hydrogen bonding effect cannot contribute to the rate. For example, the inverse electron-demand, polar cycloaddition of acridizinium bromide (**29**) with cyclopentadiene (**12**) (Scheme 8.11) occurs via an AC with hydrogen bonding interactions that are negligibly different from those of the reactants. This has been confirmed by the absence of a significant rate increase in trifluoroethanol compared with other organic solvents [47]. In water, the second-order rate constant is increased by only a factor of five as compared with ethanol, and this effect can now be (almost) fully ascribed to enforced hydrophobic interactions. However, if a $-CO_2CH_3$ substituent is introduced at the 9-position in the cation, there will be a slight increase in negative charge density on the ester carbonyl in the AC and, indeed, now the rate acceleration in water compared to ethanol is doubled.

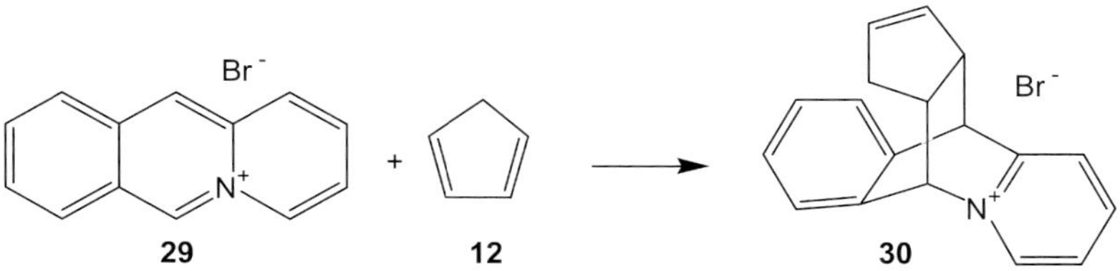

**Scheme 8.11** Diels–Alder reaction of two reactants lacking hydrogen bonding ability.

Interestingly, if the hydrophobicity in addition to the hydrogen bonding capacity of the dienophile is increased in the Diels–Alder reaction with cyclopentadiene, one could ask how much the aqueous rate acceleration can be increased. A test was conducted in 1991 [33], when rather hydrophobic and more strongly hydrogen bonding 5-substituted naphthoquinones were employed as dienophiles. It was found that for R = $OCH_3$ (**17**; Scheme 8.6), the second-order rate constant in water was increased by a factor of 353 relative to 1-propanol and by a factor of 5000 relative to *n*-hexane. These results demonstrate that huge rate increases in water are feasible for Diels–Alder reactions if substantial enforced hydrophobic interactions and hydrogen boding effects work in concert.

In sum, it can be concluded that normal and inverse electron-demand Diels–Alder reactions in water not only have an outstanding potential for green chemistry, but that the aqueous reaction environment, in comparison with organic solvents, can also induce very efficient transformations. Since the mechanism of the aqueous rate enhancements is now known in some detail, the favorable kinetic and stereochemical effects of aqueous reaction media can be tuned with reasonable accuracy by structural variations in the diene and dienophile.

### 8.2.3
### Intramolecular Diels–Alder Reactions

On the basis of the previous discussion of the mechanism behind the acceleration of *inter*molecular Diels–Alder reactions in water, it is expected that *intra*molecular Diels–Alder reactions will also be accelerated. This is borne out in practice. Thus, the

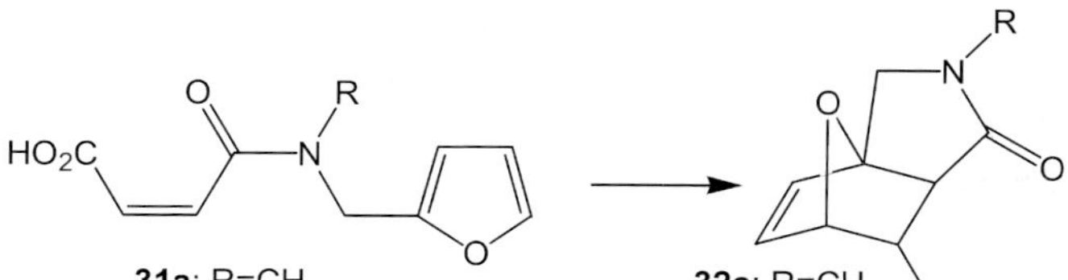

## Scheme 8.12

**31a:** R=CH$_3$
**31b:** R=C$_2$H$_5$
**31c:** R=n-C$_3$H$_7$

**32a:** R=CH$_3$
**32b:** R=C$_2$H$_5$
**32c:** R=n-C$_3$H$_7$

**Scheme 8.12** Intramolecular Diels–Alder reaction of 31a–c.

intramolecular Diels–Alder cycloaddition of N-furfuryl-N-methylmaleamic acid (**31a**) (Scheme 8.12), for which the AC is reached via a conformation in which the furan and dienophile are geometrically disposed for π-overlap, occurs 103 times faster in water than in methanol, and 153 times faster than in n-hexane [33, 48]. Also in this reaction both the decreased solvent-accessible surface area of the AC relative to the reactant and the stronger H-bonding stabilization of the carbonyl group in the AC will contribute to the rate enhancement in water.

In general, intramolecular Diels–Alder reactions, for which the diene and dienophile are linked by a bridge of three or four atoms, still have negative volumes of activation, although generally less negative than for intermolecular Diels–Alder cycloadditions.

The aqueous rate enhancements further argue against initial "hydrophobic association" as the reason for aqueous rate accelerations of bimolecular Diels–Alder reactions, since for **31a** the diene and dienophile are already close together in the reactant.

The stereochemistry of the intramolecular process depends crucially on the conformation of the diene, which can be cis or trans. Compound **31a** exists in solution both in the s-trans and in the (reactive) s-cis conformation. NMR studies have shown that there is no significant increase in the concentration of the s-cis conformation in deuterium oxide ([s-cis]: [s-trans] = 1.05 in chloroform, 1.03 in methanol, and 1.20 in water). Apparently, substrate **31a** does not "coil" intramolecularly in water [49].

The same intramolecular Diels–Alder reaction was studied for dienophiles in which the N-methyl substituent was replaced by an ethyl (**31b**) or an n-propyl group (**31c**). Rate accelerations in water relative to 1-propanol were 86 (**31a**), 81 (**31b**), and 41 (**31c**). The somewhat smaller effect for **31c** is caused by both a slower reaction in water and a faster reaction in 1-propanol [49].

Other examples of water-accelerated intramolecular Diels–Alder reactions have been reported and played an important role in synthetic protocols in both homogeneous and heterogeneous aqueous reaction media [30].

### 8.2.4
### Retro-Diels–Alder Reactions

Retro-cycloaddition reactions are synthetically useful and are expected to occur at elevated temperatures because of the relatively favorable entropy of activation for the unimolecular process. The principle of microscopic reversibility tells us that the mechanistic pathway is just the reverse of that of the forward reaction. What is the effect of an aqueous reaction medium on these cycloreversions? An early study [50] examined the homo-retro-Diels–Alder reaction of **33a** and **b** to give the corresponding naphthoquinones and cyclopentadiene, which can be kinetically studied at relatively low temperatures. Rate measurements in the temperature range 30–49 °C showed that the reaction of **33a** is dominated by the enthalpy of activation ($\Delta^\ddagger H^\circ = 109.8 \pm 1.9$ kJ mol$^{-1}$) with the $T\Delta^\ddagger S^\circ$ term being only $-0.3 \pm 1.8$ kJ mol$^{-1}$. In organic solvents, ranging from *n*-hexane to 2-propanol, rates vary only by a factor of about 10, similar to what is usually found for forward Diels–Alder reactions. However, in water the reaction is almost 140 times faster than in *n*-hexane. All evidence points to an AC in which the solvent-accessible surface area has hardly changed, but with a significantly increased polarization of the carbonyl groups. Therefore, enforced hydrophobic interactions will not importantly contribute to the aqueous rate acceleration, and the aqueous medium effect will be governed by the increased hydrogen bonding interactions in the AC.

The fast reaction of **33a** (Scheme 8.13) in the very strong hydrogen-bonding solvent 1,1,1,3,3,3-hexafluoroisopropanol (HFIP, 1.3 times faster than in water) further illustrates the strong H-bond stabilization of the AC. Interestingly, the rather acidic trifluoroethanol cannot compete with water (2.2 times slower), demonstrating the special H-bonding capability of liquid water composed of the small, not sterically demanding, water molecules. Gibbs energies of activation for the reaction of **33a** in eight solvents (Figure 8.4), including water, showed a remarkable linear correlation with the $E_T(30)$ values of the solvents, suggesting that both polarity and hydrogen bond donating capacity of these solvents govern the reaction rates.

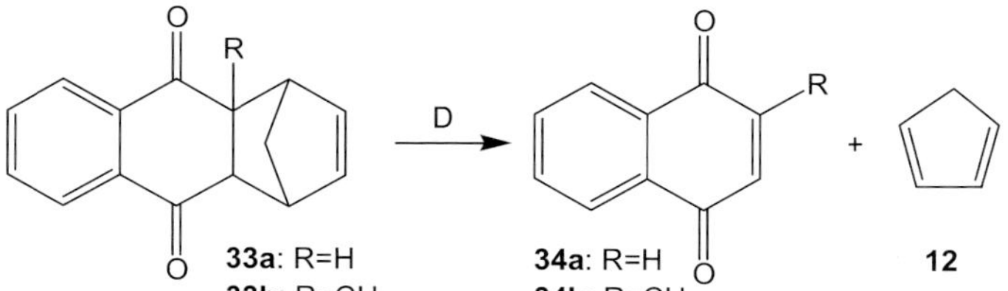

**Scheme 8.13** Reverse Diels–Alder reaction of **33a** and **b**.

## 8.2.5
### Forward and Retro-hetero-Diels–Alder Reactions

Diels–Alder cycloadditions in which one or more of the diene or dienophile carbons are replaced by heteroatoms are called hetero-Diels–Alder reactions. They can be of the normal electron-demand or inverse electron-demand type, and they have found numerous applications in synthetic organic chemistry. The synthetic procedures have also been successfully extended to aqueous media, particularly by the groups of Grieco, Lubineau and Breslow.

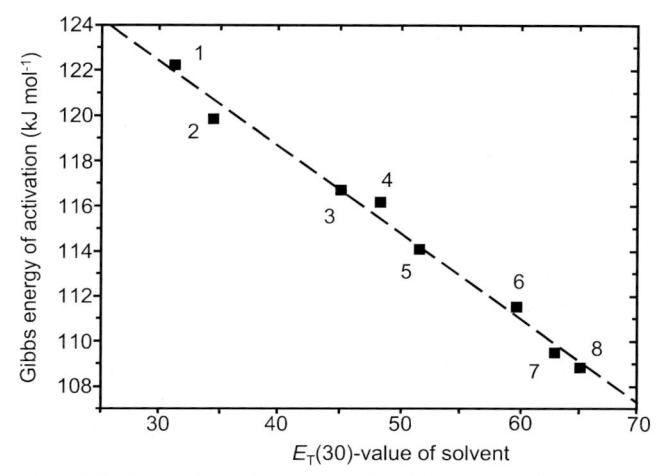

**Figure 8.4** Linear relationship between the Gibbs energy of activation for the reaction of **33a** and the $E_T(30)$ value of the solvent. n-Hexane (1); benzene (2); DMSO (3); i-PrOH (4); acetic acid (5); TFE (6); water (7); HFIP (8). Reproduced with permission from *J. Org. Chem.* 1997, 62, 2039–2044.

A detailed kinetic study [42] has been made of the inverse electron-demand hetero-Diels–Alder reaction of di(2-pyridyl)-1,2,4,5-tetrazine (**35**) with five 5-substituted styrenes (**36a**, R = OMe; **36b**, R = Me; **36c**, R = H; **36d**, R = Cl; **36e**, R = NO$_2$) (Scheme 8.14). Apart from an attempt to relate the aqueous rate effects to interactions with the solvent determining the activation process, the use of the five substituents in the styrene components would allow an analysis of medium effects on the substituent effects as expressed in the Hammett ρ-values.

**Scheme 8.14** Example of an inverse electron-demand hetero-Diels–Alder reaction.

Cycloadditions involving (substituted) tetrazines had been already investigated in a number of previous studies. It was established that the cycloaddition, with the tetrazine acting as a diene, was a two-step process. The first step is a rate-determining

[4 + 2] cycloaddition leading to **37**, which is followed by a rapid retro-Diels–Alder reaction to yield the final dihydropyridazine **38** [51].

The reaction is characterized by a dominant HOMO(dienophile)–LUMO(diene) interaction, in accord with the observed sequence of reactivities, **36a** > **36b** > **36c** > **36d** > **36e**. Because of solubility constraints, the kinetics could not be reliably monitored in pure water and the reaction was therefore studied in water containing 5 mol% of *t*-BuOH. This highly aqueous mixture is still representative of the reaction in pure water and even several examples have been recorded of a somewhat higher rate in this mixture than in pure water [52]. Comparison of the second-order rate constants in $H_2O$–*t*-BuOH ($X_w = 0.95$) and in five organic solvents again shows a strong rate acceleration in the aqueous medium [$k_2(H_2O$–*t*-BuOH ($X_w = 0.95$))/$k_2$(THF) = 143 (**36a**), 136 (**36b**), 123 (**36c**), 75 (**36d**) and 26 (**36e**)], particularly for the dienophile with the electron-releasing substituents in the phenyl ring. The rate constants in the six solvents show a rough correlation with both the $E_T(30)$ parameter and the acceptor number (AN) of the solvent, but the plots exhibit considerable scattering.

The substituent effects can be correlated with the $\sigma^+$-values for the aprotic solvents, but for the protic solvents the normal $\sigma$-values are more satisfactory. A more sophisticated treatment in terms of the CR equation and electron-demand parameter was illuminating [53, 54]. Overall, the picture emerged that hydrogen bonding of water (and other protic solvents) to the lone pairs of the tetrazine decreases the electron density and accelerates the reaction. The higher rate of reaction in the aqueous medium than in THF can be reasonably explained by the operation of enforced hydrophobic interactions.

The cheap commercial aqueous solution of glyoxylic acid (**39**) was employed by Lubineau *et al.* [55] for Diels–Alder reactions (Scheme 8.15) with, for example, cyclopentadiene (**12**) and 1,3-cyclohexadiene (**27**) to yield the corresponding epimeric lactones (**41**, **42**) via the intermediate **40**.

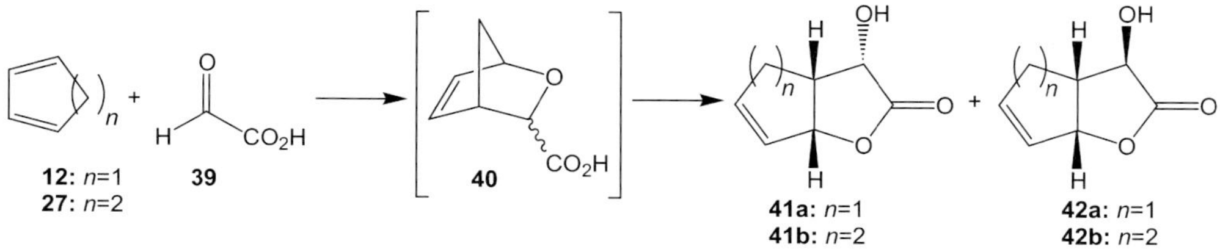

**Scheme 8.15** Glyoxylic acid as the dienophile in hetero-Diels–Alder reactions.

It can be concluded that green aqueous reaction media are equally effective for both homo- and hetero-Diels–Alder reactions.

The beneficial effect of water on a retro-hetero-Diels–Alder reaction was reported in 1987 by Grieco *et al.* [56]. They found that azanorbornene and its N-alkyl derivatives (Scheme 8.16) undergo a dramatically accelerated, acid-catalyzed retro-hetero-cycloaddition in water to yield **44** and **45**. This reaction, which needs extreme conditions (temperatures in the range 400–600 °C) and does not proceed at all in organic solvents at 50 °C, could now be performed at ambient temperature in water in

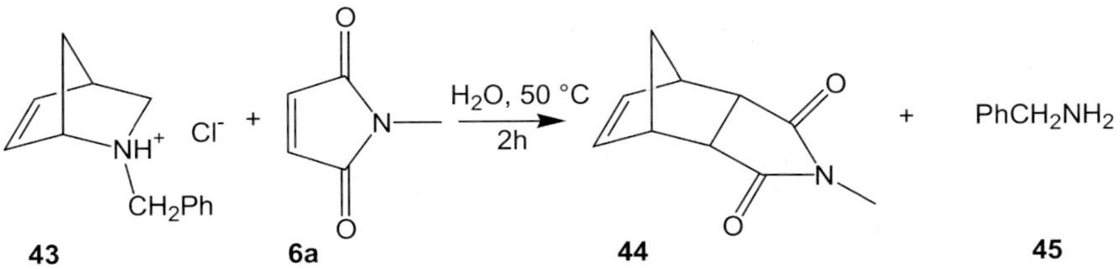

**Scheme 8.16** The retro-Diels–Alder reaction of **44** in the presence of the trapping reagent **6a**.

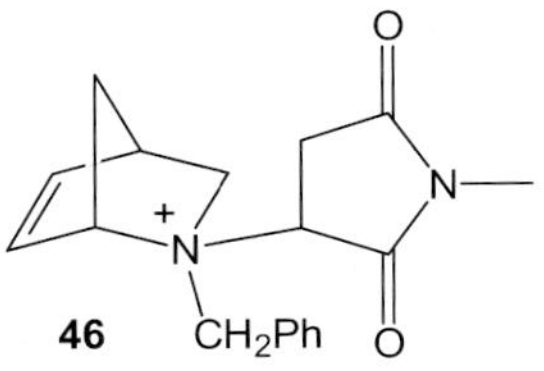

**Scheme 8.17** Structure of the proposed intermediate.

the presence of the dienophile N-methylmaleimide acting as a trapping agent. It has been suggested that the cycloreversion occurs through an intermediate ammonium salt **46** (Scheme 8.17).

The heterocycloreversion was later employed for unmasking of primary amines via catalysis by Cu(II) salts or sulfonic acid, but these protocols need binary ethanol–water mixtures instead of pure water [57].

The retro-hetero-Diels–Alder reaction of **47** to give cyclopentadiene and nitrosobenzene (Scheme 8.18) has been examined quantitatively in water and in a number of organic solvents [58]. In fact, the adduct **47** cycloreverts in solution to yield an equilibrium mixture of reactants and adduct. The equilibrium constant $K$ could be easily determined by UV/Vis spectroscopy. Because of the reverse reaction, the second-order rate constants for the addition of cyclopentadiene to nitrosobenzene could not be reliably measured, and were estimated from the addition of cyclopentadiene to 1,3-cyclohexadiene (to give a stable adduct) as a model substrate. A spectacular increase in $K$ was found on going from n-hexane ($K = 6.6\,M^{-1}$) to water ($K = 5775\,M^{-1}$). This is a result of both an *acceleration* of the forward Diels–Alder reaction and an *inhibition* of the cycloreversion in water relative to organic solvents. The rate enhancement for the forward reaction is as expected, taking into account the smaller water-accessible surface area of the AC compared with that of the reactants and the much stronger hydrogen bond acceptor capability of the isoxazolidine relative to that of nitrosobenzene. The inhibition of the rate of the

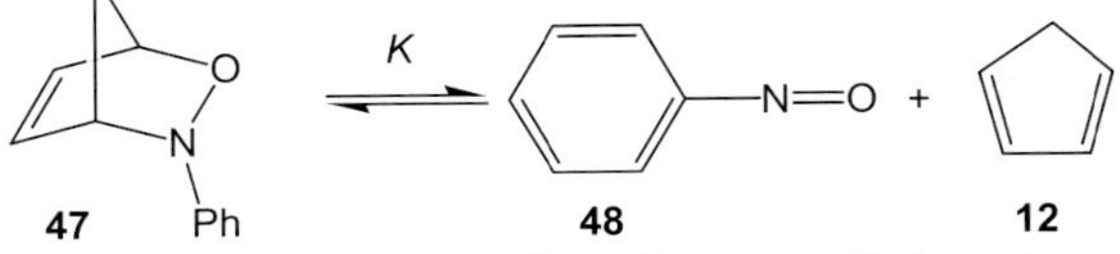

**Scheme 8.18** Cycloreversion of **47** yielding an equilibrium mixture of diene (**12**) and dienophile (**48**).

cycloreversion in water shows that hydrogen bonding can also *slow* a pericyclic reaction. This is in accord with the inhibition of the 1,3-cycloaddition of aromatic nitrones or nitrile oxides to electron-poor dipolarophiles in water (Section 8.2.7).

### 8.2.6
### Photocycloadditions

A [2 + 2] cycloaddition does not occur under thermal conditions because the required suprafacial/suprafacial approach is forbidden. The process can be turned into an allowed suprafacial/antarafacial reaction by electronic excitation of one of the reactants. Such photochemical cycloadditions may also benefit from water as the reaction medium. For example, the photodimerization of *cis*- and *trans*-stilbenes is greatly enhanced in a heterogeneous aqueous medium, whereas in benzene the dominant reaction for *cis*-stilbene is *cis*–*trans* isomerization (Scheme 8.19) [59]. After 2 months of irradiation of *trans*-stilbene in benzene (0.75 M) only 27% of dimer was formed.

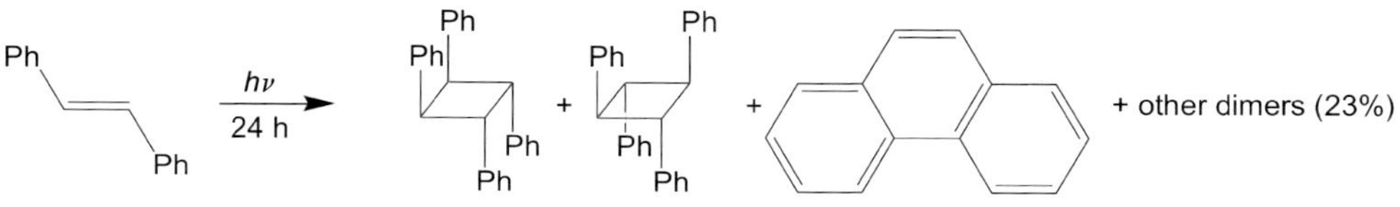

**Scheme 8.19** Photodimerization of *trans*-stilbene.

The authors assumed that preassociative hydrophobic interaction between the reactants in water (sometimes leading to a fluorescent aggregate) facilitates the aqueous dimerization, even at concentrations of $10^{-6}$–$10^{-4}$ M. The reaction of *cis*-stilbene occurs via initial isomerization to the *trans* isomer. Other photodimerizations were also found to have increased quantum efficiencies in water [60].

Photodimerization of the coumarin **53** at room temperature has been investigated in pure water and the adduct **54** was formed exclusively (Scheme 8.20) [61]. In EtOH and in benzene, the light-induced cycloaddition led to a high preference for the stereoisomer **55**, and the quantum yield was much lower than in water.

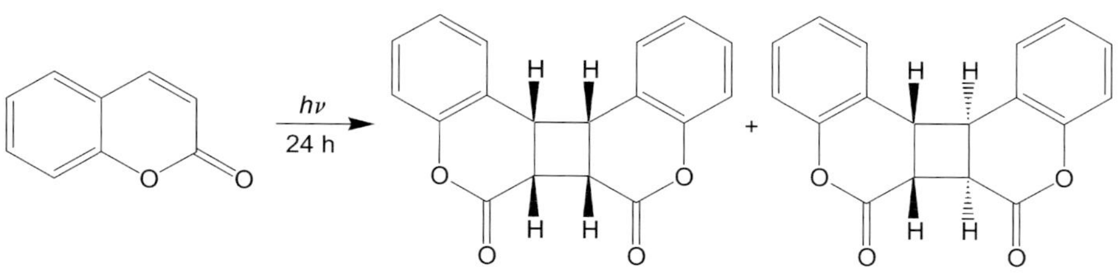

**Scheme 8.20** Photodimerization of coumarin **53**.

## 8.2.7
## 1,3-Dipolar Cycloadditions

This class of pericyclic reactions is of considerable synthetic importance, particularly for the synthesis of five-membered heterocycles [62]. Therefore, the development of green protocols has significant relevance. Mechanistically, the [$2\pi + 4\pi$] cycloaddition is akin to Diels–Alder reactions, as illustrated by its reversibility, orbital symmetry-controlled stereochemistry, and concertedness leading to stereospecificity. The negative volume of activation is indicative for the bimolecular rate-determining step. An example has already been given in Section 8.1.1. A 1,3-dipole reacts with a dipolarophile, and the orbital overlap in the activated complex can be rationalized by frontier molecular orbital (FMO) theory. For an electron-rich dipolarophile the dominating interaction is LUMO(dipole)–HOMO(dipolarophile) overlap, whereas for an electron-poor dipolarophile it is LUMO(dipolarophile)–HOMO(dipole) overlap that governs the cycloaddition. Just as for Diels–Alder reactions, solvent effects in organic media are modest, with rate constants either increasing or decreasing with solvent polarity depending on the nature of both reactants. Some reactions even hardly respond to changes of the medium. In the later part of the previous century, examples were reported of a beneficial kinetic effect of water on 1,3-dipolar cycloadditions. Moreover, product precipitation in highly aqueous media was a practical advantage, but prevented reliable kinetic measurements in pure water.

An extensive kinetic study of the cycloaddition of norbornene (**56**) with phenyl azide (**57**) to give **58** in 11 organic solvents and in five binary aqueous mixtures showed a marked acceleration in the aqueous media (Scheme 8.21) [63]. The increase in rate on going from *n*-hexane to water containing 1 mol% of 1-cyclohexyl-2-pyrrolidone by a factor of 53 was unprecedented and enforced hydrophobic interactions were held responsible for this huge kinetic effect for a 1,3-dipolar cycloaddition.

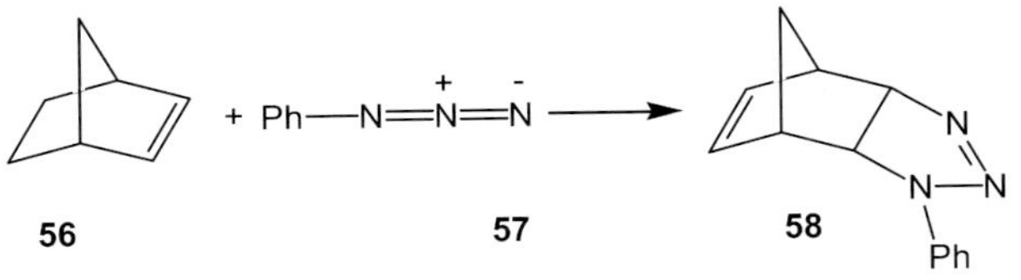

**Scheme 8.21** 1,3-Dipolar cycloaddition of **56** with **57**.

A completely different water-induced kinetic effect was found in a kinetic study of the 1,3-dipolar cycloaddition of benzonitrile oxide (**5**) with five different dipolarophiles [cyclopentene (**59**), 2,3-dihydrofuran (**60**), methyl vinyl ketone (**19a**), acrylonitrile (**61**), and *N*-methylmaleimide (**6a**)] in six organic solvents and in water (Scheme 8.22) [64]. The low substrate concentration allowed accurate kinetic measurements in pure water. An interesting phenomenon was observed: cycloadditions with electron-rich dipolarophiles (**59**, **60**) were *accelerated* in water, whereas the reactions with the electron-poor dipolarophiles (**6a**, **19a**, **61**) were slightly *retarded* in water. The importance of hydrogen bonding interactions was indicated by the fact

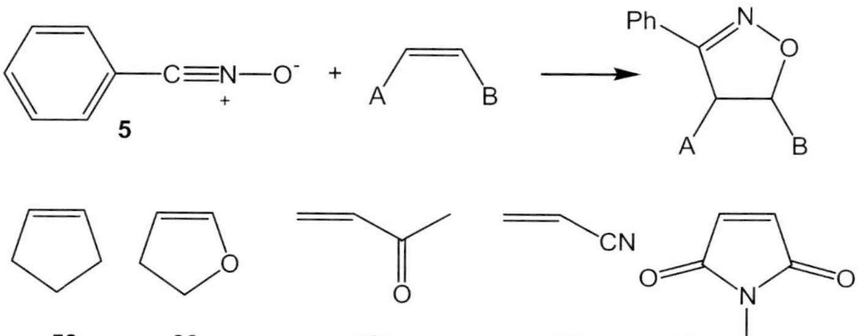

**Scheme 8.22** 1,3-Dipolar cycloaddition of benzonitrile oxide **5** with a series of dipolarophiles **6a**, **19a**, and **59–61**.

that, taking n-hexane as the reference solvent, the reactions with **59** and **60** were speeded up in 2,2,2-trifluoroethanol (TFE), in contrast with the rate decrease found in TFE for **6a**, **19a** and **61**. However, in all cases the reaction was faster in water than in TFE, in accord with a beneficial kinetic effect of enforced hydrophobic interactions. The different hydrogen bonding effects for electron-rich and electron-poor dipolarophiles could be explained in terms of the FMO theory. The interesting issue is that in the case of the electron-poor dipolarophiles, hydrogen bond interactions and enforced hydrophobic interactions *counteract* each other. This behavior is not unique: the Diels–Alder reaction of p-nitrostyrene (**36e**) with di(2-pyridyl)-1,2,4,5-tetrazine (**35**) is also enhanced by enforced hydrophobic interactions and inhibited by hydrogen bonding (as indicated by the slow reaction in TFE) [42].

A more systematic and extensive kinetic study [65] has been made of the 1,3-dipolar cycloaddition of benzonitrile oxide (**5**) with several N-alkylmaleimides (**6**) and cyclopentene (**59**) (Scheme 8.23). The second-order rate constants and the thermodynamic activation parameters ($\Delta^\ddagger H^\circ$, $\Delta^\ddagger S^\circ$) reveal the more complex aqueous solvent effects on 1,3-dipolar cycloadditions compared with Diels–Alder reactions. Although all reactions are accelerated by enforced hydrophobic interactions, the overall aqueous rate enhancements are not impressive since solvent polarity and

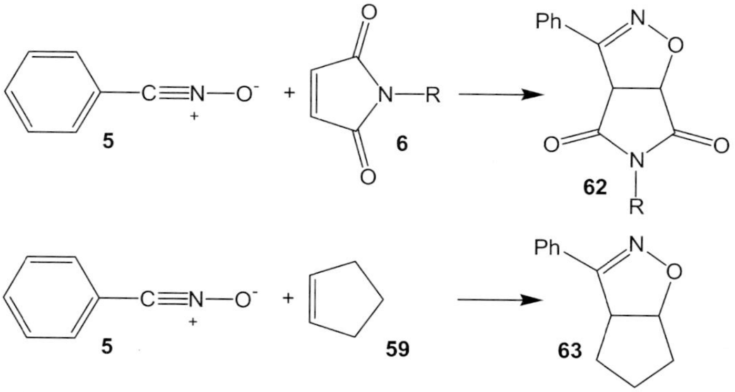

**Scheme 8.23** Two 1,3-dipolar cycloadditions responding differently to an aqueous medium.

hydrogen bonding interactions slow these cycloadditions in water. For example, the reaction of **5** with N-ethylmaleimide (**6b**) is only 1.05 times faster (hardly outside experimental error) in water than in n-hexane, but for cyclopentene as the dipolarophile there is a 2.9 times rate increase. The rather complex dependence of the rate constants on the reaction medium can, however, be rationalized with the FMO theory [65]. Interestingly, the $\Delta^{\ddagger}H°$ and $\Delta^{\ddagger}S°$ values signaled significant differences in solvation in different solvents, which are not revealed in the rate constants.

Another example illustrating the complex medium effects is the 1,3-dipolar cycloaddition of C,N-diphenylnitrone (**64**) to di-n-butyl fumarate (**65**) (Scheme 8.24) [66]. Second-order rate constants are similar in dimethylformamide, EtOH and 1-PrOH, 10 times higher in n-hexane, and 125 times higher in water.

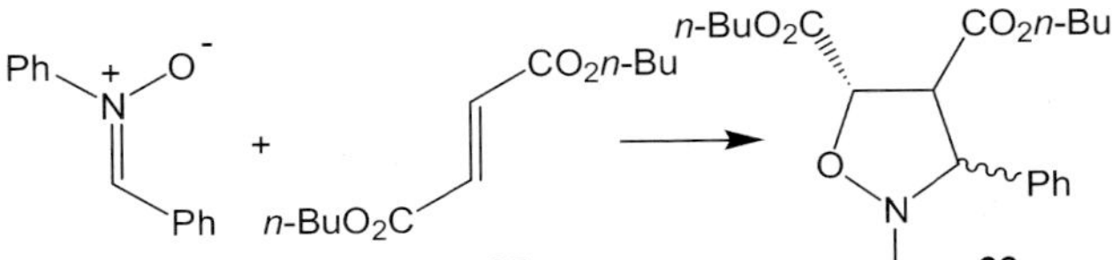

**Scheme 8.24** A 1,3-dipolar cycloaddition exhibiting complex medium effects.

An example of an aqueous, Lewis acid-catalyzed 1,3-dipolar cycloaddition is the reaction of diazocarbonyl compounds with alkynes in the presence of $InCl_3$ (Scheme 8.25) [67].

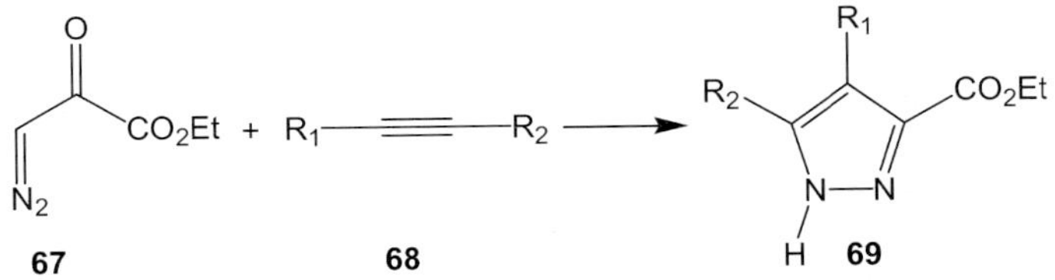

**Scheme 8.25** Reaction of diazocarbonyl **67** with an alkyne (**68**).

Apart from the obvious advantage of employing water as a green reaction medium, 1,3-dipolar cycloadditions benefit less from being performed in water than the majority of the Diels–Alder reactions discussed in Section 8.2.

### 8.2.8
### Claisen Rearrangements

Inspired by the important biochemical [3]-sigmatropic rearrangement of chorismate **70** to prephenate **71**, which is enzymatically catalyzed by chorismate mutase in water (Scheme 8.26) [68, 69], and by the usefulness of Claisen rearrangements in natural product synthesis, solvent effects on this pericyclic reactions have been examined in some detail [70, 71]. Reactions that previously needed extreme thermal conditions or led to elimination as a side reaction could now be conveniently performed in aqueous media. The rearrangement of the bis(tetra-n-butylammonium) chorismate occurs

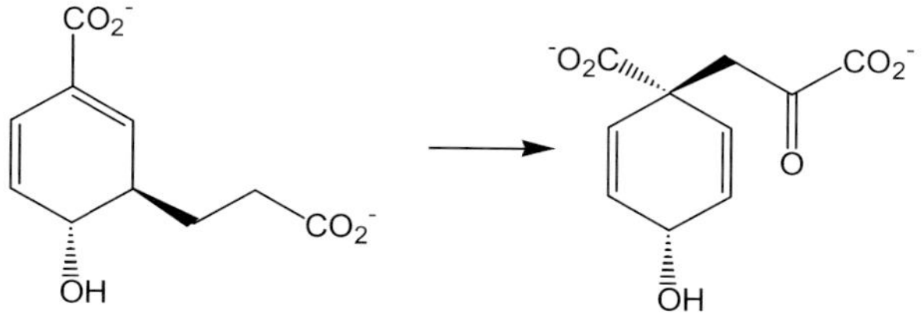

**Scheme 8.26** Claisen rearrangement of the chorismate anion.

100 times faster in water than in methanol (50 °C), whereas chorismic acid reacts 11 times faster in water than in methanol [72].

Solvent effects on this unimolecular pericyclic reaction are complex and the structural details of the AC are rather sensitive to the substituents in the reactants. A simple case involves the methyl carboxylate **72a** with the allyl vinyl ether unit as the reactive group (Scheme 8.27) [73]. Relative to cyclohexane, the first-order rate constant is increased by a factor of 2.1 in acetone, 8.6 in MeOH, and 43 in MeOH–$H_2O$ (1:1). The corresponding sodium carboxylate **72b** reacts 23 times faster in water than in methanol and 66 times faster than in DMSO–$H_2O$ (9:1).

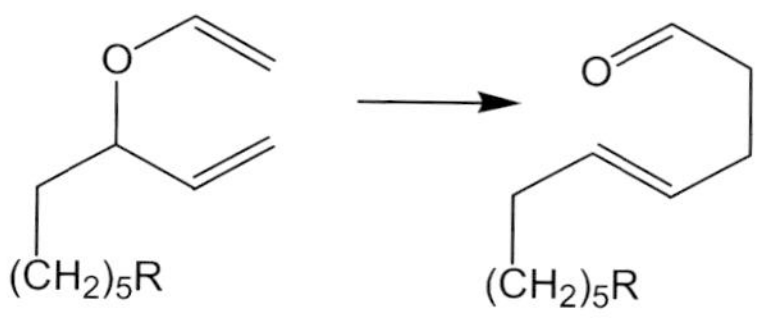

**72a**: R=$CO_2$Me
**72b**: R=$CO_2^-$ $Na^+$

**73a**: R=$CO_2$Me
**73b**: R=$CO_2^-$ $Na^+$

**Scheme 8.27** Claisen rearrangement of **72a** and **b**.

Water is clearly a favorable medium for Claisen rearrangements, as shown, for example, by the substantial rate enhancement found by Butler et al. [74], and also by Nicolaou et al. [75] in their biomimetic synthesis of gambogin. Unfortunately, there appear to be insufficient kinetic data to establish the relative contributions of hydrogen bonding and hydrophobic interactions to the observed aqueous rate accelerations.

## 8.2.9
### Mixed Aqueous Binary Mixtures

Mixtures of water with organic cosolvents are extremely popular reaction media, which can often solve the solubility problems that occur if pure water is used as the solvent. They serve as media for synthetic reactions, but also for mechanistic studies, and there is a huge literature on this topic. However, of course, from the viewpoint of green chemistry they are much less attractive and their attractiveness decreases with increasing mole fraction of the organic cosolvent. Despite this fact, a brief discussion is warranted here, since pericyclic reactions have frequently been performed in these media.

An extensive, comparative study [52] has recently been performed for four types of pericyclic processes: (i) a normal electron-demand (**12** + **6d**, Scheme 8.10) and (ii) an inverse electron-demand Diels–Alder reaction (**29** + **12**, Scheme 8.11), (iii) a retro-Diels–Alder reaction (**33b**, Scheme 8.13), and (iv) a 1,3-dipolar cycloaddition (**5** + **6d**, Scheme 8.2) in mixtures of water with methanol, acetonitrile and poly(ethylene glycol) (MW 1000) over the whole water mole fraction range. In most cases, the addition of organic cosolvents leads to rate inhibitions, which become particularly strong in the more water-poor mixtures. Remarkably, in a few cases small rate maxima were found in highly aqueous alcohol–water mixtures (around 40 M water). The latter could be explained in terms of hydrophobicity effects.

The solvent dependence of the second-order rate constants was analyzed by employing two different approaches. First, a linear multi-parameter Abraham–Kamlett–Taft (AKT) model [76] was used to correlate solvent polarity, hydrogen bonding capacities and hydrophobic effects with the kinetic data. The AKT model employs the following parameters: $\Pi^*$ (dipolarity, polarizability), $\alpha$ (hydrogen bond acidity) and $\beta$ (hydrogen bond basicity). "Solvophobicity" (whatever that exactly may be) could be included by adding the Hildebrand solubility parameter ($\delta 2$) or the solvophobicity parameter ($Sp$) [32]. The physical significance of this AKT approach is not sufficiently clear, but such correlations at least provide some insights into the factors determining the solvent dependences of the rate constants. For the pericyclic reactions i–iv, the application of AKT was not very satisfactory.

The second approach was couched in terms of an analysis of the transfer parameters of reactants and ACs starting from a reference solvent (see also Section 8.2). The Gibbs energy, enthalpy, and entropy changes of the AC can be easily calculated from the corresponding parameters of the reactants in combination with the isobaric thermodynamic activation parameters. The physical meaning of this approach is clear, and it appears to work well for the four types of pericyclic processes. However, the challenge is the interpretation of the thermodynamic data in terms of specific interactions of reactants and AC with the solvent. In many cases, computational studies (particularly MD simulations) are helpful here.

It was found that plots of $\log k$ against the molar concentration or volume fraction of water were approximately linear, but in most cases with a characteristic break at about 40 M water. The break is caused by hydrophobic effects in the highly aqueous region, consistent with the observation that the break did not occur for the retro-Diels–Alder reaction, which is not sensitive to hydrophobic effects (Section 8.2.4). Quite generally, the kinetic solvent effects can be interpreted using the insights discussed in the previous sections for the different types of pericyclic reactions in pure water. For the more hydrophobic cosolvents such as 1-propanol, deviations from the linear trends are found, indicative of preferential solvation in the water-rich mixtures where hydrophobic interactions are operative. These cosolvent effects can be quantitatively described by using a combined thermodynamic and kinetic analysis, taking into account the difference in Gibbs energy of interaction between the cosolvent and the reactants and between the cosolvent and the AC, respectively. For details, the interested reader is referred to the literature [52].

## 8.2.10
### "On Water" Pericyclic Reactions

In Section 8.1.1, we noted that for a long time water was not considered a useful solvent for pericyclic reactions and generally for synthetic organic chemistry. A major drawback was the (often very) limited solubility of apolar organic substrates in water. Certainly, heterogeneous media have been used previously with some success [77], and it was particularly after Rideout and Breslow's paper [4] in 1980 that a number of synthetic organic reactions were performed in heterogeneous aqueous emulsions and sometimes with satisfactory results [78]. However, organic solvents remained the first choice, although it was recognized that from the viewpoint of green chemistry this was an unfortunate situation. Then came a big surprise. In 2005, Sharpless and colleagues published a revolutionary paper [8] in which it was shown that solubility constraints can be turned into an advantage. They showed that several uni- and bimolecular reactions carried out in adequately stirred aqueous suspensions are substantially, and sometimes dramatically, accelerated compared with the reactions in homogeneous aqueous media or organic solvents or performed solvent-free. The showcase so far is the $[2\sigma + 2\sigma + 2\pi]$ cycloaddition of quadricyclane (**74**) with dimethyl azodicarboxylate (**75**) to yield the substituted 1,2-diazetidine **76** (Scheme 8.28). The reaction time decreased from 48 h under neat conditions to 10 min in the aqueous emulsion (23 °C). In alcohol–water mixtures, the reaction times were in the order of a few hours, and in toluene 120 h. Several other reactions were also carried out "on water", including a Diels–Alder reaction and a Claisen rearrangement, and although the rate enhancements were usually somewhat less impressive, single products were obtained in high yields with high stereo- and regioselectivity. Synthetically useful substrate concentrations could be used and the synthetic protocol was simple and highly effective. It is clear that these "on water" conditions provide great perspectives for green organic synthesis in water. For a detailed review, the reader is referred to Chapter 7.

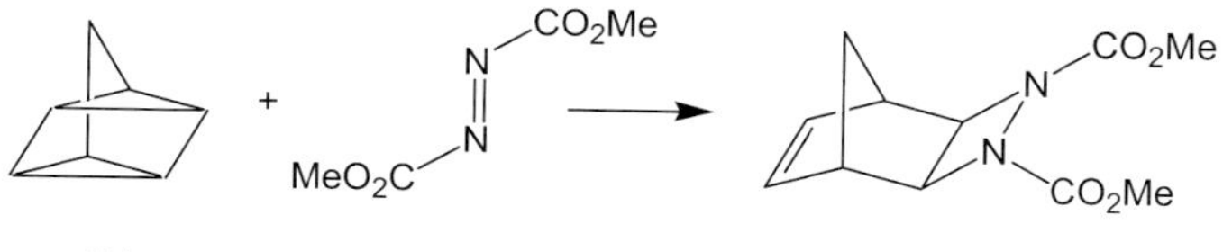

**Scheme 8.28** Cycloaddition of **74** to **75** performed under "on water" conditions.

Of course, the question was, what is the explanation? As argued by Sharpless and co-workers [8], the data suggested that the secret would lie in the droplet/water interface. Based on this assumption and the previous knowledge about water-accelerated pericyclic processes, many of the main issues were resolved in a beautifully detailed study by Jung and Marcus [10]. First, it was necessary to get a realistic and quantitative picture of the differences in rate constants for the reaction of **74** with **75** under homogeneous, neat, and "on water" conditions. Therefore, the experimental rate constants for these conditions were reduced to the same units.

These theoretical rate constants for neat, homogeneous, and "on water" conditions were $5 \times 10^{-7}$, $2 \times 10^{-4}$, and $0.2\,\mathrm{s}^{-1}$, respectively. Taking into account that for large hydrophobic surfaces, such as those for the oil droplets, no tangentially oriented waters are possible, but that in the first hydration layer there are many (in the order of 25%) dangling OH groups, protruding into the droplet containing the hydrophobic reactants. These OH groups can stabilize the AC for the reaction through hydrogen bonding interactions, just as discussed previously for cycloadditions in pure water. However, there is an important difference: these protruding OH groups do not need to be dehydrated before they undergo the interaction with the hydrogen bonding acceptors in the substrate and AC, and are therefore catalytically much more effective. Another factor contributing to their larger catalytic effect is the low polarity of the droplet interior, which is favorable for the largely electrostatic hydrogen bond interaction.

It is expected that in the near future several aspects will receive further attention, such as the effect of varying the droplet size.

We have little doubt that "on water" chemistry will develop into an important preparative-scale procedure for green aqueous organic chemistry.

## 8.2.11
### (Bio)catalysis, Cyclodextrins, Surfactant Aggregates, Molecular Cages, Microwaves, Supercritical Water

Catalytic reactions are specifically mentioned in the 12 "guidelines" for green chemistry [18]:

- *Catalytic reagents (as selective as possible) are superior to stoichiometric reagents.*

Reasons include the increased rate of product formation (less energy required for heating of reaction mixtures), higher yields, and more specific reactions with less necessity for separation and purification of the desired reaction product(s), and often better stereochemistry. A disadvantage can be the need to remove the (often toxic) catalyst from the reaction mixture.

Enormous efforts have been made to maximize the advantages of catalytic procedures. A brief discussion will be given of major possibilities for catalyzing pericyclic reactions in water.

Pericyclic reactions can be accelerated by several essentially different catalytic mechanisms. First, the catalyst may bind covalently or non-covalently to a functional group in one of the reagents, thereby promoting the formation of the AC. An example is Lewis acid catalysis [79]. Second, the reaction partners may bind non-covalently to a specific binding site of the catalyst, which may be a favorable area in the Stern region of a surfactant aggregate, a cavity, or a molecular cage offered by a catalytic guest molecule. For bimolecular reactions, this will lead to a reduced reaction volume and an accompanying enhanced rate, even under conditions such that the AC is less strongly bound than the reactants with an associated smaller second-order rate constant than that in bulk water. Furthermore, the binding process of the reactants may also result in a favorable orientation of the reactants for the pericyclic reaction.

A third possibility that we will briefly consider is the use of special aqueous reaction conditions, such as the application of microwave irradiation or the use of supercritical water. Among the different pericyclic transformations, these possibilities have been most frequently examined for Diels–Alder reactions.

### 8.2.11.1 (Bio)catalysis

In the previous discussion, much emphasis was placed on hydrogen bond interactions as an important reason for the rate enhancement of many pericyclic reactions in water. Proton catalysis could, of course, be more effective, but often the relevant hydrogen bond acceptor sites have a low thermodynamic basicity in water. In organic solvents, acid catalysis of pericyclic reactions has been employed previously, but there is little information on aqueous reaction media [80, 81]. In contrast, Lewis acid catalysis of normal electron-demand Diels–Alder reactions was reported by Yates and Eaton [82] in 1960. Coordination of the Lewis acid to the activating group of the dienophile speeds the cycloaddition by lowering the LUMO(dienophile)–HOMO-(diene) barrier. Problems involved aqueous deactivation and decomposition of the Lewis acid. Also, sufficiently efficient binding of the Lewis acid is an important problem (see below). As a consequence, often an excess of Lewis acid is required for efficient catalysis, thereby reducing the potential for green chemistry. Several water-tolerant Lewis acids have been developed, which, in some cases, were employed in mixed aqueous solutions (see below) [83–85]. Taking advantage of the aqueous rate accelerations of Diels–Alder reactions, water is often the best solvent for Lewis acid catalysis. Solid protic acids [86] and surfactant-assisted Brønsted acids [87] have been employed to cope with the problem of catalyst removal.

In 2004, a kinetic study appeared in which a comparison was made between specific acid catalysis (HCl) and Lewis acid catalysis [Cu(II) nitrate] of Diels–Alder reactions in water [79]. Cyclopentadiene (**12**) was reacted with three types of dienophiles, **77a–c**, **79**, and **34a** and **81** (Scheme 8.29). The inverse primary kinetic deuterium isotope effects observed for the acid-catalyzed reactions of **77a** and **b** indicated that specific-acid catalysis was involved. For **77b**, a $pK_a$ of 2.55 was estimated for $N$-protonation. At 32 °C, the reaction of **77b** with **12** in 0.01 M aqueous HCl is 21 times faster than the uncatalyzed reaction in water. The corresponding reaction with **79** under the same conditions is only six times faster. The reaction of **34a** with **12** is strongly accelerated in water compared with organic solvents (Section 8.2.2), but did not exhibit any acid catalysis even at pH values as low as 1.

In contrast to **79**, the bidentate dienophiles **77a–c** are prone to highly efficient Cu (II) cation catalysis for cycloaddition to **12** [80, 81]. Catalysis by $Ni^{2+}$, $Co^{2+}$, and $Zn^{2+}$ is less effective, which can be explained by the preferred four-coordination of $Cu^{2+}$ ions due to the Jahn–Teller effect. In aqueous solution, 0.010 M Cu(II) nitrate accelerates the reaction of **77** about 1000-fold. The Lewis acids do not affect the *endo/exo* ratios. A likely mechanism is shown in Scheme 8.30. Comparison of the second-order rate constants for catalysis by 0.015 M HCl and 0.015 M Cu(II) nitrate shows [79] that Cu(II) catalysis is 40 times faster than proton catalysis for **77a** and 50 times faster for **77b**. The crucial role [79] of the bidentate character of the dienophile is illustrated

**Scheme 8.29** Diels–Alder reactions examined for proton and Lewis acid catalysis in water.

by the fact that no Cu(II) catalysis in water is found for **79**. The dehydration of the Cu(II) cation necessary for binding to the dienophile cannot be sufficiently compensated by binding to a *mono*dentate substrate. Specific-acid catalysis is also more efficient for **77a** and **b** than for **79**, which can be explained by assuming that the proton attached to the pyridyl nitrogen of **77a** and **b** can form an intramolecular hydrogen bond with the carbonyl oxygen atom, thereby further lowering the energy of the

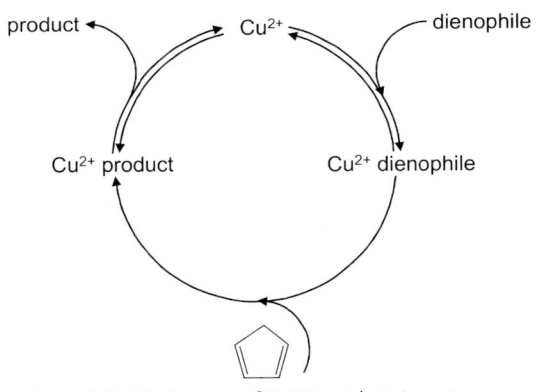

**Scheme 8.30** Mechanism of Cu(II) catalysis in water.

LUMO and enhancing the rate of the cycloaddition. Intramolecular electrostatic interactions may also play a role.

Cu(II) catalysis for the reaction of **34a/81** with **12** was not examined in any detail, but there is evidence that the observed small rate accelerations are most likely just salt effects.

Enantioselective Lewis acid-catalyzed Diels–Alder reactions were reported for the first time in 1998 [88]. Using aromatic α-amino acids as chiral ligands for the $Cu^{2+}$ ions gave up to 74% enantioselectivity in case of L-abrine (N-methyl-L-tryptophan) for the reaction of **77a** with **12**. The enantioselectivities appear to be water enhanced since lower enantiomeric excess (*ee*) values (17–40%) were found in organic solvents. Several of these reactions are also ligand accelerated, which most likely results from arene–arene hydrophobic interactions between the pyridine ring of the nucleophile and the aromatic ring of the α-amino acid ligand during the activation process (Scheme 8.31) [89]. Using chiral hosts, enantioselective cycloaddition products were also obtained in aqueous suspensions [90].

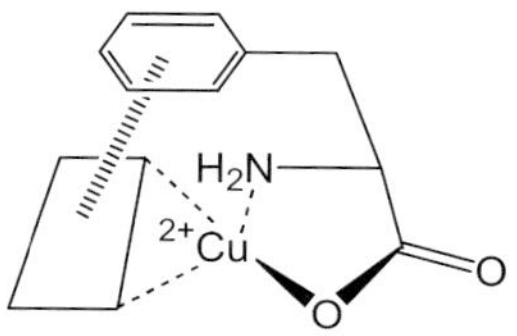

**Scheme 8.31** General structure of the activated catalyst.

Lanthanide triflates (La, Ce, Pr, Nd, Sm, Eu, Gd, Yb, Lu, and Bi, Sc, Y) are water-tolerant Lewis acids that have been extensively studied as Lewis acid catalysts for several types of Diels–Alder reactions. Often mixed aqueous reaction media were employed, but pure water is also possible. The area has been adequately reviewed by Fringuelli *et al.* [91] and Li [30].

Diels–Alder reactions can also be catalyzed by enzymes and antibodies in aqueous buffer solutions. This is superb green chemistry, but applications so far remain limited to specific cases and their applicability in synthetic organic chemistry awaits further studies. An early example is the cycloaddition of 5-methoxynaphthoquinone (**17**) with 1-methoxy-1,3-cyclohexadiene (**83**) (Scheme 8.32) catalyzed by bovine

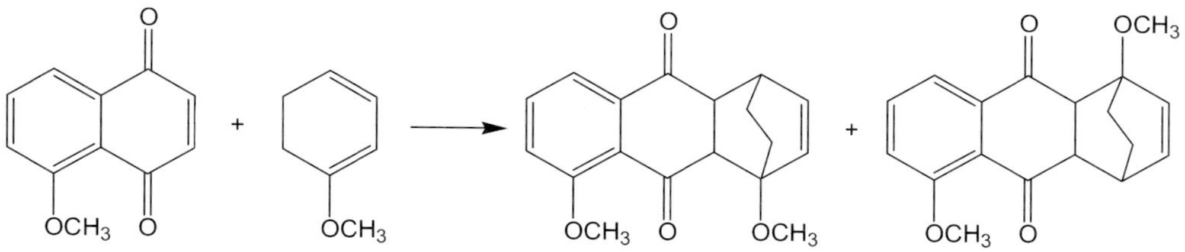

**Scheme 8.32** A BSA-catalyzed Diels–Alder reaction.

serum albumin (BSA) to yield the regioisomers **84** and **85** in a yield of 76% but with a low *ee* [92]. Other examples of enzyme- and antibody-catalyzed Diels–Alder reactions have recently been reviewed by Fringuelli *et al.* [93]. High diastereo-selectivities and enantioselectivities have occasionally been obtained.

A highly interesting novel development is the DNA-induced enantioselectivity of aqueous Diels–Alder reactions as reported by Roelfes and Feringa [94]. Employing the $Cu^{2+}$-catalyzed cycloaddition of bidentate chelating dienophiles **77** with cyclopentadiene (**12**) [80, 81], the achiral or racemic Cu(II)–dienophile complex was brought into intimate contact with the chiral environment of the DNA double helix, thereby inducing enantioselectivity in the Diels–Alder reaction [95].

Two classes of catalysts have been used: (1) ligands in which the DNA binding moiety is linked to the metal binding domain via a short spacer and (2) ligands in which the metal binding and DNA binding domains are combined in a single moiety, eliminating the necessity for a spacer [96]. Both methods lead to significant *ee* values but the second approach is superior since the catalytic metal center is positioned in closer proximity to the DNA helix. The best results were found for the 4,4′-dimethyl-2,2″-bipyridine ligand (**86**, Scheme 8.33) which gave complete *endo* selectivity and up to 99% *ee* for the (+)-enantiomer. The protocol was recently extended to α,β-unsaturated 2-acylimidazoles as highly practical dienophiles in organic synthesis [97]. Interestingly, both methods gave similar results regardless of the source of the DNA employed.

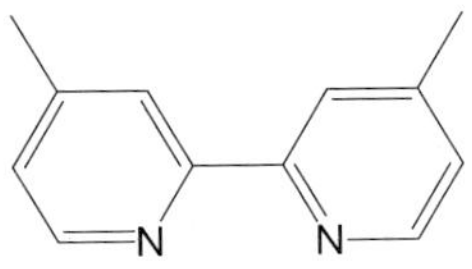

**86**

**Scheme 8.33** The 4,4′-dimethyl-2,2″-bipyridine ligand.

Further mechanistic studies are being carried out, aimed at understanding the specific structural features determining the kinetics and the enantioselectivities. A realistic prospect for highly stereoselective aqueous, green pericyclic transformations is emerging, also because relatively cheap DNA can do the job here.

### 8.2.11.2 Catalysis by Cyclodextrins

We now switch our attention to water-soluble systems that provide cavities for organic substrates, leading not only to locally increased substrate concentrations (in case more than one molecule can fit into the cavity), but also to reaction environments different from that of the bulk aqueous solution. Among these systems, cyclodextrins already have a long tradition [98]. The most important natural cyclodextrins are α-, β-, and γ-cyclodextrins, cyclic oligomers of α-D-glucose linked by α-(1 → 4) bonds, which possess six, seven, or eight α-D-glucose units, respectively. The morphology of these molecules is characterized by the presence of a hollow cone, which can bind "guest" molecules of a suitable size, shape, and polarity by non-covalent interactions. Roughly, α-, β-, and γ-cyclodextrins can act comfortably as a "host" for benzene, naphthalene, and anthracene, respectively. At least three issues always need attention, but are not frequently addressed. The first is the exact positioning of the "guest": fully or only partly bound inside the cavity, or even at the surface of the cyclodextrin, not entering the cavity. Second, often (1:1) association is assumed, without considering a different stoichiometry. Finally, the possibility of both intra- and extra-cavity reactions has to be considered, depending on the binding modes of the substrates.

In their classic 1980 paper [4], Rideout and Breslow showed that the Diels–Alder reactions of cyclopentadiene (**12**) with methyl vinyl ketone (**19a**) and acrylonitrile (**61**) at 30 °C in water are further accelerated by 10 mM of β-cyclodextrin by factors of about 2.5 and 9.1, respectively, even under conditions of incomplete binding of the substrates to the "host" molecule. It was suggested that the diene and dienophile reacted *within* the cavity with a corresponding favorable fit of the AC. Interestingly, the same reactions were slightly retarded by 5–10 mM of α-cyclodextrin, which was attributed to the inability of the dienophile to enter the smaller cavity. Moreover, the aqueous Diels–Alder reaction of anthracene-9-carbinol (**15**) with N-ethylmaleimide (**6b**) at 45 °C is inhibited by β-cyclodextrin (10 mM) by a factor of ~1.6. The increased size of particularly the diene was held responsible for this observation. Of course, inhibition does not indicate exclusion of both reactants from the cavity. Removal of just one substrate from the aqueous reaction environment can perfectly explain the rate retardation. Therefore, separate measurements of substrate binding constants lead to valuable insights into the mechanism of kinetic effects induced by cyclodextrins.

A molecular mechanics and dynamics study of β-cyclodextrin-catalyzed Diels–Alder reactions revealed interesting data regarding the kinetics and thermodynamics of the cycloadditions, with special emphasis on favorable entropic effects [99].

Perhaps rather unexpectedly, Sternbach and Rossana [100] observed a catalytic effect by 1 equiv. of β-cyclodextrin for the *intra*molecular Diels–Alder reaction of the furan derivatives **87a–d** to yield the epimers **88a–d** and **89a–d** (89 °C) (Scheme 8.34). There was also a small change in the epimeric selectivity (**88/89**) from (1:2) to (1:1.5). Mechanistic details require further investigation.

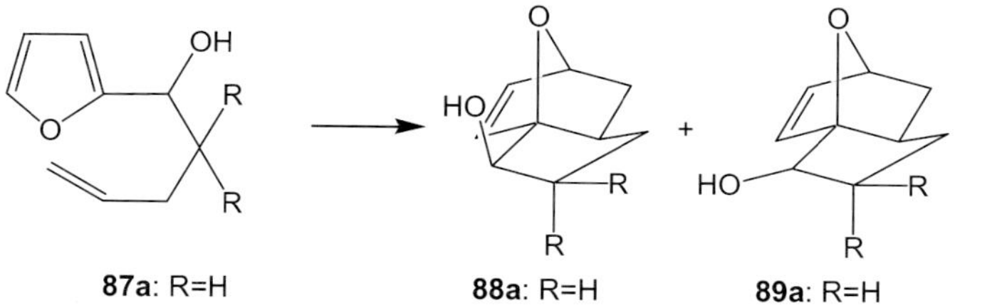

**87a**: R=H
**87b**: R=CH$_3$
**87c**: R=-S(CH$_2$)$_3$S-
**87d**: OEt

**88a**: R=H
**88b**: R=CH$_3$
**88c**: R=-S(CH$_2$)$_3$S-
**88d**: OEt

**89a**: R=H
**89b**: R=CH$_3$
**89c**: R=-S(CH$_2$)$_3$S-
**89d**: OEt

**Scheme 8.34** Intramolecular Diels–Alder reaction of **87** catalyzed by β-cyclodextrin.

Despite the fundamental interest in catalysis by cyclodextrins for aqueous organic reactions, the results obtained so far do not appear to suggest important prospects for green pericyclic reactions in water.

Finally, we note that Diels–Alder reactions can also be accelerated by incorporation of the reactants into the cavity of a coordination cage formed by self-assembly of six metal ions and four ligands [101]. The reaction of naphthoquinone with 1,3-cyclohexadiene is accelerated in an aqueous suspension of the cage 21-fold relative to the medium not containing the cage, and the reaction with 2-methyl-1,3-butadiene 113-fold. It was suggested that the binding in the cage reduces the entropy of activation for the cycloaddition.

### 8.2.11.3 Catalysis by Surfactant Aggregates

Surfactant aggregates (mainly spherical and worm-like micelles, vesicles) [102] are among the most frequently used systems for aqueous dissolution of apolar substrates for pericyclic reactions. We will largely concentrate our attention on micelles, since the thermodynamically unstable vesicles have rarely been employed.

Above their critical micelle concentration (cmc), micelles bind sufficiently hydrophobic substrates primarily by hydrophobic interaction with segments of the surfactant's tails close to the headgroup. The more hydrophobic the substrate, the deeper is the penetration into the micellar core and the stronger the binding.

The rates and stereochemistry of pericyclic reactions are influenced by added micellar aggregates [103]. Both rate accelerations and decelerations, mostly modest, have been observed. From the viewpoint of green pericyclic chemistry, reaction environments containing surfactants are not desirable, except in cases where the surfactant can be decomposed into non-surface-active molecules once the catalyzed reaction has been performed. However, the use of surfactants also has definite advantages and over the years micellar catalysis and inhibition have been investigated in some detail [104, 105]. We will briefly review some of the most important features.

For a first-order reaction, such as the monomolecular Claisen rearrangement, micelles exert a kinetic effect by supplying a specific, local reaction environment, in most cases positioned in the Stern region of the micelles. These reaction sites have a reduced polarity when compared with water, and the local water concentration is smaller. Both factors usually retard pericyclic reactions.

For a bimolecular reaction with second-order kinetics, this factor also plays a role, but now the concentration of both reactants in the micellar reaction volume is a second significant factor. In the case that both reactants bind to the micelles, and the average binding sites are sufficiently close to each other, the *enhanced* local substrate concentrations, relative to those in the bulk aqueous solution, will enhance the reaction rate. Hence the overall kinetic effect is now determined by the outcome of both counteracting factors [106]. As a consequence, micellar catalysis is not very effective for most pericyclic reactions.

The pseudo-phase model, largely developed and tested by Berezin et al. [106] and Romsted [107], is a popular and successful vehicle to quantify micellar rate effects. Using a pseudo-phase approach, the total reaction volume is taken as the sum of (1) an aqueous (w) phase and (2) a micellar (m) pseudo-phase. Partitioning of both reactants over these phases is described with partition coefficients $P_x = [X]_m/[X]_w$, and they react with different rate constants $k_m$ and $k_w$. The observed rate constant is given by the following equation:

$$k_{app} = \frac{k_m P_A P_B CV + k_w(1-CV)}{[1+(P_A-1)CV][1+(P_B-1)CV]} \tag{8.2}$$

where $C$ is the total surfactant concentration minus the cmc, and $V$ is the partial molar volume of the surfactant.

For two neutral organic substrates A and B, illustrative examples of the observed second-order rate constants, as calculated with the pseudo-phase model, are given for different values of $k_m$, $P_A$, and $P_B$, and with $V_m = 0.25$ M$^{-1}$ in Figure 8.5. The micellar rate effect is then given by $k_{rel} = k_m/k_w$.

For bimolecular reactions with one neutral substrate and one ionic substrate (acting as a counterion), the pseudo-phase ion-exchange model has been developed [107].

A systematic, comparative study of Diels–Alder reactions in micellar media was published in 2002 [108]. Using three N-substituted maleimides, **6b**, **6d**, and **6e** (R = CH$_2$Ph), as the dienophiles, cycloadditions were examined for an apolar diene

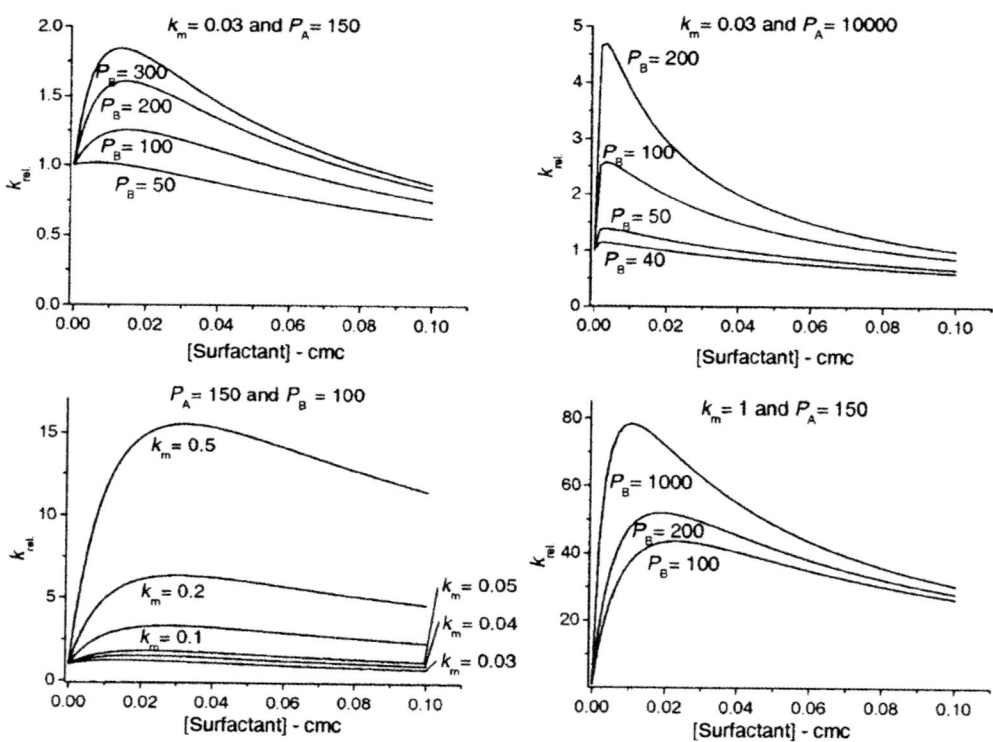

Figure 8.5 Illustrative examples of predictions of the observed second-order rate constants by the pseudo-phase model for various choices of $k_m$, $P_A$, and $P_B$ and with $V_m = 0.25$ M$^{-1}$. $k_{rel} = k_m/k_w$. Reproduced with permission from *J. Org. Chem.*, **2002** 67, 7369–7377.

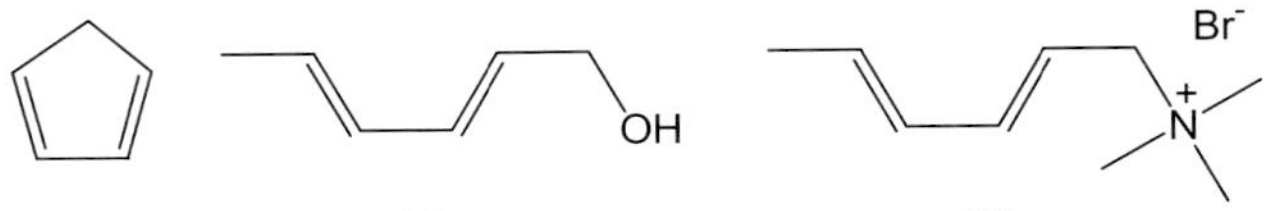

**Scheme 8.35** The dienes of different polarity.

(cyclopentadiene, **12**), a polar diene (sorbyl alcohol, **90**) and an ionic diene (sorbyl-trimethylammonium bromide, **91**) (Scheme 8.35).

These dienes were expected to bind with different binding constants and at different locations to micelles formed from an anionic surfactant [sodium n-dodecyl sulfate (SDS)] and a cationic surfactant [cetyltrimethylammonium bromide (CTAB)]. Rate constants for the cycloadditions of the three types of dienes in SDS solutions were analyzed with the use of the pseudophase model. In all cases, the micellar rate constants ($k_m$) were lower than the rate constants in water ($k_w$) and differences in $k_m/k_w$ were not primarily governed by effects due to possible "mismatch" of the substrates. Although **90** binds more strongly ($P_A = 10^4$) to the SDS micelles than **12** ($P_A \approx 75$) and **91** ($P_A \approx 100$), the $k_m/k_w$ values (0.023–0.055) differ only slightly. Previous evidence that apolar **12** binds significantly deeper in the micelles [109], thereby hampering efficient reactive encounters with the dienophiles, was not substantiated in these studies. These conclusions were confirmed by measurements of paramagnetic ion-induced relaxation enhancements of the $^1$H NMR signals of the solubilizates.

For the dienes **12** and **90**, the trend in micellar rate constants is $k_{m,6e} > k_{m,6b} > k_{m,6d}$, similar to the trend in 1-propanol–water containing 18 M water. This finding suggests that the reaction environment in the micelle is relatively water poor.

In CTAB micelles, there is a more pronounced overall rate retardation compared with SDS micelles. This is probably related to stronger binding of **12**, **90**, and the dienophiles, leading to reactions in less polar and less water-rich environments.

Looking over all the kinetic data, the conclusion is that, depending on the binding characteristics of the diene and dienophile, *observed* rate constants can be enhanced, relative to water, by a factor of ∼4–5. The micellar rate constants $k_m$ are often 20–40 times lower than $k_w$, in accord with micellar reaction sites in the region between the micellar core and the Stern region.

Although micelles provide more favorable solubilization of apolar dienes and dienophiles in water, the micellar catalytic activities are depressingly low. However, there are notable exceptions.

As discussed above, the Diels–Alder reactions of **77a–c** with **12** are effectively accelerated by catalytically active transition metal cations, and particularly by Cu(II) ions. In the absence of these Lewis acid catalysts, the cycloadditions are retarded by cationic (CTAB), anionic (SDS) and uncharged surfactant micelles (n-dodecyl heptaoxyethylene ether). However, if Cu(II) ions are combined with n-dodecylsulfate (DS) anions to form Cu(DS)$_2$ micelles (cmc 1.11 mM), enormous catalytic effects are observed [109]. Compared with the uncatalyzed reaction in acetonitrile, the reaction of **77a** with **12** at 2.4 mM Cu(DS)$_2$ is accelerated by a factor of $1.8 \times 10^6$, resulting from a combination of the beneficial aqueous solvent effect, Lewis acid catalysis by

Cu(II) ions and micellar catalysis. This catalytic efficiency approaches that of conventional enzymes, and primarily stems from the strong complexation of the dienophile to the Cu(II) ions at the micellar surface. Interestingly, **12** binds more strongly to Cu(DS)$_2$ micelles than to micelles formed from SDS and CTAB, which has been rationalized in terms of the higher counterion binding of the divalent Cu(II) ions relative to the monovalent sodium ions, leading to less water penetration into the micelles and more hydrophobic binding sites for **12** at the micellar surface.

The Diels–Alder reaction of the bidentate dienophiles **77a–c** with **12** has also been examined in aqueous solutions containing the vesicle-forming amphiphile **92** (Scheme 8.36) with Cu(II) counterions [110]. These vesicles have diameters of ~40 nm and, as expected, show a fusogenic tendency. Again, strong catalysis is found, although slightly smaller (about 1.5–2 times) than for the micelle-forming Cu(DS)$_2$. However, green chemistry benefits from the fact that catalysis is already effective at 10–20 times lower surfactant concentrations for the metallo-vesicles as a consequence of the low critical vesicle concentration compared with the cmc for Cu(DS)$_2$. Cu(II)–dienophile complexation was estimated to be about seven times stronger than that in bulk water. The slight instability of the vesicles should be taken into account in possible applications. Rate constants were found to be somewhat dependent on the exact reaction conditions and the method used for vesicle preparation also affected the kinetics.

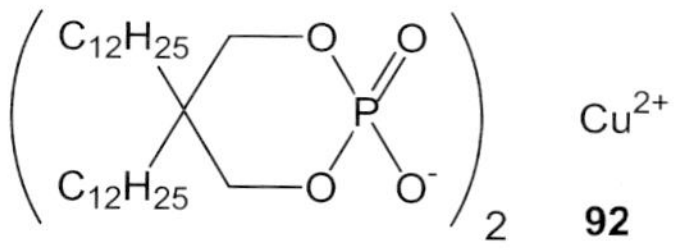

**Scheme 8.36** The double-tailed vesicle-forming amphiphile **92**.

1,3-Dipolar cycloadditions have also been studied in micellar media and their behavior is significantly different from that of Diels–Alder reactions. A detailed study has been made of the cycloaddition of benzonitrile oxide (**5**) with three N-substituted maleimides (**6b, 6d**, and **6e**) in aqueous solutions of SDS, CTAB, and C$_x$E$_y$ [alkyl poly (ethylene) oxide, $x,y = $ 12,8; 12,23; 16,10; and 16,20] surfactants. The second-order rate constants were analyzed using the pseudophase model. Binding of **5** to SDS micelles ($P_B \approx 180$) is stronger than to CTAB micelles ($P_B \approx 60$) as expected for the charge distribution in the molecule, that allows a better fit into the anionic micellar surface. Dienophile **6b** binds about equally strongly to SDS and CTAB micelles, but the more hydrophobic **6d** and **6e** have a clear preference for CTAB aggregates, as expected for an amphiphile with longer alkyl tails.

Although $k_m$ is again smaller than $k_w$, much larger overall micellar rate accelerations (up to a factor of 17) were found than for the Diels–Alder reactions, as a result of the higher $k_m/k_w$ values (~0.25–0.45) for the three types of micelles compared with those for the Diels–Alder cycloadditions (~0.02–0.05). This is obviously due to the smaller solvent sensitivities and aqueous rate accelerations for the 1,3-dipolar cycloadditions (Section 8.2.6), thereby allowing a larger contribution of the enhanced substrate concentrations in the micellar reaction volume.

n-Propanol–water mixtures ($[H_2O] \approx 15$ M) are good mimics for the micellar reaction environments. This was strongly substantiated by the finding that isobaric activation parameters ($\Delta^\ddagger H^\circ$, $\Delta^\ddagger S^\circ$) determined for the reaction of **5** with **6b** using *micellar* rate constants $k_m$ had a close resemblance to those obtained for the reaction in this binary aqueous mixture.

We also note that attempts to employ micellar media to induce favorable regio- and enantioselectivities have had only moderate success so far. An example is the Diels–Alder reaction of **12** with *n*-nonyl acrylate **93** in aqueous solutions of micelles formed from the chiral (*S*)-leucine-derived amphiphile **94** (Scheme 8.37) [111].

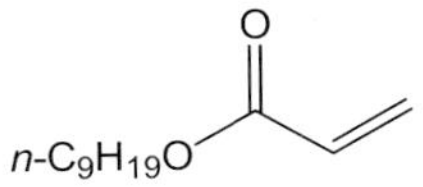

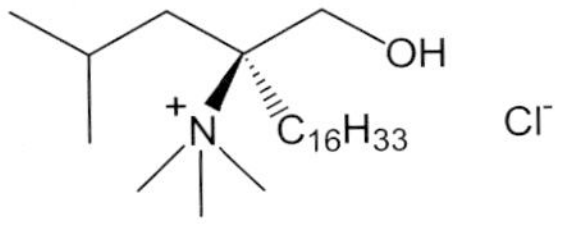

**Scheme 8.37** The chiral cationic amphiphile **94**.

The *endo/exo* ratio increased from 1.7 in water to 2.2 in an aqueous solution of micellar **94** in the presence of 4.86 M LiCl and a small preference for the *R*-isomer (15%) was found. Regioselectivities have been studied in some detail, particularly by Jaeger and Su [112]. An interesting case involves the Diels–Alder reaction of a cationic diene (**95**) and a dienophile (**96**) that form mixed micelles in water (Scheme 8.38) [112]. Two reaction products are formed (**97** and **98**), and at concentrations above their cmc values a significant preference for product **97** (**97** : **98** = 6.6 : 1) exemplifies the micellar effect on the regioselectivity.

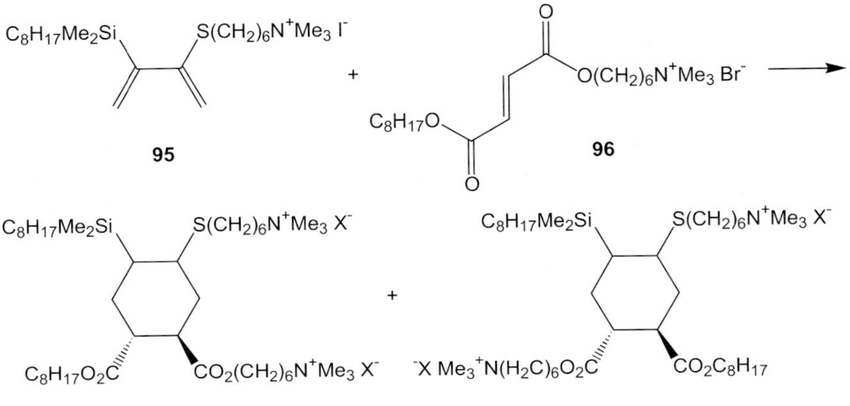

**Scheme 8.38** Diels–Alder reaction of a cationic diene (**95**) and a dienophile (**96**).

The 1,3-dipolar cycloaddition of **5** with **6b** has also been investigated in AOT–isooctane–water microemulsions [AOT is sodium bis(2-ethylhexyl)sulfosuccinate] at 25 °C. These reaction media can hardly be designated as green, but interesting rate accelerations have been found. The about six times faster reaction in these media compared with water was ascribed to an increase in the local reactant concentrations at the innermost region of the microemulsion interface and specific electrostatic interactions with the headgroups of the surfactant.

The Diels–Alder reaction of **12** with **6b** in the same water-in-oil microemulsions was also examined in kinetic detail [113]. The second-order rate constant was 2.3–16 times higher than that in isooctane, depending on the water content in the microemulsion, but much lower than that in water.

### 8.2.11.4 Microwave-assisted Aqueous Pericyclic Reactions

The popularity of microwave-assisted organic reactions has increased substantially in recent years [114]. Microwave heating may lead to rate accelerations and increases in yield compared with traditional heating techniques. A debate is continuing on the exact reasons for these observations. Of course, one could ask about the energy requirements. Could it be called green chemistry? Perhaps somewhat unexpectedly, the answer is yes. Compared with traditional oil-bath heating, large energy savings have been noted. However, for large-scale processes, the comparison has still to be examined in sufficient detail. However, as argued by Dallinger and Kappe [114], the prospects for energy and time saving appear bright.

At the moment, little sound information is available for aqueous pericyclic reactions. A detailed study has been made by Yu and co-workers [115], who examined aqueous Diels–Alder reactions of six dienes (**12, 27, 25, 99–101**) with six dienophiles (**34a, 102–106**) (Scheme 8.39) catalyzed by 3 mol% of the highly water-soluble Lewis acid $[O=P(2\text{-py})_3W(CO)(NO)_2](BF_4)_2$ using microwave heating conditions. Short reaction times (compared with nitromethane as the medium), good to excellent yields, and high *endo/exo* ratios were found. However, problems involve the usually low substrate solubility and possible competing polymerization of the diene.

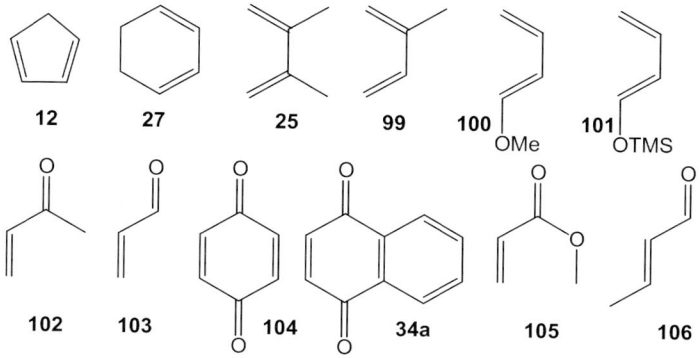

**Scheme 8.39** Dienes and dienophiles used in the study by Yu and co-workers [115].

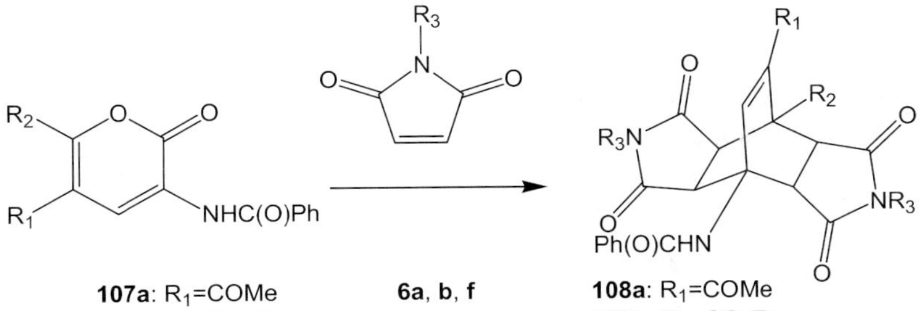

**Scheme 8.40** Diels–Alder reaction of 2H-pyran-2-ones with maleimides.

107a: $R_1$=COMe
107b: $R_1$=CO$_2$Et
107c: $R_1$=p-PhOMe

6a, b, f

108a: $R_1$=COMe
108b: $R_1$=CO$_2$Et
108c: $R_1$=p-PhOMe

Another example is the Diels–Alder reaction of 2H-pyran-2-ones (**107a–c**, $R_1$ = COMe, CO$_2$Et, p-MeOPh; $R_2$ = Me) with several maleimides (**6a, b, f**, $R_3$ = Me, Et, Ph) to give bicyclo[2.2.2]octenes (**108a–c**) (Scheme 8.40) [116]. In water at 150 °C, and with microwave irradiation for 10–45 min, the products were obtained in yields of 86–94%. Conventional heating in decalin required higher temperatures (190 °C) and longer reaction times (1.5–2 h). Further studies are eagerly awaited to see the synthetic potential of microwave-assisted, green pericyclic reactions in water.

### 8.2.11.5 Supercritical Water

At high temperatures and pressures, the solvent properties of water undergo drastic changes, largely caused by thermal destruction of the extensive three-dimensional hydrogen bond network in the liquid. Water can be compressed from gas-like to liquid-like densities above a critical temperature of 647 K. The relative dielectric permittivity of such "hydrothermal" fluids is reduced from 78 to 10–25. Furthermore, the autodissociation of water is greatly increased, providing possibilities for spontaneous acid catalysis.

Diels–Alder reactions in these media are accelerated [117, 118] and substrate solubilities are increased relative to those in water at room temperature and ambient pressure. The *endo/exo* ratios appear to be unchanged. Of course, the use of water gives these reactions a green flavor, but the energy needed to create these unusual aqueous media is a clear drawback. For a detailed discussion, see Chapter 11.

## 8.3
## Conclusion

Pericyclic reactions are important transformations in broad areas of organic chemistry. In this chapter, we have made an attempt to show that there are challenging perspectives for performing these reactions in aqueous media under conditions that can be qualified as green chemistry. The reasons are that many cycloadditions are accelerated in water relative to organic solvents, and exhibit more favorable stereochemical features. Moreover, practical factors, such as work-up of the desired product, are often improved when using aqueous reaction media. For each pericyclic process, both in the (university) research laboratory and in the industrial laboratory, a creative choice has to be made as to what the best reaction conditions are. A broad variety of issues have to be considered. These include environmental and safety factors, reaction yields and rates, stereochemical features, possibilities for catalysis, isolation of the product, economic factors, and several more.

We have summarized for different types of forward and reverse pericyclic reactions the available insights into the effects of aqueous reaction media, and we hope that this information can be employed as useful background for planning green and aqueous pericyclic reactions in the coming years.

## References

1 Houk, K.N., Li, Y. and Evanseck, J.D. (1992) *Angew. Chem. Int. Ed. Engl.*, **31**, 682–708.
2 Sauer, J. and Sustmann, R. (1980) *Angew. Chem. Int. Ed. Engl.*, **19**, 779–807.
3 Anslyn, E.V. and Dougherty, D.A. (2006) *Modern Physical Organic Chemistry*, University Science Books, Sausalito, CA.
4 Rideout, D.C. and Breslow, R. (1980) *J. Am. Chem. Soc.*, **102**, 7816–7817.
5 Grieco, P.A., Yoshida, K. and Garner, P. (1983) *J. Org. Chem.*, **48**, 3137–3139.
6 Grieco, P.A., Garner, P. and He, Z. (1983) *Tetrahedron Lett.*, **24**, 1897–1900.
7 Itami, K., Nokami, T. and Yoshida, J.-I. (2002) *Adv. Synth. Catal.*, **344**, 441–451.
8 Narayan, S., Muldoon, J., Finn, M.G., Fokin, V.V., Kolb, H.C. and Sharpless, K.B. (2005) *Angew. Chem. Int. Ed.*, **44**, 3275–3279.
9 Klijn, J.E. and Engberts, J.B.F.N. (2005) *Nature*, **435**, 746–747.
10 Jung, Y.S. and Marcus, R.A. (2007) *J. Am. Chem. Soc.*, **129**, 5492–5502.
11 Leitner, W. (1999) *Science*, **284**, 1780–1781.
12 Tundo, P., Anastas, P., Black, D.S., Breen, J., Collins, T., Memoli, S., Miyamoto, J., Polyakoff, M. and Tumas, W. (2000) *Pure Appl. Chem.*, **72**, 1207–1228.
13 *Green Chemistry: Designing Chemistry for the Environment* (eds P.T. Anastas and T.C., Williamson), (1996) American Chemical Society, Washington, DC.
14 Cathcart, C. (1990) *Chem. Ind. (London)*, 684–687.
15 Sheldon, R.A. (1996) *J. Mol. Catal. A*, **107**, 75–83.
16 Linthorst, J.A., submitted for publication.
17 Horvath I.T. and Anastas P.T.(eds) (2007) *Chem. Rev.* 107 (Special Issue on Green Chemistry), 2167–2820.
18 Anastas, P.T. and Warner, J.C. (1998) *Green Chemistry: Theory and Practise*, Oxford University Press, New York. Chapter 1, p. 8.
19 Ball, P. (2008) *Chem. Rev.*, **108**, 74–108.
20 Engberts, J.B.F.N. (2007) in *Organic Reactions in Water* (ed. U.M. Lindström), Blackwell, Oxford. Chapter 2, pp. 29–59.
21 Graziano, G. (2006) *J. Phys. Chem. B*, **110**, 11421–11426.
22 Blokzijl, W. and Engberts, J.B.F.N. (1993) *Angew. Chem. Int. Ed. Engl.*, **32**, 1545–1579.
23 Southall, N.T., Dill, K.A. and Haymet, A.D.J. (2002) *J. Phys. Chem. B*, **106**, 521–533.
24 Kolb, H.C., Finn, M.G. and Sharpless, K.B. (2001) *Angew. Chem. Int. Ed. Engl.*, **40**, 2004–2021.
25 Li, Z., Held, M., Panke, S., Schmid, A., Mathys, R. and Withat, B. (2007) *Methods and Reagents for Green*

*Chemistry*, Wiley-Interscience, Hoboken, NJ. Chapter 15, pp. 281–296.
26 Lubineau, A., Augé, J. and Queneau, Y. (1994) *Synthesis*, 741–760.
27 Li, C.J. and Chan, T.H. (1997) *Organic Reactions in Aqueous Media*, John Wiley & Sons, Inc., New York.
28 Kobayashi, S. (1998) in *Organic Synthesis in Water* (ed. P.A. Grieco), Blackie Academic & Professional, London, Chapter 8, pp. 262–306.
29 Lindstrom, U.M. (2002) *Chem. Rev.*, **102**, 2751–2771.
30 Li, C.J. (2005) *Chem. Rev.*, **105**, 3095–3165.
31 Li, C.J. and Chen, L. (2006) *Chem. Soc. Rev.*, **35**, 68–82.
32 Schneider, H.J. and Sangwan, N.K. (1987) *Angew. Chem. Int. Ed. Engl.*, **26**, 896–897.
33 Blokzijl, W., Blandamer, M.J. and Engberts, J.B.F.N. (1991) *J. Am. Chem. Soc.*, **113**, 4241–4246.
34 Otto, S., Blokzijl, W. and Engberts, J.B.F.N. (1994) *J. Org. Chem.*, **59**, 5372–5376.
35 Blokzijl, W. and Engberts, J.B.F.N. (1992) *J. Am. Chem. Soc.*, **114**, 5440–5442.
36 Jorgensen, W.L. (1989) *Acc. Chem. Res.*, **22**, 184–189.
37 Furlani, T.R. and Gao, J.L. (1996) *J. Org. Chem.*, **61**, 5492–5497.
38 Blake, J.F. and Jorgensen, W.L. (1991) *J. Am. Chem. Soc.*, **113**, 7430–7432.
39 Chandrasekhar, J., Shariffskul, S. and Jorgensen, W.L. (2002) *J. Phys. Chem. B*, **106**, 8078–8085.
40 Blake, J.F., Lim, D. and Jorgensen, W.L. (1994) *J. Org. Chem.*, **59**, 803–805.
41 Acevedo, O. and Jorgensen, W.L. (2007) *J. Chem. Theory Comput.*, **3**, 1412–1419.
42 Wijnen, J.W., Zavarise, S., Engberts, J.B.F.N. and Charton, M. (1996) *J. Org. Chem.*, **61**, 2001–2005.
43 Tiwari, S. and Kumar, A. (2006) *Angew. Chem. Int. Ed.*, **45**, 4824–4825.
44 Meijer, A., Otto, S. and Engberts, J.B.F.N. (1998) *J. Org. Chem.*, **63**, 8989–8994.
45 Lubineau, A., Bienayme, H., Queneau, Y. and Scherrmann, M.C. (1994) *New J. Chem.*, **18**, 279–285.
46 Hunt, I., and Johnson, C.D. (1991) *J. Chem. Soc., Perkin Trans.*, **2**, 1051–1056.
47 Van der Wel, G.K., Wijnen, J.W. and Engberts, J.B.F.N. (1996) *J. Org. Chem.*, **61**, 9001–9005.
48 Blokzijl, W. (1991) *PhD Dissertation*, University of Groningen.
49 Engberts, J.B.F.N. (1995) *Pure Appl. Chem.*, **67**, 823–828.
50 Wijnen, J.W. and Engberts, J.B.F.N. (1997) *J. Org. Chem.*, **62**, 2039–2044.
51 Sauer, J., Mielert, A., Lang, D. and Peter, D. (1965) *Chem. Ber.*, **98**, 1435–1445.
52 Rispens, T. and Engberts, J.B.F.N. (2005) *J. Phys. Org. Chem.*, **18**, 725–736.
53 Charton, M. (1987) *Prog. Phys. Org. Chem.*, **16**, 287–315.
54 Charton, M. (1989) in *The Chemistry of Functional Groups* (eds S. Patai and Z. Rappoport), John Wiley & Sons, Ltd, Chichester, pp. 239–298.
55 Lubineau, A., Augé, J. and Lubin, N. (1991) *Tetrahedron Lett.*, **32**, 7529–7530.
56 Grieco, P.A., Parker, D.T., Fobare, W.F. and Ruckle, R. (1987) *J. Am. Chem. Soc.*, **109**, 5859–5861.
57 Grieco, P.A. and Clark, J.D. (1990) *J. Org. Chem.*, **55**, 2271–2272.
58 Wijnen, J.W. and Engberts, J.B.F.N. (1997) *Liebigs Ann. Recl.*, 1085–1088.
59 Syamala, M.S. and Ramamurthy, V. (1986) *J. Org. Chem.*, **51**, 3712–3715.
60 Ramamurthy, V. (1986) *Tetrahedron*, **42**, 5753–5839.
61 Muthuramu, K. and Ramamurthy, V. (1982) *J. Org. Chem.*, **47**, 3976–3979.
62 Padwa, A. (ed.) (1987) *1,3-Dipolar Cycloaddition Chemistry*, Wiley-Interscience, New York.
63 Wijnen, J.W., Steiner, R.A. and Engberts, J.B.F.N. (1995) *Tetrahedron Lett.*, **36**, 5389–5392.
64 van Mersbergen, D., Wijnen, J.W. and Engberts, J.B.F.N. (1998) *J. Org. Chem.*, **63**, 8801–8805.
65 Rispens, T. and Engberts, J.B.F.N. (2005) *J. Phys. Org. Chem.*, **18**, 908–917.

66 Gholami, M.R. and Yangjeh, A.H. (1999) *J. Chem. Res.*, 226–227.
67 Jiang, N. and Li, C.J. (2004) *Chem. Commun.*, 394–395.
68 Andrews, P.R., Smith, G.D. and Young, I.G. (1973) *Biochemistry*, **12**, 3492–3498.
69 Ganem, B. (1996) *Angew. Chem. Int. Ed. Engl.*, **35**, 937–945.
70 Grieco, P.A., Brandes, E.B., Mccann, S. and Clark, J.D. (1989) *J. Org. Chem.*, **54**, 5849–5851.
71 Gajewski, J.J. (1997) *Acc. Chem. Res.*, **30**, 219–225.
72 Copley, S.D. and Knowles, J.R. (1987) *J. Am. Chem. Soc.*, **109**, 5008–5013.
73 Brandes, E., Grieco, P.A. and Gajewski, J.J. (1989) *J. Org. Chem.*, **54**, 515–516.
74 Butler, R.N., Cunningham, W.J., Coyne, A.G. and Burke, L.A. (2004) *J. Am. Chem. Soc.*, **126**, 11923–11929.
75 Nicolaou, K.C., Xu, H. and Wartmann, M. (2005) *Angew. Chem. Int. Ed.*, **44**, 756–761.
76 Abraham, M.H., Grellier, P.L., Abboud, J.L.M., Doherty, R.M. and Taft, R.W. (1988) *Can. J. Chem. Rev.*, **66**, 2673–2686.
77 Amantini, D., Fringuelli, F., Piermatti, O., Pizzo, F. and Vaccaro, L. (2003) *J. Org. Chem.*, **68**, 9263–9268.
78 Breslow, R., Maitra, U. and Rideout, D. (1983) *Tetrahedron Lett.*, **24**, 1901–1904.
79 Mubofu, E.B. and Engberts, J.B.F.N. (2004) *J. Phys. Org. Chem.*, **17**, 180–186.
80 Otto, S. and Engberts, J.B.F.N. (1995) *Tetrahedron Lett.*, **36**, 2645–2648.
81 Otto, S., Bertoncin, F. and Engberts, J.B.F.N. (1996) *J. Am. Chem. Soc.*, **118**, 7702–7707.
82 Yates, P. and Eaton, P. (1960) *J. Am. Chem. Soc.*, **82**, 4436–4437.
83 Kobayashi, S., Hachiya, I., Araki, M. and Ishitani, H. (1993) *Tetrahedron Lett.*, **34**, 3755–3758.
84 Yang, Y. and Chan, T.H. (2000) *J. Am. Chem. Soc.*, **122**, 402–403.
85 Tucker, C.J., Welker, M.E., Day, C.S. and Wright, M.W. (2004) *Organometallics*, **23**, 2257–2262.
86 Olah, G.A., Meidar, D. and Fung, A.P. (1979) *Synthesis*, 270–271.
87 Manabe, K., Mori, Y., Wakabayashi, T., Nagayama, S. and Kobayashi, S. (2000) *J. Am. Chem. Soc.*, **122**, 7202–7207.
88 Otto, S., Boccaletti, G. and Engberts, J.B.F.N. (1998) *J. Am. Chem. Soc.*, **120**, 4238–4239.
89 Otto, S. and Engberts, J.B.F.N. (1999) *J. Am. Chem. Soc.*, **121**, 6798–6806.
90 Miyamoto, H., Kimura, T., Daikawa, N. and Tanaka, K. (2003) *Green Chem.*, **5**, 57–59.
91 Fringuelli, F., Piermatti, O., Pizzo, F. and Vaccaro, L. (2001) *Eur. J. Org. Chem.*, 439–455.
92 Colonna, S., Manfredi, A. and Annunziata, R. (1988) *Tetrahedron Lett.*, **29**, 3347–3350.
93 Fringuelli, F., Piermatti, O., Pizzo, F. and Vaccaro, L. (2007) in *Organic Reactions in Water* (ed. U.M. Lindström), Blackwell, Oxford. Chapter 5, pp. 146–184.
94 Roelfes, G. and Feringa, B.L. (2005) *Angew. Chem. Int. Ed.*, **44**, 3230–3232.
95 For a review of DNA- and RNA-based asymmetric catalysis, see: Roelfes, G., (2007) *Mol. Biol. Syst.*, **3**, 126–135.
96 Roelfes, G., Boersma, A.J. and Feringa, B.L. (2006) *Chem. Commun.*, 635–637.
97 Boersma, A.J., Feringa, B.L. and Roelfes, G. (2007) *Org. Lett.*, **9**, 3647–3650.
98 Fringuelli, F. and Taticchi, A. (2002) *The Diels–Alder Reaction: Selected Practical Methods*, John Wiley & Sons, Inc., Chichester.
99 Kim, S.P., Leach, A.G. and Houk, K.N. (2002) *J. Org. Chem.*, **67**, 4250–4260.
100 Sternbach, D.D. and Rossana, D.M. (1982) *J. Am. Chem. Soc.*, **104**, 5853–5854.
101 Kusukawa, T., Nakai, T., Okano, T. and Fujita, M. (2003) *Chem. Lett.*, **32**, 284–285.
102 Israelachvili, J.N. (1992) *Intermolecular and Surface Forces*, Academic Press, London.
103 For a review, see: Otto, S. and Engberts, J.B.F.N., (2001) in *Reactions and Synthesis in Micellar Media* (ed. J. Texter), Marcel Dekker, New York. Chapter 9, pp. 247–263.

104 Menger, F.M. and Portnoy, C.E. (1967) *J. Am. Chem. Soc.*, **89**, 4698–4703.
105 Bunton, C.A., Nome, F., Quina, F.H. and Romsted, L.S. (1991) *Acc. Chem. Res.*, **24**, 357–364.
106 Berezin, I.V., Martinek, K. and Yatsimirskii, A.K. (1973) *Russ. Chem. Rev. (Engl. Transl.)*, **42**, 787–802.
107 Romsted, L.S. (1977) in *Micellization, Solubilization and Microemulsions* (ed. K.L. Mittal), Plenum Press, New York. pp. 509–530.
108 Rispens, T. and Engberts, J.B.F.N. (2002) *J. Org. Chem.*, **67**, 7369–7377.
109 Otto, S., Engberts, J.B.F.N. and Kwak, J.C.T. (1998) *J. Am. Chem. Soc.*, **120**, 9517–9525.
110 Rispens, T. and Engberts, J.B.F.N. (2001) *Org. Lett.*, **3**, 941–943.
111 Diego-Castro, M.J. and Hailes, H.C. (1998) *Chem. Commun.*, 1549–1550.
112 Jaeger, D.A. and Su, D. (1999) *Tetrahedron Lett.*, **40**, 257–260.
113 Engberts, J.B.F.N., Fernandez, E., Garcia-Rio, L. and Leis, J.R. (2006) *J. Org. Chem.*, **71**, 6118–6123.
114 Dallinger, D. and Kappe, C.O. (2007) *Chem. Rev.*, **107**, 2563–2591.
115 Chen, I.H., Young, J.N. and Yu, S.J. (2004) *Tetrahedron*, **60**, 11903–11909.
116 Kranjc, K., Kocevar, M., Iosif, F., Coman, S.M., Parvulescu, V.I., Genin, E., Genet, J.P. and Michelet, V. (2006) *Synlett*, 1075–1079.
117 Korzenski, M.B. and Kolis, J.W. (1997) *Tetrahedron Lett.*, **38**, 5611–5614.
118 Harano, Y., Sato, H. and Hirata, F. (2000) *Chem. Phys.*, **258**, 151–161.

# 9
# Non-conventional Energy Sources for Green Synthesis in Water (Microwave, Ultrasound, and Photo)
*Vivek Polshettiwar and Rajender S. Varma*

## 9.1
## Introduction

The synthetic chemicals community has been under increasing pressure to produce, in an environmentally benign fashion, the myriad of substances required by society in short periods of time, and one of the best options to accelerate these synthetic processes is to use microwave (MW) technology. The efficiency of MW flash heating has resulted in dramatic reductions in reaction times (reduced from days and hours to minutes and seconds) [1]. The time saved by using the MW heating approach is potentially important in traditional organic synthesis and assembly of heterocyclic systems [2]. Ultrasound and photochemical reactions are also well recognized and proceed via the formation and adiabatic collapse of the transient cavitation bubbles [3]. These techniques can also be utilized as alternative sources of energy for various organic transformations.

These non-conventional energy sources have been widely recognized as important enabling technologies in organic synthesis due to a range of benefits such as improved yields, shorter reaction times, and enhanced selectivity. Although it has been widely presumed that ultrasound and MW provide differential assistance to competing reaction pathways, there has been a dearth of reports wherein the comparative efficacy of ultrasound and MW towards selective formation of competing products has been clearly discerned [4].

Organic reactions in water or aqueous media have recently attracted increasing interest because of the environmental issues and an increased understanding of biochemical processes. In the context of green chemistry, there are several issues which influence the choice of solvent. It should be relatively non-toxic and relatively non-hazardous, for example, not flammable or corrosive. The solvent should also be contained, that is, it should not be released to the environment. All these traits are ideally fulfilled by water, which is non-toxic, non-flammable, abundantly available, and inexpensive [5, 6].

One of the most attractive concepts for the development of rapid and benign synthetic protocols is the combination of an MW or ultrasound technique and a benign aqueous solvent [7, 8]. To illustrate the advantages of MW and ultrasound irradiation, and the utility of photochemical reactions, in aqueous media for rapid and greener organic synthesis, in this chapter we summarize various synthetic pathways developed in recent years.

## 9.2
## MW-assisted Organic Transformations in Aqueous Media

### 9.2.1
### Carbon–Carbon Coupling Reactions

Carbon–carbon cross-coupling reactions are among the most important processes in organic chemistry [9]. The Heck [10] and Suzuki [11] reactions are among the most widely used reactions for the formation of carbon–carbon bonds. These reactions are generally catalyzed by soluble palladium complexes with various ligands [12]. However, the efficient separation and subsequent recycling of homogeneous transition metal catalysts remain a scientific challenge and are aspects of economic and ecological relevance. Although heterogeneous Pd catalyst systems were found to be highly effective to overcome some of these issues [13], the MW-assisted coupling reaction in aqueous media is a recent choice of chemists.

Leadbeater and Marco [14] and Wang and co-workers [15] studied Suzuki reactions in aqueous medium using MW irradiation (Scheme 9.1), and prepared various biaryl derivatives from aryl halides and phenylboronic acid.

$$R^1\text{-}C_6H_4\text{-}X + C_6H_5\text{-}B(OH)_2 \xrightarrow[H_2O, \text{ MW-150 °C, 5 min}]{Pd(OAc)_2, Na_2CO_3, TBAB} R^1\text{-}C_6H_4\text{-}C_6H_5$$

X - Cl, Br, I
R$^1$ - Me, OMe, COMe

Yield = 66 - 96%

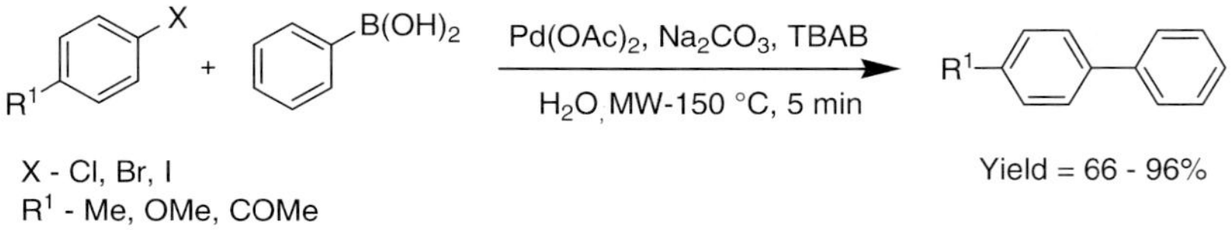

**Scheme 9.1**

Although the above process is generally good in terms of yield and reaction time, the yield when using aryl chloride is poor compared with other halides. This has been overcome by carrying out the reaction of aryl chlorides and phenylboronic acid with a Pd/C catalyst in aqueous medium using a cooling technique in conjugation with MW heating [16]. It was observed that the use of this technique prolongs the life of the aryl chloride; as a result, the yields of desired biaryls were increased.

The Suzuki reaction has been shown to be extremely versatile and has found extensive use in natural products and heterocyclic synthesis. Vanelle and co-workers studied the Suzuki reaction of imidazo[1,2-a]pyridines with arylboronic acid in aqueous medium under MW irradiation (Scheme 9.2). This is an excellent protocol which works more efficiently and rapidly compared with the conventional reaction conditions [17].

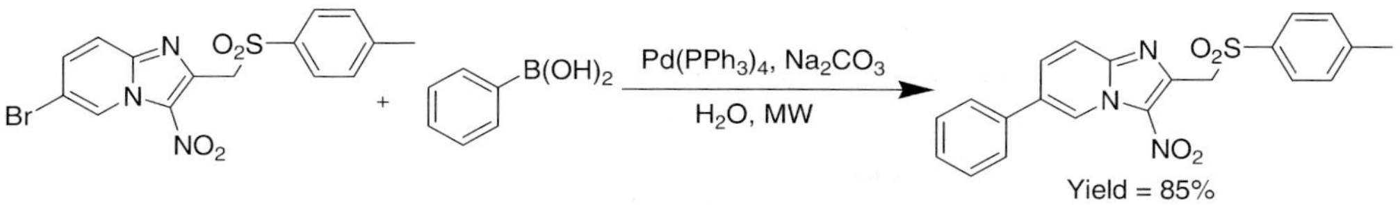

**Scheme 9.2**

5-Aryltriazole acyclonucleosides, with various aromatic groups on the triazole ring, occupy a pivotal position in the arsenal of drug candidates for combating various viruses. These compounds were synthesized via the Suzuki coupling reaction in aqueous solution and promoted by MW irradiation (Scheme 9.3) [18].

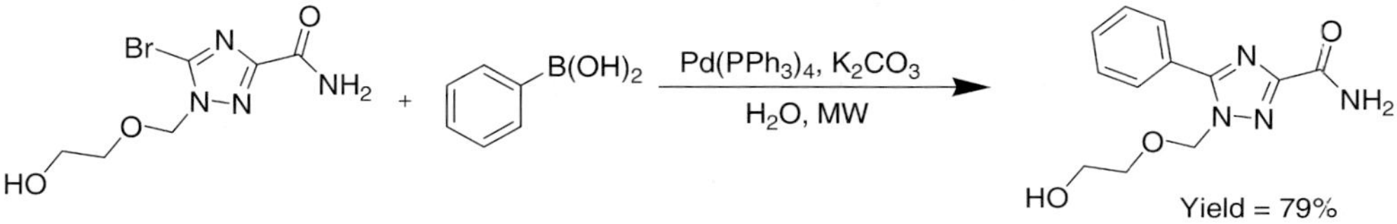

**Scheme 9.3**

The coupling reaction was significantly promoted in aqueous solution under MW irradiation, giving the corresponding compounds in good to excellent yields. This one-step method directly synthesized the aryltriazole acyclonucleosides in an aqueous solvent and involved no protection and deprotection steps.

Heck reactions between aryl halides and alkenes continue to attract attention within the chemistry community because of the versatility of the reaction [10]. Arvela and Leadbeater performed the Heck coupling reaction in water using MW heating (Scheme 9.4) [19]. They observed that Pd catalyst concentrations as low as 500 ppb were sufficient for these reactions, with moderate to good product yield.

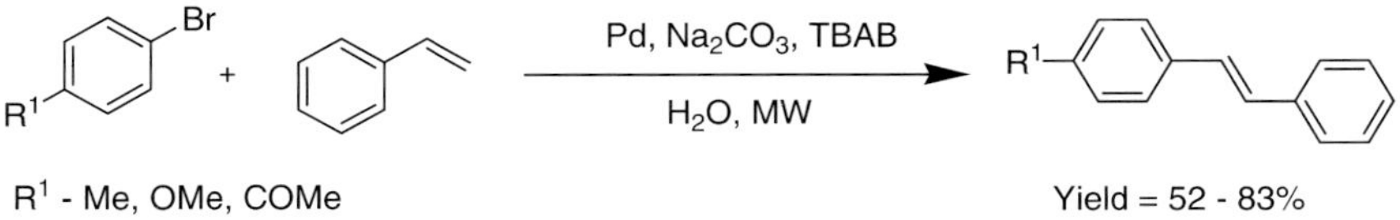

$R^1$ - Me, OMe, COMe       Yield = 52 - 83%

**Scheme 9.4**

Recently, Larhead and co-workers reported highly regioselective and fast Pd(0)-catalyzed internal R-arylation of ethylene glycol vinyl ether with aryl halides in aqueous medium (Scheme 9.5) [20].

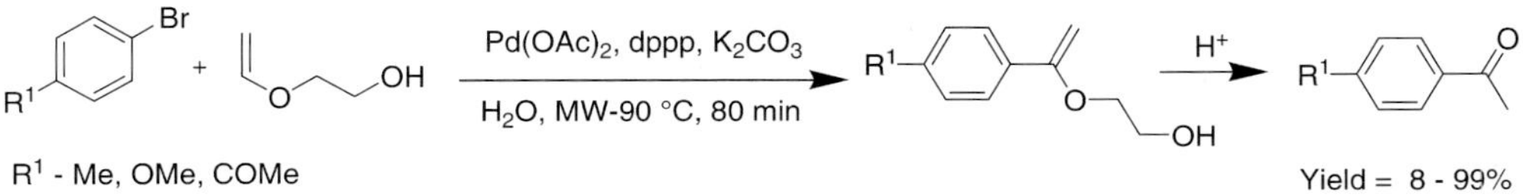

$R^1$ - Me, OMe, COMe       Yield = 8 - 99%

**Scheme 9.5**

Aryl bromides and iodides were efficiently converted to corresponding acetophenones in high yields in water using ethylene glycol vinyl ether as the olefin and potassium carbonate as the base. This Pd(0)-catalyzed method is highly advantageous as no heavy metal additives or ionic liquids are necessary, it proceeds cleanly without any noticeable byproduct formation, it avoids the need for an inert atmosphere, and it allows for easy purification of the products. Also, MW irradiation was shown to be beneficial in activation of aryl chlorides toward the internal Heck arylation.

The Sonogashira cross-coupling reaction of terminal acetylenes with aryl or vinyl halides has proved to be a powerful method for the creation of carbon–carbon bonds. An aqueous Sonogashira-type coupling reaction was studied by Van der Eycken and co-workers under MW irradiation (Scheme 9.6) [21]. The reaction proceeded in water as the sole solvent, without the need for copper(I) or any transition metal–phosphane complex. This eliminated the problem of the intrinsic toxicity and air sensitivity of transition metal complexes, and also the use of expensive phosphane ligands.

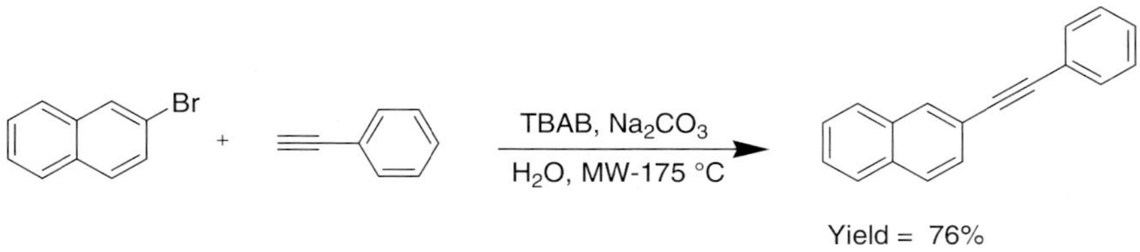

$$\text{Naphthyl-Br} + \text{HC≡C-Ph} \xrightarrow[\text{H}_2\text{O, MW-175 °C}]{\text{TBAB, Na}_2\text{CO}_3} \text{Naphthyl-C≡C-Ph}$$

Yield = 76%

**Scheme 9.6**

Kremsner and Kappe investigated MW-assisted organic synthesis in near-critical water (NCW) in the 270–300 °C temperature range. Several different known transformations, such as the hydrolysis of esters or amides, the hydration of alkynes, Diels–Alder cycloadditions, pinacol rearrangements, and the Fischer indole synthesis, were successfully performed in MW-generated NCW without the addition of an acid or base catalyst [22]. This study showed that it is technically feasible to perform MW synthesis in water on scales from 15 to 400 ml at temperatures up to 300 °C and 80 bar pressure in a multimode MW reactor.

The Hiyama cross-coupling reaction between vinylalkoxysilanes and aryl halides promoted by aqueous sodium hydroxide under fluoride-free conditions was carried out under MW irradiation (Scheme 9.7). This reaction was catalyzed by palladium(II) acetate or a 4-hydroxyacetophenone oxime-derived palladacycle at 120 °C with low catalyst loading in the presence of tetrabutyl ammonium bromide (TBAB) as additive in air [23].

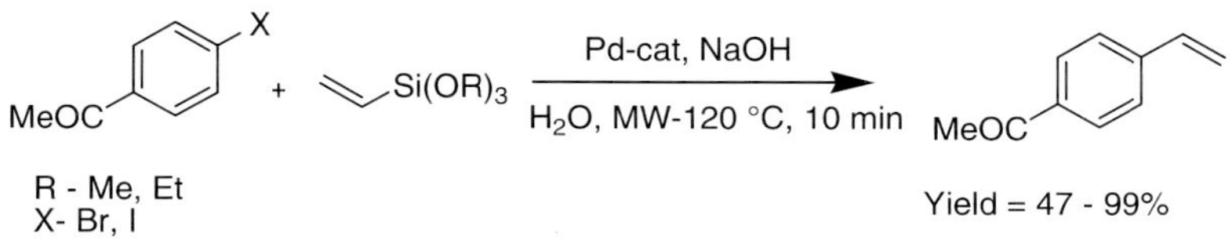

$$\text{MeOC-C}_6\text{H}_4\text{-X} + \text{CH}_2\text{=CH-Si(OR)}_3 \xrightarrow[\text{H}_2\text{O, MW-120 °C, 10 min}]{\text{Pd-cat, NaOH}} \text{MeOC-C}_6\text{H}_4\text{-CH=CH}_2$$

R - Me, Et
X - Br, I

Yield = 47 - 99%

**Scheme 9.7**

Aryl bromides were rapidly converted to the corresponding secondary and tertiary benzamides in water using $Mo(CO)_6$ as the source of carbon monoxide and MW

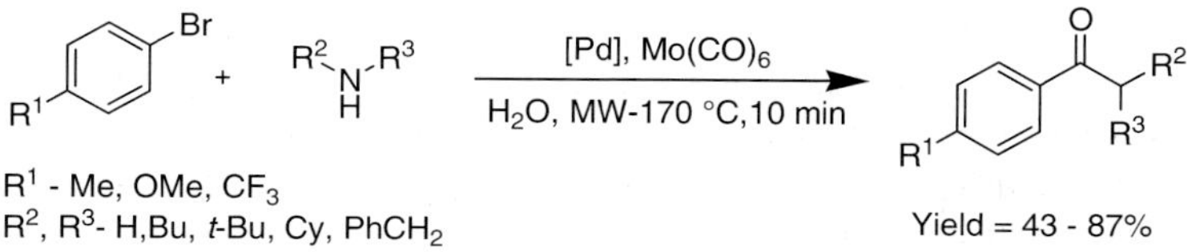

## Scheme 9.8

R¹—C₆H₄—Br + R²R³NH → [Pd], Mo(CO)₆, H₂O, MW-170 °C, 10 min → R¹—C₆H₄—C(O)—C(R³)(R²)

R¹ - Me, OMe, CF$_3$
R², R³ - H, Bu, t-Bu, Cy, PhCH$_2$

Yield = 43 - 87%

heating (Scheme 9.8) [24]. It is important to note that, despite the use of water as solvent, aminocarbonylation strongly dominated over hydroxycarbonylation, providing good yields of both secondary and tertiary model benzamides.

### 9.2.2
### Nitrogen-containing Heterocycles

Nitrogen heterocycles are abundant in Nature and are of great significance to life because their structural subunits exist in many natural products, such as vitamins, hormones, antibiotics, and alkaloids, and also in pharmaceuticals, herbicides, dyes, and many more compounds [25].

The synthesis of nitrogen-containing heterocycles such as substituted azetidines, pyrrolidines, piperidines, azepanes, N-substituted 2,3-dihydro-1H-isoindoles, 4,5-dihydropyrazoles, pyrazolidines, and 1,2-dihydrophthalazines has been accomplished in basic aqueous media under the influence of MW irradiation; the reactions proceed via double N-alkylation of primary amines and hydrazine derivatives (Scheme 9.9) with readily available alkyl dihalides (or ditosylates), thus providing facile entry to important classes of building blocks in natural products and pharmaceuticals [26–28].

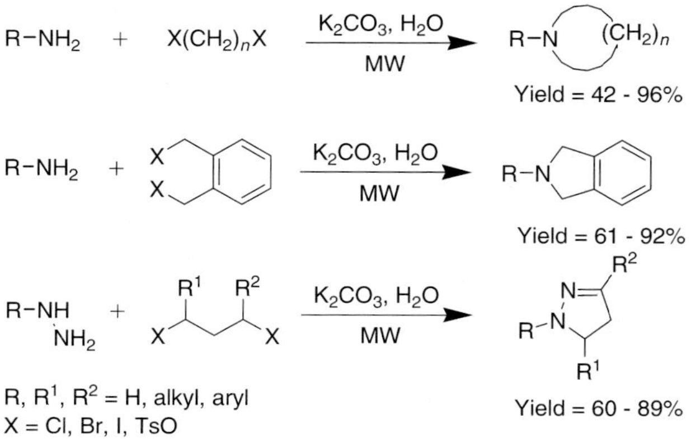

R—NH$_2$ + X(CH$_2$)$_n$X →(K$_2$CO$_3$, H$_2$O, MW)→ R—N(CH$_2$)$_n$ (cyclic)

Yield = 42 - 96%

R—NH$_2$ + o-C$_6$H$_4$(CH$_2$X)$_2$ →(K$_2$CO$_3$, H$_2$O, MW)→ R—N(isoindoline)

Yield = 61 - 92%

R—NH—NH$_2$ + X—CH(R¹)—CH(R²)—X →(K$_2$CO$_3$, H$_2$O, MW)→ pyrazoline with R², R¹, R-N

Yield = 60 - 89%

R, R¹, R² = H, alkyl, aryl
X = Cl, Br, I, TsO

## Scheme 9.9

This MW-accelerated general approach shortened the reaction time significantly and utilized readily available amines and hydrazines with alkyl dihalides or ditosylates to assemble two C—N bonds in a simple $S_N2$-like sequential heterocyclization experimental protocol which has never been fully realized under conventional reaction conditions. The strategy circumvents multi-step reactions and functional group protection/deprotection sequences, and eliminates the use of expensive phase-transfer and transition metal catalysts.

It is noteworthy that this reaction is not a homogeneous single-phase system as neither reactant is soluble in aqueous alkaline reaction media. It is believed that the selective absorption of MWs by polar molecules and intermediates in a multiphase system could substitute as a phase-transfer catalyst (PTC) without using any phase-transfer reagent, thereby providing the acceleration that has also been observed for ultrasonication [29].

The double alkylation of hydrazine was favored by MW irradiation (Scheme 9.10) because the increased polarity of **II** and **IV** drives the reaction to form 1-phenylpyrazolidine (**V**). The formation of a carbon–nitrogen double bond to afford 1-phenyl-4,5-dihydro-1H-pyrazole (**VI**) was explored in two control experiments by introducing oxygen into the system or by using activated palladium on carbon as a dehydrogenation catalyst. It was found that oxygen in the air plays a critical role in promoting the formation of a C–N double bond.

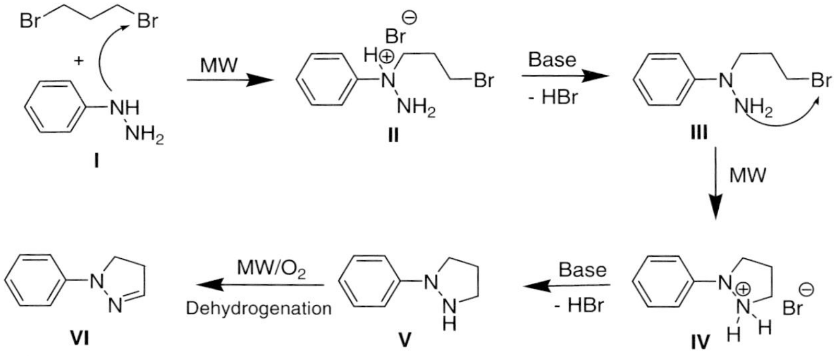

Scheme 9.10

The experimental observation is consistent with the mechanistic postulation wherein the polar transition state of the reaction is favored by MW irradiation with respect to the dielectric polarization nature of MW energy transfer [5a]. The phase separation of the desired product in either solid or liquid form from the aqueous medium can facilitate product purification by simple filtration or decantation instead of tedious column chromatography, distillation, or extraction processes. This eventually reduces the use of volatile organic solvent required for extraction or column chromatography.

A variety of nitrogen heterocycles have been synthesized by the condensation of hydrazine, hydrazide, and diamines with diketones and β-keto esters, respectively (Scheme 9.11) [30].

This N-alkylation of nitrogen heterocycles has also been achieved in aqueous media under MW irradiation conditions (Scheme 9.12) [31]. Shorter reaction times and higher product yields are some of the advantages that render this procedure a greener alternative to conventional chemical synthesis.

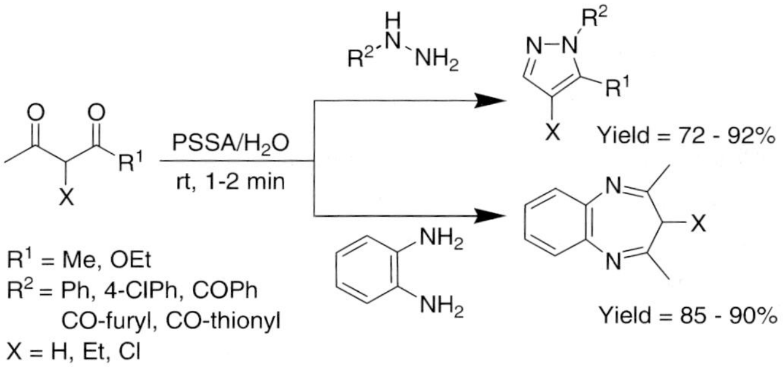

## 9.2 MW-assisted Organic Transformations in Aqueous Media

**Scheme 9.11**

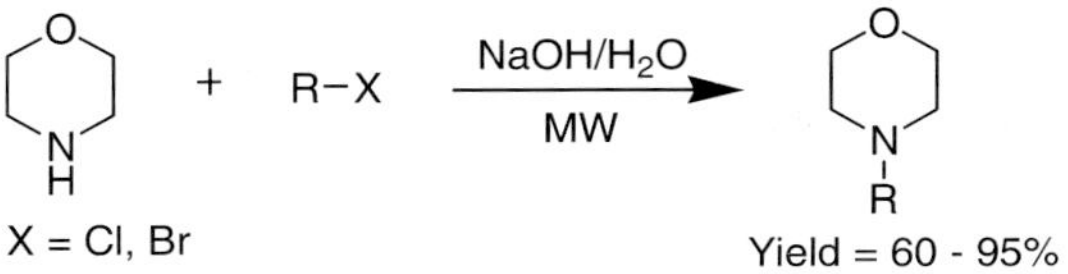

$R^1$ = Me, OEt
$R^2$ = Ph, 4-ClPh, COPh, CO-furyl, CO-thionyl
X = H, Et, Cl

Reagents: PSSA/H$_2$O, rt, 1-2 min
Yield = 72 - 92% (pyrazole)
Yield = 85 - 90% (benzodiazepine)

**Scheme 9.12**

X = Cl, Br

Reagents: NaOH/H$_2$O, MW
Yield = 60 - 95%

Dihydropyrimidinones are an important class of organic compounds which show prominent biological activity, and were synthesized by an environmentally benign aqueous Biginelli protocol using poly(styrenesulfonic acid) (PSSA) as a catalyst. (Scheme 9.13) [32]. This three-component MW protocol proceeds efficiently in water without the use of any organic solvent. Also, the use of polymer-supported, low-toxic, and inexpensive PSSA as a catalyst renders this method eco-friendly, with a very simple isolation procedure that entails the filtration of the precipitated products.

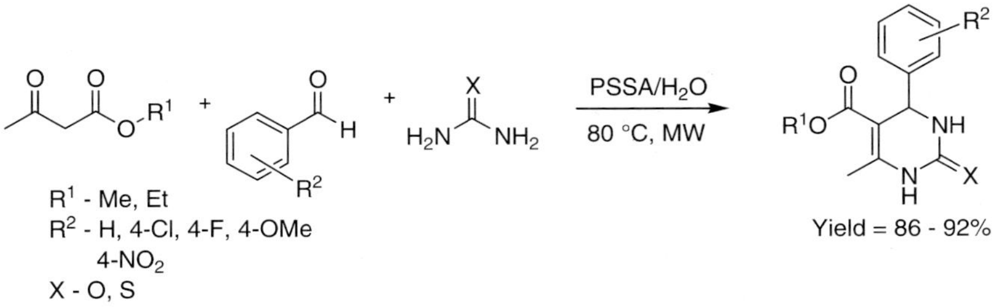

**Scheme 9.13**

$R^1$ - Me, Et
$R^2$ - H, 4-Cl, 4-F, 4-OMe, 4-NO$_2$
X - O, S

Reagents: PSSA/H$_2$O, 80 °C, MW
Yield = 86 - 92%

Another three-component reaction involving an aldehyde, 2,6-diaminopyrimidin-4(3H)-one, and either tetronic acid or indane-1,3-dione, for the synthesis of furo[3′,4′ : 5,6]pyrido[2,3-d]pyrimidine derivatives in water under MW irradiation was developed by Tu et al. (Scheme 9.14) [33]. Valuable features of this method include the excellent product yields, reduced reaction time and environmental impact, and the simple reaction procedure.

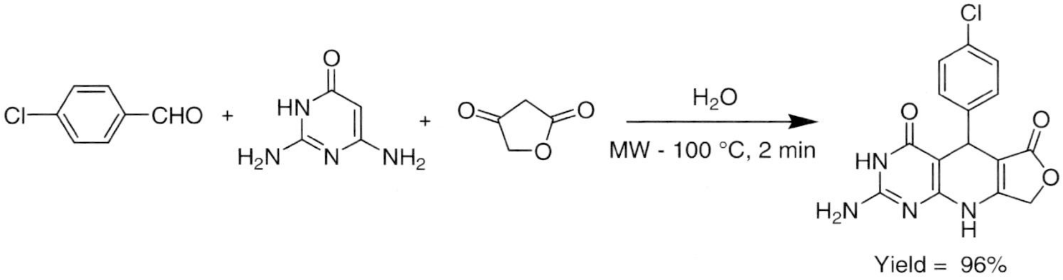

**Scheme 9.14**

Benzoxazines, an important class of bioactive heterocycles, were also synthesized using MW irradiation (Scheme 9.15), in an operationally simple and safe aqueous protocol with fair to good yields [34].

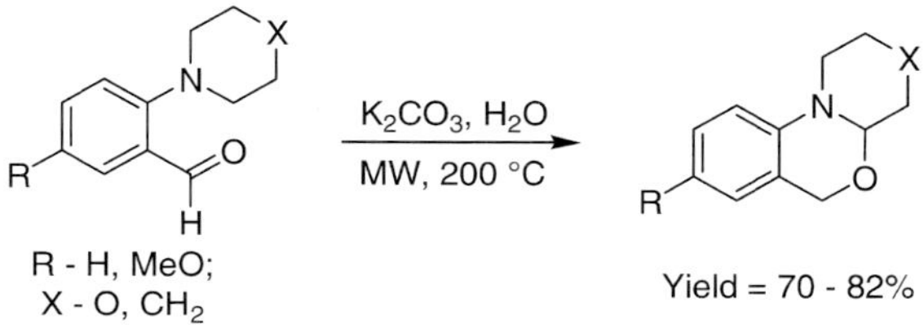

**Scheme 9.15**

### 9.2.3
### Oxygen-containing Heterocycles

Oxygen-containing heterocycles are important classes of building blocks in organic synthesis and several derivatives of these heterocycles have attracted much attention from medicinal chemists over the years.

Although dioxanes have great potential as drug candidates, the synthetic protocols for these important molecules have been largely untapped. A novel tandem bis-aldol reaction of ketones with paraformaldehyde in aqueous media catalyzed by PSSA under MW irradiation conditions to produce 1,3-dioxanes has been developed (Scheme 9.16) [35].

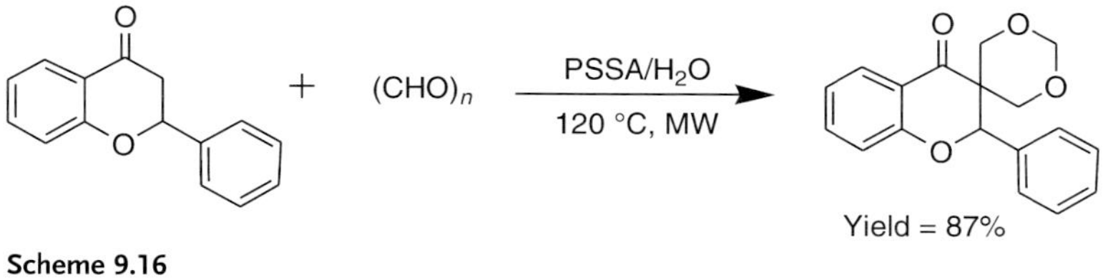

**Scheme 9.16**

Various ketones reacted efficiently with paraformaldehyde in water to afford the desired 1,3-dioxanes in good yield. This approach establishes a convenient and flexible method to attach functional arms to indanone and flavanone for further elaboration in synthetic design. Also, it is worth mentioning that these reactions proceed in an aqueous medium without using any PTC. This may be due to selective absorption of MWs by reactants, intermediates, and the polar aqueous medium [36], which accelerates the reaction even in the absence of PTC.

This PSSA-catalyzed tandem bis-aldol reaction of ketones with paraformaldehyde in water may proceed via the mechanism shown in Scheme 9.17.

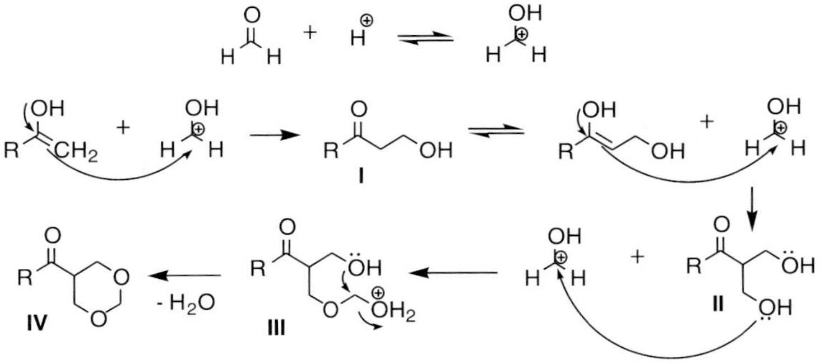

**Scheme 9.17**

The reaction involves the addition of a protonated formaldehyde molecule (generated by MW exposure of paraformaldehyde with PSSA–water) to ketone (enol) to form β-hydroxy ketone **I**. This is followed by the addition of another protonated formaldehyde molecule to **I** to yield diol **II**, which in turn attacks the third formaldehyde molecule to give adduct **III**, which after dehydration yields the final product 1,3-dioxane **IV**.

A nano-sized magnesium oxide-catalyzed three-component condensation reaction of aldehyde, malononitrile, and α-naphthol proceeded rapidly in water–poly(ethylene glycol) (PEG) to afford the corresponding 2-amino-2-chromenes in high yields at room temperature (Scheme 9.18). This greener protocol was found to be fairly general and the catalyst was reused in subsequent reactions with consistent activity [37].

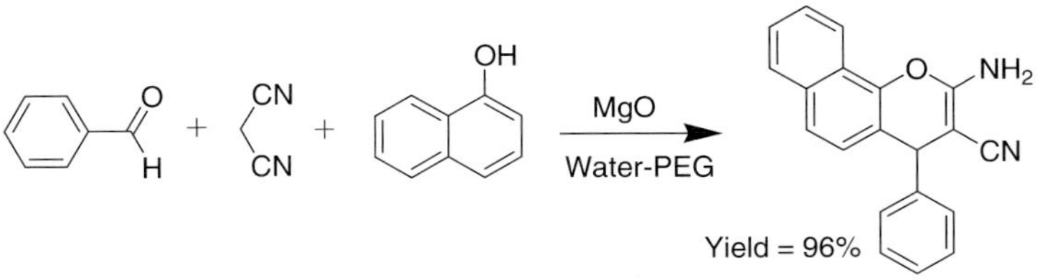

**Scheme 9.18**

Nano-sized magnesium oxide has been employed for the first time as a novel and efficient catalyst for the benign synthesis of various substituted 2-amino-2-chromenes in a three-component condensation approach. The attractive features of this protocol are the simple experimentation procedure, the use of benign reaction solvents, cost effectiveness, the recyclability of catalysts, and its adaptability for the synthesis of a diverse set of 2-amino-2-chromenes.

## 9.2.4
### Heterocyclic Hydrazones

Heterocyclic hydrazones constitute an important class of compounds in organic chemistry and recently have also been found useful as anti-malaria drugs and as inhibitors of macrophage migration inhibitory factor (MIF) tautomerase activity [38].

An environmentally benign aqueous protocol for the synthesis of these heterocyclic hydrazones using PSSA as a catalyst has been developed (Scheme 9.19). The simple reaction proceeds efficiently in water in the absence of any organic solvent under MW irradiation and involves basic filtration as the product isolation step [39].

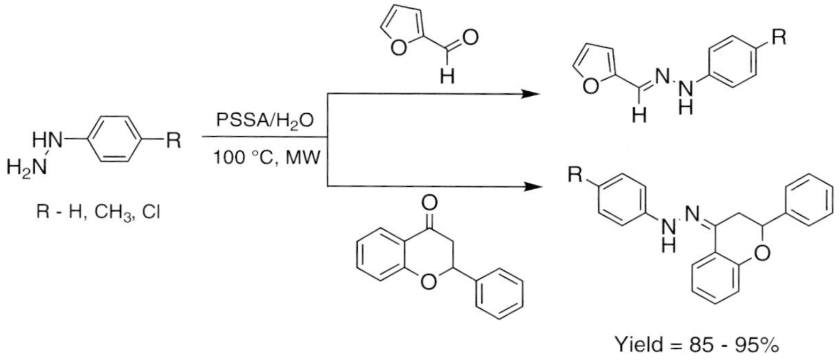

Scheme 9.19

## 9.2.5
### Other Miscellaneous Reactions

β-Hydroxy sulfides and β-hydroxy sulfoxides are important intermediates in organic synthesis. An efficient and environmentally friendly one-pot synthesis of these compounds in water under MW irradiation has been reported (Scheme 9.20) [40].

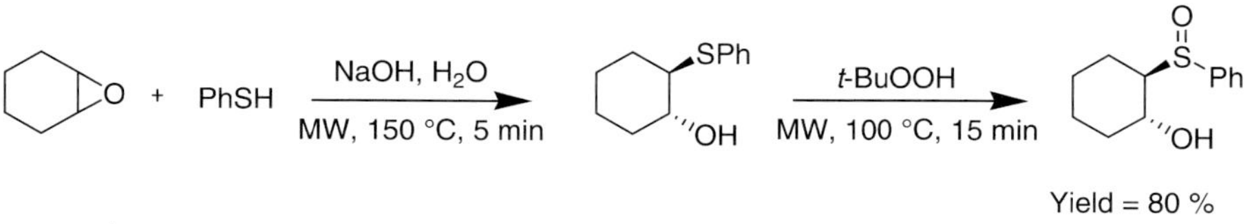

Scheme 9.20

This thiolysis of several epoxides in water, under controlled MW irradiation, was performed in a short time and with quantitative yields without the use of any metal catalyst. By coupling this procedure with the *in situ* oxidation of sulfide mediated by *tert*-butyl hydroperoxide, β-hydroxy sulfoxides were also synthesized.

4′-Aryl-2,2′,6′,2″-terpyridines were synthesized by a clean aqueous Krohnke reaction process via a one-pot reaction of 2-acetylpyridine with an aromatic aldehyde and ammonium acetate under MW irradiation in aqueous medium (Scheme 9.21) [41].

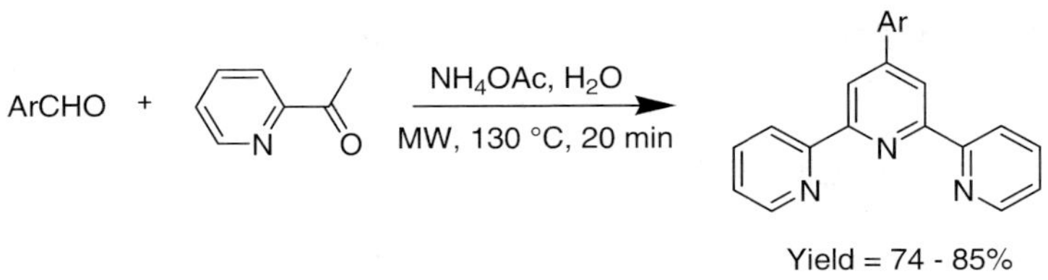

**Scheme 9.21**

Decarboxylation of substituted α-phenylcinnamic acid derivatives in aqueous media were achieved, wherein a remarkable synergism between methylimidazole and aqueous $NaHCO_3$ in PEG under MW irradiation furnished the corresponding *para/ortho*-hydroxylated (*E*)-stilbenes in a mild and efficient manner (Scheme 9.22). The critical role of water in facilitating the decarboxylation imparts an interesting facet to the synthetic utility of water-mediated organic transformations [42].

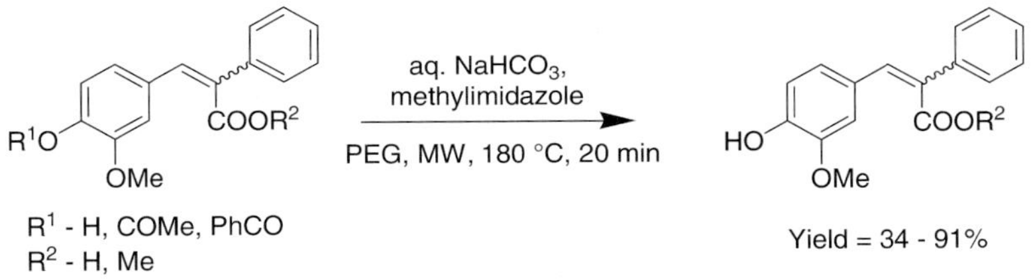

**Scheme 9.22**

An efficient and environmentally benign tandem bis-aza-Michael addition of amines catalyzed by PSSA was developed (Scheme 9.23). This operationally simple, high-yielding MW-assisted synthetic protocol proceeded in water in the absence of any organic solvent [43].

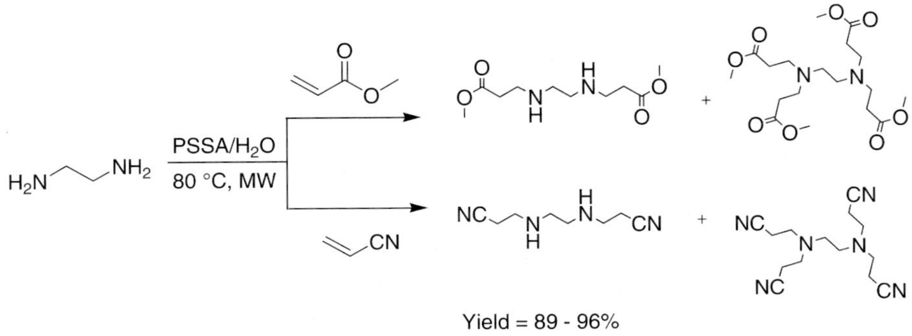

**Scheme 9.23**

A variety of bis-aza-Michael addition products of diamines were obtained readily depending on the relative mole ratio of the reactants. One mole equivalent of diamine with two mole equivalents of Michael acceptor afforded disubstituted diamines, without any tri- or tetrasubstituted products, whereas using four mole equivalents of Michael acceptor gave exclusively the tetrasubstituted diamine product.

Transition metal-mediated C–N bond-forming processes are important fundamental transformations and are extensively utilized for N-arylations. Copper-promoted aqueous N-arylations of amines, amides, imides, and β-lactams with aryl halides under MW irradiation conditions were reported (Scheme 9.24) [44]. These reactions can be performed at 85–90 °C in aqueous media, and also under solvent-free conditions, to give good yields. However, under solvent-free conditions, lower yields were obtained.

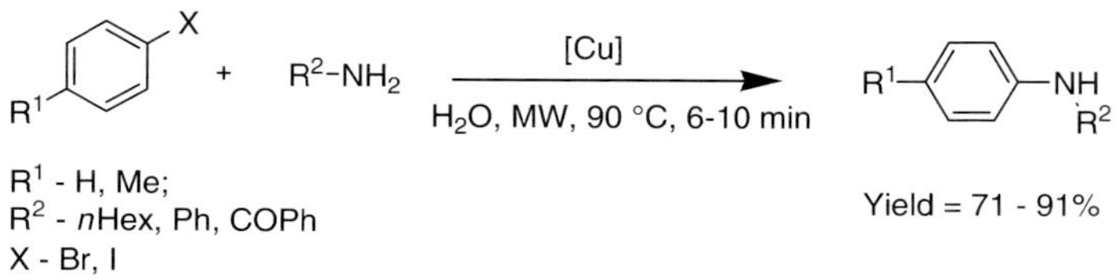

$R^1$ - H, Me;
$R^2$ - nHex, Ph, COPh
X - Br, I

Yield = 71 - 91%

**Scheme 9.24**

Interestingly, this method was also successfully applied to the intramolecular N-arylation of β-lactam derivatives (Scheme 9.25); most other methods were unsuccessful and led to the decomposition of the starting material because of the presence of a base.

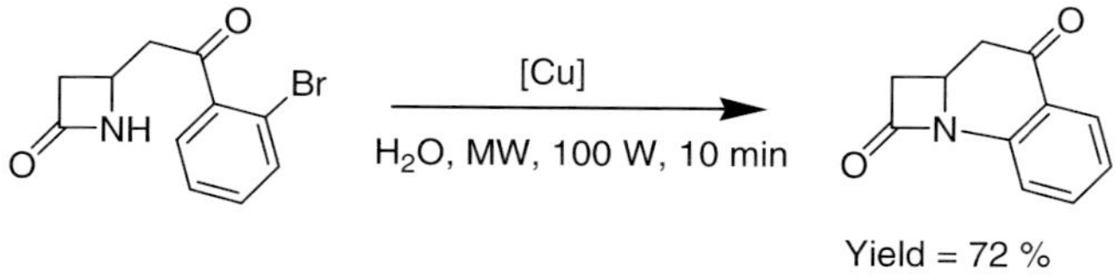

Yield = 72 %

**Scheme 9.25**

An MW-assisted catalytic transfer hydrogenation process was developed to reduce various α,β-unsaturated carbonyl compounds to the corresponding saturated carbonyl compounds in the presence of silica-supported palladium chloride as catalyst and a combination of MeOH–HCOOH–$H_2O$ (1 : 2 : 3) as hydrogen source (Scheme 9.26) [45].

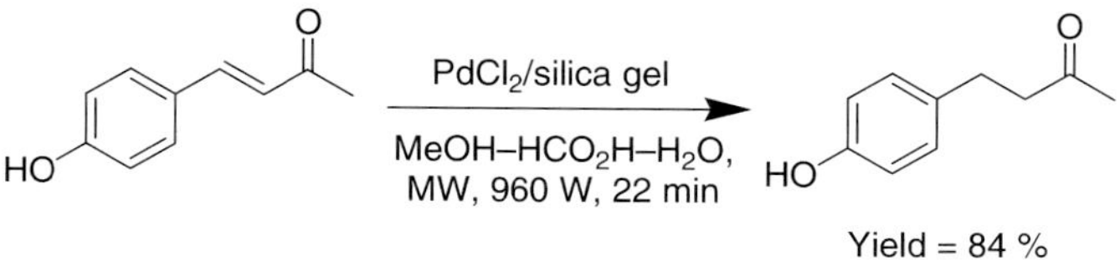

Yield = 84 %

**Scheme 9.26**

## 9.3
## Sonochemical Organic Transformations in Aqueous Media

Ultrasound techniques have increasingly been used in organic synthesis, as these techniques are more convenient and easily controllable than traditional methods. These are important techniques for increasing the rates of chemical reactions under milder reaction conditions. They are excellent tools for the fabrication of nanomaterials [46], the formation of polymers [47], and their controlled degradation [48] in green technologies [49], and more recently for promoting supramolecular aggregation [50].

### 9.3.1
### Synthesis of Heterocycles

The regioselective synthesis of 1,4-di-substituted-1,2,3-triazoles has been accomplished using ultrasound. This three-component reaction of alkyl/aryl halides, terminal alkynes, and sodium azide, catalyzed by CuI, was efficiently carried out in water (Scheme 9.27); ultrasonication dramatically shortened the reaction time. The operationally simple protocol using water as a solvent and pure product formation make this method attractive for large-scale synthesis and also for small-scale drug molecule library synthesis [51].

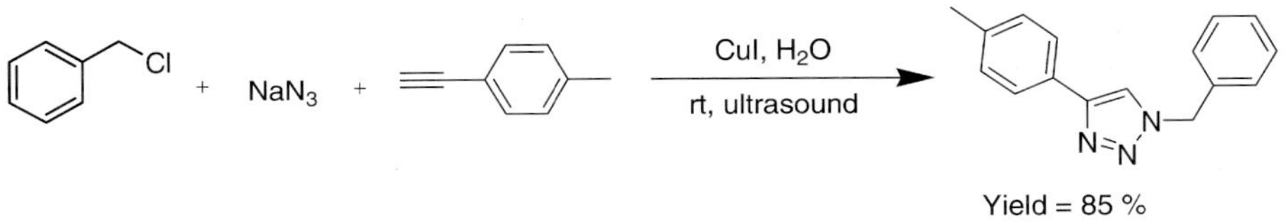

Scheme 9.27

1,2-Dihydroisoquinoline derivatives have also been synthesized by three-component coupling of 2-alkynylbenzaldehyde, an amine, and a nucleophile in aqueous medium (Scheme 9.28). This reaction was catalyzed by a Lewis acid–surfactant combined catalyst (LASC) and was accelerated by ultrasonic conditions [52].

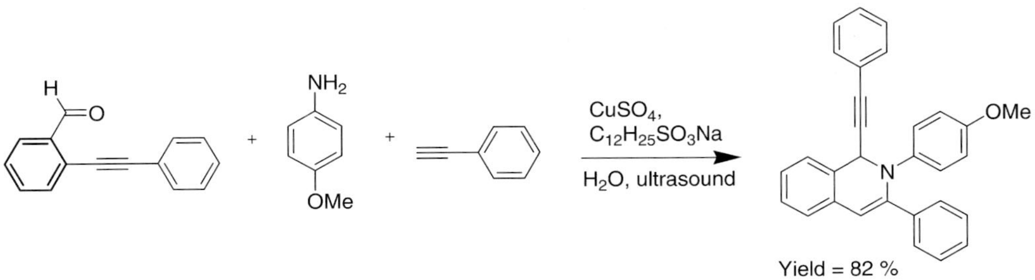

Scheme 9.28

## 9.3.2
### Pinacol Coupling Reaction

Ultrasound irradiation can efficiently prompt pinacol coupling of aromatic aldehydes in acidic aqueous media (Scheme 9.29) [53]. Pinacol coupling of aromatic aldehydes by Zn powder in aqueous $H_2NSO_3H$ or $H_3PO_4$ under ultrasound irradiation can lead to the corresponding pinacols in 14–88% yields. The main advantages of this procedure are the milder reaction conditions and the use of inexpensive reagent.

Li and co-workers also studied these pinacol coupling reactions using ultrasound irradiation [54]. Aqueous vanadium(II) solution efficiently catalyzed this reaction resulting in the pinacol product in 54–93% yield within 15–30 min.

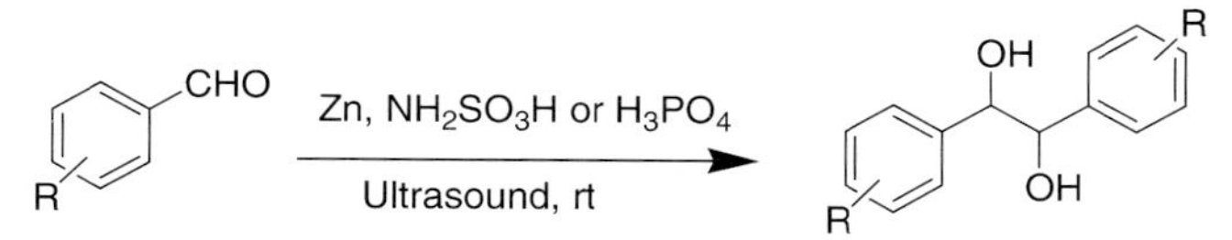

R - H, 2-Cl, 3-Cl, 4-Cl
2,4-Cl, 4-Me, 3-Br

Yield = 14 - 88%

**Scheme 9.29**

An allylation reaction between an allyl halide and carbonyl compounds is an important and powerful method for constructing carbon–carbon bonds. This reaction can also be conducted using ultrasound irradiation conditions catalyzed by La–SnCl$_2$ in aqueous medium (Scheme 9.30) [55].

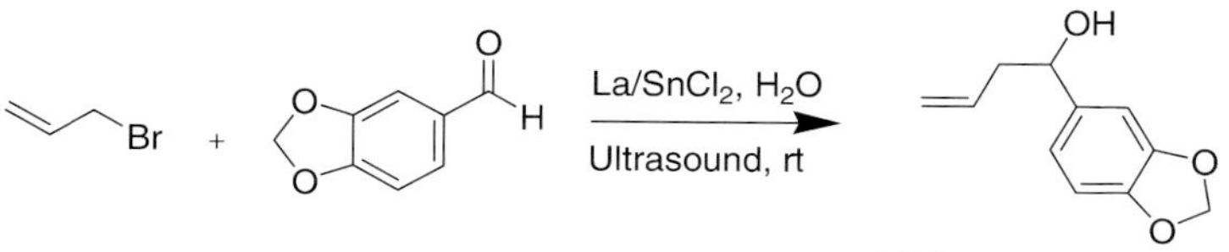

Yield = 98%

**Scheme 9.30**

Sonochemical Ullmann–Goldberg reaction of o-chlorobenzoic acid and substituted anilines for the synthesis of N-phenylanthranilic acids was achieved by Palacios and Comdom in aqueous medium (Scheme 9.31) [56]. The use of ultrasound irradiation accelerated the synthesis of N-phenylanthranilic acid using water as solvent. A number of N-phenylanthranilic acid derivatives were prepared in good yield in a very short reaction time. This protocol can also be used for the synthesis of 2-carboxy-substituted diphenyl ethers [57].

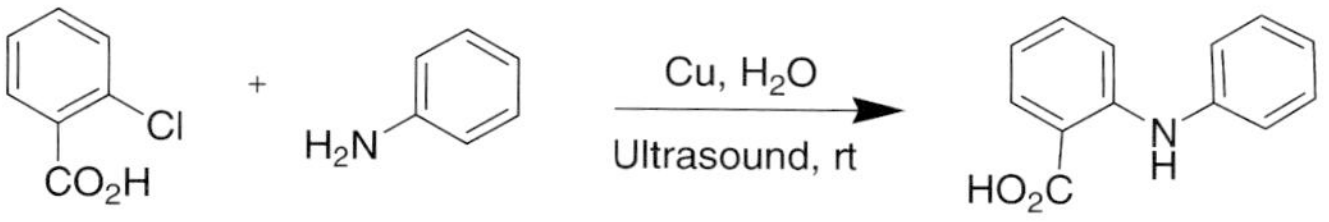

Yield = 88 %

**Scheme 9.31**

## 9.4
## Photochemical Transformations in Aqueous Media

The photocatalyzed reaction of organic molecules has become a subject of serious study as it shows promise of a viable commercial technology. It may be used as an alternative method for selective organic synthesis under environmentally benign conditions and achieve pollution reduction by eliminating potential wastes at their source. Such photocatalyzed reactions have been carried out both in organic solvents and in aqueous media. Among all explorations for potential application, the most active area of the past decade has been the photocatalysis for environmental remediation due to the emerging concern for the environment and the successful destruction of a wide range of pollutants [58].

The synthesis of a variety of cyclic peptides from N-phthaloyl-protected di-, tri-, tetra-, and pentapeptides with different aminocarboxylic acids by photodecarboxylation initiated by intramolecular electron transfer has been accomplished in aqueous media (Scheme 9.32) [59].

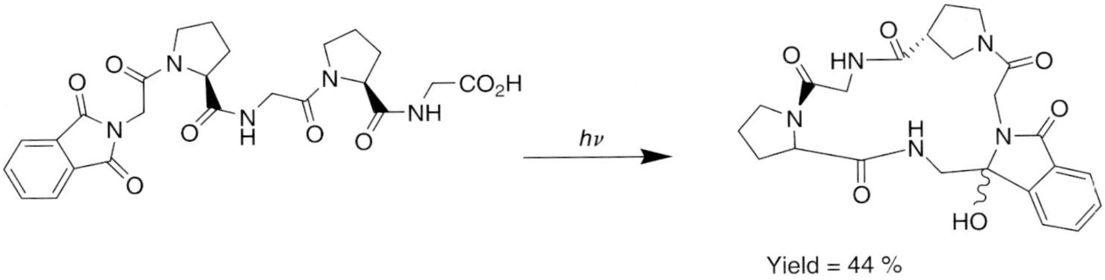

Yield = 44 %

**Scheme 9.32**

The combination of ultraviolet (UV) light and hydrogen peroxide is potentially applicable for the conversion of many types of organic contaminants in water, at concentrations ranging from ppb to ppm, into innocuous inorganic compounds. Chen and co-workers studied the destruction of 1,3-dinitrobenzene (1,3-DNB) in aqueous solution under UV irradiation in the presence of hydrogen peroxide [60]. The combination of UV radiation and hydrogen peroxide was significantly effective in degrading 1,3-DNB, and this photodegradation process can be influenced to a certain extent by increasing the content of hydrogen peroxide and the acidity of the reaction matrices. It was found that a variety of phenolic intermediates and inorganic acids were formed via hydroxyl radicals attacking the parent compound.

A similar system was also used for the degradation reaction of dichloroacetic acid [61]. This study showed that in an aqueous solution, no stable reaction intermediates were formed and at all times during the reaction, 2 mol of hydrochloric acid were formed for each mole of dichloroacetic acid that decomposed.

Photochemical reactions of acidic iron(III) solutions have been studied since 1949 and have been widely used to achieve the efficient degradation of several organic pollutants in aqueous solution. The degradation of cyclohexanol, which can be photoinduced by Fe(III) derivatives (e.g. nitrate, pyrophosphate, and sulfate), has been investigated in aqueous media at room temperature and under neutral to alkaline conditions with continuous irradiation at 254, 366 and around 500 nm for 24 h [62]. During the oxidation of cyclohexanol, cyclohexanone was formed in reasonable amounts in the presence of the nitrate salt, whereas with the other two iron salts its yield was negligible, and instead carbon dioxide was the dominant reaction product. This is an efficient cyclohexanol removal process by oxidative photocatalysis that can be carried out in neutral to alkaline aqueous media.

Perfluorocarboxylic acids have recently attracted much attention because they are recognized as ubiquitous environmental contaminants. Hori *et al.* studied the decomposition of these acids in water mediated by iron(II)/(III) and photocatalytically decomposed to $F^-$ and $CO_2$ in the presence of a small amount of Fe(III). Photocatalysis was achieved by UV irradiation (220–460 nm) under oxygen [63].

Aromatic nucleophilic photosubstitution reactions have been studied widely as major research topics in organic photochemistry. Reductive photocyanation of 1,10-phenanthroline was carried out in aqueous NaCN solution, yielding 5-cyano-5,6-dihydro-1,10-phenanthroline. The photoreaction in the presence of $Mg^{2+}$ improved the reaction quantum yield compared with that without $Mg^{2+}$ [64].

## 9.5
## Conclusion

The demands for new bioactive molecules in the field of healthcare, combined with the pressure to produce these substances expeditiously and in an environmentally benign fashion, pose significant challenges to the synthetic chemicals community. These challenges can be met by the development of wide variety of synthetic protocols using various greener techniques. The use of non-conventional energy sources in safer solvents such as water is one of the best ways to develop greener and sustainable protocols. It should also be noted that the rapid development in this field is due to the recognition that environmentally friendly products and processes will be economical in the long term and more scientific innovations are needed to develop "green organic synthesis" further.

## References

1 Varma, R.S. (2006) *Kirk-Othmer On-line Encyclopedia of Chemical Technology*, 5th edn., John Wiley & Sons, Inc., Hoboken, NJ, Vol. 16, pp. 538–594.

2 Polshettiwar, V. and Varma, R.S. (2008) *Pure Appl. Chem.*, **80**, 777–790.

3 Cravotto, G. and Cintas, P. (2007) *Angew. Chem. Int. Ed.*, **46**, 5476–5478.

4 Cravotto, G. and Cintas, P. (2007) *Chem. Eur. J.*, **13**, 1902–1909.

5 (a) Polshettiwar, V. and Varma, R.S. (2008) *Chem. Soc. Rev.*, **37**, 1546–1557.

(b) Li, C.J. and Chen, L. (2006) *Chem. Soc. Rev.*, **35**, 68–82.

6 Varma, R.S. (2007) Clean chemical synthesis in water.Org. Chem. Highlights, URL: http://www.organic-chemistry.org/Highlights/2007/01February.shtm.

7 Polshettiwar, V. and Varma, R.S. (2007) *Curr. Opin. Drug Discov.*, **10**, 723–737.

8 Polshettiwar, V. and Varma, R.S. (2008) *Acc. Chem. Res.*, **41**, 629–639.

9 (a) Yin, L. and Liebsher, J. (2007) *Chem. Rev.*, **107**, 133–173; (b) Tsuji, J. (2004) *Palladium Reagents and Catalysts: New Perspectives for the 21st Century*, John Wiley & Sons, Ltd, Chichester.

10 (a) Mizoroki, T., Mori, K. and Ozaki, A. (1971) *Bull. Chem. Soc. Jpn.*, **44**, 581–1581; (b) Heck, R.F. and Nolly, J.P. (1972) *J. Org. Chem.*, **37**, 2320–2322; (c) Trzeciak, A.M. and Ziolkowski, J.J. (2005) *Coord. Chem. Rev.*, **249**, 2308–2322.

11 (a) Miyaura, N., Yamada, K. and Suzuki, A. (1979) *Tetrahedron Lett.*, **36**, 3437–3440; (b) Bellina, F., Carpita, A. and Rossi, R. (2004) *Synthesis*, **15**, 2419–2440; (c) Miyaura, N. (2002) *Top. Curr. Chem.*, **219**, 11–59.

12 (a) Kotha, S., Lahiri, K. and Kashinath, D. (2002) *Tetrahedron*, **58**, 9633–9695; (b) Herrmann, W.A. and Cornils, B. (1997) *Angew. Chem. Int. Ed. Engl.*, **36**, 1048–1067.

13 Polshettiwar, V. and Molnar, A. (2007) *Tetrahedron*, **63**, 6949–6976.

14 Leadbeater, N.E. and Marco, M. (2003) *J. Org. Chem.*, **68**, 888–892.

15 Bai, L., Wang, J.-X. and Zhang, Y. (2003) *Green Chem.*, **5**, 615–617.

16 Arvela, R.K. and Leadbeater, N.E. (2005) *Org. Lett.*, **7**, 2101–2104.

17 Crozet, M.D., Castera-Ducros, C. and Vanelle, P. (2006) *Tetrahedron Lett.*, **47**, 7061–7065.

18 Zhu, R., Qu, F., Quéléverb, G. and Peng, L. (2007) *Tetrahedron Lett.*, **48**, 2389–2393.

19 Arvela, R.K. and Leadbeater, N.E. (2005) *J. Org. Chem.*, **70**, 1786–1790.

20 Arvela, R.K., Pasquini, S. and Larhed, M. (2007) *J. Org. Chem.*, **72**, 6390–6396.

21 Appukkuttan, P., Dehaen, W. and Van der Eycken, E. (2003) *Eur. J. Org. Chem.*, 4713–4716.

22 Kremsner, J.M. and Kappe, C.O. (2005) *Eur. J. Org. Chem.*, 3672–3679.

23 Alacida, E. and Nájera, C. (2006) *Adv. Synth. Catal.*, **348**, 2085–2091.

24 Wu, X. and Larhed, M. (2005) *Org. Lett.*, **7**, 3327–3329.

25 Padwa, A. and Bur, S. (2004) *Chem. Rev.*, **104**, 2401–2433.

26 Ju, Y. and Varma, R.S. (2005) *Tetrahedron Lett.*, **46**, 6011–6014.

27 Ju, Y. and Varma, R.S. (2005) *Org. Lett.*, **7**, 2409–2411.

28 Ju, Y. and Varma, R.S. (2006) *J. Org. Chem.*, **71**, 135–141.

29 Varma, R.S., Naicker, K.P. and Kumar, D. (1999) *J. Mol. Catal. A: Chemical*, **149**, 153–160.

30 Polshettiwar, V. and Varma, R.S. (2008) *Tetrahedron Lett.*, **49**, 397–400.

31 Ju, Y. and Varma, R.S. (2004) *Green. Chem.*, **6**, 219–221.

32 Polshettiwar, V. and Varma, R.S. (2007) *Tetrahedron Lett.*, **48**, 7343.

33 Tu, S.-J., Zhang, Y., Jiang, H., Jiang, B., Zhang, J.-Y., Jia, R.-H. and Shi, F. (2007) *Eur. J. Org. Chem.*, 1522–1528.

34 Kaval, N., Halasz-Dajka, B., Vo-Thanh, G., Dehaen, W., Van der Eycken, J., Matyus, P., Loupy, A. and Van der Eycken, E. (2005) *Tetrahedron*, **61**, 9052–9057.

35 Polshettiwar, V. and Varma, R.S. (2007) *J. Org. Chem.*, **72**, 7420–7422.

36 Loupy, A. and Varma, R.S. (2006) *Chim. Oggi*, **24**, 36–40.

37 Kumar, D., Reddy, V.B., Mishra, B.G., Rana, R.K., Nadagouda, M.N. and Varma, R.S. (2007) *Tetrahedron*, **63**, 3093–3097.

38 Dabideen, D.R., Cheng, K.F., Aljabari, B., Miller, E.J., Pavlov, V.A. and Al-Abed, Y. (2007) *J. Med. Chem.*, **50**, 1993–1997.

39 Polshettiwar, V. and Varma, R.S. (2007) *Tetrahedron Lett.*, **48**, 5649–5652.

40 Pironti, V. and Colonna, S. (2005) *Green Chem.*, **7**, 43–45.

41 Tu, S., Jia, R., Jiang, B., Zhang, J., Zhang, Y., Yaoa, C. and Jib, S. (2007) *Tetrahedron*, **63**, 381–388.
42 Kumar, V., Sharma, A., Sharma, A. and Sinha, A.K. (2007) *Tetrahedron*, **63**, 7640–7646.
43 Polshettiwar, V. and Varma, R.S. (2007) *Tetrahedron Lett.*, **48**, 8735–8738.
44 Yadav, L.D.S., Yadav, B.S. and Rai, V.K. (2006) *Synthesis*, 1868–1872.
45 Sharma, A., Kumar, V. and Sinha, A.K. (2006) *Adv. Synth. Catal.*, **348**, 354–360.
46 (a) Gedanken, A. and Mastai, Y. (2004) in *The Chemistry of Nanomaterials: Synthesis, Properties and Applications*, Vol. 1 (eds C.N.R. Rao, A. Müller and A.K. Cheetham), Wiley-VCH Verlag GmbH, Weinheim, pp. 113–169; (b) Gedanken, A. (2007) *Ultrason. Sonochem.*, **14**, 418–430.
47 Kruus, P. (1991) in *Advances in Sonochemistry*, Vol. 2, (ed. T.J. Mason), JAI Press, Stamford, CT, pp. 1–21.
48 Price, G.J. (1990) in *Advances in Sonochemistry*, Vol. 1, (ed. T.J. Mason), JAI Press, Stamford, CT, pp. 231–287.
49 Mason, T.J. and Cintas, P. (2002) in *Handbook of Green Chemistry* (eds J. Clark and D. Macquarrie), Blackwell, Oxford, pp. 372–396.
50 Paulusse, J.M.J. and Sijbesma, R.P. (2006) *Angew. Chem. Int. Ed.*, **45**, 2334–2337.
51 Sreedhar, B. and Reddy, P.S. (2007) *Synth. Commun.*, **37**, 805–812.
52 Ye, Y., Ding, Q. and Wu, J. (2008) *Tetrahedron*, **64**, 1378–1382.
53 Yang, J.-H., Li, J.-T., Zhao, J.-L. and Li, T.-S. (2004) *Synth. Commun.*, **34**, 993–1000.
54 Wang, S.-X., Wang, K. and Li, J.-T. (2005) *Synth. Commun.*, **35**, 2387–2394.
55 Bian, Y.-J., Zhang, J.-Q. and Xia, J.-P. (2006) *Synth. Commun.*, **36**, 2475–2481.
56 Palacios, M.L.D. and Comdom, R.F.P. (2003) *Synth. Commun.*, **33**, 1771–1775.
57 Comdom, R.F.P. and Palacios, M.L.D. (2003) *Synth. Commun.*, **33**, 921–926.
58 Tariq, M.A., Faisal, M., Muneera, M. and Bahnemannc, D. (2007) *J. Mol. Catal. A: Chemical*, **265**, 231–236.
59 Griesbeck, A.G., Heinrich, T., Oelgemller, M., Molis, A. and Heidtmann, A. (2002) *Helv. Chim. Acta*, **85**, 4561–4578.
60 Chen, Q.M., Yang, C., Goh, N.K., Teo, K.C. and Chen, B. (2004) *Chemosphere*, **55**, 339–344.
61 Zalazar, C.S., Labas, M.D., Brandi, R.J. and Cassano, A.E. (2007) *Chemosphere*, **66**, 808–815.
62 Garrel, L., Bonetti, M., Tonucci, L., d'Alessandro, N. and Bressan, M. (2006) *J. Photochem. Photobiol. A*, **179**, 193–199.
63 Hori, H., Yamamoto, A., Koike, K., Kutsuna, S., Osaka, I. and Arakawa, R. (2007) *Chemosphere*, **66**, 572–578.
64 Kitamura, N., Kita, A. and Kitagawa, F. (2005) *J. Photochem. Photobiol. A*, **174**, 149–155.

# 10
# Functionalization of Carbohydrates in Water

*Marie-Christine Scherrmann, André Lubineau, and Yves Queneau*

## 10.1
## Introduction

*Carbohydrate chemistry* and *chemistry in water* go well with each other for three main reasons. First, the synthesis of carbohydrate derivatives of biological relevance has profited from the advances in the use of aqueous media as solvents, leading to new compounds with improved reactivity and selectivity. Second, carbohydrates, being water soluble, have been used as water-solubility inducers for water-promoted reactions. Third, part of the chemistry devoted to the use of carbohydrate as raw materials, as an alternative to fossil resources, is performed in water. Indeed, it is now strategic to combine different key principles of green chemistry, namely the use of clean solvents and the use of renewable starting materials in synthetic chemistry. Moreover, it can be connected with the design of new derivatives with improved environmental properties.

The chapter focuses on the chemistry of small carbohydrates (not polysaccharides), reviewing C–C bond formation reactions involved in multi-step syntheses of complex structures or in direct processes towards functional derivatives, C–N and C–O bond formation reactions. The latter aspect, which deals with the reactivity of sugars as polyols, is an extremely large topic, and for reasons of conciseness the main issues of this chemistry will be illustrated by focusing on the example of sucrose, the most available small carbohydrate.

Also, some approaches in which a carbohydrate serves as a polar appendage which provides the water solubility necessary to take advantage of hydrophobic effects are described, in addition to the use of aqueous solutions of sugars as media in which hydrophobic effects are intensified.

## 10.2
## C—C Bond Formation Reactions

### 10.2.1
### Knoevenagel Condensations

Galbis Perez and co-workers reported in 1986 a method for the synthesis of C-glycosyl barbiturates, involving a one-step reaction between aldoses and barbituric acids **1a** or **1b** in water in the presence of sodium carbonate [1]. This general method was applied to D-glucose, D-galactose, D-mannose, D-xylose, D-ribose, D-arabinose and 2-amino-2-deoxy-D-glucose to give 5-D-glycopyranosyl-1,3-dimethylbarbiturates **2a–5a** or 5-D-glycopyranosylbarbiturates **2b–5b** in good yields (Scheme 10.1) [1, 2]. The mechanism of the transformation presumably involves condensation of the carbanion of the barbituric acid with the formyl group of the starting sugar, β-elimination of water and then cyclization to the β-C-glycoside through a Michael-type addition.

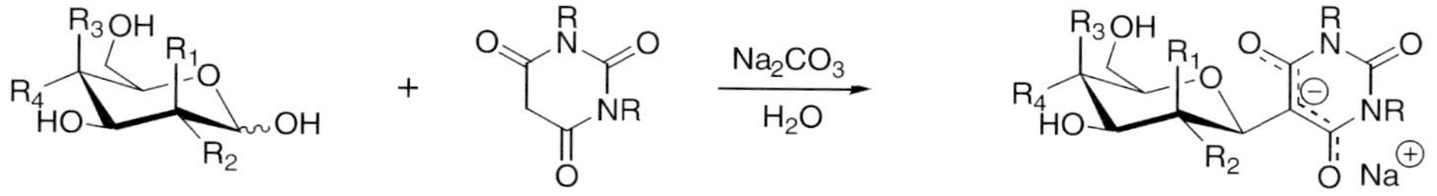

**1a** R = Me
**1b** R = H

**2a** $R_1 = R_3 = H$, $R_2 = R_4 = OH$, R = Me   78%
**2b** $R_1 = R_3 = H$, $R_2 = R_4 = OH$, R = H    80%
**3a** $R_1 = R_4 = H$, $R_2 = R_3 = OH$, R = Me   78%
**3b** $R_1 = R_4 = H$, $R_2 = R_3 = OH$, R = H    73%
**4a** $R_2 = R_3 = H$, $R_1 = R_4 = OH$, R = Me   80%
**4b** $R_2 = R_3 = H$, $R_1 = R_4 = OH$, R = H    74%
**5a** $R_1 = R_3 = H$, $R_2 = NH_2$, $R_4 = OH$, R = Me   85%
**5b** $R_1 = R_3 = H$, $R_2 = NH_2$, $R_4 = OH$, R = H    70%

**Scheme 10.1** Reaction of hexoses with barbituric acids.

The reaction was exploited later on by Wulff and Clarkson to synthesize monomers devoted to the preparation of polyvinylsaccharides [3]. The barbiturate moiety was alkylated in water at the C-5 position, using allylic or benzylic bromides under phase-transfer catalyst conditions promoted by ultrasound, in 63–75% yield. The resulting C-5 alkylated C-glycosylbarbiturates **6** were then converted into glutarimides **7** designed to prepare glycopolymers (Scheme 10.2). Surprisingly, the same reaction conducted in aqueous medium with Meldrum's acid, the acidity of which is similar to that of barbituric acids, gave a complex mixture instead of the expected adduct [4].

The reaction of 1,3-diketone with reducing sugars had been studied in acidic media or under mildly basic conditions. When treated with 1,3-diketones or β-keto esters in the presence of acids, unprotected aldoses gave C-glycosylfurans or polyhydroxyalkylfuran derivatives. Garcia Gonzalez presented the first report on this topic in 1956, describing the condensation of D-glucose with acetylacetone promoted by zinc chloride [5]. The product was proved to be a tetrahydrofuranylfuran, consistently with an other investigation of the reaction 30 years later [6]. Using milder conditions, Misra and Agnihotri obtained an excellent yield of furyl C-glycosides **9** from hexoses

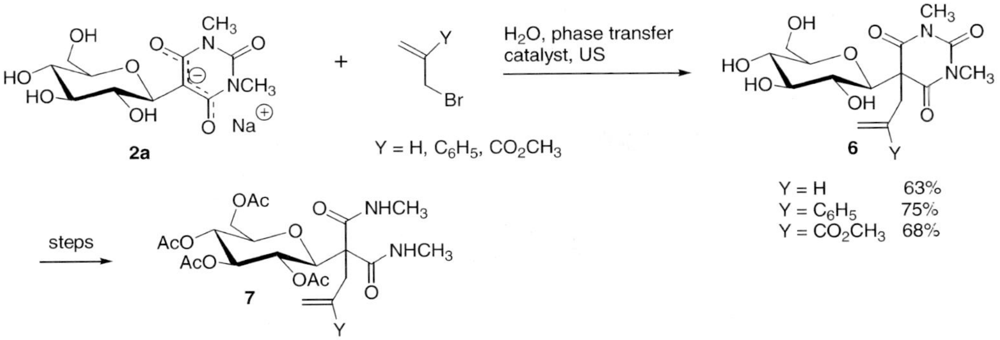

**Scheme 10.2** Synthesis of potential monomers for the preparation of polyvinylsaccharides.

and trihydroxyalkyl-substituted furans **8** from pentose using cerium(III) chloride as promoter in aqueous solution [7]. Later, Yadav et al. reported the same reaction catalyzed by indium(III) chloride [8]; in both cases, yields between 81 and 93% were obtained (Scheme 10.3).

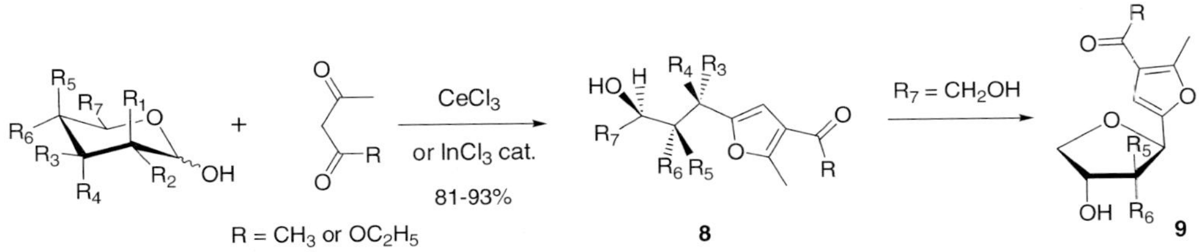

**Scheme 10.3** Synthesis of trihydroxyalkyl-substituted furan derivatives from pentose and furyl C-glycosides from hexoses.

The acid-promoted Knoevenagel condensation of pentane-2,4-dione or ethyl acetoacetate is followed by acyclization involving attack of OH-2 of the sugar moiety (i.e. $R_2$ or $R_3$) on the ketone and then aromatization to furan to afford **8**. In the case of pentoses ($R_7 = H$), the reaction does not proceed further. Since only the configurations at C-3 and C-4 are retained, D-ribose and D-arabinose gave the same polyhydroxylated furan derivatives **8** ($R_3 = R_5 = H$, $R_4 = R_6 = OH$, $R_7 = H$), which is a diastereoisomer of the product obtained from D-xylose ($R_4 = R_5 = H$, $R_3 = R_6 = OH$, $R_7 = H$). In the case of hexoses ($R_7 = CH_2OH$), the intermediate polyhydroxylated alkylfurans **8** undergoes cyclization to β-tetrahydrofuranylfuran derivatives **9** regardless of the stereochemistry at C-3 of the starting sugar, so that only the configurations at C-4 and C-5 of the starting sugar are maintained. As a consequence, D-glucose and D-mannose afforded the same product **9** ($R_6 = OH$, $R_5 = H$), diastereoisomer of the product obtained from D-galactose **9** ($R_5 = OH$, $R_6 = H$). The reaction was also applied with success to unprotected reducing disaccharides [7, 8]. By the same mechanism, the condensation in aqueous acetone in the presence of sodium carbonate with reducing amino sugars such as D-glucosamine afforded the pyrrole derivative **10** in 80% yield [9] (Scheme 10.4).

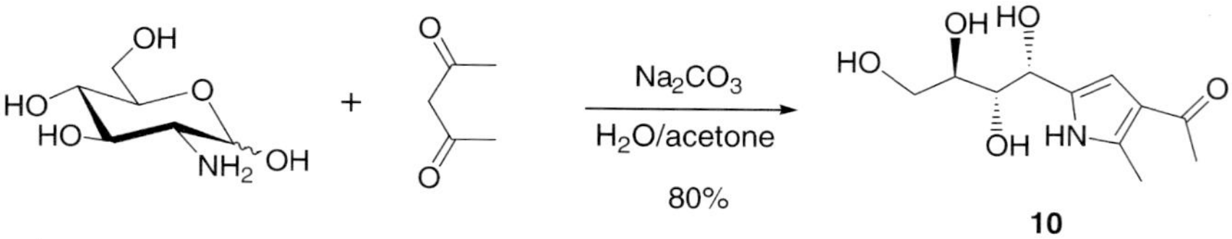

**Scheme 10.4** Reaction of D-glucosamine with pentane-2,4-dione in basic aqueous medium.

The reaction between dimedone **11** and unprotected sugars was studied by Sato's group [10]. Different products were obtained, depending on the promoter: the use of Sc(OTf)$_3$ resulted in the production of hydroxyalkyl-6,7-dihydrobenzofuran-4(5H)-one derivatives **12** in 86% yield, whereas scandium cation-exchanged montmorillonite gave hydroxyalky-3,3,6,6,-tetramethyl-3,4,5,6,7,9-hexahydro-1H-xanthene-1,8(2H)-dione **13** as the sole product in good yield (Scheme 10.5).

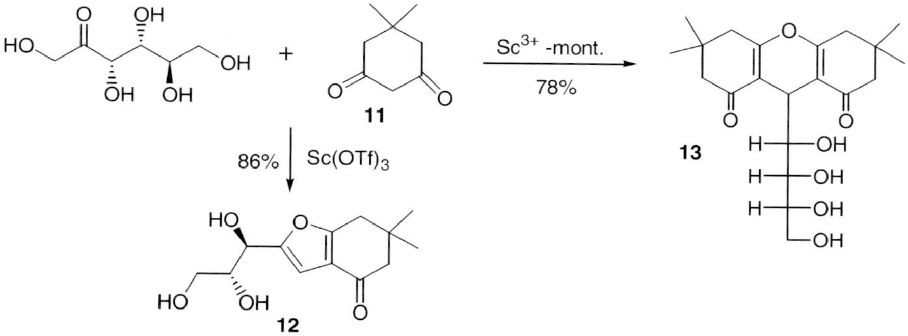

**Scheme 10.5** Reaction of D-ribose with dimedone; use of scandium cation-exchanged montmorillonite or scandium triflate as catalyst.

The quantitative one-step synthesis of β-D-C-glycosidic ketones by condensation of pentane-2,4-dione with unprotected sugars in alkaline aqueous media (Scheme 10.6) was explored by Lubineau's group [11]. The reaction in water of pentane-2,4-dione with D-glucose, D-mannose, D-galactose, D-maltose and D-cellobiose, in the presence of sodium hydrogencarbonate gave quantitatively in one-step the β-C-glycosidic

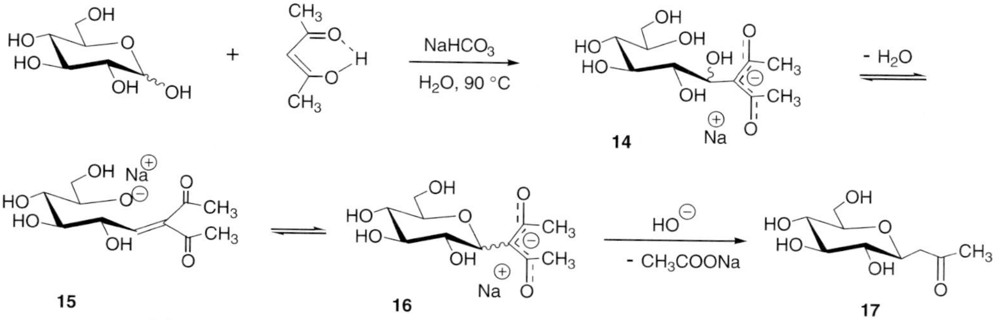

**Scheme 10.6** One step synthesis of a β-C-glucosidic ketone in aqueous medium.

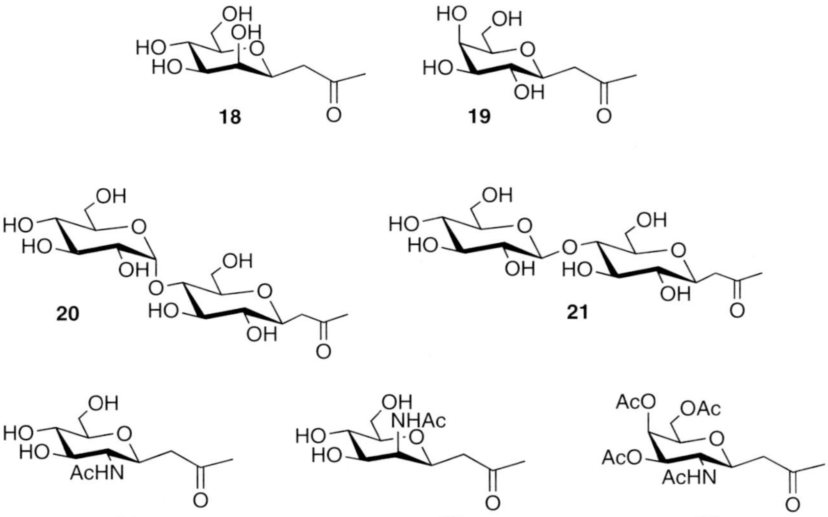

**Scheme 10.7** Various β-D-C-glycosidic ketones obtained from unprotected sugar through an aqueous Knoevenagel condensation.

ketones (Scheme 10.7, Table 10.1). It is worth pointing out that the D-glucosyl and D-cellobiosyl ketone derivatives **17** and **21** were previously prepared in seven and eight steps, respectively, and in low overall yields from commercial 2,3,4,6-tetra-O-benzyl-D-glucopyranose [12].

Also in this process, the reaction started with the Knoevenagel condensation of the β-diketone with the hemiacetalic sugar. Then, the β-elimination of water was followed by cyclization to the intermediate C-glycoside **16**, which underwent a retro-Claisen aldolization under basic conditions with concomitant sodium acetate elimination to give **17** (Scheme 10.6).

The selectivity for the formation of the β-D-pyranoside (equatorial) stereoisomer in the reaction originated from thermodynamic control. Indeed, after 24 h at room temperature, a mixture of the four possible α,β-furanoside and α,β-pyranoside stereoisomers, in which the α-furanoside predominated, was obtained. Under further equilibration in basic medium, almost exclusive formation of the β-glucopyranoside isomer was observed. This reaction applied to xylose is the first step in the industrial preparation by L'Oréal of Pro-Xylane, the first active cosmetic synthesized following the principles of green chemistry [13–15]. Studies on the use of other 1,3-dicarbonyl compounds as nucleophilic species were reported by Riemann et al. [16].

The condensation of pentane-2,4-dione with unprotected N-acetyl-D-gluco-, -manno-, and -galactosamine in alkaline aqueous media was also explored [17]. N-Acetyl-D- gluco- and -mannosamine gave, in good yield, a mixture of the two *gluco-* and *manno-C-*glycosidic ketones **22** and **23**, which were separated by crystallization after acetylation, whereas N-acetyl-D-galactosamine afforded the *galacto-C-*glycosidic ketone, which was isolated as its acetylated derivative **24** in 50% yield (Scheme 10.7; Table 10.1).

**Table 10.1** Reaction conditions and results for the condensation of unprotected sugars with pentane-2,4-dione in aqueous medium.

| Substrate | Conditions | Product | Yield (%) | Ref. |
|---|---|---|---|---|
| D-Glucose | H$_2$O, pentane-2,4-dione (1.2 equiv.) NaHCO$_3$ (1.5 equiv.), 6 h, 90 °C | 17 | 100 | [11] |
| D-Mannose | H$_2$O, pentane-2,4-dione (1.2 equiv.) NaHCO$_3$ (1.5 equiv.), 12 h, 90 °C | 18 | 100 | [11] |
| D-Galactose | H$_2$O–THF (2:1), pentane-2,4-dione (2 equiv.) NaHCO$_3$ (4 equiv.), 24 h, 90 °C | 19 | 90 | [18] |
| D-Maltose | H$_2$O, pentane-2,4-dione (1.2 equiv.) NaHCO$_3$ (1.5 equiv.), 12 h, 90 °C | 20 | 91 | [18] |
| D-Cellobiose | H$_2$O, pentane-2,4-dione (1.2 equiv.) NaHCO$_3$ (1.5 equiv.), 12 h, 90 °C | 21 | 100 | [11] |
| N-Acetyl-D-glucosamine | H$_2$O–THF (2:1), pentane-2,4-dione (2 equiv.) NaHCO$_3$ (4 equiv.), 24 h, 90 °C | 22, 23 (4:6) | 83 | [17] |
| N-Acetyl-D-mannosamine | H$_2$O–THF (2:1), pentane-2,4-dione (2 equiv.) NaHCO$_3$ (4 equiv.), 24 h, 90 °C | 22, 23 (1:3) | 40 | [17] |
| N-Acetyl-D-galactosamine | H$_2$O–THF (2:1), pentane-2,4-dione (2 equiv.) NaHCO$_3$ (4 equiv.), 2 h, 90 °C then Ac$_2$O, pyridine | 24 | 50 | [17] |

This straightforward method was applied to the preparation of C-glycosides related to D- and L-*glycero*-β-D-*manno*-heptoses in order to prepare isosteric analogs of nucleotide-activated sugars [19]. Lichtenthaler's group studied the application of this reaction to D-fructose. Surprisingly, a bicyclic product **26** was obtained in 27% yield by a double cycloketalization of the intermediate **25** [20] (Scheme 10.8).

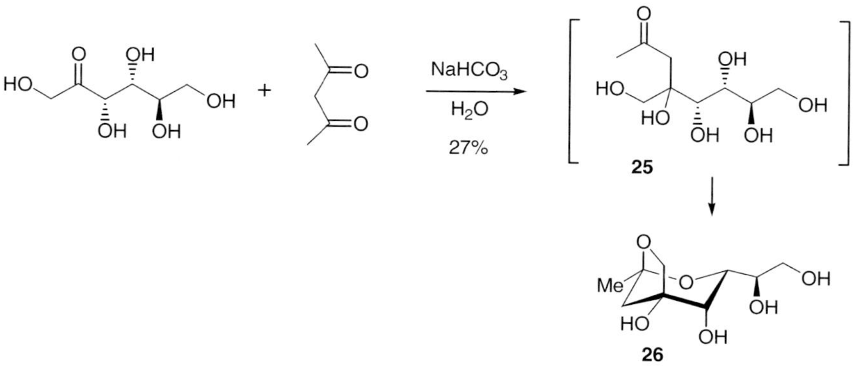

**Scheme 10.8** Aqueous condensation of D-fructose with pentane-2,4-dione.

In order to prepare carbohydrate-based amphiphiles by this efficient methodology, the use of non-symmetrical ketones **27** was tested in the condensation. It was assumed that the acetate elimination from the intermediate **28** should be favored towards the elimination of the shorter chain carboxylate residue, affording C-glycolipids rather than propanone C-glycoside [18] (Scheme 10.9).

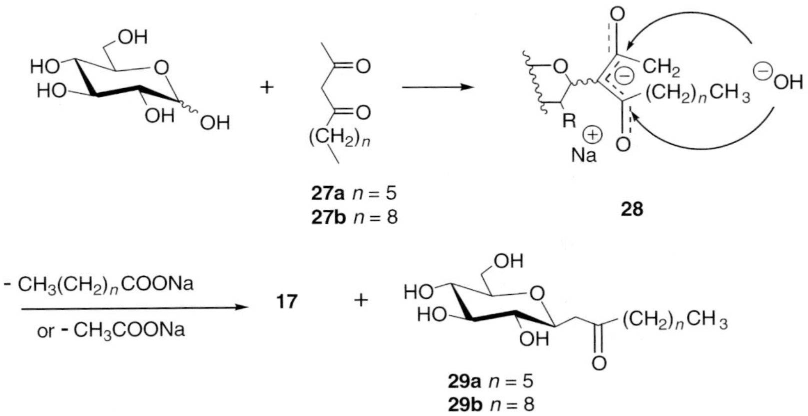

**Scheme 10.9** Use of unsymmetrical diketones in the preparation of C-glycolipids.

The reaction of glucose with decane-2,4-dione **27a** and NaHCO$_3$ in water at 90 °C for 48 h gave the C-glycolipid **29a** and the propanone C-glycoside **17** in a 2 : 1 ratio and 56% total yield, whereas the reaction with the less soluble tridecane-2,4-dione **27b** led to a complex mixture from which no major products could be isolated. Use of EtOH as a cosolvent gave a clean reaction (90–94% yield), presumably by making the long-chain diketones more soluble. Unfortunately, under these conditions no selectivity of the elimination of sodium acetate rather than sodium heptanoate or sodium decanoate was obtained. Therefore, symmetrical β-diketones **30a** and **30b** were used. The reactions of these symmetrical diketones with glucose in EtOH–H$_2$O as the solvent and NaHCO$_3$ as the base were complete and the C-glycolipids were isolated in 75% yield ($n = 5$, **29a**) and in 52% yield for the less soluble diketone ($n = 8$, **29b**). The condensation was also applied to D-maltose to obtain **31a** or **31b** [18] (Scheme 10.10).

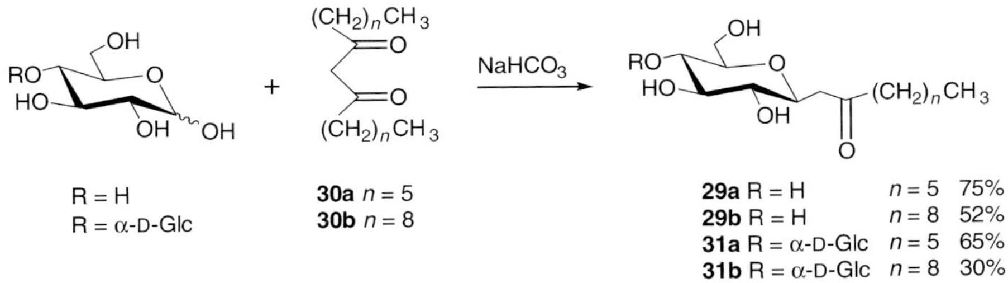

**Scheme 10.10** Synthesis of C-glycolipids using symmetrical fatty diketones.

## 10.2.2
### Barbier-type Reactions

The Barbier-type reaction has been largely studied in aqueous medium with various metals such as antimony [21], bismuth [22], cobalt [23], gallium [24], iron [25], magnesium [26], manganese [27] and mercury [28], although zinc- [29–33],

Table 10.2 First ionization potential of some metals.

| Metal | First ionization potential (eV) |
|---|---|
| In | 5.79 |
| Mg | 7.65 |
| Zn | 9.39 |
| Sn | 7.43 |
| Li | 5.39 |
| Na | 5.12 |

tin- [34–37] and indium-mediated [38, 39] allylations have received more attention, as outlined by Li [40].

The first report dealing with the application of the Barbier reaction to carbohydrate was published by Schmid and Whitesides in 1991 [41]. The condensation of the carbonyl group of the sugars and allyl bromide promoted by tin powder proceeded with high diastereoselectivities and good yields. At the same time, Li and Chan reported on the use of the more environmentally friendly indium to mediate Barbier-type reactions [42]. A comparative study of tin- and indium-mediated allylation of carbohydrates in aqueous media revealed that indium metal is more reactive than tin, giving fewer byproducts and higher diastereoselectivity [43]. The explanation of the superiority of indium with respect to tin comes from the comparison of the first ionization potential of the elements, as proposed by Li and Chan [42]. Compared with the elements near it in the periodic table, indium has the lowest first ionization potential, allowing very mild reaction conditions (Table 10.2).

Mediated by tin or indium, the stereochemical outcome of the allylation is *syn* when applied to unprotected sugars, since the free α-hydroxy group favors the chelation control model (Scheme 10.11).

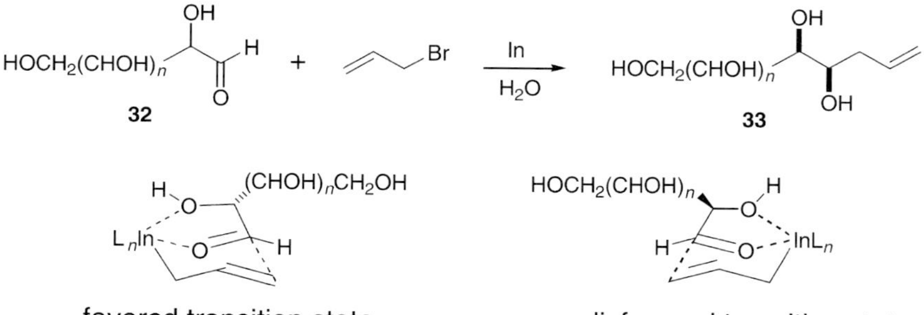

Scheme 10.11 *Syn* selectivity of the indium-mediated allylation of sugars in water through a chelating transition state.

Interestingly, protection of the α-hydroxy group as an alkoxyl will give the non-chelating transition state, allowing the reversal of the diastereoselectivity of the addition, as illustrated by the addition of allyl bromide to glyceraldehyde [44]. As

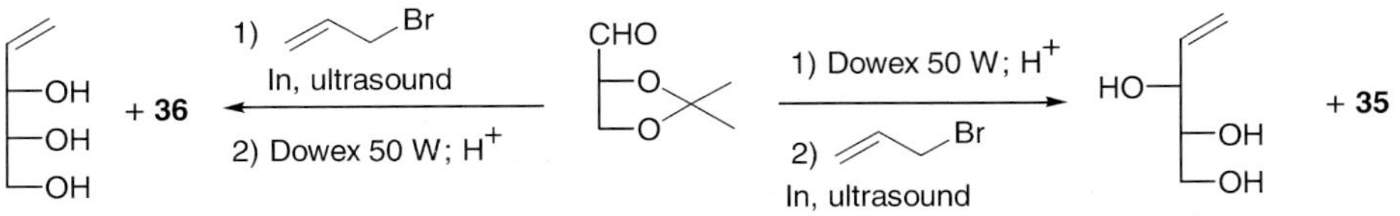

**35** **34** **36**

(**35** : **36** = 6.5 : 1) 85% (**35** : **36** = 1 : 2) 68%

**Scheme 10.12** Reversal of the diastereoselectivity in the indium-mediated addition by protection of the α-hydroxy group.

the reaction of 2,3-*O*-isopropylidene-D-glyceraldehyde **34** led to the *trans* isomer **35** as the main product after deprotection, the reaction carried out after deprotection of the isopropylidene afforded mainly the *syn* isomer **36** (Scheme 10.12).

The three-carbon elongated polyols **33** were further converted to 2-deoxyaldoses [41, 43, 44] (Scheme 10.13).

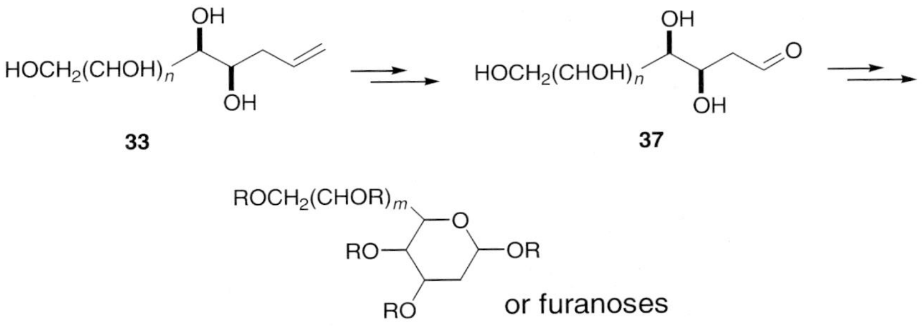

**33** **37** **38**

**Scheme 10.13** Synthesis of 2-deoxyaldoses from chain-elongated sugars.

Pamelund and Madsen described the acyloxyallylation of unprotected aldoses [45]. The reaction, carried out in ethanol or dioxane–water mixture with **40a** or **40b**, proceeded with good yields when mediated by indium, whereas other metals such as zinc, tin, bismuth, antimony, magnesium, or aluminum gave only traces of the elongated aldoses. Starting from pentoses, only two diastereoisomers were obtained among the four possible, the selectivity being better when the more bulky benzoate ester was used (Scheme 10.14; Table 10.3). The major diastereoisomers **41a–d** (Scheme 10.15) were subjected to ozonolysis to afford the corresponding heptoses **42a–d** (Scheme 10.16).

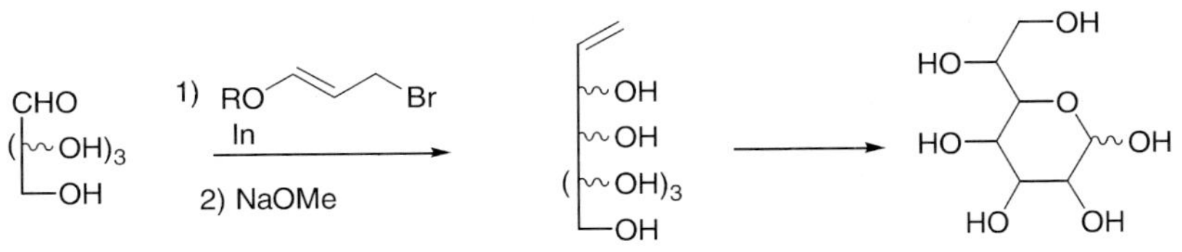

**39** **40a** R = Ac **40b** R = Bz **41a-d** **42a-d**

**Scheme 10.14** Use of 3-bromopropenyl esters in the indium-mediated chain elongation of aldoses.

**Table 10.3** Yields and selectivity of the chain elongation of aldoses using 3-bromopropenyl acetate or benzoate.

| Pentose 39 | R | Yield (selectivity) | Major diastereoisomer | Heptose |
|---|---|---|---|---|
| D-Arabinose | Ac | 85% (2.5:1) | 41a | 42a |
|  | Bz | 67% (4.0:1) |  |  |
| D-Lyxose | Ac | 60% (4.0:1) | 41b | 42b |
|  | Bz | 60% (8.5:1) |  |  |
| D-Ribose | Ac | 82% (1.5:1) | 41c | 42c |
|  | Bz | 77% (3.5:1) |  |  |
| D-Xylose | Ac | 78% (2.5:1) | 41d | 42d |
|  | Bz | 75% (4.5:1) |  |  |

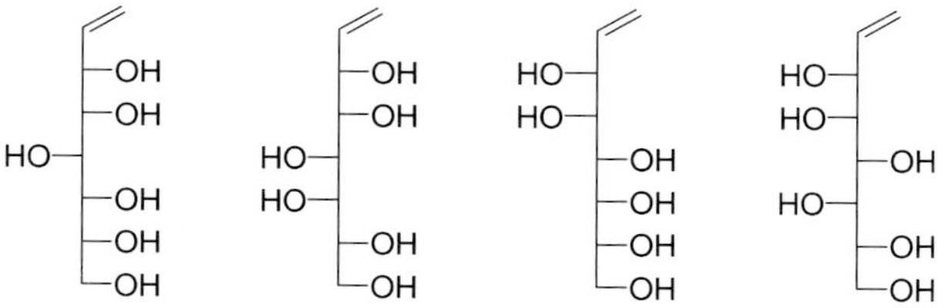

**Scheme 10.15** Major diastereoisomers obtained from various sugars (see Table 10.3).

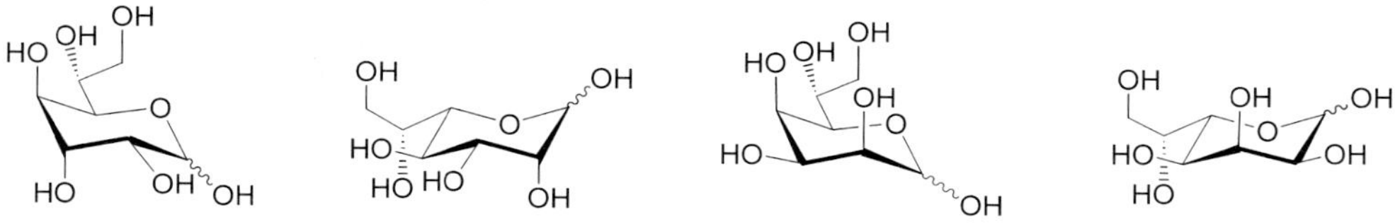

**Scheme 10.16** Heptoses obtained after ozonolysis of the chain-elongated sugars (see Table 10.3).

The major isomers had always a *syn* orientation between the hydroxy group previously located on C-2 of the pentose and the one originating from the attack of the aldehyde. Moreover, a *trans* orientation was always observed between the latter and the ester coming from the allylic derivatives. Hence, after further transformations, aldoses having *lyxo* configurations at C-2, C-3 and C-4 were obtained. Unfortunately, the minor isomer was not characterized. The condensation was also carried out using hexoses as starting material to afford, after ozonolysis of the major isomers, the octoses **43**, **44**, and **45** from D-galactose, D-glucose, and D-mannose, respectively (Scheme 10.17).

Ulosonic acids were obtained using 2-(bromomethyl)acrylates as allylic compounds. Indium-mediated addition of methyl 2-(bromomethyl)acrylate to unprotected D-mannose gave the homoallylic alcohol precursor of 3-deoxy-D-*glycero*-D-*galacto*-2-

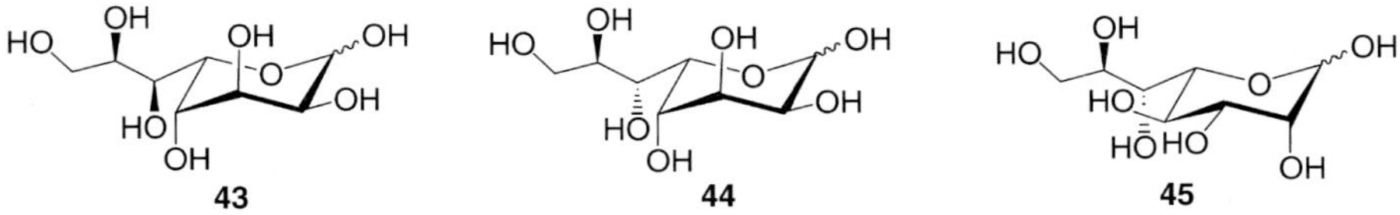

**Scheme 10.17** Octoses obtained from D-galactose, D-glucose, and D-mannose through indium-mediated chain elongation using 3-bromopropenyl benzoate and further ozonolysis.

nonulosonic acid (KDN) with a 6:1 diastereoselectivity (Scheme 10.18) [46]. Use of 2-(bromomethyl)acrylic acid instead of the methyl ester simplified the overall sequence, avoiding ester saponification [47]. Further improvements were reported by Warwel and Fessner by conducting the reaction under acidic conditions, giving exclusive formation of the *threo* product, and by developing a new purification method based on the conversion of the promoter derivatives into phosphate salts easily removed by filtration [48].

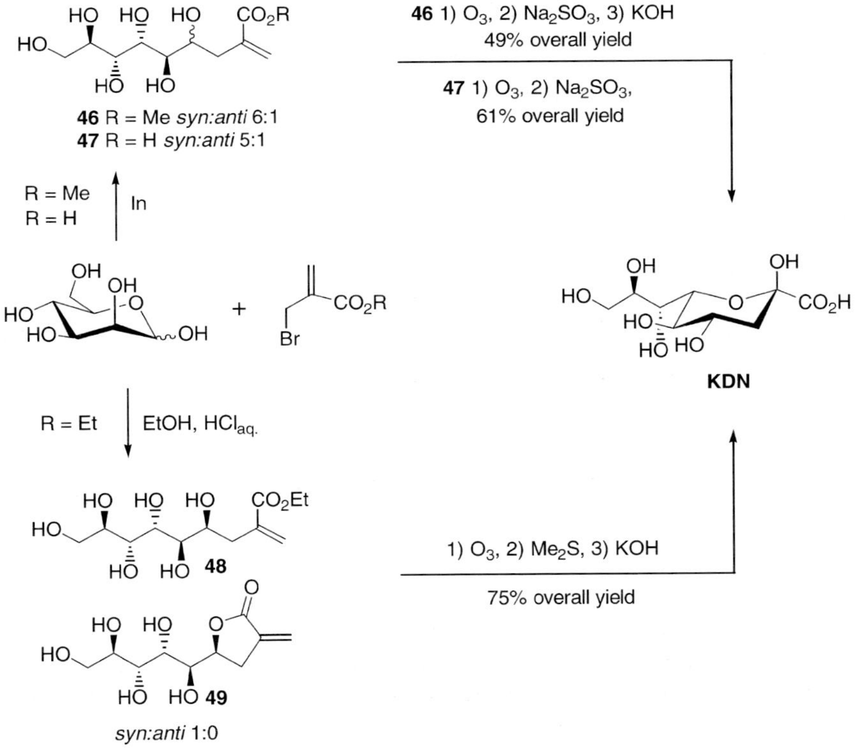

**Scheme 10.18** Preparation of KDN using 2-(bromomethyl)acrylates or acrylic acid.

Similarly, N-acetyl-D-mannosamine was converted into N-acetylneuraminic acid (Neu5Ac) [47, 49, 50] (Scheme 10.19), whereas the synthesis of 3-deoxy-D-manno-2-octulosonic acid (KDO) was achieved using the protected arabinose **50** to obtain the required configuration through a non-chelated transition state [51] (Scheme 10.20).

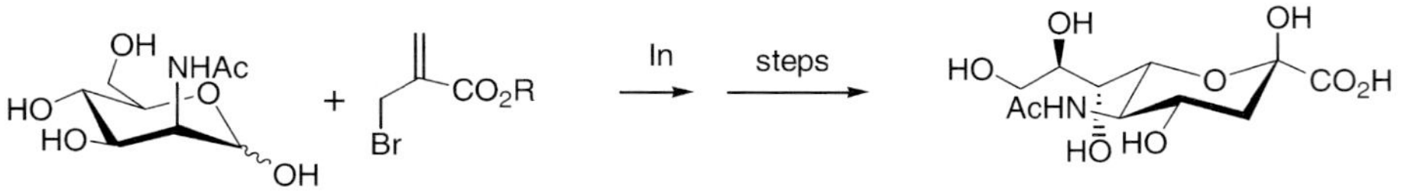

Neu5Ac

Scheme 10.19 Synthesis of Neu5Ac in aqueous medium.

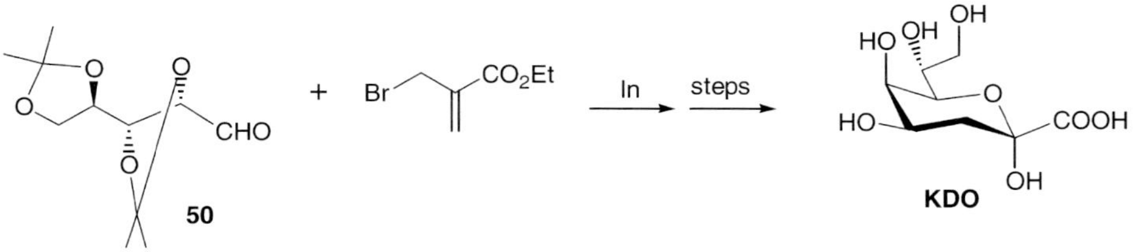

50 KDO

Scheme 10.20 Synthesis of KDO in aqueous medium.

Moreover, the use of dimethyl (3-bromopropen-2-yl)phosphonate as an allylic compound allowed the synthesis of **51** and **52**, phosphonic acid analogs of Neu5Ac and KDN, from *N*-acetyl-D-mannosamine and D-mannose, respectively [52] (Scheme 10.21).

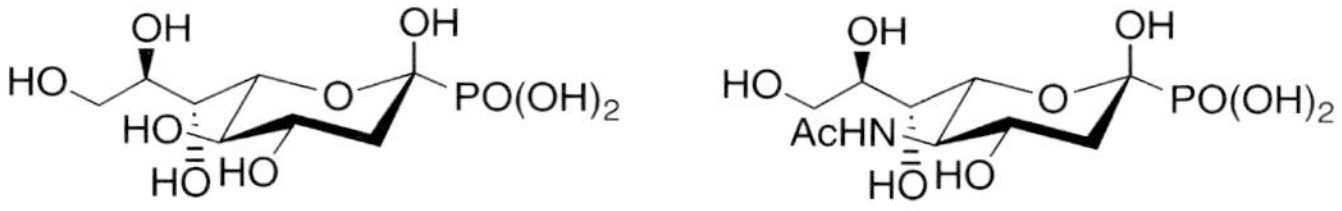

51 52

Scheme 10.21 Phosphonic acid analogs of Neu5Ac and KDN.

Lubineau's group described a very convenient protocol for the preparation of *C*-branched monosaccharides and *C*-disaccharides involving indium-promoted Barbier-type allylations in aqueous media [53–55]. The condensation of bromoenopyranosides **53–58** (Scheme 10.22) with aldehydes was studied.

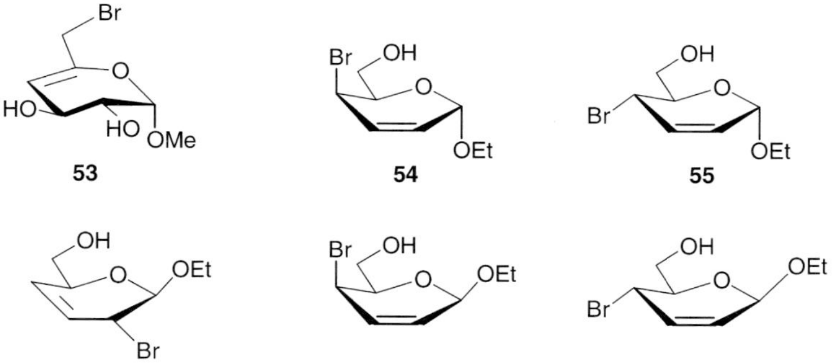

53 54 55 56 57 58

Scheme 10.22 Bromoenopyranosides used in indium-promoted allylation in aqueous media.

## 10.2 C—C Bond Formation Reactions

The reaction involving the 6-bromo derivative **53** and benzaldehyde was totally regio- and diastereoselective and afforded **59** in a good yield (Scheme 10.23).

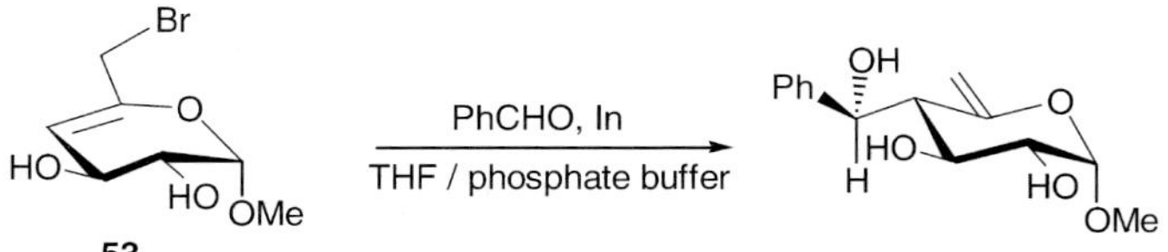

**Scheme 10.23** Example of a totally regio- and diastereoselective reaction involving a 6-bromoenopyranoside.

Using primary bromides, the formation of the allylindium was always stereospecific with retention of configuration. In the case of an axial bromine atom, the product of γ-alkylation *syn* relative to the bromine atom was obtained through the currently accepted six-membered cyclic transition state. Compound **54** gave **60** as the transition state in which the phenyl group is in an equatorial position is preferred (Scheme 10.24). In the case of **57**, an unfavorable steric interaction between the phenyl substituent in the equatorial position and the β-ethoxy group favors the formation of the isomer **61** resulting from the transition state in which the phenyl occupied the axial position (Scheme 10.25).

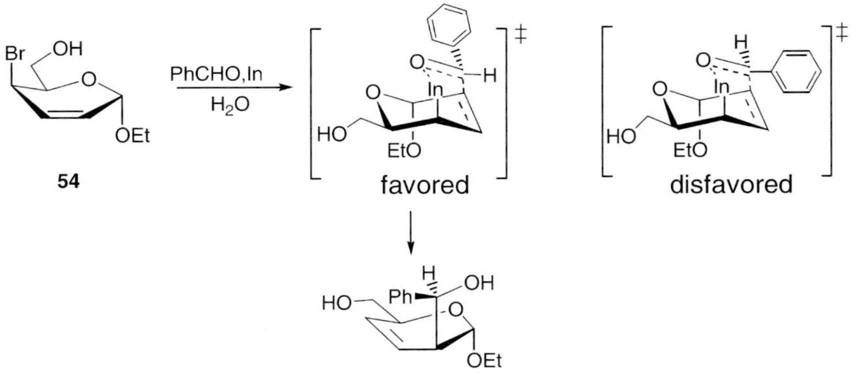

**Scheme 10.24** Control of diastereoselectivity through a cyclic transition state with the phenyl group in an equatorial position.

When the bromine is in an equatorial position, the allylic indium intermediate must undergo a half-chair inversion to react with the aldehyde. This energetic process competes with 1,3-allylindium migration or inversion of configuration of the indium species through elimination–recombination, so that α or γ selectivities could be obtained, depending on the allylindium 1,3-migration. As a result, **56** gave a 1:1 mixture of **60** and **62** (60%), and **55** a 1:4.5 mixture of **60** and **63**, respectively (70%), whereas **58** gave only **62** in a modest 22% yield (Scheme 10.26).

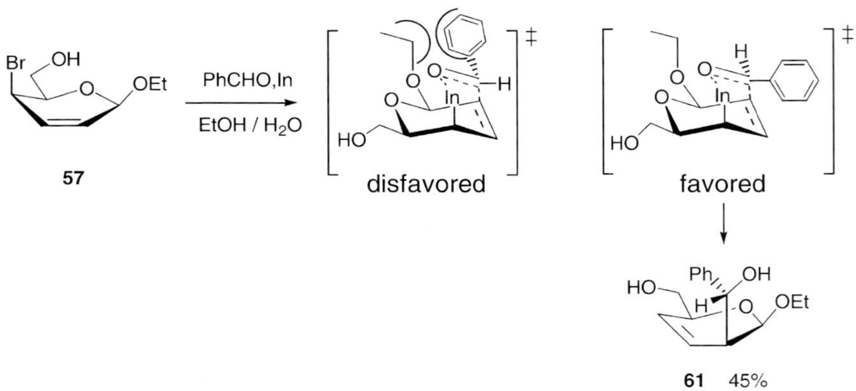

**Scheme 10.25** Control of diastereoselectivity through a cyclic transition state with the phenyl group in an axial position.

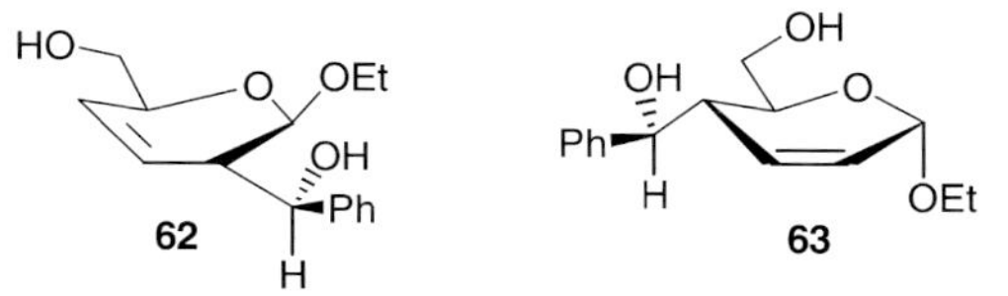

**Scheme 10.26** Compounds obtained from bromoenopyranosides by indium-promoted allylation in aqueous media.

This methodology was applied to sugar aldehydes. The reaction between **54** and 3-*O*-benzyl-1,2-*O*-isopropylidene-α-D-*xylo*-dialdose **64** gave the C-2 axial adduct **65** as a single diastereoisomer in 94% yield (Scheme 10.27).

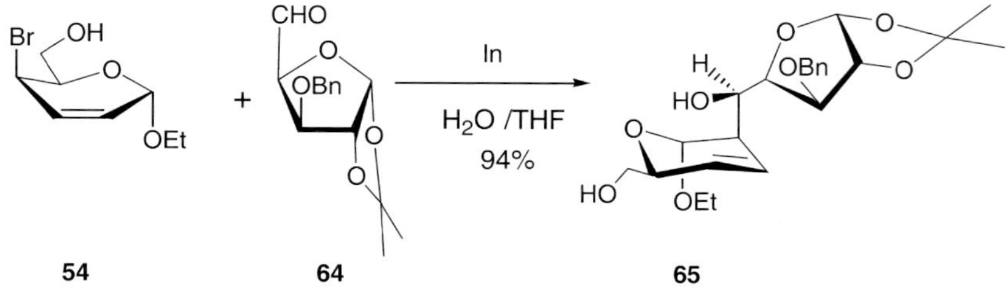

**Scheme 10.27** Indium-mediated condensation between a sugar aldehyde and a sugar allyl derivative.

The indium-mediated coupling of the formyl C-glycoside **66** with **53** was the key step in the synthesis of the C-disaccharides **68a** and **68b** (Scheme 10.28).

Another type of C-glycoside could be obtained in a two-step and one-pot procedure by indium-mediated addition of benzaldehyde to the allylic bromide **69** affording the *anti* allylation product **70**, which upon treatment with $K_2CO_3$ afforded the tetrasubstituted tetrahydrofuran **71** (Scheme 10.29) [56].

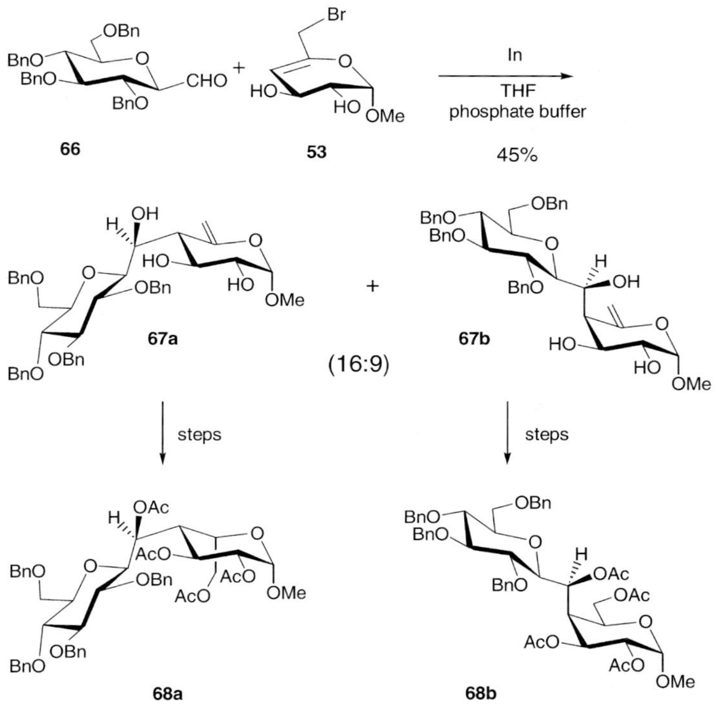

**Scheme 10.28** Access to C-disaccharides through an aqueous Barbier-type reaction as the key step.

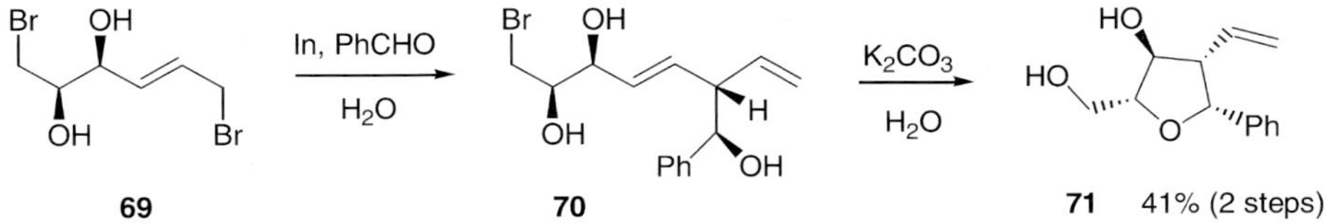

**Scheme 10.29** Synthesis of C-glycosidic tetrasubstituted THF in water.

## 10.2.3
## Baylis–Hillman Reactions

The Baylis–Hillman reaction, the condensation between an activated alkene and a carbonyl group, was found to be greatly accelerated in water compared with usual organic solvents. [57]. The reaction of various aldehydes with methyl acrylate was investigated in aqueous medium by Hu and co-workers [58]. They proposed efficient conditions such as the use of a greater loading of DABCO catalyst and a water–1,4-dioxane mixture as the solvent to improve the yields and rates of the reaction. Radha Krishna *et al.* applied these conditions to the reaction of sugar aldehydes **72a–c** with activated olefins **73** to obtain compounds **74** with diastereoisomeric excess varying from 36 to 86% and yields between 56 and 82% [59] (Scheme 10.30).

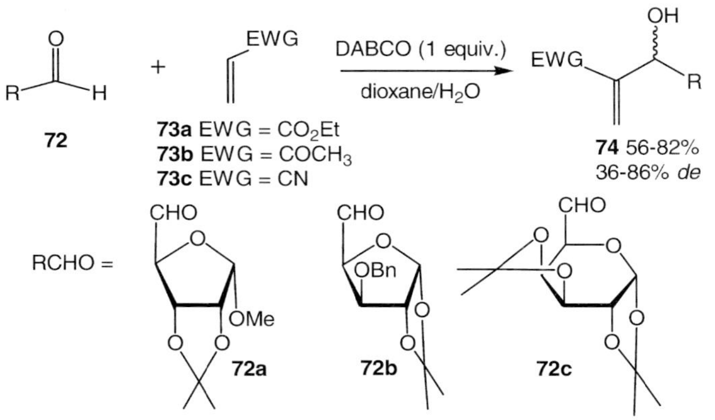

**Scheme 10.30** Baylis–Hillman reactions of sugar aldehydes with activated olefins.

## 10.2.4
### Electrophilic Aromatic Substitution Reactions

Electron-rich aromatics can react in water with unprotected sugars in the presence of an acid catalyst to give aryl C-glycosides (Scheme 10.31). The aqueous electrophilic aromatic substitution reactions of 3,5-dimethoxyphenol **75** with 2-deoxy-D-aldoses **74a–d** were reported by Toshima et al. using montmorillonite K-10 as insoluble acid [60].

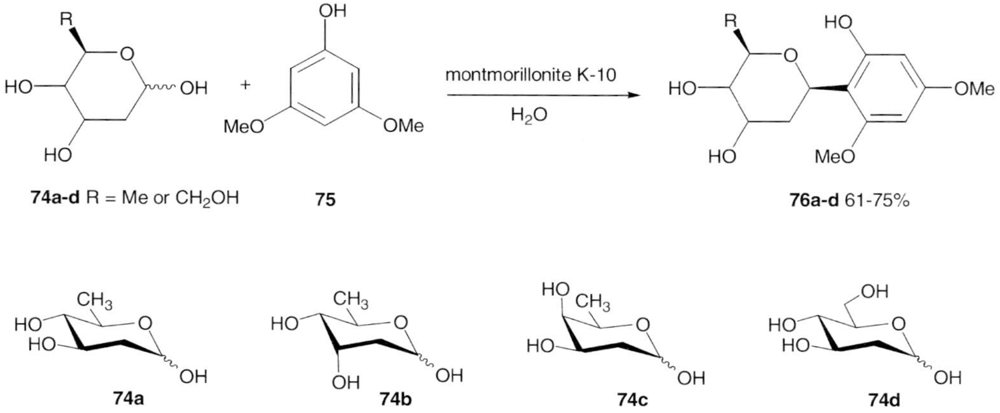

**Scheme 10.31** Montmorillonite-catalyzed C-glycosidation of electron-rich aromatics using 2-deoxy sugars.

The aryl C-glycosides **76a–d** were obtained with high β-D stereoselectivities (>99%) and in yields varying from 61 to 75%, comparable to those of the glycosidation using dry $CHCl_3$ as the solvent. Later, Sato's group reported the aqueous scandium(III) trifluoromethanesulfonate-catalyzed reaction of phloroacetophenone with unprotected D-glucose [61]. Mono- and bis-β-C-glucosides **78** and **79** were obtained in yields

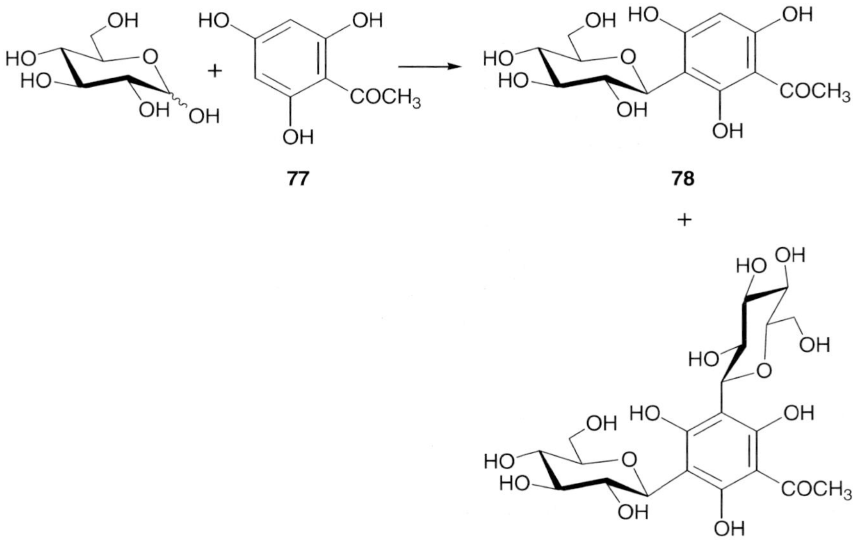

**Scheme 10.32** Mono-and bis-β-D-C-glucosides of phloroacetophenone (see Table 10.4 for conditions).

and selectivity which proved to depend on the conditions used (cosolvent and amount of promoter). Some examples are shown in Scheme 10.32 and Table 10.4.

As the conditions tested were unsuccessful with phenol and 2,4-O-dimethyl-protected phloroacetophenone, the authors claimed that a 1,3-diphenol structure is required and that the reaction was a "carbon–carbon bond-forming reaction of the 1,3-diketone with an aldehyde." Taking into account the above-mentioned results of Toshima et al. (see Scheme 10.31), it should rather be envisaged that the promoter, Sc(OTf)$_3$, is not strong enough to allow the reaction involving the phenol or the dimethyl-protected phloroacetophenone derivatives, since these compounds are less activated towards the electrophilic aromatic substitution. The reaction was applied by the same group to the synthesis of another C-glucosylflavonoid [62].

**Table 10.4** Conditions and yields for the C-glucosidation of phloroacetophenone.

| Sc(OTf)$_3$ (equiv.) | Solvent | Time (h) | Total yield (%) (78/79) |
|---|---|---|---|
| 0.2 | CH$_3$CN–H$_2$O 2:1 | 8 | 72 (60:40) |
| 0.1 | CH$_3$CN–H$_2$O 2:1 | 8 | 62 (77:23) |
| 0.2 | CH$_3$CN–H$_2$O 1:2 | 8 | 71 (66:34) |
| 0.2 | EtOH–H$_2$O 2:1 | 9 | 81 (53:47) |
| 0.4 | EtOH–H$_2$O 2:1 | 6.5 | 79 (49:51) |

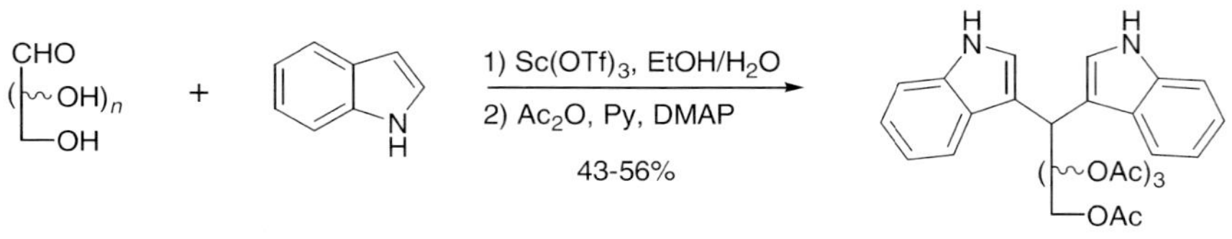

**Scheme 10.33** Scandium triflate-catalyzed reaction of indole with aldoses.

Sato's group also reported the use of indole in a similar reaction [63, 64]. In this case, the π-electron-excessive heterocycle gave the products of double addition **80**, as already observed by Wang and co-workers for the lanthanide-catalyzed reaction of indole derivatives with aldehydes or ketones in aqueous solution (Scheme 10.33) [65].

## 10.2.5
### Mukaiyama Aldol Reaction

The reaction between a silyl enol ether and an aldehyde in an organic solvent requires the presence of a stoichiometric amount of titanium tetrachloride [66] (Mukaiyama's conditions), leading mainly to the *anti* addition product, or the use of high pressure [67] (without acidic promoter) affording preferentially, in this case, the *syn* hydroxy ketone. In aqueous medium the reaction proceeds under atmospheric pressure without any catalyst, with the same *syn* selectivity as under high pressure [68, 69]. Kobayashi and co-workers developed water-tolerant Lewis acids in order to improve the yields and therefore the scope of this aqueous aldolization [70–72]. The use of Lewis acid–surfactant combined catalysts promotes the reaction only in water and avoids the need for an organic cosolvent [72]. Lubineau's group developed a synthesis of C-glycosides and C-disaccharides based on this aqueous condensation as the key step [73]. The reaction with of the formyl C-glycoside **66** and the trimethylsilyl enol ether **81** derived from acetophenone was found to give the best results when carried out in 2 : 1 $H_2O$–THF with $Yb(OTf)_3$ as the catalyst. The two C-glycosides **82** and **83** were isolated in a 60 : 40 ratio and 90% total yield (Scheme 10.34).

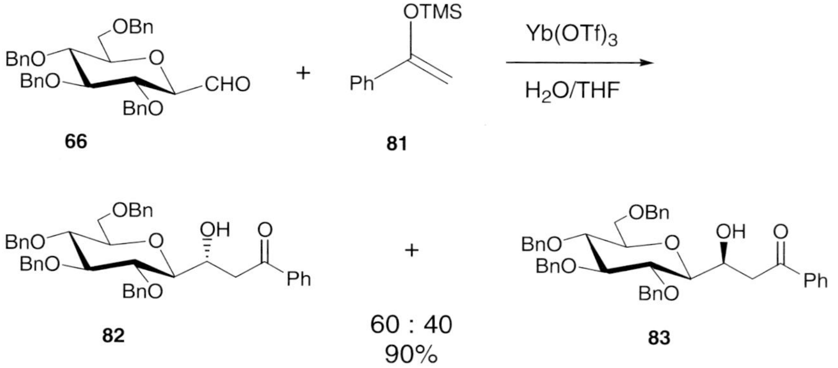

**Scheme 10.34** Mukaiyama aldol reaction involving a formyl C-glucoside.

When applied to the sugar silyl enol ether **84**, the condensation led to **85** and **86**, two of the four possible diastereoisomers, in a 59 : 41 ratio and 95% yield (Scheme 10.35).

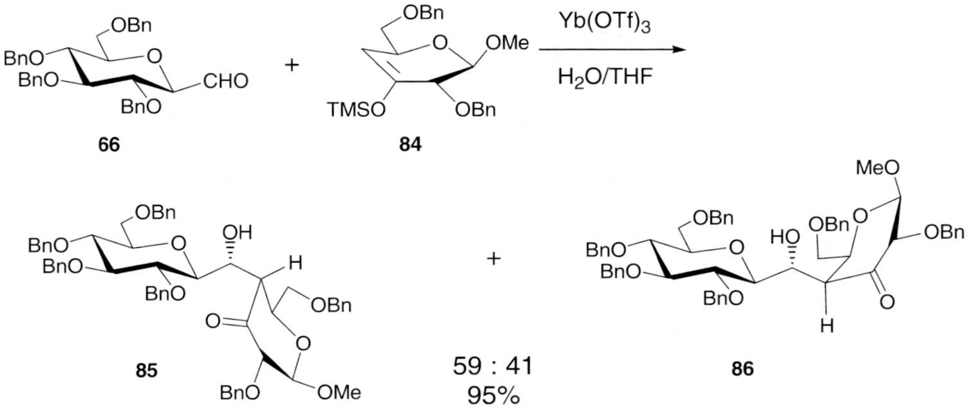

**Scheme 10.35** Synthesis of C-disaccharides using Mukaiyama aldol reaction in aqueous medium.

## 10.3
## C—N Bond Formation Reactions

### 10.3.1
### Glycosylamines and Glycamines

The reaction of aldoses with ammonia or primary amines gives glycosylamines. These compounds are generally not stable in neutral or slightly acidic aqueous media but can be efficiently prepared in aqueous ammonia in the presence of 1 equiv. of ammonium hydrogencarbonate [74]. The reaction pathway involves the formation of the imine **87** that gives rise to a ring closure, affording the N-glycosylamine **88**. In the presence of ammonium hydrogencarbonate, the latter can lead to β-D-glycosylcarbamate **89**, or to the di-β-D-glycosamine **90**, when reacting with ammonia at a temperature above 50 °C (Scheme 10.36). The equilibrium shown in Scheme 10.36 being totally shifted to the formation of β-D-glycosamine **88** by lyophilization, a quantitative yield was obtained. This method was also applied to D-galactose, D-lactose, D-cellobiose, and D-maltose, giving quantitatively the corresponding β-D-glycosamines [74].

The N-glycosylamines were transformed into N-acylglycosylamines. The reaction of β-D-glucosylamine with acetic anhydride was immediate at 0 °C and afforded N-acetyl–β-D-glucopyranosylamine **91**, which was further acetylated to give peracetylated β-D-glucosamine **92** in 78% yield from glucose on a 100 g scale

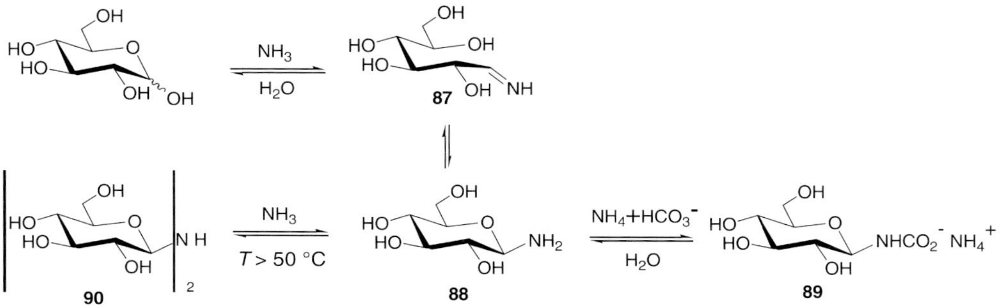

# 10 Functionalization of Carbohydrates in Water

**Scheme 10.36** Reaction of glucose with aqueous ammonia in the presence of ammonium hydrogencarbonate.

(Scheme 10.37). This compound turned out to be an excellent bleaching booster for detergents [75].

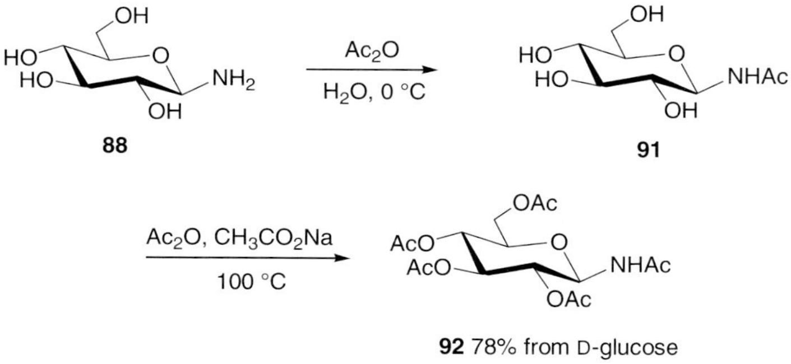

**Scheme 10.37** Synthesis of peracetylated β-D-glucosamine, an efficient bleaching booster for detergents.

Glycosylamines were also reacted with fatty acyl chlorides in EtOH–$H_2O$ to prepare N-octanoyl-, N-decanoyl-, N-lauryl-, and N-myristoyl-β-D-glucopyranosylamine in 40–56% yield [74]. Alternatively, reductive amination followed by either acylation or alkylation is a widely used method for the synthesis of amphiphilic glucamide-type surfactants from many different sugars such as glucose and lactose [76–80]. From the readily available disaccharide isomaltulose, produced by bioconversion from sucrose on the ton scale, isomaltamides having good surfactant properties were prepared [78] (Scheme 10.38).

The reaction of unprotected aldoses with aniline derivatives in aqueous media was investigated by Maugard and co-workers [81]. Under optimized conditions, they prepared aryl-β-D-N-glycosides in yields varying from 30 to 55% (Scheme 10.39, Table 10.5).

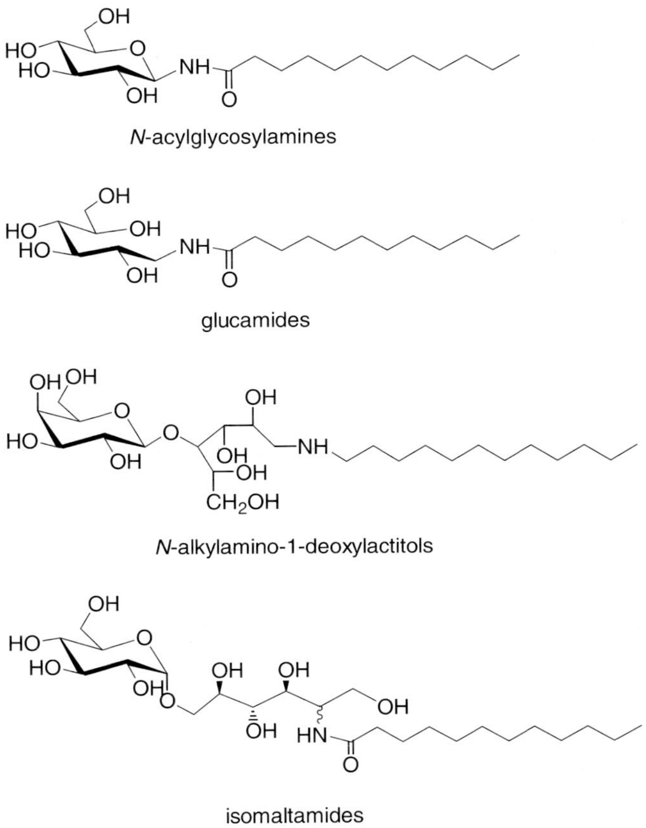

**Scheme 10.38** Glycosylamine- and glycamine-based surfactants.

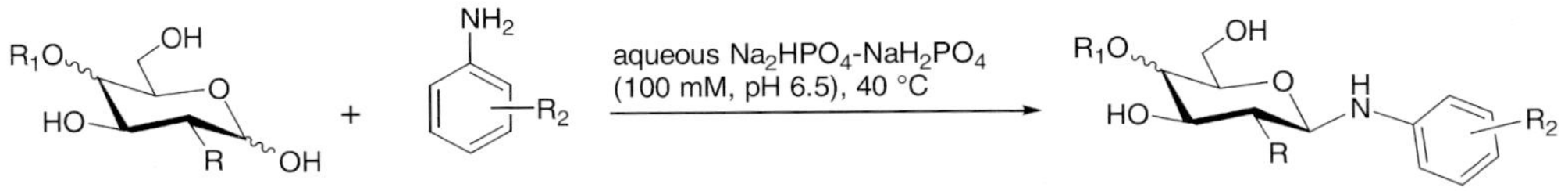

**Scheme 10.39** N-Glycosylation of various compounds (see Table 10.5).

## 10.3.2
## Aza Sugars

A very efficient asymmetric synthesis of the aza sugar **93e** was reported by Lindström et al. [82]. The strategy involved four steps, all carried out in water as the solvent, without the need for a protecting group. An asymmetric dihydroxylation afforded diol **93b** in good yield and enantioselectivity. This compound was then transformed into the epoxide **93d** via the allylic alcohol **93c**. The last step involved the nucleophilic displacement of the bromine with ammonia followed by intramolecular ring opening of the

**Table 10.5** Yields of N-glycosylation of various compounds.

| R | R$_1$ (orientation) | R$_2$ | Yield (%) |
|---|---|---|---|
| OH | H (equatorial) | H | 55 |
| OH | H (axial) | H | 51 |
| OH | Gal (β-1) | H | 52 |
| NHAc | H (equatorial) | H | 37 |
| OH | H (equatorial) | m-CH$_2$OH | 40 |
| OH | H (axial) | m-CH$_2$OH | 38 |
| OH | H (equatorial) | p-CH$_2$CH(CO$_2$H)NH$_2$ | 54 |
| OH | H (axial) | p-CH$_2$CH(CO$_2$H)NH$_2$ | 34 |
| OH | Gal (β-1) | p-CH$_2$CH(CO$_2$H)NH$_2$ | 40 |
| NHAc | H (equatorial) | p-CH$_2$CH(CO$_2$H)NH$_2$ | 41 |
| NHAc | H (axial) | p-CH$_2$CH(CO$_2$H)NH$_2$ | 41 |
| NHAc | Gal (β-1) | p-CH$_2$CH(CO$_2$H)NH$_2$ | 30 |

epoxide giving the aza sugar **93e** in 60% overall yield from diene **93a** (Scheme 10.40). This synthesis was a great improvement on a previously described strategy involving seven steps from a protected galactosides and affording **92e** in 21% overall yield [83].

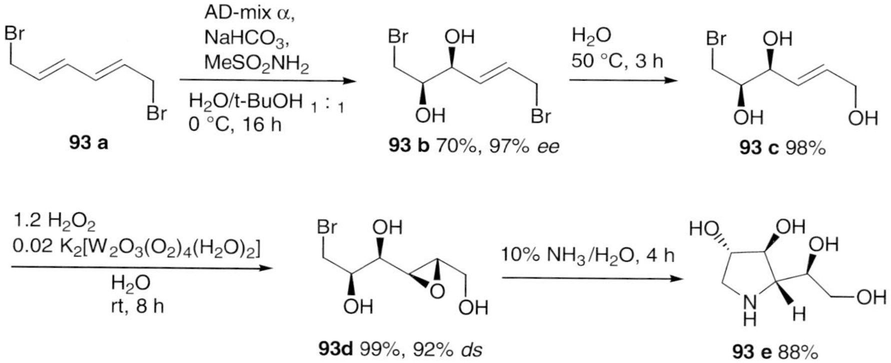

**Scheme 10.40** A four-step efficient asymmetric synthesis of an aza sugar in water.

## 10.4
## Functionalization of Hydroxy Groups

Carbohydrates are indeed a considerable reserve of organic molecules, and therefore their chemistry has attracted intense efforts for the synthesis of a variety of compounds such as surfactants and polymerizable derivatives [84–90]. The interest in sugars as substrates for the preparation of derivatives of high biological relevance has led to elegant multi-step syntheses, mostly based on subtle protection–deprotection strategies and the use of specialized catalysts and reagents. In parallel with this, some research has been carried out in the context of simple direct transformations

## 10.4 Functionalization of Hydroxy Groups

towards functional derivatives, notably in the field of surfactants, but not exclusively since polymers, complexing agents, and other applications have also been considered. Unprotected carbohydrates being polyols with generally good solubility in water, it seems reasonable that water, which is known to be an interesting medium for chemistry because of its non-toxicity, its good solvating properties, and the peculiar activation which can be promoted by hydrophobic effects, should be a solvent of choice for their chemistry. Also, water is the solvent of living systems, and is therefore the medium in which bioconversions can be performed. Important processes dealing with this latter aspect will be briefly mentioned.

### 10.4.1
### Esterification, Etherification, Carbamation: the Example of Sucrose

Sucrose (β-D-fructofuranosyl α-D-glucopyranoside, Scheme 10.41), with its glucose and fructose moieties connecting both anomeric carbon atoms, is a typical example of "non-reducing" sugar and its chemistry is mostly connected with the transformations of some of its eight hydroxy groups (three primary and five secondary). Many products can arise from a simple reaction, resulting from different degrees of substitution (mono-, di-, etc.) and from different positions of the substituents on the disaccharidic skeleton [91–94]. The control of the two types of selectivity, that is, the degree of substitution and the regiochemistry, is therefore the main issue, with influence due to the use of water in some cases, as will be seen in the following. In sucrose, carbon atoms C-2, C-1′, and C-3′ are linked to either one or the other anomeric centers of sucrose. The electron-withdrawing effects and the hydrogen bonds connecting OH–2, OH–1″ and OH–3′ make these latter more reactive. In solution, a hydrogen bond between O-2 and H-O-1′ or O-H-3′ [95–99] has been shown to monitor the conformation of the molecule. The conformational behavior of sucrose in water, involving a bridging water molecule between O-2 and OH-1′, has also been determined by NMR spectroscopy and molecular modeling [100–102] (Scheme 10.41). This results in an increased acidity for sucrose ($pK_a = 12.6$) compared with "normal" alcohols, and even compared with water [103–105]. Some scales of relative acidity of the hydroxy groups of sucrose, converging on the highest acidity for OH-2, can either be established by semiempirical calculations [106] or be deduced from the distribution of the regioisomers after selective substitution, as illustrated by many examples in the following

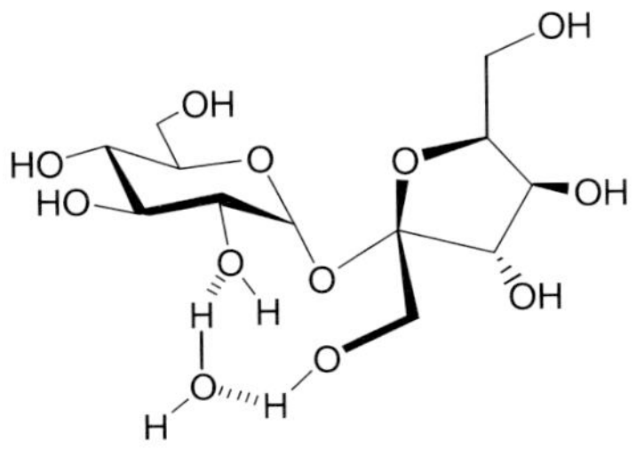

Sucrose

**Scheme 10.41** Conformational behavior of sucrose in water [101, 102].

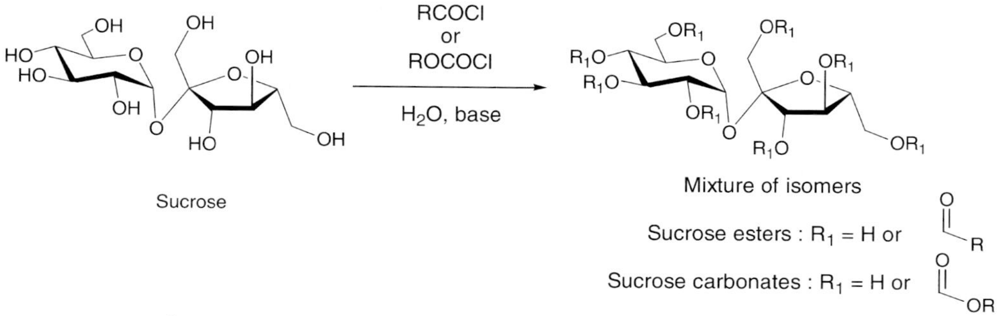

**Scheme 10.42** Preparation of sucrose esters and carbonates in aqueous basic medium.

sections. The acidity of sucrose has also been discussed in various reviews dealing with the relationships between structure and properties (such as acidity and reactivity) of carbohydrates that point out the higher reactivity of an equatorial hydroxy group vicinal to an axial oxygen atom, and pointing out the "α-glucoside" behavior for sucrose [107].

The possibility of using water as an alternative solvent for the esterification of sucrose instead of organic solvents such as DMF, Me$_2$SO, or pyridine has been studied. In the case of short esters such as methacryloyl esters, good yields of sucrose esters can be obtained in an aqueous or hydro-organic medium [108, 109] (Scheme 10.42). High concentrations of sugar, thus limiting the amount of water, decrease the competitive hydrolysis of the acid chloride. For acid chlorides having a longer alkyl chain, such as octanoyl chloride, the outcome of the reaction in basic aqueous medium proved to depend on the pH and the starting concentration. At low water content, fair yields of sucrose esters, mostly polysubstituted, are obtained. Sucrose strengthens the water structure, thus increasing the polysubstitution pathway driven by the hydrophobic effect [110]. When stoichiometric 4-(dimethylamino)pyridine (DMAP) is used as the base, polysubstitution was not observed upon increasing the starting sucrose concentration, likely because the acylpyridinium intermediate might be less sensitive to the entropy-driven aggregation effect. For higher sucrose concentrations, very high conversion of the fatty chain towards polysubstituted sucrose fatty esters were observed. Using NaOH (at pH 10) as the stoichiometric base with a catalytic amount of DMAP led to very good yields even in dilute solutions, with extensive polysubstitution for high sucrose concentrations. This tendency was even stronger when higher fatty acid chlorides were used. In the presence of an organic cosolvent (1 : 1 water–THF or water–2-propanol mixture), laurates and palmitates having a low average degree of substitution were obtained due to lower hydrophobic aggregation, showing that it is possible to direct the reaction towards more or less substituted products by choosing the medium with respect to its ability to develop hydrophobic effects [111, 112]. (Table 10.6)

Chloroformates, known to provide either cyclic carbonates or alkyloxycarbonyl derivatives [113–116], can also be used in aqueous media. For example, addition of allyl chloroformate to an aqueous sucrose solution led to monosubstituted allyloxycarbonylsucrose derivatives. A 40% (w/w) sucrose solution in water at pH 10 and a reaction temperature of 2 °C was the best balance between esterification and hydrolysis of the products or the starting chloroformate, with no effect of the

**Table 10.6** Esterification of sucrose in aqueous medium with fatty acid chlorides[a].

| | | Sucrose ester (%) | | | |
|---|---|---|---|---|---|
| R | Solvent | Mono- | Di- | Poly | Total |
| $C_7H_{15}$ | Water | 52 | 11 | 9 | 72 |
| $C_{11}H_{23}$ | Water | 2 | 6 | 12 | 20 |
| $C_7H_{15}$ | Water–THF (1:1) | 67 | 2 | 0 | 69 |
| $C_{11}H_{23}$ | Water–THF (1:1) | 58 | 2 | 0 | 60 |
| $C_{15}H_{31}$ | Water–THF (1:1) | 45 | 15 | 0 | 60 |

[a] 10% (w/w) sucrose solution, 0.1 equiv. of acid chloride, pH 10 (NaOH), 0.01 equiv of DMAP.

**Table 10.7** Formation of monosubstituted sucrose carbonates from chloroformates (0.25 equiv) in aqueous medium.

| R | Sucrose concentration (% w/w) | Solvent | pH | DMAP equiv./chloroformate | Temperature (°C) | Yield (%) |
|---|---|---|---|---|---|---|
| Allyl | 40 | Water | 7 | – | 23 | 55 |
| Allyl | 40 | Water | 10 | – | 2 | 77 |
| n-Octyl | 10 | Water | 10 | 0.1 | 20 | 8 |
| n-Octyl | 10 | Water | 9 | 0.01 | 0 | 13 |
| n-Octyl | 10 | Water–THF (1:1) | 10 | 0.1 | 1 | 72 |
| n-Octyl | 10 | Water–2-propanol (1:1) | 10 | 0.1 | 1 | 76 |

concentration of sucrose in water on the degree of substitution (Table 10.7). In the case of octyl chloroformate, polysubstitution was observed as for the esterification of sucrose with fatty acid chlorides, and again, using a cosolvent and catalytic amounts of DMAP allowed the above tendency to be counteracted and gave acceptable yields of mono-O-octyloxycarbonylsucrose [117].

Isocyanates were shown to lead to sucrose carbamates under similar conditions (Scheme 10.43). Reaction with octyl isocyanate in pure water led to polysubstituted derivatives, whereas in water–THF or water–alcohol mixtures, good yields of low-substituted sucrose octyl carbamates are obtained [118]. Alcohol carbamation and isocyanate hydrolysis (and also subsequent carbamic acid and urea derivative formation) were observed as side reactions, with variations depending on the nature of the alcohol cosolvent. Hindered alcohols such as 2-propanol and *tert*-butanol gave better conversion of low-substituted products (Table 10.8).

In the case of the synthesis of amphiphilic hydroxyalkyl ethers by addition of sucrose to a fatty epoxide (Scheme 10.44) [119–121], moderate to fair yields were obtained. A key parameter was found to be the heterogeneity of the medium, which changes during the reaction because of the surfactant properties of the compounds. Tertiary amines appeared to be good basic catalysts for this reaction, and the presence of cetyltrimethylammonium bromide (CTAB) as phase-transfer catalyst was beneficial to the formation of the desired low-substituted products, which are

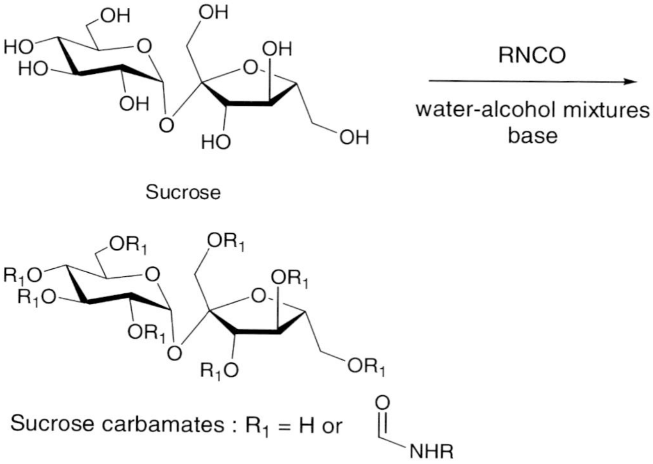

**Scheme 10.43** Reaction of sucrose with alkyl isocyanates.

Sucrose carbamates: $R_1$ = H or $\overset{O}{\underset{}{\|}}$C-NHR

**Table 10.8** Reaction of sucrose and octyl isocyanate in water–cosolvent media[a].

| Co-solvent | Mono- + di- (%) |
| --- | --- |
| – | 9 |
| THF | 58 |
| 2-Propanol | 58 |
| tert-Butanol | 58 |
| EtOH | 45 |
| MeOH | 12 |
| 2-Butanol | 22 |
| 1-Butanol | 10 |

[a][Sucrose] = 10% (w/w) in the mixture of solvents; [isocyanate] = [NaOH] = 0.5 equiv.; room temperature; cosolvent (1 : 1) with water.

surfactants with increased stability, and therefore with wider applications, compared with esters [122, 123] (Table 10.9). The strongly basic quaternary ammonium type 1 anion-exchange resin A26 (OH$^-$ form), associated with CTAB, was shown also to produce sucrose ethers efficiently. Mixtures of water and dimethyl sulfoxide (DMSO) were also used, at lower reaction temperatures. which permitted better stability and recyclability of the catalyst [124]. Hydroxypropyl ethers designed for food applications were obtained by etherification of sucrose with propylene oxide in aqueous basic medium [125]. From diepoxides, cross-linked polymers having hydrogel properties were prepared [123]. Cyanoethylated sucrose derivatives and the corresponding carboxylated compounds were also prepared [126].

An interesting alternative method for the synthesis of carbohydrate ethers is the use of the palladium-catalyzed telomerization of butadiene, a very satisfactory

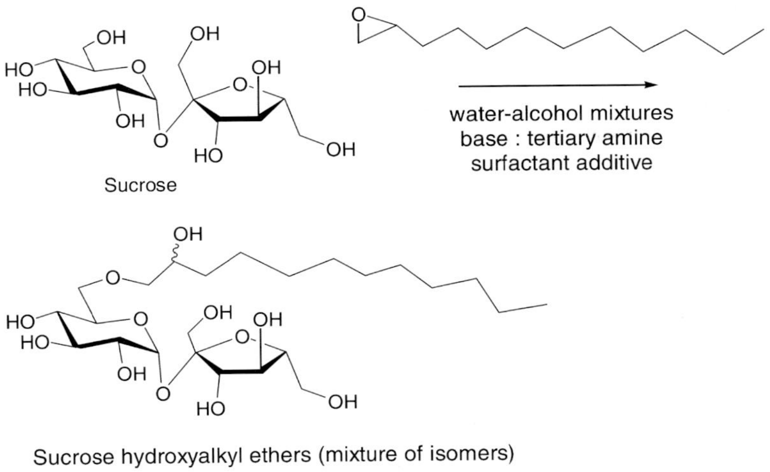

## 10.4 Functionalization of Hydroxy Groups

**Scheme 10.44** Reaction of sucrose with fatty epoxides.

Sucrose + fatty epoxide → Sucrose hydroxyalkyl ethers (mixture of isomers)

Conditions: water-alcohol mixtures, base: tertiary amine, surfactant additive.

**Table 10.9** Sucrose etherification versus epoxide hydrolysis.

| Base[a] | Additive | Sugar/epoxide (mol/mol) | Mono- + diethers (%) | Dodecanediol (%) |
|---|---|---|---|---|
| KOH | None | 2/1 | 0 | 0 |
| KOH | CTAB | 2/1 | – | 56 |
| NMM | None | 1/1 | 38 | 14 |
| NMM | None | 2/1 | 44 | 11 |
| NMM | None | 4/1 | 36 | 12 |
| Me$_2$NBu | None | 2/1 | 47 | 7 |
| NMM | CTAB | 2/1 | 55 | 7 |
| NMM | CTAB | 4/1 | 53 | 6 |
| Me$_2$NBu | CTAB | 2/1 | 52 | 6 |

[a]NMM = N-methylmorpholine.

method with respect to the green context. Among other examples, sucrose octadienyl ethers were obtained in aqueous media using sulfonated phosphines as water-soluble ligands (Scheme 10.45) [127–132].

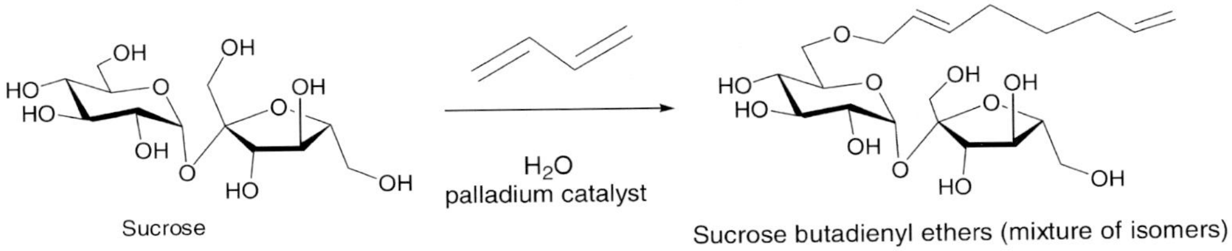

**Scheme 10.45** Formation of sucrose butadienyl ethers by palladium-catalyzed telomerization of butadiene.

Sucrose + butadiene →(H$_2$O, palladium catalyst)→ Sucrose butadienyl ethers (mixture of isomers)

## 10.4.2
## Oxidation Reactions

Aqueous media are more conventional for oxidations reactions, since many oxidizing systems are stable in water. Taking into account the natural good or high solubility of carbohydrates in water, it is thus obvious that many examples of oxidations reactions of carbohydrates have been reported, with important contributions in the field of metal-catalyzed processes [133–137]. The oxidation of sucrose by oxygen catalyzed by platinum provides derivatives with carboxy groups at the primary positions (a mixture of mono-, di-, and tricarboxy derivatives (**94**, Scheme 10.46). Yields vary with the amount of catalyst, the tricarboxylated compounds being more difficult to obtain because of catalyst poisoning, but excellent conversion to monocarboxylated sucrose could be achieved using electrodialysis [138–142]. Alternatively, 2,2,6,6-tetramethylpiperidine-1-oxyl (TEMPO)-mediated oxidation using hypochlorite as stoichiometric oxidant also led to tricarboxysucrose in yields up to 80% [143]. This method has been applied to many carbohydrates with success. With respect to environmental concerns, many variations of the method have been reported, notably heterogeneous TEMPO analogs [144–150] and bromide-free methods [151–153] (see also Chapter 4).This included a sonochemical version in which the reaction could also be performed in the absence of sodium bromide at a slower rate but with yields in the same range, providing new elements in the search for the mechanistic role of bromide ions [143, 154, 155].

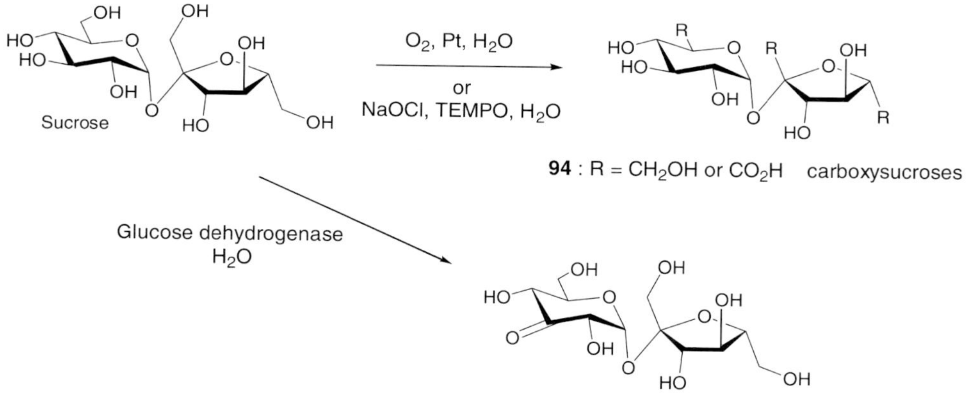

**94** : R = $CH_2OH$ or $CO_2H$   carboxysucroses

**95** : 3-oxosucrose

**Scheme 10.46** Examples of oxidation methods in water applied to sucrose.

Reaction of reducing sugars with hydrogen peroxide has also been studied. It normally leads to compounds with are shortened after C—C bond cleavage, and is often difficult to control because the immediate products can undergo further degradations, often leading to totally destructured materials. In a number of cases, the reaction can, however, be of preparative interest [156, 157]. For example, carboxy methylglucoside, obtained in a very simple manner from isomaltulose in 35% yield (a moderate yield but competitive compared with multi-step sequences), is an intermediate towards a bicyclic lactone which proved to be a useful synthon for grafting carbohydrate on nucleophilic species [158–162].

Many microbial carbohydrate oxidations have been described [163–167]. For example, conversions of sucrose to "3-ketosucrose" **95** in 60% yield by action of the D-glucoside 3-dehydrogenase (the active species in *Agrobacterium tumefaciens*) has been extensively studied [168–170] (Scheme 10.46).

### 10.4.3
### Bioconversions

Most of the natural chemical processes take place in aqueous media, and it is not the purpose of this chapter to review all kinds of bioconversions. However, it might be useful to mention a few of them which have connections with carbohydrate chemistry, just in order to give the reader a few general references in this field. Two very important reactions should be noted first, the fermentation to ethanol by action of *Saccharomyces cerevisiae* or *Zymomonas mobilis* [171] and the glycoside hydrolases catalyzed hydrolysis of oligosaccharides followed by isomerization reactions [172]. Other fermentations use sucrose and molasses as carbon sources for the preparation of polyhydroxyalkanoates [173]. Some microorganisms, yeasts, and bacteria also convert sucrose or other carbohydrates into alcohols, organic acids, amino acids, or vitamins [174–176]. Bio-hydrogen is also an interesting target for which examples of sugar biocracking have been reported [177–179].

Oligosaccharide isomerizations and oligo- and polymerizations are other examples of transformations achieved in water. For example, isomaltulose, glucose α-(1–6)-fructose is obtained from sucrose by action of *Protoaminobacter rubrum* and analogous processes provide other disaccharides from different disaccharides [180–183]. Transglycosidases or glycosyltransferases using sucrose either as the receiving sugar or as the provider lead to oligo- and polysaccharides [184], such as inulin-like polymers [185]. Oxidations using biocatalysts were mentioned in the preceding paragraph.

### 10.5
### Glyco-organic Substrates and Reactions in Aqueous Sugar Solutions

After having described diverse aspects of the chemistry of sugars in water, we want to recall briefly in this section our encounter with the field of the use of water as an unusual solvent for organic chemistry, which led us to build "glyco-organic" substrates. This came shortly after Rideout and Breslow [186], and then Grieco *et al.* [187], reported their key achievements with aqueous Diels–Alder reactions. Being previously involved with asymmetric Diels–Alder reactions of chiral dienes in which the chiral appendage was a carbohydrate, we developed the project further based on the idea that the carbohydrate, if most of its hydroxy functions were unprotected, would provide sufficient water solubility to the dienes in order to bring them within the water lattice, a property indispensable for being subjected to the hydrophobic effect. This was an alternative to the use of carboxylates or ammonium ions as water solubility inducers. In glyco-organic substrates, the carbohydrate was attached on the functional moiety at the anomeric position, allowing easy removal of the solubility inducer by acidic or enzymatic hydrolysis. The concept, applied to dienyl glycosides for [4 + 2] and [4 + 6] cycloadditions and to allyl vinyl ethers for Claisen rearrangements, is only briefly described in this section, as Chapters 1 and 8 cover the fundamental aspects of the hydrophobic effects in both aqueous and pericyclic reactions.

Dienyl glycosides were prepared by Wittig olefination of unsaturated aldehydes obtained either by reaction of the sodium salt of malonaldehyde with acetobromoglucose or anomeric alkylation with tosylacrolein. Using methacrolein as dienophile, a first comparison between the reaction of unprotected and the acetylated dienes **96** and **97** in water or toluene, respectively, showed a dramatic rate enhancement for the aqueous cycloaddition (3.5 h at room temperature vs 144 h at 80 °C) (Scheme 10.47). Exclusively *endo* products were obtained in water, in yields up to 90%, whereas a 87 : 13 *endo–exo* mixture was obtained from the acetylated diene in toluene. From the preparative viewpoint, separations of the diastereoisomers was possible, leading to the enantiomerically pure new cyclohexane derivative **98** after acidic or enzymatic removal of the carbohydrate moiety. Selective orientation of the reaction towards one or the other major enantiomer could be obtained after structural variations on the carbohydrate (α- or β-dienes, substitutions at various positions of the carbohydrate) [188–190]. Dienyl glycosides were also employed in the first example of aqueous [4 + 6] cycloaddition reaction using tropone leading to the [4 + 6] adducts in 62–66% yields [191].

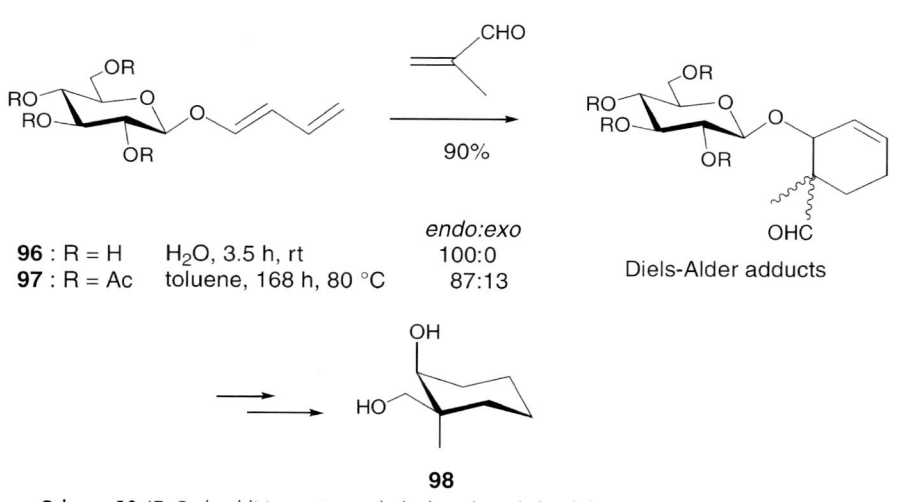

**Scheme 10.47** Cycloaddition using carbohydrate-based chiral dienes.

A detailed study of the thermodynamic outcome of the reaction allowed the determination of the activation parameters in various media, and confirmed the entropic origin of the activation. Insights into carbohydrate–water interactions were provided by comparing the dependence of the activation parameters, not only on the nature of the

Table 10.10 Activation parameters for the aqueous cycloaddition of diene 96 with methacrolein the presence of additives[a]

| Additive | $k_2 \times 10^5$ (M$^{-1}$ s$^{-1}$) | $\Delta H^{\ddagger}$ (kJ mol$^{-1}$) | $\Delta S^{\ddagger}$ (J mol$^{-1}$ K$^{-1}$) | $\Delta(\Delta H^{\ddagger})$ (kJ mol$^{-1}$) | $\Delta(T\Delta S^{\ddagger})$ (kJ mol$^{-1}$) |
|---|---|---|---|---|---|
| None | 28.5 | 40.0 ± 0.6 | −178 ± 2.1 | | |
| MeOH (50%) | 8.5 | 33.6 ± 0.8 | −211.1 ± 2.6 | −6.4 | −9.63 |
| LiCl (2.6 m) | 57.8 | 39.3 ± 1.7 | −175.1 ± 5.4 | −0.7 | +1.10 |
| Glucose (2.6 m) | 45.0 | 39.2 ± 0.3 | −177.4 ± 1.1 | −0.8 | +0.41 |
| Ribose (2.6 m) | 35.0 | 36.7 ± 1.5 | −188.3 ± 4.9 | −3.3 | −2.83 |

[a] At 25 °C.

sugar connected to the diene, but also on the presence of sugar additives in the solvent. Similarly to the modulations of hydrophobic effects due to the presence of salts (structure making or breaking, salting-out or salting-in, antichaotropic or chaotropic, pro- or antihydrophobic) [192–195] in water, glucose and sucrose were shown to strengthen the water structure [196, 197], unlike ribose, consistent with their effect on protein stability [198–201].

For concentrated glucose and sucrose solutions, the rate enhancement was correlated with a favorable change of both enthalpic and entropic terms of the activation energy, and likewise the influence of lithium chloride, with a quantitative connection of the rate variation with the sugar concentration (Table 10.10). The effect of the presence of sucrose, or glucose, proved to be larger than that of β-cyclodextrin. The latter actually excludes the substrates from the water solution by incorporating them in its hydrophobic pocket, whereas for sugar additives, it is the effect of the solutes on the water structure which affects the thermodynamic and kinetic behavior of the reaction.

Wittig carbon–carbon double bond formation performed on the carbonyl group of shorter carbohydrates is an alternative sequence towards carbohydrate-based dienes. From glyceraldehyde, the simplest diene 99 derived from a carbohydrate is obtained, for which the reaction with acrolein is 50 times faster in water than in toluene and leads to total *endo* selectivity. In this case, it was also observed that facial selectivity was increased when water was the medium, possibly due to an attack of the dienophile on the more hydrophobic face [202] (Scheme 10.48).

The glyceraldehyde-derived diene 99 and similar ones prepared from erythrose 100, or its shorter analog pentadienol 101, were used for asymmetric aqueous Diels–Alder reactions with glyoxylic acid, leading to ulosonic acids such as KDH, KDO and KDN and analogs [203–205] (Scheme 10.49). C-Glycosidic analogs of the disaccharide trehalose could be obtained in 68% yield by the same strategy from a C-dienylglycoside analog of diene 96 [206]. Other aqueous hetero-Diels–Alder reactions using a series of simple carbonyl compounds which are available as aqueous solutions led to dihydropyrans, which are precursors of pseudo-sugars [207, 208].

Glyceraldehyde-derived iminium salts were also used in aqueous aza-hetero-Diels–Alder reactions which exhibited good stereoselectively towards chiral

# 10 Functionalization of Carbohydrates in Water

**Scheme 10.48** Facial discrimination in aqueous cycloaddition.

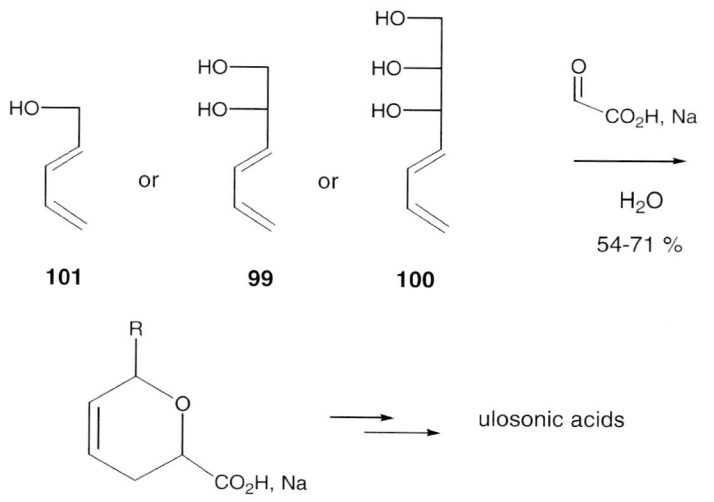

**Scheme 10.49** Aqueous hetero-Diels–Alder reactions towards ulosonic acids.

heterocyclic compounds subsequently transformed into aza sugars [209, 210] (Scheme 10.50).

The "glyco-organic substrates" concept was further applied to the study of aqueous Claisen rearrangements. This reaction also displays a negative activation volume, and is therefore also accelerated in water (see Chapter 8). The aqueous rate-enhanced Claisen rearrangement of glyco-organic compounds **102** and **104** proceeded in 80% to quantitative yields [211, 212], leading to, for example, enantiomerically pure (R)- and (S)-1,3-diols **103** and **105** after reduction, separation of diastereoisomers, and enzymatic hydrolysis (Scheme 10.51).

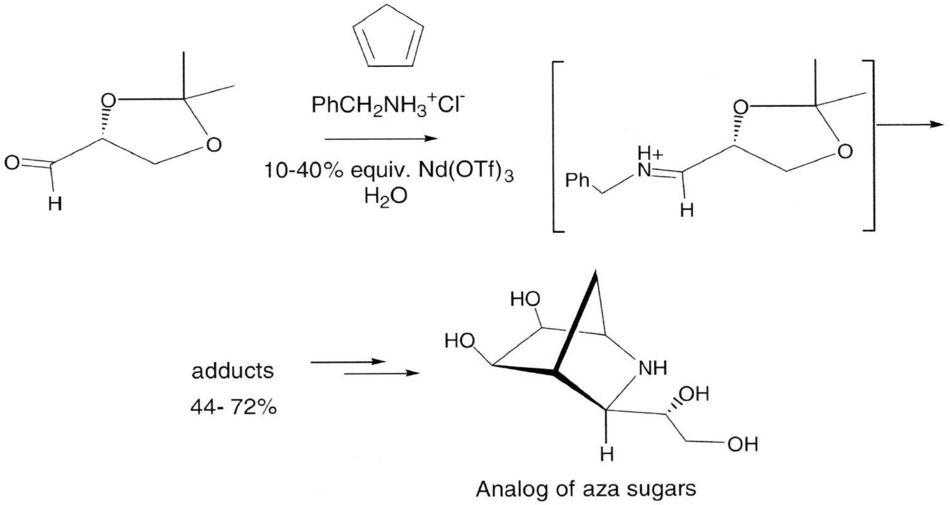

**Scheme 10.50** Lanthanide-catalyzed aza-Diels–Alder reactions.

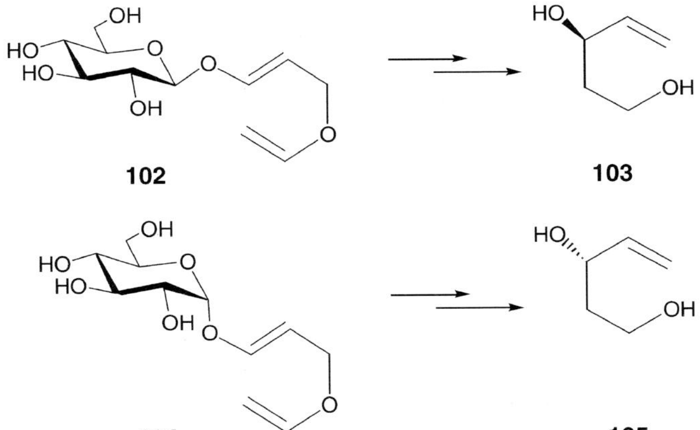

**Scheme 10.51** Preparation of enantiomerically pure (R)- and (S)-1,3-diols through aqueous Claisen rearrangements.

The use of sugar solutions as the medium, briefly mentioned above, was investigated in the case of various other reactions. As for Diels–Alder reactions, the Claisen rearrangement of glucosides **102** or **104** was also found to be accelerated. Aldolization of silylenol ethers, which was reported to be accelerated in water because of the hydrophobic effect [68, 69], was also performed in sugar solutions, and led to improved yields because of a decrease in the undesired competitive hydrolysis of the substrate due to lower water activity [196]. In the case of the Michael addition reaction of nitromethane with methyl vinyl ketone, a rate acceleration was also observed. The reduction of α,β-unsaturated ketones was studied under these conditions, and variations of the ratio between 1,2- and 1,4-addition were observed [196, 213]. It had previously been reported that the selectivity of the reduction of cyclohexen-2-one by sodium borohydride was modified when additives such as cyclodextrin or amylase were present in the medium [214, 215]. Finally, efficient epoxidation of allylic alcohols by hydrogen peroxide together with molybdenum or tungsten salts in aqueous solutions of carbohydrates was reported [216].

## 10.6 Conclusion

An important issue in modern chemistry is to design new products having improved environmental properties, prepared from renewable starting materials, and produced using clean processes. This chapter has described many examples which follow this philosophy, combining carbohydrate chemistry with the use of water for bringing new reactivity and selectivity. Also, carbohydrates were shown to serve as water solubility inducers, allowing benefit to be gained from hydrophobic effects.

It is predicted that carbohydrate chemistry will develop as a major field of organic synthesis in the future, due, on one hand, to the high biological relevance of many carbohydrate derivatives and, on the other, to the necessity for finding alternatives to fossil resources. Because of the high polarity and water solubility of most carbohydrates, aqueous media are the *"natural medium"* for performing their chemistry.

## References

1 Avalos Gonzales, M., Jimenez Requejo, J.L., Palacios Albarran, J.C. and Galbis Perez, J.A. (1986) *Carbohydr. Res.*, **158**, 53–66.

2 Bueno Martinez, M., Zamara Mata, F., Ruiz, A.M., Galbis Perez, J.A. and Cardiel, C.J. (1990) *Carbohydr. Res.*, **199**, 235–238.

3 Wulff, G. and Clarkson, G. (1994) *Carbohydr. Res.*, **257**, 81–95.

4 Zamara Mata, F., Bueno Martinez, M. and Galbis Perez, J.A. (1990) *Carbohydr. Res.*, **201**, 223–231.

5 Garcia Gonzalez, F. (1956) *Adv. Carbohydr. Chem.*, **11**, 97–143.

6 Kosikowski, A.K., Lin, G.Q. and Springer, J.P. (1987) *Tetrahedron Lett.*, **28**, 2211–2214.

7 Misra, A.K. and Agnihotri, G. (2004) *Carbohydr. Res.*, **339**, 1381–1387.

8 Yadav, J.S., Reddy, B.V.S., Sreenivas, M. and Satheesh, S. (2007) *Synthesis*, **11**, 1712–1716.

9 García González, F., Gómez Sánchez, A. and Goni De Rey, M.I. (1965) *Carbohydr. Res.*, **1**, 261–273.

10 Sato, S., Naito, Y. and Aoki, K. (2007) *Carbohydr. Res.*, **342**, 913–918.

11 Rodrigues, F., Canac, A. and Lubineau, Y. (2000) *Chem. Commun.*, 2049–2050.

12 Howard, S. and Withers, S.G. (1998) *J. Am. Chem. Soc.*, **120**, 10326–10331.

13 Philippe, M. and Semeria, D., Patent WO 2002051803; (2002) *Chem. Abstr.*, **137**, 63422.

14 Dalko, M. and Breton, L., Patent WO 2002051828; (2002) *Chem. Abstr.*, **137**, 79179.

15 Dalko-Csiba, M., Cavezza, A., Pichaud, P., Trouille, S., Pineau, N. and Breton, L.

(2007) Abstracts of Papers, 234th ACS National Meeting, Boston, MA, 19–23 August 2007, CARB-117.

16. Riemann, I., Papadopoulos, M.A., Knorst, M. and Fessner, W.D. (2002) *Aust. J. Chem.*, **55**, 147–154.

17. Bragnier, N. and Scherrmann, M.C. (2005) *Synthesis*, **5**, 814–818.

18. Hersant, Y., Abou-Jneid, R., Canac, Y., Lubineau, A., Philippe, M., Semeria, D., Radisson, X. and Scherrmann, M.C. (2004) *Carbohydr. Res.*, **339**, 741–745.

19. Graziani, A., Amer, H., Zamyatina, A., Hofinger, A. and Kosma, P. (2005) *Tetrahedron: Asymmetry*, **16**, 167–175.

20. Peters, S., Lichtenthaler, F.W. and Lindner, H.J. (2003) *Tetrahedron: Asymmetry*, **14**, 2475–2479.

21. Li, L.H. and Chan, T.H. (2000) *Tetrahedron Lett.*, **41**, 5009–5012.

22. Smith, K., Lock, S., El-Hiti, G.A., Wada, M. and Miyoshi, N. (2004) *Org. Biomol. Chem.*, **2**, 935–938.

23. Khan, R.H. and Rao, T.S.R.P. (1998) *J. Chem. Res. (S)*, 202–203.

24. Wang, Z., Yuan, S. and Li, C.J. (2002) *Tetrahedron Lett.*, 5097–5099.

25. Chan, T.C., Lau, C.P. and Chan, T.C. (2004) *Tetrahedron Lett.*, **45**, 4189–4191.

26. Li, C.J. and Zhang, W.-C. (1998) *J. Am. Chem. Soc.*, **120**, 9102–9103.

27. Li, C.J., Meng, Y., Yi, X.-H., Ma, J. and Chan, T.H. (1998) *J. Org. Chem.*, **63**, 7498–7504.

28. Chan, T.H. and Yang, Y. (1999) *Tetrahedron Lett.*, **40**, 3863–3866.

29. Hanessian, S., Park, H. and Yang, R.Y. (1997) *Synlett*, 353–354.

30. Cho, Y.S., Lee, J.E., Pae, A.N., Choi, K.I. and Kho, H.Y. (1999) *Tetrahedron Lett.*, **40**, 1725–1728.

31. Lu, W. and Chan, T.H. (2000) *J. Org. Chem.*, **65**, 8589–8594.

32. Alcaide, B., Almendros, P., Aragoncillo, C. and Rodríguez-Acebes, R. (2001) *J. Org. Chem.*, **66**, 5208–5216.

33. Márquez, F., Montoro, R., Llebaria, A., Lago, E., Molins, E. and Delgado, A. (2002) *J. Org. Chem.*, **67**, 308–311.

34. Loh, T.P. and Li, X.R. (1999) *Eur. J. Org. Chem.*, 1893–1899.

35. Chan, T.H., Yang, Y. and Li, C.J. (1999) *J. Org. Chem.*, **64**, 4452–4455.

36. Chang, H.-M. and Cheng, C.-H. (2000) *Org. Lett.*, **2**, 3439–3442.

37. Tan, X.-H., Shen, B., Deng, W., Zhao, H., Liu, L. and Guo, Q.-X. (2003) *Org. Lett.*, **5**, 1833–1835.

38. Li, C.J. and Chan, T.H. (1999) *Tetrahedron*, **55**, 11149–11176.

39. Podlech, J. and Maier, T.C. (2003) *Synthesis*, 633–655.

40. Li, C.J. (1996) *Tetrahedron*, **52**, 5643–5668.

41. Schmid, W. and Whitesides, G.M. (1991) *J. Am. Chem. Soc.*, **113**, 6674–6675.

42. Li, C.J. and Chan, T.H. (1991) *Tetrahedron Lett.*, **32**, 7017–7020.

43. Kim, E., Gordon, D.M., Schmid, W. and Whitesides, G.M. (1993) *J. Org. Chem.*, **58**, 5500–5507.

44. Binder, W.H., Prenner, R.H. and Schmid, W. (1994) *Tetrahedron*, **50**, 749–758.

45. Palmelund, A. and Madsen, R. (2005) *J. Org. Chem.*, **70**, 8248–8251.

46. Chan, T.H. and Li, C.J. (1992) *J. Chem. Soc., Chem. Commun.*, 747–748.

47. Chan, T.H. and Lee, M.C. (1995) *J. Org. Chem.*, **60**, 4228–4232.

48. Warwel, M. and Fessner, W.-D. (2000) *Synlett*, 865–867.

49. Gordon, D.M. and Whitesides, G.M. (1993) *J. Org. Chem.*, **58**, 7937–7938.

50. Choi, S.K., Lee, S. and Whitesides, G.M. (1996) *J. Org. Chem.*, **61**, 8739–8745.

51. Gao, J., Härter, R., Gordon, D.M. and Whitesides, G.M. (1994) *J. Org. Chem.*, **59**, 3714–3715.

52. Chan, T.H. and Xin, Y.C. (1997) *J. Org Chem.*, **62**, 3500–3504.

53. Canac, Y., Levoirier, E. and Lubineau, A. (2001) *J. Org. Chem.*, **66**, 3206–3210.

54. Lubineau, A., Canac, Y. and Le Goff, N. (2002) *Adv. Synth. Catal.*, **344**, 319–327.

55. Levoirier, E., Canac, Y., Norsikian, S. and Lubineau, A. (2004) *Carbohydr. Res.*, **339**, 2737–2747.

56. Hidelstål, O., Ding, R. and Lindström, U.M. (2005) *Green Chem.*, **7**, 259–261.

57 Augé, J., Lubin, N. and Lubineau, A. (1994) *Tetrahedron Lett.*, **35**, 7947–7948.
58 Yu, C., Liu, B. and Hu, L. (2001) *J. Org. Chem.*, **66**, 5413–5418.
59 Radha Krishna, P., Kannan, V., Sharma, G.V.M. and Ramana Rao, M.H.V. (2003) *Synlett*, 888–890.
60 Toshima, K., Ushiki, Y., Matsuo, G. and Matsumura, S. (1997) *Tetrahedron Lett.*, **42**, 7375–7378.
61 Sato, S., Akiya, T., Suzuki, T. and Onodera, J. (2004) *Carbohydr. Res.*, **339**, 2611–2614.
62 Sato, S., Nojiri, T. and Onodera, J. (2005) *Carbohydr. Res.*, **340**, 389–393.
63 Sato, S. and Sato, T. (2005) *Carbohydr. Res.*, **340**, 2251–2255.
64 Sato, S., Masukawa, H. and Sato, T. (2006) *Carbohydr. Res.*, **341**, 2731–2736.
65 Chen, D., Yu, L. and Wang, P.G. (1996) *Tetrahedron Lett.*, **37**, 4467–4470.
66 Mukaiyama, T., Banno, K. and Narasaka, K. (1974) *J. Am. Chem. Soc.*, **96**, 7503–7504.
67 Yamamoto, Y., Maruyama, K. and Mataumoto, K. (1983) *J. Am. Chem. Soc.*, **105**, 6963–6965.
68 Lubineau, A. (1986) *J. Org. Chem.*, **51**, 2142–2144.
69 Lubineau, A. and Meyer, E. (1988) *Tetrahedron*, **44**, 6065–6070.
70 Kobayashi, S. and Manabe, K. (2000) *Pure Appl. Chem.*, **72**, 1373–1380.
71 Kobayashi, S., Manabe, K. and Nagayama, S., (2000) in *Modern Carbonyl Chemistry*, (ed. J. Otera,), Wiley-VCH Verlag GmbH, Weinheim, pp. 539–562.
72 Kobayashi, S. and Manabe, K. (2002) *Acc. Chem. Res.*, **35**, 209–217.
73 Zeitouni, J., Norsikian, S., Merlet, D. and Lubineau, A. (2006) *Adv. Synth. Catal.*, **348**, 1662–1670; corrigendum: (2006) *Adv. Synth. Catal.*, **348**, 2009.
74 Lubineau, A., Augé, J. and Drouillat, B. (1995) *Carbohydr. Res.*, **266**, 211–219.
75 Burzio, F., Beck, R., Lubineau, A., Augé, J., Drouillat, B. and Mentech, J. (1994) European Patent Application, EP 0600359, (1995) *Chem. Abstr.*, **122**, 58871.
76 Burczykn, B. (2003) in *Novel Surfactants*, (ed. K. Holmberg), Surfactant Science Series, Vol. 114, Marcel Dekker, New York. pp. 129–192.
77 Kelkenberg, H. (1988) *Tenside Surfact. Deterg.*, **25**, 8–13.
78 Laughlin, R.G., Fu, Y.C., Wireko, F.C., Scheibel, J.J. and Munyon, R.L. (2003) in *Novel Surfactants* (ed. K. Holmberg), Surfactant Science Series, Vol. 114, Marcel Dekker, New York, pp. 1–33.
79 Guderjahn, L., Kunz, M. and Schüttenhelm, M. (1994) *Tenside Surfact. Deterg.*, **31**, 146–150.
80 Rico-Lattes, I. and Lattes, A. (1997) *Colloids Surf. A*, **123–124**, 37–48.
81 Bridiau, N., Benmansour, M., Legoy, M.D. and Maugard, T. (2007) *Tetrahedron Lett.*, **63**, 4178–4183.
82 Lindström, U.M., Ding, R. and Hidelstäl, O. (2005) *Chem. Commun.*, 1773–1774.
83 Bernotas, R.C. (1990) *Tetrahedron Lett.*, **31**, 469–472.
84 Dodds, D.R. and Gross, R.A. (2007) *Science*, **318**, 1250–1251.
85 Lichtenthaler, F.W. (2007) in *Methods and Reagents for Green Chemistry* (eds P. Tundo, A. Perosa and F. Zecchini), John Wiley & Sons, Inc., Hoboken, NJ, pp. 23–63.
86 Lichtenthaler, F.W. (ed.) (1991) *Carbohydrates as Organic Raw Materials*, Vol. 1, VCH Verlag GmbH, Weinheim.
87 Descotes, G. (ed.) (1993) *Carbohydrates as Organic Raw Materials*, Vol. 2, VCH Verlag GmbH, Weinheim.
88 van Bekkum, H., Röper, H., and Voragen, A.G.J. (eds) (1996) *Carbohydrates as Organic Raw Materials*, Vol. 3, VCH Verlag GmbH, Weinheim.
89 Praznik, W. and Huber, H. (eds) (1998) *Carbohydrates as Organic Raw Materials*, Vol. 4, WUV-Universitätsverlag, Vienna.
90 Godshall, M.A. (2001) *Int. Sugar J.*, **103**, 378–384.
91 Hickson, J.L. (ed.) (1977) *Sucrochemistry*, ACS Symposium Series, Vol. 41, American Chemical Society, Washington, DC.

92 Mathlouthi, M. and Reiser, P. (eds) (1995) *Sucrose: Properties and Applications*, Blackie, Glasgow.

93 Khan, R. (1984) *Pure Appl. Chem.*, **56**, 883–844.

94 Queneau, Y., Jarosz, S., Lewandowski, B. and Fitremann, J. (2007) *Adv. Carbohydr. Chem. Biochem.*, **61**, 217–292.

95 Bock, K. and Lemieux, R.U. (1982) *Carbohydr. Res.*, **100**, 63–74.

96 McCain, D.C. and Markley, J.L. (1986) *Carbohydr. Res.*, **152**, 73–80.

97 Christofides, J.C. and Davies, D.B. (1985) *J. Chem. Soc., Chem. Commun.*, 1533–1534.

98 Pérez, S. (1995) in *Sucrose: Properties and Applications* (eds M. Mathlouthi and P. Reiser), Blackie, Glasgow, pp. 11–32.

99 Bernet, B. and Vasella, A. (2000) *Helv. Chim. Acta*, **83**, 2055–2071.

100 Hervé du Penhoat, C., Imberty, A., Roques, N., Michon, V., Mentech, J., Descotes, G. and Pérez, S. (1991) *J. Am. Chem. Soc.*, **113**, 3720–3727.

101 Engelsen, S.B., Hervé du Penhoat, C. and Pérez, S. (1995) *J. Phys. Chem.*, **99**, 13334–13351.

102 Immel, S. and Lichtenthaler, F.W. (1995) *Liebigs. Ann. Chem.*, 1925–1937.

103 Wooley, E.M., Tomkins, J. and Hepler, L.G. (1972) *J. Solution Chem.*, **1**, 341–351.

104 Coccioli, F. and Vicedomini, M. (1974) *Ann. Chim.(Rome)*, **64**, 369–375.

105 Fang, X.M., Gong, F.Y., Ye, J.N. and Fang, Y.Z. (1997) *Chromatographia*, **46**, 137–140.

106 Houdier, S. and Perez, S. (1995) *J. Carbohydr. Chem.*, **14**, 1117–1132.

107 Haines, A.H. (1976) *Adv. Carbohydr. Chem. Biochem.*, **33**, 11–109.

108 Mentech, J., Betremieux, I., and Legger, B. (1992) European Patent Application, EP 467 762, (1992) *Chem. Abs.*, **116**, 256232v.

109 Jhurry, D., Deffieux, A., Fontanille, M., Betremieux, I., Mentech, J. and Descotes, G. (1992) *Makromol. Chem.*, **193**, 2997–3007.

110 Lubineau, A., Bienaymé, H., Queneau, Y. and Scherrmann, M.-C. (1994) *New J. Chem.*, **18**, 279–285.

111 Thévenet, S., Descotes, G., Bouchu, A. and Queneau, Y. (1997) *J. Carbohydr. Chem.*, **16**, 691–696.

112 Thévenet, S., Wernicke, A., Belniak, S., Descotes, G., Bouchu, A. and Queneau, Y. (1999) *Carbohydr. Res.*, **318**, 52–66.

113 Hough, L., Priddle, J.E. and Theobald, R.S. (1962) *J. Chem. Soc.*, 1934–1938.

114 Doane, W.M., Shasha, B.S., Stout, E.I., Russel, C.R. and Rist, C.E. (1967) *Carbohydr. Res.*, **4**, 445–451.

115 Gray, C.J., Al-Dulaimi, K. and Barker, S.A. (1976) *Carbohydr. Res.*, **47**, 321–325.

116 Theobald, R.S. (1961) *J. Chem. Soc.*, 5370–5376.

117 Wernicke, A., Belniak, S., Thévenet, S., Descotes, G., Bouchu, A. and Queneau, Y. (1998) *J. Chem. Soc., Perkin Trans.*, **1**, 1179–1181.

118 Christian, D., Fitremann, J., Bouchu, A. and Queneau, Y. (2004) *Tetrahedron Lett.*, **45**, 583–586.

119 Gagnaire, J., Toraman, G., Descotes, G., Bouchu, A. and Queneau, Y. (1999) *Tetrahedron Lett.*, **40**, 2757–2760.

120 Gagnaire, J., Cornet, A., Bouchu, A., Descotes, G. and Queneau, Y. (2000) *Colloids Surf. A*, **172**, 125–138.

121 Pierre, R., Adam, I., Fitremann, J., Jerome, F., Bouchu, A., Courtois, G., Barrault, J. and Queneau, Y. (2004) *C. R. Chimie*, **7**, 151–160.

122 Gérard, E., Götz, H., Pellegrini, S., Castanet, Y. and Mortreux, A. (1998) *Appl. Catal. A*, **170**, 297–306.

123 Spila Riera, G.C., Azurmndi, H.F., Ramia, M.E., Bertorello, H.E. and Martin, C.A. (1998) *Polymer*, **39**, 3515–3521.

124 Villandier, N., Adam, I., Jérôme, F., Barrault, J., Pierre, R., Bouchu, A., Fitremann, J. and Queneau, Y. (2006) *J. Mol. Catal. A: Chemistry*, **259**, 67–77.

125 Yadav, M.P., BeMiller, J.N. and Wu, Y. (1994) *J. Carbohydr. Chem.*, **14**, 991–1001.

126 Bazin, H., Bouchu, A. and Descotes, G. (1995) *J. Carbohydr. Chem.*, **14**, 1187–1207.
127 Hill, K., Gruber, B. and Weese, K.J. (1994) *Tetrahedron Lett.*, **35**, 4541–4542.
128 Hill, K. and Weese, K.J. (1993) Patent Ger. Offen, DE 4 242 467.
129 Pennequin, I., Meyer, J., Suisse, I. and Mortreux, A. (1997) *J. Mol. Catal. A.*, **120**, 139–142.
130 Desvergnes-Breuil, V., Pinel, C. and Gallezot, P. (2001) *Green Chem.*, **3**, 175–177.
131 Zakharkin, L.I., Guseva, V.V., Sulaimankulova, D.D. and Korneva, G.M. (1988) *Zh. Org. Khim.*, **24**, 119–121; (1988) *Chem. Abstr.*, **109**, 231376.
132 Muzart, J., Hénin, F., Estrine, B. and Bouquillon, S. (2001) French Patent 0116363.
133 Gallezot, P., Besson, M., Djakovitch, L., Perrard, A., Pinel, C. and Sorokin, A. (2006) in *Feedstocks for the Future*, ACS Symposium Series, Vol. 921, American Chemical Society, Washington, DC. pp. 52–66.
134 Varela, O. (2003) *Adv. Carbohydr. Chem. Biochem.*, **58**, 307–369.
135 Abbadi, A. and van Bekkum, H. (1996) in *Carbohydrates as Organic Raw Materials*, Vol. 3 (eds H. van Bekkum, H. Röper and A.G.J. Voragen) VCH Verlag GmbH, Weinheim, pp. 37–65.
136 Mehdi, H., Tuba, R., Mika, L.T., Bodor, A., Torkos, K. and Horvath, I.T. (2007) in *Renewable Resources and Renewable Energy*, (eds M. Graziani and P. Fornasiero), CRC Press LLC, Boca Raton, FL, pp. 55–60.
137 Kolaric, S. and Sunjic, V. (1996) *J. Mol. Catal. A*, **111**, 239–249.
138 Edye, L.A., Meehan, G.V. and Richards, G.N. (1991) *J. Carbohydr. Chem.*, **10**, 11–23.
139 Edye, L.A., Meehan, G.V. and Richards, G.N. (1994) *J. Carbohydr. Chem.*, **13**, 273–283.
140 Kunz, M., Schwarz, A. and Kowalczyk, J. (1997) European Patent EP 0775709, (1997) German Patent DE 19542287, (1997) *Chem. Abstr.*, **127**, 52504.
141 Ehrhardt, S., Kunz, M. and Munir, M. (1997) European Patent EP 0774469.
142 Kunz, M., Puke, H., Recker, C., Scheiwe, L. and Kowalczyk, J. (1994) German Patent DE 4307388; (1995) *Chem. Abstr.*, **122**, 56411q.
143 Davis, N.J. and Flitsch, S.L. (1993) *Tetrahedron Lett.*, **34**, 1181–1184.
144 de Nooy, A.E.J., Besemer, A.C. and van Bekkum, H. (1996) *Synthesis*, 1153–1174.
145 Sheldon, R.A., Arends, I.W.C.E., ten Brink, G.J. and Dijksman, A. (2002) *Acc. Chem. Res.*, **35**, 774–781.
146 Ciriminna, R., Blum, J., Avnir, D. and Pagliaro, M. (2000) *J. Chem. Soc., Chem. Commun.*, 1441–1142.
147 Brunel, D., Fajula, F., Nagy, J.B., Deroide, B., Verhoef, M.J., Veum, L., Peters, J.A. and van Bekkum, H. (2001) *Appl. Catal. A*, **213**, 73–82.
148 Sakuratani, K. and Togo, H. (2003) *Synthesis*, 21–23.
149 Tanyeli, C. and Gümüs, A. (2003) *Tetrahedron Lett.*, **44**, 1639–1642.
150 Fleche, G. (1996) EP 0798 310 A1; CA 96–2193034, (1997) *Chem. Abstr.*, **127**, 697180.
151 Bragd, P.L., Besemer, A.C. and van Bekkum, H. (2000) *Carbohydr. Res.*, **328**, 355–363.
152 Kochkar, H., Lassalle, L., Morawietz, M. and Hölderich, W.F. (2000) *J. Catal.*, **194**, 343–351.
153 Lemoine, S., Thomazeau, C., Joannard, D., Trombotto, S., Descotes, G., Bouchu, A. and Queneau, Y. (2000) *Carbohydr. Res.*, **326**, 176–184.
154 Brochette-Lemoine, S., Trombotto, S., Joannard, D., Descotes, G., Bouchu, A. and Queneau, Y. (2000) *Ultrasonics Sonochem.*, **7**, 157–161.
155 Brochette-Lemoine, S., Joannard, D., Descotes, G., Bouchu, A. and Queneau, Y. (1999) *J. Mol. Catal. A*, **150**, 31–36.
156 Arts, S.J.H.F., Mombarg, E.J.M., van Bekkum, H. and Sheldon, R.A. (1997) *Synthesis*, 597–613.

157 Isbel, H.S. and Frush, H.L. (1987) *Carbohydr. Res.*, **161**, 181–193.
158 Trombotto, S., Bouchu, A., Descotes, G. and Queneau, Y. (2000) *Tetrahedron Lett.*, **41**, 8273–8277.
159 Pierre, R., Chambert, S., Alirachedi, F., Danel, M., Trombotto, S., Doutheau, A. and Queneau, Y. (2008) *C. R. Chimie*, **11**, 61–66.
160 Trombotto, S., Danel, M., Fitremann, J., Bouchu, A. and Queneau, Y. (2003) *J. Org. Chem.*, **68**, 6672–6678.
161 Le Chevalier, A., Pierre, R., Kanso, R., Chambert, S., Doutheau, A. and Queneau, Y. (2006) *Tetrahedron. Lett.*, **47**, 2431–2434.
162 Chambert, S., Cowling, S.J., Mackenzie, G., Goodby, J.W., Doutheau, A. and Queneau, Y. (2007) *J. Carbohydr. Chem.*, **26**, 37–39.
163 Sedmera, P., Halada, P., Kubatova, E., Haltrich, D., Prikrylova, V. and Volc, J. (2006) *J. Mol. Catal. B*, **41**, 32–42.
164 Henriksson, G., Johansson, G. and Pettersson, G. (2000) *J. Biotechnol.*, **78**, 93–113.
165 Stoppok, E. and Buchholz, K. (1996) in (eds H.J. Rehm and G. Reed), *Biotechnology*, 2nd edn, VCH Verlag GmbH, Weinheim. pp. 5–29.
166 Buchholz, K. (1995) *Zuckerindustrie*, **120**, 692–699.
167 Stoppok, E., Matalla, K. and Buchholz, K. (1992) *Appl. Microbiol. Biotechnol.*, 604–610.
168 Bernaerts, M.J., Furnelle, J. and De Ley, J. (1963) *Biochim. Biophys. Acta*, **69**, 322–330.
169 Van Beeumen, J. and De Ley, J. (1968) *Eur. J. Biochem.*, **6**, 331–343.
170 Pietsch, M., Walter, M. and Buchholz, K. (1994) *Carbohydr. Res.*, **254**, 183–194.
171 Bai, F.W., Anderson, W.A. and Moo-Young, M. (2008) *Biotechnol. Adv.*, **26**, 89–105.
172 Bourne, Y. and Henrissat, B. (2001) *Curr. Opin. Struct. Biol.*, **11**, 593–600.
173 Liu, F., Li, W., Ridgway, D., Gu, T. and Shen, Z. (1998) *Biotechnol. Lett.*, **20**, 245–248.
174 Kamm, B., Gruber, P.R. and Kamm, M. (eds) (2006) *Biorefineries – Industrial Processes and Products*, Wiley-VCH Verlag GmbH, Weinheim.
175 Danner, H. and Braun, R. (1999) *Chem. Soc. Rev.*, **28**, 395–405.
176 Wilke, D. (1996) in *Carbohydrates as Organic Raw Materials*, Vol. 3 (eds H. van Bekkum, H. Röper and A.G.J. Voragen), VCH Verlag GmbH, Weinheim, pp. 115–128.
177 Hallenbeck, P.C. and Benemann, J.R. (2002) *Int. J. Hydrogen Energy*, **27**, 1185–1193.
178 Hussy, I., Hawkes, F.R., Dinsdale, R. and Hawkes, D.L. (2005) *Int. J. Hydrogen Energy*, **30**, 471–483.
179 Tao, Y., Chen, Y., Wu, Y., He, Y. and Zhou, Z. (2007) *Int. J. Hydrogen Energy*, **32**, 200–206.
180 Weidenhagen, R. and Lorenz, S. (1957) *Angew. Chem.*, **69**, 641–641.
181 Weidenhagen, R. and Lorenz, S. (1957) German Patent DE 1 049 800; (1961) *Chem. Abstr.*, **55**, 2030b.
182 Seibel, J., Moraru, R. and Gotze, S. (2005) *Tetrahedron*, **61**, 7081–7086.
183 Seibel, J., Moraru, R., Götze, S., Buchholz, K., Na'amnieh, S., Pawlowski, A. and Hecht, H.-J. (2006) *Carbohydr. Res.*, **341**, 2335–2349.
184 Remaud-Simeon, M., Albenne, C., Joucia, G., Fabre, E., Bozonnet, S., Pizzut, S., Escalier, P., Potocki-Veronese, G. and Monsan, P. (2003) in *Oligosaccharides in Food and Agriculture*, ACS Symposium Series, Vol. 849, American Chemical Society, Washington, DC, pp. 90–103.
185 Tungland, B.C. (2003) in *Oligosaccharides in Food Agriculture*, ACS Symposium Series, Vol. 849, American Chemical Society, Washington, DC, pp. 135–152.
186 Rideout, D.C. and Breslow, R. (1980) *J. Am. Chem. Soc.*, **102**, 7816–7817.
187 Grieco, P.A., Yoshida, K. and Garner, P. (1983) *J. Org. Chem.*, **48**, 3137–3139.
188 Lubineau, A. and Queneau, Y. (1985) *Tetrahedron Lett.*, **26**, 2653–2654.
189 Lubineau, A. and Queneau, Y. (1987) *J. Org. Chem.*, **52**, 1001–1007.

190 Lubineau, A. and Queneau, Y. (1989) *Tetrahedron*, **45**, 6697–6712.

191 Lubineau, A., Bouchain, G. and Queneau, et Y. (1997) *J. Chem. Soc., Perkin Trans.*, **1**, 2863–2867.

192 Breslow, R. and Rizzo, C.J. (1991) *J. Am. Chem. Soc.*, **113**, 4340–4341.

193 Breslow, R. (1991) *Acc. Chem. Res.*, **24**, 159–164.

194 Breslow, R. (2004) *Acc. Chem. Res.*, **37**, 471–478.

195 Blokzijl, W., Blandamer, M.J. and Engberts, J.B.F.N. (1991) *J. Am. Chem. Soc.*, **113**, 4241–4246.

196 Lubineau, A., Augé, J., Bienaymé, H., Queneau, Y. and Scherrmann, M.C. (1993) in *Carbohydrates as Organic Raw Materials*, Vol. 2 (ed. G. Descotes), VCH Verlag GmbH, Weinheim, pp. 99–112.

197 Lubineau, A., Bienaymé, H., Queneau, Y. and Scherrmann, M.C. (1994) *New J. Chem.*, **18**, 279–285.

198 Hofmeister, F. (1888) *Arch. Exp. Pathol. Pharmakol. (Leipzig)*, **24**, 247–260.

199 Rizzo, C.J. (1992) *J. Org. Chem.*, **57**, 6382–6384.

200 Back, J.F., Oakenfull, D. and Smith, M.B. (1979) *Biochemistry*, **18**, 5191–5196.

201 Lee, J.C. and Timasheff, S.N. (1981) *J. Biol. Chem.*, **256**, 7193–7201.

202 Lubineau, A., Augé, J. and Lubin, N. (1990) *J. Chem. Soc., Perkin Trans*, **1**, 3011–3015.

203 Lubineau, A., Augé, J. and Lubin, N. (1993) *Tetrahedron*, **49**, 4639–4650.

204 Lubineau, A. and Queneau, Y. (1995) *J. Carbohydr. Chem.*, **14**, 1295–1306.

205 Lubineau, A., Arcostanzo, H. and Queneau, Y. (1995) *J. Carbohydr. Chem.*, **14**, 1307–1328.

206 Lubineau, A., Grand, E. and Scherrmann, M.C. (1997) *Carbohydr. Res.*, **297**, 169–174.

207 Lubineau, A., Augé, J. and Lubin, N. (1991) *Tetrahedron Lett.*, **32**, 7529–7530.

208 Lubineau, A., Augé, J., Grand, E. and Lubin, N. (1994) *Tetrahedron*, **34**, 10265–10276.

209 Yu, L., Chen, D. and Wang, P.G. (1996) *Tetrahedron Lett.*, **37**, 2169–2172.

210 Yu, L., Li, J., Ramirez, J., Chen, D. and Wang, P.G. (1997) *J. Org. Chem.*, **62**, 903–907.

211 Lubineau, A., Augé, J., Bellanger, N. and Caillebourdin, S. (1990) *Tetrahedron Lett.*, **31**, 4147–4150.

212 Lubineau, A., Augé, J., Bellanger, N. and Caillebourdin, S. (1992) *J. Chem. Soc., Perkin Trans.*, **1**, 1631–1636.

213 Denis, C., Laignel, B., Plusquellec, D., Le Marouille, J.-Y. and Botrel, A. (1996) *Tetrahedron Lett.*, **37**, 53–56.

214 Chênevert, R. and Cumberland, D. (1985) *Chem. Lett.*, 1117–1118.

215 Chênevert, R. and Ampleman, G. (1985) *Chem. Lett.*, 1489–1490.

216 Denis, C., Misbahi, K., Kerbal, A., Ferrières, V. and Plusquellec, D. (2001) *Chem. Commun.*, 2461–2461.

# 11
# Water Under Extreme Conditions for Green Chemistry
*Phillip E. Savage and Natalie A. Rebacz*

## 11.1
## Introduction

Much research and development work in green chemistry and engineering has been devoted to the use of liquid water at near-ambient temperatures as a reaction medium. Progress and advances in this area are described in most of the other chapters in this volume. The present chapter has a different focus. Here, we center upon applications of water as a reaction medium at temperatures above its normal boiling point. We refer to this hot liquid water as high-temperature water (HTW), and we also use this term broadly so that it includes water above its critical temperature (374 °C) and pressure (218 atm). At times in this chapter, we will single out supercritical water (SCW) for special consideration. The system pressure and temperature must both exceed the fluid's critical values to reach the supercritical fluid state.

HTW is an important reaction medium in many different natural and engineered environments. In nature, HTW chemistry can profoundly influence the species that exist in regions surrounding ocean floor hydrothermal vents. In the energy sector, HTW is generated in nuclear and fossil fuel-fired electrical power plants, so there is interest in understanding reaction rates and equilibria in these media. Also, the need for sustainable and renewable energy sources and chemical products has prompted much research and development work on the aqueous-phase processing of biomass and its derivatives [1]. In the manufacturing industry, HTW is used in different chemical processes, and it has long been used in hydrothermal syntheses of different inorganic materials. More recently, there has been interest in using HTW as a medium for organic syntheses that are currently being done commercially in organic media.

This chapter begins with a background section and then moves on to a review of recent studies in the field. The background section is intended to be tutorial. We provide enough information about key topics so that new investigators in this field can pick up the basics that are needed to plan and execute experiments and interpret results from experiments in HTW chemical synthesis. This background section is not

*Handbook of Green Chemistry, Volume 5: Reactions in Water.* Edited by Chao-Jun Li
Copyright © 2010 WILEY-VCH Verlag GmbH & Co. KGaA, Weinheim
ISBN: 978-3-527-31591-7

intended to be exhaustive, however. Readers interested in more details about fundamentals and applications should consult our review articles on reactions in supercritical fluids [2] and in supercritical water [3, 4].

## 11.2
## Background

This background section provides basic information about HTW itself and organic chemicals and reactions in HTW. The scope of coverage includes both the science and the engineering of reactions in HTW.

### 11.2.1
### Properties of HTW

Liquid water under ambient conditions is a fluid familiar to all. It has a pH of 7 and a dielectric constant of around 80 because the polar water molecules arrange themselves in a hydrogen-bonded network. This hydrogen bonding also leads to the relatively high normal boiling point of water. Hydrocarbons are generally insoluble in liquid water at room temperature. All of these properties of liquid water change as its temperature increases. The properties of supercritical water are also different from those of liquid water and they vary with both temperature and pressure (or density).

Table 11.1 shows selected properties of water as a liquid at room temperature, as a high-temperature saturated liquid, as a supercritical fluid, and as steam [5, 6]. We see that HTW is less dense than ambient liquid water. The density of supercritical water is variable, and it can be controlled by the system pressure. Doubling the pressure at 400 °C from 250 to 500 bar more than triples the fluid density. Even higher pressures are required for SCW to reach more liquid-like densities. The properties of SCW are clearly density (or pressure) dependent. Table 11.1 shows that the dielectric constant ($\varepsilon$), ion product ($K_w$), viscosity, and thermal conductivity all increase as pressure increases at 400 °C.

Table 11.1 Some properties of water at different conditions.

| Parameter | Room temperature | HTW | SCW | SCW | Steam |
|---|---|---|---|---|---|
| Temperature (°C) | 25 | 250 | 400 | 400 | 400 |
| Pressure (bar) | 1 | 50 | 250 | 500 | 1 |
| Density (kg m$^{-3}$) | 997 | 800 | 170 | 580 | 0.3 |
| Dielectric constant | 78.5 | 27.1 | 5.9 | 10.5 | 1 |
| Ion product, $K_w$ (mol$^2$ kg$^{-2}$) | $10^{-14}$ | $10^{-11.2}$ | $10^{-19.4}$ | $10^{-11.9}$ | – |
| Viscosity (MPa s) | 890.8 | 106.1 | 29.00 | 67.89 | 24.45 |
| Thermal conductivity (mW K$^{-1}$ m$^{-1}$) | 607.2 | 622.7 | 169.3 | 451.6 | 54.76 |

Data taken from [5, 6].

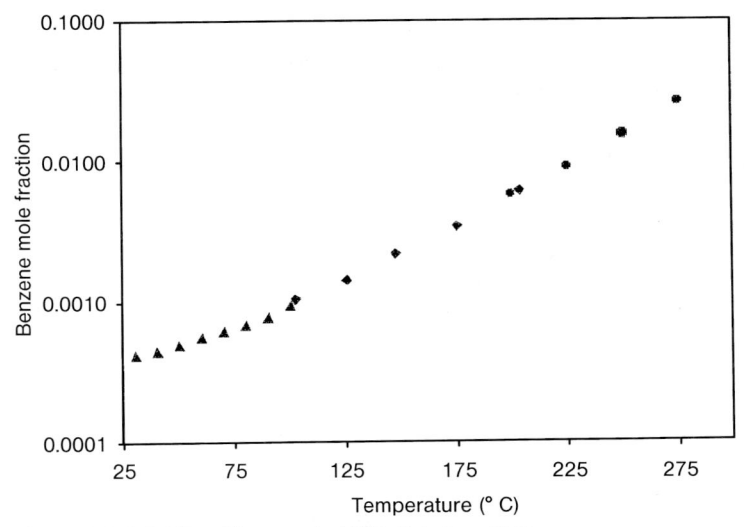

**Figure 11.1** Solubility of benzene in HTW. Data from [7–9].

The dielectric constant of liquid water decreases as temperature is increased along the saturation curve. As shown in Table 11.1, the dielectric constant at 250 °C is about 27. At 350 °C, it decreases further to 13. These reduced dielectric constants in HTW are similar in value to those of moderately polar organic liquids at room temperature. It should not be surprising, then, that small hydrocarbons become increasingly soluble in HTW as the temperature increases. Figure 11.1 displays representative solubility data, in this case for benzene in HTW [7–9]. The solubility increases exponentially with temperature and is on the order of a few mole percent around 250 °C. Of course, as the temperature increases above the critical point of the mixture, organic compounds become completely miscible in water.

Table 11.2 shows the average number of hydrogen bonds per water molecule at different temperature/density states, as determined experimentally [10]. At room temperature, each water molecule has about four hydrogen bonds, and liquid water has the familiar tetrahedral structure. At an elevated temperature (300 °C), liquid

**Table 11.2** Average number of hydrogen bonds per water molecule

| Condition | Temperature (°C) | Density (g cm$^{-3}$) | No. of hydrogen bonds |
|---|---|---|---|
| Room temperature | 25 | 1.0 | 3.9 |
| HTW | 300 | 0.72 | 2.4 |
| SCW | 400 | 0.66 | 2.0 |
| SCW | 400 | 0.60 | 1.9 |
| SCW | 400 | 0.41 | 1.4 |
| SCW | 400 | 0.29 | 1.1 |
| SCW | 400 | 0.19 | 0.7 |

Data taken from [10].

water has fewer hydrogen bonds, on average. As the temperature increases to exceed the critical point, the number of hydrogen bonds decreases. Table 11.2 also shows that at a constant supercritical temperature (400 °C) the number of hydrogen bonds per water molecule decreases as density (pressure) decreases. In addition to reducing the number of hydrogen bonds, increasing the temperature also reduces the persistence of hydrogen bonds [11]. That is, the process by which hydrogen bonds form and break is characterized by faster rates (shorter hydrogen bond lifetimes).

### 11.2.2
### Process Engineering Considerations

As discussed above, the properties of HTW make it an attractive medium for organic chemistry because small organic molecules more readily go into solution. In addition, there are often process engineering advantages associated with the use of HTW as the reaction medium. The chief advantage comes from the ability to control the solubility of many reactants and products in the reaction medium by controlling the temperature. In principle, this ability allows one to cool the reactor effluent and, in many cases, thereby generate two immiscible liquid phases or a liquid phase and a solid phase. If two liquid phases exist, the organic phase would typically contain the desired product(s) and any unconverted reactant(s). It can be easily separated from the aqueous phase by decanting. The aqueous phase can then be recycled to the reactor. Separating the solvent from the product(s) using a gravity-based method (decanting, crystallization, and filtration) is energy efficient. Separation processes such as distillation and liquid–liquid extraction, which would often be required to separate an organic solvent from organic reaction products, would typically require higher capital costs, higher operating costs, and higher energy consumption.

An objection sometimes leveled at the use of HTW for organic synthesis in industry is that doing so requires higher pressures and, often, higher temperatures than a competing synthesis in an organic solvent. These more severe conditions are thought to require much higher energy costs for the process. This need not be the case, however, in a well-designed HTW process. The pressurization that is required would be done on the water while it is in the liquid phase under ambient conditions. The work required to pressurize a liquid is much less than that required to pressurize a gas to the same pressure, because liquids have much smaller specific volumes. Additionally, the heat required to operate at an elevated temperature can be recovered and used to preheat the reactor feed stream. Thus, with proper heat integration an HTW process need not require significantly more energy than a lower temperature process. Of course, a complete process design would be required to estimate accurately the energy requirements for any proposed HTW chemical synthesis process. Figure 11.2 shows a simplified process flow diagram for a continuous chemical synthesis in HTW.

Published studies on the use of water as a reaction medium for chemical synthesis often claim or at least intimate that a process that employs water is "greener" than one that does not. On the surface, this claim may seem reasonable since water is non-toxic

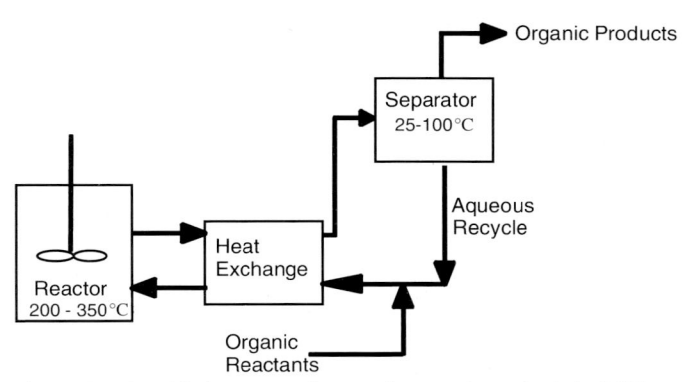

**Figure 11.2** Simplified process schematic for organic synthesis in HTW.

and naturally occurring. However, there are energy inputs and environmental emissions associated with water treatment and water recovery if it is to be recycled in a chemical process. Add to these inputs and emissions an additional amount when high-temperature liquid water or supercritical water is used as the reaction medium because elevated temperatures and pressures are required. To determine whether a water-based process is in fact less environmentally harmful than one that uses an organic reaction medium, one needs to determine and then compare the energy requirements and emissions associated with the production and use of the competing reaction media. This type of assessment is within the domain of industrial ecology, and life-cycle assessment (LCA) is a tool commonly used to make these comparisons.

To the best of our knowledge, the literature provides few, if any, LCA studies of aqueous versus organic reaction media for chemical processes. One published report that provides some quantitative information about the different trade-offs (but not a complete LCA study) is the work of Dunn and Savage [12] on terephthalic acid synthesis from *p*-xylene. There is clearly a need for more work in this area.

Demonstrating environmental benefits is just one of the hurdles that must be cleared for a proposed aqueous-phase process. A second hurdle is that of economics. The proposed process or synthetic pathway must be more effective in creating value than the existing process or pathway with which it would compete. There have been instances in the literature where new aqueous-phase syntheses that are proposed and studied actually destroy economic value rather than create it. For example, proposed syntheses wherein the raw materials have greater value than the intended product(s) will never be implemented commercially, regardless of the environmental friendliness of the process.

## 11.2.3
### Theoretical, Computational, and Experimental Methods

Theory, computation, and experiments have all led to progress in the field of chemical synthesis in HTW. Theory has been used to interpret results from experiments.

Computation has been used to provide an explicitly molecular picture of the dynamics of reactants, catalysts, and water molecules. Experiments have been used to test the feasibility of different proposed syntheses and to validate results from theory and computation.

#### 11.2.3.1 Classical Theory

Theories applied in this field are those commonly used in solution-phase chemical kinetics [13]. Classical theories exist to describe the influence of different substituents, pressure, solvent dielectric constant, solubility parameter, ionic strength, and polarity on reaction rates. Substituent effects for SCW reactions, for example, have been rationalized by use of the Hammett correlation [14, 15]. The solubility parameter has also been used to correlate the influence of SCW polarity on reaction kinetics [16].

Since the properties of HTW, and especially SCW, can be varied by changing he temperature or density (pressure), there have been several instances where classical theories have been used to correlate and rationalize the density dependence of reaction rates in SCW. Transition-state theory indicates that reaction rate constants are expected to be functions of pressure. The extent of the pressure dependence is determined by the activation volume (difference in partial molar volume between the activated complex and the reactants). Activation volumes on the order of hundreds or even thousands of $cm^3\,mol^{-1}$ have been reported (e.g. [17]) for reactions in SCW. The intrinsic activation volume due to the formation or cleavage of chemical bonds is much smaller in magnitude (tens of $cm^3\,mol^{-1}$), so the high absolute values obtained in many SCW reaction studies are due to factors in addition to the making and breaking of bonds in the transition state. Iyer and Klein [18] showed that absolute values of activation volumes for nitrile hydrolysis in HTW (not SCW) decreased from several hundred to tens of $cm^3\,mol^{-1}$ as the pressure (and water density) increased. They suggested that the apparent activation volume can be decomposed into separate contributions from electrostatic, hydrostatic, diffusion, and phase behavior effects. Subsequent modeling showed that the electrostatic contribution was the largest for nitrile hydrolysis. That is, the effect of pressure on the dielectric constant of water and the influence of the dielectric constant on the kinetics was of greater influence than the effect of the hydrostatic pressure.

The influence of the dielectric constant on the reaction rate is often modeled via simple Kirkwood theory [19]. Here, one expects a plot of $\ln k$ versus the Kirkwood polarity parameter, $(\varepsilon - 1)/(2\varepsilon + 1)$, to be linear. Figure 11.3 shows one such plot, in this case for ester hydrolysis in SCW [20].

This brief discussion shows that solution-phase reaction theories developed over the years can be applied profitably to reactions in HTW and SCW. The important distinction in the case of the aqueous medium, however, is that the solvent properties (e.g. dielectric constant, solubility parameter) can be changed simply by changing the system pressure. In conventional reaction media at room temperature, the only way to change the solvent properties is by changing the solvent. This change in the chemical identity of the solvent introduces additional perturbations into the reaction system.

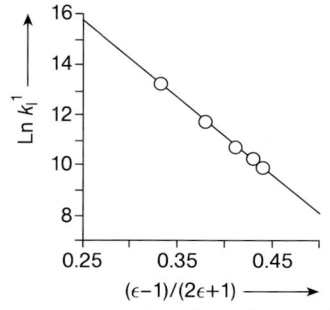

**Figure 11.3** Kirkwood plot for ester hydrolysis in SCW. Reproduced with permission from [20].

#### 11.2.3.2 Molecular and Computational Modeling

As outlined above, interpreting results by use of classical theories of reactions in solution is one approach that has been used in this field. This approach is straightforward, and the methods are well known. The approach does not provide a molecular-level description of the reaction and the solvent effects, however. To obtain more molecular details, investigators have turned to molecular modeling and simulations. The literature on molecular simulation of HTW, and especially SCW, systems is rich and voluminous. Molecular dynamics simulations have been carried out to examine the structure, properties, and hydrogen bonding characteristics of SCW (e.g. [11, 21, 22]). Molecular simulations of reactions in HTW or SCW and *ab initio* electronic structure calculations have also been performed. We discuss a few selected examples here to illustrate the various approaches used.

Nakahara and co-workers have published a series of articles on $C_1$ chemistry in SCW. The intention was to determine the fundamental mechanisms for interconversion of different functional groups. Computational approaches played an important role in this work. For example, Matubayasi and Nakahara [23] determined equilibrium constants for the transformations shown in Scheme 11.1 by molecular simulation. The SPC/E model was used for water and force fields for

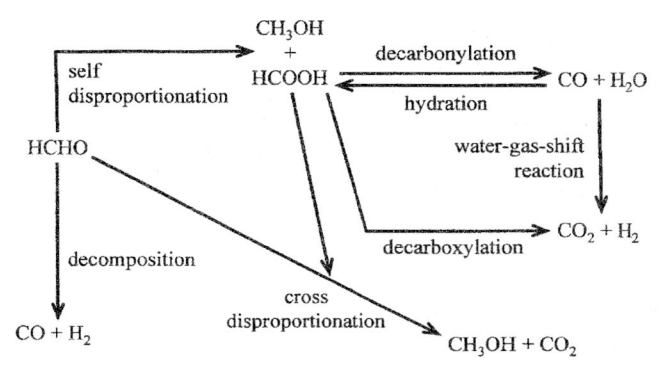

**Scheme 11.1** $C_1$ interconversion pathways in SCW. Reproduced with permission from [23].

the solutes in Scheme 11.1 were from the literature. Monte Carlo simulations were performed to determine the solvation free energy for each species and thereby the influence of SCW on the equilibrium constants for each reaction.

Studies such as these are important, but they provide information only about the equilibrium states. There have also been reports of molecular simulation of chemical reactions in SCW where attention has been paid to the progress of the reaction along the reaction coordinate. Balbuena et al. [24]. did pioneering work in this area via their analysis of an $S_N2$ reaction in SCW. Akiya and Savage [25] examined $H_2O_2$ decomposition in SCW. Boero et al. [26] used first-principles molecular dynamics to elucidate details about the mechanism of the Beckmann rearrangement of cyclohexanone oxime into ε-caprolactam in SCW. This reaction is one of the steps that can be used to synthesize nylon-6. Taylor et al. [27] examined the hydrolysis of methyl tert-butyl ether in SCW via *ab initio* calculations with a dielectric continuum rather than an explicit accounting for the water molecules. Harano et al. [28] combined an integral equation approach with *ab initio* molecular orbital theory to study the effect of SCW on a Diels-Alder reaction.

The discussion above shows that many different computational approaches have been used to obtain mechanistic and molecular-level details about chemical reactions in SCW. The methods have different advantages and disadvantages, so the method chosen must be adapted to the needs of the research project and the computational resources available. This area of computational molecular-level studies of reactions in HTW and SCW will become increasingly important as more opportunities are demonstrated for green chemistry in this medium.

### 11.2.3.3 Experimental Methods

Laboratory experiments have probably been the source of most of the advances in synthesis in HTW. The experimental techniques used include batch and continuous flow reactor systems. The reactors are typically made of metal (stainless steel, high-nickel alloys, and titanium are common) to withstand the elevated pressures required, but there are reports, increasing in frequency, of the use of quartz capillary tubes as batch reactors. Quartz solves the problem of unintentional metal catalysis by reactor walls, which is a concern with metal reactors. Analysis of products is often made by examining samples withdrawn from the reactor. The samples are typically quenched to ambient conditions prior to analysis, so any chemistry occurring during this cooling period will affect the results reported. *In situ* measurement techniques, such as Fourier transform infrared (FTIR), Raman, and NMR spectroscopy and microscopy, can remedy this situation for those cases where it is important.

Nakahara and co-workers have made extensive studies with NMR detection in SCW. For example, they examined the proton chemical shift of water under different supercritical conditions to draw inferences about the state of hydrogen bonding in supercritical water [10]. They also used NMR for the analysis of reactants and products during numerous reaction studies in HTW and SCW (e.g. [29, 30]).

Brill and co-workers have reported numerous results from *in situ* FTIR and Raman studies of reactions in HTW and SCW (e.g. [31]). The detection and quantification of molecular intermediates in reactions provides tremendous insight into reaction networks and mechanisms. Brill indicated that diamond and sapphire are appropriate window materials for *in situ* cells, and that Au, Ti, Ta, and high-Ni steel alloys are appropriate for the body of the cell. The windows must be IR transparent and stable under SCW conditions. The cell material must be able to withstand that chemical environment, which can be very corrosive if acids are present, and also be machinable. This group reported on the hydrothermal reactions of many different compounds, including amines, nitriles, organic acids, and amides. Ikushima *et al.* have also developed and utilized *in situ* IR cells for measuring the progress of reactions in SCW [32]. For example, they showed that Beckmann and pinacol rearrangements, although normally acid catalyzed, occurred in SCW with no added acid.

Another experimental technique used to study reactions in SCW *in situ* is the diamond anvil cell. This apparatus generates the high pressure by pressing together two diamond anvils. The sample cell is typically heated electrically. This instrument includes an optical microscope, so images of the sample can be obtained, and often FTIR or Raman spectroscopy for chemical measurements. This method has been used to examine the behavior of solid materials in HTW and SCW. Reports on poly(ethylene terephthalate) and benzo[*a*]pyrene serve as examples for the interested reader [33, 34].

Koda and co-workers have used a different method for studying the behavior of solids in SCW (e.g. [35]). His laboratory has developed a shadowgraph apparatus that provides visual images of solid materials. The focus of Koda's work has been on oxidation reactions in SCW, but the technique can also be used for other reactions of interest.

Finally, in this discussion of specialized reactors for HTW/SCW studies, we mention electrochemical cells. These cells are useful for determining some properties of solutes in SCW, for developing electrochemical sensors, and for studying electron transfer reactions in SCW. Important contributions have come from several research laboratories (e.g. [36–39]).

## 11.2.4
## pH Effects

Pure HTW becomes more acidic (and basic) as its temperature increases to about 250 °C because the ion product, $K_w$, increases with temperature up to this point (see Table 11.1). The higher ion product means that an elevated level of hydronium (and hydroxide) ions will exist in HTW. There have been many demonstrations of reactions that require acid catalysis at room temperature proceeding with no added acid in HTW. Therefore, HTW provides an opportunity to perform acid-catalyzed syntheses with less environmental impact from the production and post-reaction neutralization of acid catalysts. The reasons why acid- or base-catalyzed reactions can proceed in HTW with no added catalyst are both the elevated levels of $H^+$ and $OH^-$ naturally present in HTW because of the elevated $K_w$ value and also the activity of a water-induced or water-catalyzed path that becomes operative because of the thermal

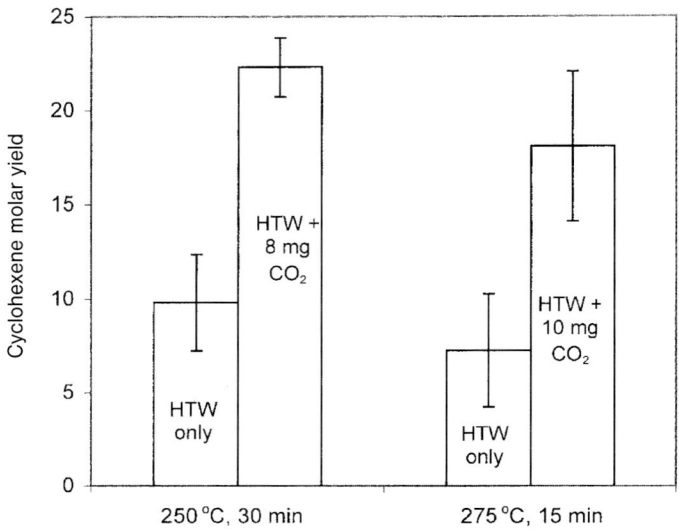

**Figure 11.4** Enhanced cyclohexene yields from cyclohexanol dehydration in $CO_2$-enriched HTW. Data from [41].

energy in the system. Recent work [40] suggests that it is this second route that is the more important.

This elevated level of hydronium ions in HTW can be enhanced even further through the addition of $CO_2$ to the reaction medium. $CO_2$ is often used as an environmentally benign substitute for organic solvents in reaction, extraction, and cleaning processes. $CO_2$ addition to water generates carbonic acid, which then dissociates to generate even more hydronium ions in the solution. The advantage of using HTW, either with or without added $CO_2$, is that the acidity can be "neutralized" after the reaction without adding a base, so no waste salt is generated. The acidity can be easily reduced by reducing temperature and/or the $CO_2$ partial pressure. Figure 11.4 shows that adding $CO_2$ can increase yields from some acid-catalyzed reactions in HTW. These data are for the dehydration of cyclohexanol to cyclohexene [41].

Hunter and Savage [42] used the thermodynamics of the $CO_2$–$H_2O$ system to explore the influence of $CO_2$ on the pH of HTW. Figure 11.5, which summarizes some of these results, shows that a reduction of at least two pH units can be achieved at moderate $CO_2$ pressures. Adding $CO_2$ has the same effect on HTW reactions as adding a mineral acid, namely it reduces the pH. Figure 11.6 shows the kinetics of dibenzyl ether hydrolysis in HTW with added $CO_2$ and added mineral acid. The data from $CO_2$ addition follow the same trend as those from addition of mineral acid. Therefore, $CO_2$–$H_2O$ mixtures at elevated temperatures may be a more environmentally friendly option for acid catalysis than mineral acids at lower temperatures. One can control the pH of the medium by controlling temperature and by adding $CO_2$ to the medium.

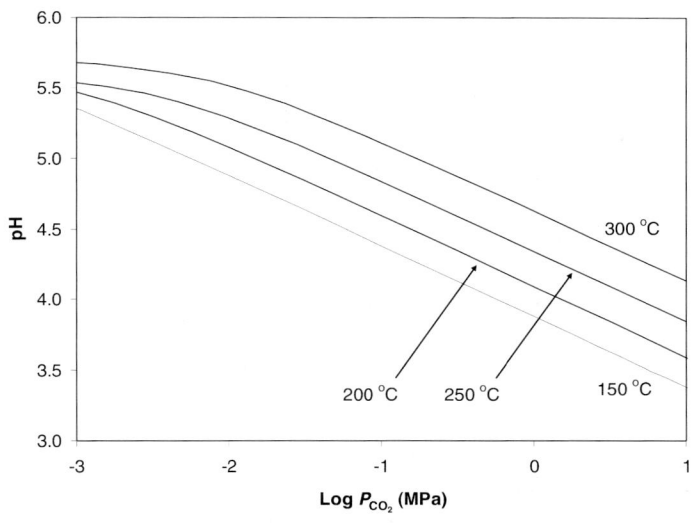

**Figure 11.5** pH of $CO_2$-enriched HTW at different temperatures and $CO_2$ partial pressures. Reproduced with permission from [42].

Since acid- and base-catalyzed reactions constitute a class that has often been considered for green chemistry applications in HTW, it is instructive to consider the influence of pH on such reactions. The responses of reactions in HTW to pH are similar to those encountered in water near room temperature. Figure 11.7 shows

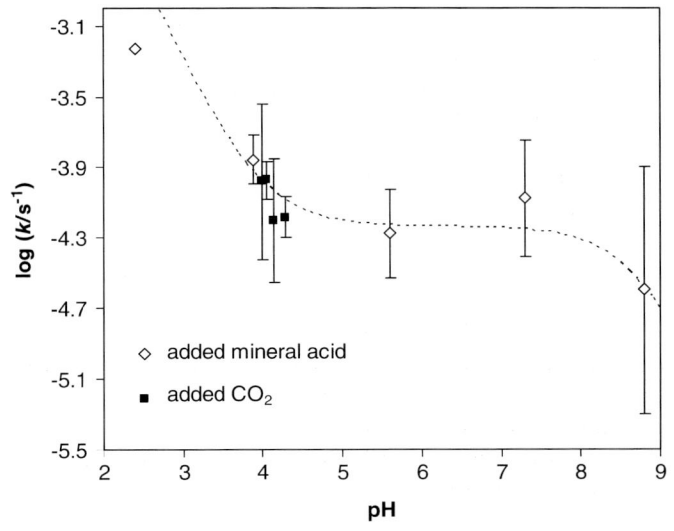

**Figure 11.6** Effect of pH on dibenzyl ether hydrolysis kinetics in HTW at 250 °C. Filled squares are from experiments with added $CO_2$. Open diamonds are from experiments with added mineral acid or base. The dashed curve is the fit of a mechanistic model to the data. Reproduced with permission from [42].

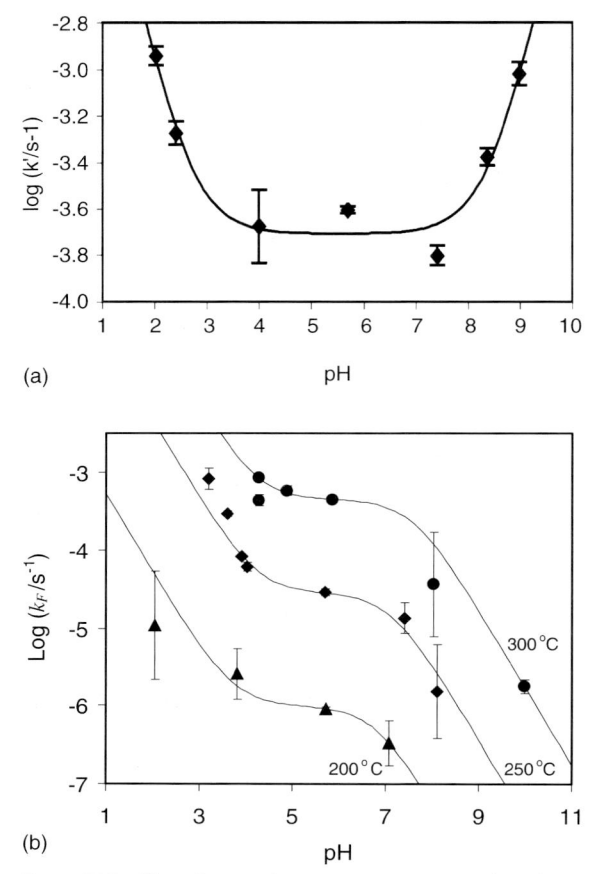

**Figure 11.7** Effect of pH on kinetics in HTW. (a) Bisphenol A decomposition. Reproduced with permission from [43]. (b) Tetrahydrofuran synthesis. Reproduced with permission from [44] Copyright 2006 American Chemical Society.

the influence of pH on the rate constants for isopropenylphenol synthesis via bisphenol A cleavage in HTW [43] and for tetrahydrofuran synthesis from butanediol dehydration [44]. Similar curves were observed for other reactions, including aldol condensation, benzil rearrangement, and numerous hydrolyses [40, 45, 46].

What is striking is that most of the reactions examined to date in HTW do not show a simple linear semilogarithmic dependence of the rate constant on pH, as one would expect if the reactions were first order in $H^+$ (or $OH^-$ for the case of bases). Therefore, it is advised that healthy skepticism be applied when mechanisms in the literature, which had been elucidated for reactions in water near room temperature, are extrapolated to HTW and SCW conditions.

## 11.3
## Recent Progress in HTW Synthesis

This section updates previous reviews [3, 4] on organic synthesis in high-temperature water. Studies of kinetic observables, for example, rate constant, activation energy, and frequency factor, are included, as are physical organic investigations concerned more with mechanistic pathways. However, SCW gasification and thermal decomposition studies are mostly omitted unless they are meant to feature some model transformation, as for example, "decomposition" of anisole to phenol as a model hydrolysis reaction.

### 11.3.1
### Hydrogenation

Crittendon and Parsons first demonstrated the applicability of SCW for hydrogenation over Pd catalysts [47, 48]. Subsequently, decomposition of formic acid or formate salts was developed as a means of producing hydrogen *in situ* for HTW reductions [49, 50]. Poliakoff and co-workers investigated the reduction of simple aromatic nitro, ketone, and aldehyde compounds by the $H_2$ generated *in situ* [50]. Using $HCO_2H$ as the hydrogen source, nitrobenzene was hydrogenated to aniline in up to 75% yield within a residence time of around 11 s. Aniline was the only product observed at temperatures above 200 °C. Surprisingly, formic acid provided little to no yield for the reductions of benzaldehyde and cyclohexanone. Sodium formate, however, successfully reduced benzaldehyde to benzyl alcohol in 65% yield after 19 s (247 °C, 15.7 MPa) and reduced cyclohexanone to cyclohexanol in 28% yield in 24 s (250 °C, 15.5 MPa). The researchers surmised that CO poisoning of the stainless-steel reactor surface is to blame for reaction failure with formic acid decomposition. CO was produced in much larger quantities from formic acid (31.0 mol% at 15 MPa) than from sodium formate (2.5 mol% at 15 MPa). As an alternative explanation, the researchers proposed direct H-transfer from $HCO_2^-$, rather than hydrogenation by $H_2$. Unfortunately, no further work was presented to test these hypotheses, and the actual mechanism explaining these results is still poorly understood. Finally, acetophenone was successfully hydrogenated to 1-phenylethanol in 78% yield at 15.6 MPa, 250 °C, 20 s residence. A large molar ratio of sodium formate to acetophenone (at least 10:1) was needed to achieve reasonable yields. However, at these high loadings of sodium formate, no further reduction of 1-phenylethanol was observed, perhaps illustrating the typical relationship between selectivity and reactivity. No added catalyst was used in this study, but the researchers suspected that the 316 stainless-steel reactor wall (surface area $= 5.7 \text{ cm}^2$, flow reactor volume $= 0.23 \text{ cm}^3$) or stainless-steel components that leached into solution were catalyzing the reaction. Although the transformations in this study proceeded quickly and easily, the authors noted that the large quantity of reducing agent (the ratio of HCOOH or HCOONa to substrate is as much as 10:1) limits the practicality of this procedure to bench-scale preparations.

## 11.3.2
### C–C Bond Formation

The formation of new carbon–carbon bonds is the foundation of most synthetic routes in every chemical industry. The Friedel–Crafts reaction is probably the most important example within this class. Heck coupling is also useful, especially when Friedel–Crafts conditions cannot be tolerated. Both Friedel–Crafts alkylation and Heck coupling have been shown to be successful in HTW.

#### 11.3.2.1 Friedel–Crafts Alkylation

Friedel–Crafts alkylation is by far the most important method of attaching alkyl side chains to aromatic rings [51]. A general example of this transformation is shown in Scheme 11.2.

R–X + C$_6$H$_6$ ⇌ C$_6$H$_5$–R + HX

**Scheme 11.2** The most general representation of a Friedel–Crafts alkylation reaction.

Among the first Friedel–Crafts reactions carried out in HTW were the addition of *tert*-butanol to phenol and *p*-cresol to achieve about 20% yield in both cases after 1–2 days [52]. This result was promising because the reaction proceeded without a Lewis acid catalyst, but the working procedure was hardly practical owing to its low yields and long reaction times. Arai and co-workers built upon the promising results of Chandler *et al.* with a suite of investigations [53–58].

Sato *et al.* showed that as much as 58% yield of *o*-isopropylphenol was possible at 400 °C for the alkylation of phenol with 2-propanol [53]. Regioselectivity was demonstrated for the addition of propionaldehyde to phenol; low water densities at 400 °C favored *ortho* addition [54]. High water densities seemed to increase the rate of both *ortho* and *para* substitution, but *o*-propylphenol was lost to dehydrogenation to form 2,3-dihydro-2-methylbenzofuran. Hence the *ortho/para* ratio decreased with increasing water density. In a more thorough kinetic study of 2-propanol addition to phenol, the researchers again noted that alkylation rates are increased with increasing water density, but this acceleration cannot entirely be attributed to $[H^+]$ [55]. A study of the reverse reaction showed that the rate of dealkylation increased with increasing water density whereas the rate of rearrangement was independent of water density [56]. It then follows that phenol alkylation by 2-propanol to *o*-isopropylphenol is optimized at high water densities (0.5 g ml$^{-1}$) [57]. Alkylation of phenol by *t*-BuOH at the *para* position increases with increasing water density; at the *ortho* position, it is independent of water density [58].

Hunter and Savage demonstrated that the addition of $CO_2$ to an HTW reaction mixture can increase the rate of *tert*-butylation of *p*-cresol to achieve more than a 10% yield of 2-*tert*-butylmethylphenol in 120 min with $CO_2$ enrichment, compared with a 7% yield without $CO_2$ [41, 59]. The researchers attribute the rate enhancement to the lower pH achieved by formation of carbonic acid.

### 11.3.2.2 Heck Coupling

Palladium-catalyzed alkene–arene coupling reactions in HTW and SCW were studied by Parsons and co-workers [47, 60]. The general reaction studied was coupling of halobenzene with an alkene, as shown in Scheme 11.3.

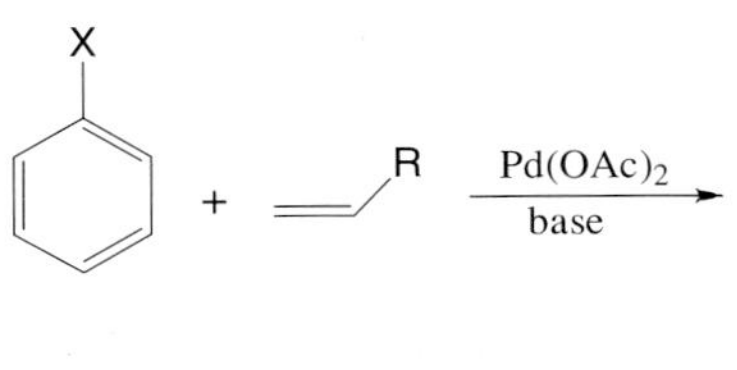

**Scheme 11.3** Heck coupling of a halobenzene with an alkene.

Gron and Tinsley further investigated Heck coupling of p-iodophenol with medium-sized cycloalkenes, producing, for instance, phenylcyclohexenes in 21% yield in 20 min at 225 °C under saturated liquid conditions in the presence of $Pd(OAc)_2$ [61]. As is typical for Heck reactions, electron-withdrawing groups on iodobenzene accelerated the reaction. Addition of LiCl depressed the reaction rate. Addition of n-$Bu_4$NCl enhanced it and modified the distribution of isomers for the addition of cyclohexene to iodobenzene. Interestingly, cyclopentene gave the lowest yield compared with cyclooctene, -heptene, and -hexene. Coupling in DMF at room temperature typically achieves higher reactivity for cyclopentene than for the other cyclic alkenes. The authors suspect that hydrophobic effects explain the reactivity behavior in HTW.

Gron et al. went on to investigate the effects of batch reactor filling factor on the Heck coupling of iodobenzene with cyclohexene at 225 °C and observed that higher filling factors lead to increased yields [62]. However, decreasing the filling factor below the level necessary to achieve a saturated liquid at reaction conditions leads to complicated phase behavior; reactants and products will partition between liquid and vapor phases, making the results more difficult to interpret correctly. The best yield, 47%, was achieved at a water density of 0.84 g $ml^{-1}$ (which achieved saturated liquid water at 225 °C) in the presence of 1 M n-$Bu_4$NBr. The presence of tetrabutylammonium bromide altered the product isomer ratio for 1-: 3-: 4-phenylcyclohexene from 8:31:62 without the additive to 55:16:28. This effect is also seen in organic solvents [62].

Zhang et al. investigated non-Pd-catalyzed Heck couplings in SCW with the more reactive olefin styrene, which polymerizes somewhat during reaction [63]. In the presence of the weak base potassium acetate at 650 K and 25 MPa, a 55.6% yield of stilbene was achieved after 10 min with moderate selectivity: $E: Z = 45:10$. The researchers found that when stronger bases such as $K_2CO_3$, $Na_2CO_3$, $NaHCO_3$, and NaOH were used, hydrogenation of iodobenzene to phenol was the chief transformation. The coupling reaction yielded the same $E: Z$ stereoselectivity regardless of base. Increasing water density accelerated the reaction, in accord with expectations for a simple bimolecular reaction with a negative volume of activation [64]. Water density, however, has no effect upon the $E: Z$ selectivity. Hence it was hypothesized that the transition state structure for $(E)$- versus $(Z)$-stilbene must have similar partial molar volumes. Two postulates regarding the role of water in the reaction were offered. One is formation of carbanion by proton abstraction for the β-C of styrene. The other is a water-catalyzed proton transfer and Ar–C(sp$^2$) bond formation through an eight-membered ring involving two water molecules [64].

### 11.3.2.3 Nazarov Cyclization

Leikoski et al. [65] discovered that the Nazarov reaction of *trans,trans*-dibenzylideneacetone, which normally produces 3,4-diphenyl-2-cyclopentanone in conventional media (acidic, chlorinated solvent at elevated temperatures), instead produces the "abnormal" [66, 67] Nazarov cyclization product 2,3-diphenyl-2-cyclopentenone in as much as 38% yield in $CO_2$-enriched HTW. The unusual product is thought to arise because of water or $CO_2$ addition to the intermediate oxyallyl cation formed after ring closure. Direct deprotonation of the intermediate leads to the ordinary 3,4-disubstituted-2-cyclopentanone. Without $CO_2$, yields of 2,3-diphenyl-2-cyclopentenone are only around 20%.

## 11.3.3
### Condensation

Eckert et al. investigated the Claisen–Schmidt condensation of benzaldehyde with 2-butanone [68]. Most asymmetric ketones would attack benzaldehyde with the alpha carbon attached to the most hydrogens, but 2-butanone behaves differently as both 2-buten-1-ol and 1-buten-2-ol may add to benzaldehyde to form 4-phenyl-3-methyl-3-buten-2-one and 1-phenyl-1-penten-3-one, respectively (Scheme 11.4). Experiments were carried out between 250 and 300 °C, where the ion product of water is roughly $10^{-11}$ mol$^2$ kg$^{-2}$. Condensation to both products occurred at each temperature tested. Interestingly, the selectivity of the butenone was independent of temperature over the range tested, remaining between 10 and 15%, whereas the selectivity of the pentenone decreased with increasing temperature. Other products included second

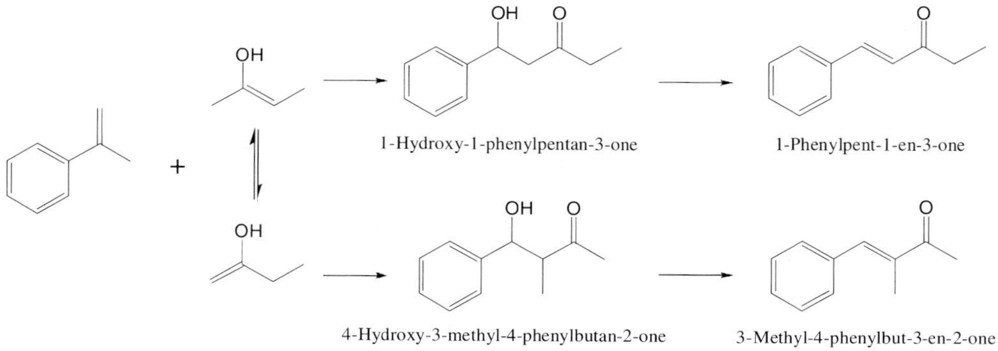

**Scheme 11.4** Claisen–Schmidt condensation of benzaldehyde with 2-butanone yields a pair of enones, and a respective pair of ketol intermediates [68].

additions and oligomerization products. As expected, addition of HCl increased the yield of butenone after 15 min from 0% without acid to 30%. The rate of formation of pentenone also increased slightly from 1 to 10%, as did the formation of side products.

Poliakoff and co-workers developed a means of synthesizing simple benzimidazoles in near- and supercritical water from a 1,2-phenylenediamine and benzoic acid (Scheme 11.5) [69].

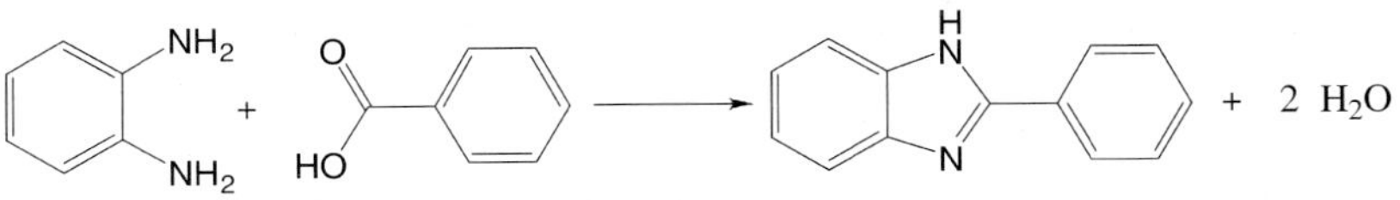

**Scheme 11.5** Condensation of 1,2-phenylenediamine with benzoic acid yields 2-phenylbenzimidazole [69].

This reaction is normally carried out under highly acidic conditions, often coupled with great thermal excitation. It was found that reaction in HTW removes the need for acid and simplifies the purification while still achieving good yields (near 90%) in reasonable reaction times (4 h) when operating at 350 °C and 21.7 MPa. 1,2-Phenylenediamine was then reacted with various simple aliphatic carboxylic acids, diacids, and aromatic dicarboxylic acids at temperatures ranging from 210 to 350 °C for 1–4 h, depending on reactant stability and reactivity. Moderate to excellent yields were achieved [69]. This work was followed by the synthesis of phthalimide derivatives in water–ethanol (1 : 1 v/v) between 260 and 380 °C at 1240–4800 psi [70].

Continuing with condensation reactions, Comisar and Savage [45] investigated the synthesis of benzalacetone from benzaldehyde and acetone and the related synthesis of chalcone from benzaldehyde and acetophenone. Maximum yields achieved were 24% for benzalacetone after 5 h at 250 °C and 21% for chalcone after 15 h at 250 °C. A reaction network (Scheme 11.6) was proposed to account for the various products

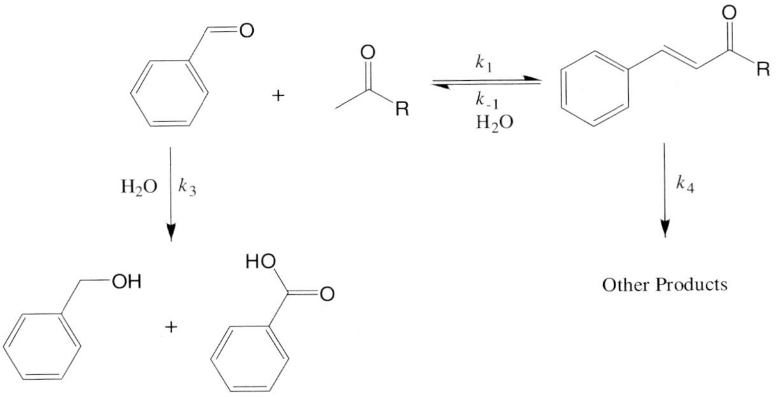

**Scheme 11.6** Condensation of benzaldehyde with ketones yields α,β-unsaturated ketones such as benzalacetone and chalcone [45].

formed. Data were fit to the proposed model and the relevant kinetic parameters determined [45].

A study of the effects of pH on this reaction was also carried out by adding HCl or NaOH to the reaction vessel. Rate acceleration was observed in both cases, but more so with added acid. At pH 4.25, after 1 h at 250 °C, the yield of benzalacetone was 16%, compared with 6% with no added acid.

A portion of the paper [45] treated the degradation of benzalacetone via Michael addition followed by intramolecular Meerwein–Pondorf–Verley reduction followed by retro-aldol condensation to form benzaldehyde and acetaldehyde. Lu *et al.* were very interested in this degradation reaction and studied the related retro-aldol condensation of cinnamaldehyde to benzaldehyde and acetaldehyde [71]. They found that the addition of $NH_3$ to the reaction mixture promoted the reaction rate.

The dehydration of 1,4-butanediol to form tetrahydrofuran is an industrially important reaction, the ease of which may be improved significantly by HTW. The most common process, the Reppe process, is typically carried out at temperatures above 100 °C, pressures near atmospheric, and with catalysis by either mineral acids which must later be neutralized, or by aluminum silicates or ion-exchange resins which must be regenerated. Use of HTW may minimize the cost of waste removal for this process. Richter and Vogel studied the reaction in HTW in a flow reactor between 300 and 400 °C and achieved selectivities of nearly 100%. They showed that the correlation between $K_w$ and reaction rate is weak [72]. Nagai *et al.* [73] showed that alone, the $K_w$ of water could not explain the reactivity of 1,4-butanediol entirely. Hence the "water-induced" pathway must be significant in neutral water. Hunter *et al.* found that the addition of $CO_2$ to the reaction mixture did not increase yield as significantly as one would expect from an acid-catalyzed mechanism [44]. They conducted experiments over a wide range of pH values (between 2 and 10) with the addition of HCl or NaOH at 200, 250, and 300 °C, and proposed that under acidic conditions $H^+$ is the dominant proton donor, but under neutral or basic conditions $H_2O$ serves in this role. Further, under acidic or neutral conditions the protonated oxonium ion of tetrahydrofuran is deprotonated predominantly by water; whereas under basic conditions $OH^-$ serves this role [44]. Of course, the exact pH of "neutral" water changes with temperature, as does the onset or cut-off of specific-acid catalysis for butanediol dehydration. Accounting for these effects, Hunter *et al.* explained kinetically why addition of acid increases the rate more at lower temperatures than a commensurate addition of acid at higher temperatures.

### 11.3.4
#### Hydrolysis

Researchers studying benzoic acid ester hydrolysis in near-critical water demonstrated autocatalysis with an inflection point in a plot of the conversion of *n*-propyl benzoate to benzoic acid versus time [74]. The Hammett reaction constant ρ was determined as nearly zero, signifying that the transition state of the rate-determining step maintains the charge of the ground state. This is consistent with acid-catalyzed hydrolysis of esters, and is contrary to what would be expected if the reaction were base catalyzed. Hence the ordinary $A_{AC}2$ mechanism that is normally expected for acid-catalyzed hydrolysis also dominates in near-critical water.

Eckert and co-workers later published a more complete study with experiments converting methoxy- and ethoxybenzene derivatives to the alcohols [14]. Hydrolysis of anisole may proceed by three mechanisms: acid-catalyzed, base-catalyzed, or $S_N2$ addition. Experimentally, a positive ρ value was determined, which is inconsistent with an acid-catalyzed mechanism. This is in agreement with Klein et al. [15], who more carefully decoupled solvent and substituent effects in their Hammett plot analysis to determine $ρ_H = 1.8$. Eckert and co-workers basified their reaction medium to estimate the rate of the base-catalyzed pathway. With this parameter known, the rate of $S_N2$ addition was then determined. Nucleophilic attack by water was found to be the dominant pathway for hydrolysis of anisole in neutral HTW, mainly because the concentration of water is much greater than that of hydroxide ions. The earlier analysis of Klein et al. also favored the $S_N2$ pathway.

The value of this study is in the physical understanding that it achieved of how the reaction proceeded in HTW. It is important to note, however, that Hammett plots only allow one to conclude whether a particular mechanism with a particular rate-determining step is either consistent or inconsistent with an experimentally determined ρ value. The explaining potential of a ρ value is limited in this way because all that a Hammett plot really measures is the accumulation or depletion of electron density in the transition state relative to the ground state. This is an important distinction to make because the expected ρ value for a particular mechanism changes based on which step is rate determining.

Nitrile hydrolysis was first studied in HTW by Iyer and Klein, who showed that butyronitrile autocatalytically hydrolyzed first to butanamide and then to butyric acid [18]. Similarly, acetonitrile was shown to hydrolyze first to acetamide and then to acetic acid in SCW [75, 76]. Kinetic observables for the hydrolyses of benzonitrile, benzamide, and iminodiacetonitrile were also studied in HTW [77, 78].

In a large study of the hydrolysis of dibenzyl ether, benzyl tert-butyl ether, methyl tert-butyl ether, methyl benzoate, and diphenyl carbonate in HTW, both with and without added acid or base, it was discovered that the apparent reaction order in [H$^+$] did not exceed 0.2 for any example ether, ester, or carbonate [40]. This was contrary to the widespread notion that such hydrolysis reactions were specific-acid catalyzed in HTW, a scenario in which the apparent reaction order in [H$^+$] is expected to be one. This result is consistent with catalysis by water molecules.

Ogawa et al. studied the synthesis of polyorganosiloxanes from alkoxysilanes by hydrolysis and condensation without added catalyst [79]. At 300 °C, an average MW of 1550 was achieved within 15 min. The effects of monomer to water ratio, pressure, temperature, and substrate upon MW were broadly studied. The authors believe that supercritical water could be shown to be an advantageous medium for the formation of polyorganosiloxanes [79].

The hydrolysis of 6-aminocapronitrile followed by cyclization to form ε-caprolactam was studied in HTW [80]. Caprolactam finds importance as an intermediate in the production of nylon-6. Under the proper conditions, conversion of

6-aminocapronitrile reached 94% and the yield of ε-caprolactam reached 90% in under 2 min in a flow reactor (400 °C, 400 bar, 30% by volume initial concentration of starting material). The researchers studied the effect of temperature, pressure, and residence time upon yield and selectivity.

### 11.3.5
**Rearrangements**

The Beckmann and pinacol rearrangements occur in SCW without any added catalyst. Of the temperatures examined, Beckmann rearrangement of cyclohexanone oxime into ε-caprolactam occurred fastest at 380 °C and 22.1 MPa with an observed first-order rate constant of $8160 \pm 750\, s^{-1}$. Observed first-order rate constants for pinacol rearrangement of 2,3-dimethyl-2,3-dihydroxybutane to 3,3-dimethyl-2-butanone were determined by IR measurements to be on the order of tens of thousands of $s^{-1}$. The researchers attributed the observed rate enhancements for these systems to an increased local proton concentration about the substrates. A theoretical study of neopentyl and pinacol rearrangements showed that the mechanism does not involve the formation of carbocations [81]. In particular, pinacol rearrangement of the model substrates 2,3-dimethyl-2,3-butanediol and 2,3-diphenyl-2,3-butanediol proceeded by a concerted process where proton transfer was promoted by an H-bonded relay. The reaction was modeled with interaction with 12 water molecules and one hydronium molecule [81].

A quantum mechanical/molecular mechanical method calculation with energy representation was carried out to study the thermodynamics of Beckmann rearrangement of acetone oxime in the supercritical state [82]. The activation energy was reduced by $12.3\, kcal\, mol^{-1}$ if two molecules of water participated in the reaction. Also, the transition state was stabilized in comparison with the reactant state by $2.7\, kcal\, mol^{-1}$ with the participation of two water molecules [82].

In their study of the rearrangement of benzil, Comisar and Savage [46] observed diphenylketene, benzophenone, benzhydrol, and diphenylmethane, but not the expected product, benzilic acid, which was presumed to react away more quickly than it was formed. It was discovered that whereas base catalysis is the sole mechanism for the rearrangement in the conventional water–dioxane system at 100 °C, the reaction is acid, base, and water catalyzed in HTW, depending on pH. New mechanisms for the acid- and water-catalyzed routes were posited to account for this behavior. Further, selectivities to rearrangement products and decomposition products were pH dependent. While decomposition was favored at a near-neutral pH, rearrangement rates were lowest at pH 3 and increased with increasing pH. A later study [83] of the benzil–benzilic acid rearrangement with no added catalyst tested the proposed kinetic pathways by first reacting proposed intermediates to verify an overall reaction network, and by fitting kinetic data to the proposed reaction network with favorable results.

The work of Wang et al. [84] confirmed many of the observations of Comisar and Savage in their study of benzil rearrangement. They studied the oxidation of benzhydrol and benzoin in pure HTW. After 3 h at 460 °C, benzhydrol was nearly completely consumed (>99% conversion) to benzophenone (63% yield) and diphenylmethane (10% yield). At 440 °C with no water present, pyrolysis of benzhydrol gave a 48% yield of benzophenone and a 53% yield of diphenylmethane, likely through disproportionation. The higher ratio of oxidation product to reduction product achieved in the presence of water led the authors to believe that oxidation of benzhydrol occurs through some mechanism other than disproportionation in HTW. Hydrogen gas was detected in some experiments. Hydrogen evolution is consistent with the proposed hydrogen-bonded eight-membered ring transition state structure illustrated here, which was proposed by Wang et al. [84]

Benzoin reacted more readily than benzhydrol to form a mixture of products: benzil, benzyl phenyl ketone, benzaldehyde, benzhydrol, benzophenone, and diphenylmethane. In their studies, the researchers were careful to exclude molecular oxygen from their reactors by purging the water with nitrogen and sealing the reactors under nitrogen.

### 11.3.6
**Hydration/Dehydration**

Akiya and Savage [85] found that cyclohexanol dehydration to the alkene occurred without catalyst present. They determined that two pathways for cyclohexanol dehydration are probable in HTW: unimolecular and bimolecular elimination. Bimolecular elimination seemed to be the only mechanism at work at low temperatures of 250 and 275 °C. However, increasing the temperature of the system beyond 275 °C allowed for the formation of methylcyclopentanes, which form by rearrangement of cyclohexyl carbocation. This suggested unimolecular elimination. Increasing the water density under supercritical conditions (380 °C) further improved the yields of methylcyclopentanes, suggesting further enhancement of $E1$.

The $E2$ mechanism's requirement of planarity allows for only two possible transition states, the most unfavorable of which is synperiplanar elimination from the high-energy and ephemeral boat configuration. The lower energy antiperiplanar elimination from the chair structure with oxygen axial constitutes a configuration which costs roughly 0.5 kcal mol$^{-1}$ – nothing unmanageable in HTW.

Hunter and Savage showed that addition of $CO_2$ to HTW could increase the reaction rates of acid-catalyzed reactions such as cyclohexanol dehydration to cyclohexene [41]. As shown in Figure 11.5, the addition of 8 mg of dry-ice to cyclohexanol in saturated HTW at 250 °C for 30 min increased the yield of cyclohexene from $10 \pm 3$ to $22 \pm 2\%$. At 275 °C, addition of 10 mg of dry-ice increased yields from $7 \pm 3$ to $18 \pm 4\%$.

The dimerization of hemiterpene alcohols prenol and 2-methyl-3-buten-2-ol to form a suite of monoterpene alcohols was carried out without added catalyst in HTW. The most promising result from this study was the formation of linalool, geraniol, nerol, lavandulol, and α-terpineol in yields of 9, 10, 1, 24, and 10%, respectively, at 450 °C and 40 MPa in under 5 s [86].

The synthesis of phthalimide derivatives was studied in a batch reactor in HTW [70]. Nucleophilic attack by an amine on o-phthalic acid followed by ring closure yielded a substituted phthalimide plus two molecules of water. Rather than cyclization, bis-amidation may occur, but would be followed by intramolecular deamination to yield the same substituted phthalimide. Formation of an ammonium salt with any of the carboxylic acids detracted from the phthalimide yield. Purification of phthalimides is normally done by recrystallization from an ethanol–water solvent system. With the intention of developing a reaction/purification process, the researchers therefore chose an ethanol–water mixture as the solvent system. Various ratios of water to ethanol were tested using N-phenylphthalimide and N-benzylphthalimide as model nucleophiles. A 1 : 1 volume ratio of water to ethanol gave high-purity crystals in good yields for both substrates, and was therefore chosen as the solvent system for most of the further experiments with different substrates.

A variety of amines were tested, all with reaction times between 5 and 12 min. Other functional groups (halogen, nitro) and heteroaromatics (pyridine) were found to perform poorly in this reaction due to increased salt formation with the acid, or lack of stability under the reaction conditions. The researchers expect such problems to persist for other functional groups (cyano, ester, ether, *etc.*). Three amino acids were applied as amines – 4-aminobenzoic acid, L-phenylalanine, and glycine – and all were decarboxylated in the phthalimide product [70].

Aida *et al.* studied the dehydration of D-glucose in HTW water at 40, 70, and 80 MPa [87]. Previous work elucidated the major products and reactive pathways of sugar molecules in HTW. The authors added to these studies by focusing on changes in yield and selectivity due to changes in pressure at subcritical (350 °C) and supercritical (400 °C) temperatures. Although the yields of the desired product 5-hydroxymethylfurfural never rose above 8%, it was concluded that high temperatures and pressures and short residence times increased the selectivity of 5-hydroxymethylfurfural. Short residence times are needed to prevent further reaction to 1,2,4-benzenetriol. High temperatures and pressures with long residence times, of course, increase the yields of 1,2,4-benzenetriol and furfural, which is a decomposition product of 5-hydroxymethylfurfural. The researchers suggested a new mechanism for D-glucose decomposition which they feel better represents their product distribution vs time data [87].

This work was pursued with other sugars, the decomposition products of D-glucose being compared with those of D-fructose. D-Fructose gave higher yields of 5-hydroxymethylfurfural whereas D-glucose gave higher yields of furfural. The latter is contrary to expectations based on the current reaction pathways in the literature, so another reaction scheme was proposed [88].

HTW has been found to be a useful medium for forming cyclic dipeptides such as cyclo(Gly–Gly) by the dehydration of linear dipeptides in HTW at 240–300 °C and 20 MPa [89]. Hydrolysis of the linear dipeptides formed amino acids. The cyclic peptides could also hydrolyze back to the linear dimer.

The conversion of propylene glycol to 2-methyl-2-pentenal was studied in HTW at 300 °C with and without salt additives. Without additives, the reaction yielded 1.8 wt% aldehyde in 2 h. In the presence of 1 wt% $ZnCl_2$, 59 wt% 2-methyl-2-pentenal was produced. No reaction took place with the addition of $Na_2CO_3$. Presumably, the transformation begins with dehydration of propylene glycol to propionaldehyde. Aldol condensation with another propylene glycol molecule produces 3-hydroxy-2-methylpentanal, which itself undergoes dehydration to 2-methyl-2-pentenal [90].

## 11.3.7
### Elimination

The thermal energy of HTW makes it an attractive medium for decomposition reactions. Although total decomposition is not covered in this review, some decomposition reactions are productive of small molecules. In particular, there is a trend to use HTW for "producing" amino acids and organic acids from waste protein streams, as in the seafood industry in Japan [91].

Dunn *et al.* [92]. studied the decarboxylation of various aromatic carboxylic acids as a contribution to the use of HTW in the purification and synthesis of aromatic diacids. Benzoic acid was the most stable aromatic acid tested. Kinetic analysis revealed that terephthalic acid and trimellitic anhydride decarboxylated with autocatalysis. The $CO_2$ formed during decarboxylation formed carbonic acid in solution, which lowered the pH of the reaction medium and catalyzed further decarboxylation.

The oxidative decarboxylation of benzoic acid to phenol was studied over heterogeneous catalysts (NiO, CuO, Carulite, $MnO_2$ and $Al_2O_3$) in HTW in a flow reactor [93]. Carulite 300 was found to be the best catalyst of those tested for the experimental conditions explored. As high as a 65% yield of phenol can be achieved at 134 min with this catalyst (340 °C, 14 MPa, in the presence of 18 mol% NaOH), but the activity decreases substantially shortly thereafter. However, after the catalyst was regenerated with oxygen, which was fed to the reactor in the form of hydrogen peroxide, the yield rebounded to a respectable 70%. Without NaOH, which forms the promoter sodium benzoate *in situ*, the highest yield observed was only 46%. The researchers also investigated temperature, pressure, and oxygen addition methods. One drawback of the approach is that organic material adsorbs on the catalyst and oxidizes to form $CO_x$ gases during catalyst regeneration with $O_2$ ($H_2O_2$).

The decarboxylative oxidation of benzoic acid to phenol was pursued for its potential in improving the Dow process, which produces phenol on an industrial scale in two stages. The first converts toluene to benzoic acid, and the second benzoic acid to phenol. Recently, other researchers have produced phenol in 20% yield by direct oxidation of benzene [93, 94].

## 11.3.8
### Partial Oxidation to Form Carboxylic Acids

First studied by Holliday *et al.* [95], partial oxidation of alkyl aromatics to form carboxylic acids is one of the best studied systems in HTW. In particular, the oxidation of *p*-xylene to form terephthalic acid has received a great deal of research attention for its success and industrial significance [96–98]. For example, Dunn and Savage explored the effects of various process variables upon the yield and selectivity of terephthalic acid from *p*-xylene [96]. Of seven different catalysts tested [$MnBr_2$, $CoBr_2$, Mn–Co–Br, Mn–Ni–Zr–Br, Mn–Co–Hf–Br, Mn–(OAc)–Br, and Mn–Hf–Br], $MnBr_2$ produced the highest yield of terephthalic acid (49 ± 8%) [96, 98, 99]. This was a peculiar result because Mn–Co systems delivered higher yields than $MnBr_2$ alone in acetic acid reactions. Other researchers confirmed that $MnBr_2$ achieved better yield and selectivity than $NiBr_2$ and $CoBr_2$ [100].

Temperatures between 250 and 400 °C were tested. At 250 °C the yields were low, but many intermediates were produced. At 350 °C, decarboxylation destroyed a significant portion of the terephthalic acid. The best yields were delivered at 300 °C. Experiments in a 440 ml autoclave reactor allowed the use of air as the oxidant and quantification of CO and $CO_2$ yields. An 80% yield of terephthalic acid was demonstrated at 300 °C; yields at both lower and higher (including supercritical) temperatures were much lower [99]. Osada and Savage [115, 116] discovered that the manner of oxygen addition significantly affected *p*-xylene conversion and terephthalic acid selectivity. Small, quick, discrete bursts of oxygen addition led to high selectivities (>90%) of terephthalic acid, whereas continuous oxygen feed did not. In their experiments, Osada and Savage also used a much higher loading of *p*-xylene (0.2 mol l$^{-1}$) than was previously used in research on this system (typically less than 0.05 mol l$^{-1}$).

Dunn and Savage examined the economic and environmental impact of the terephthalic acid synthesis in HTW and SCW [12]. They compared four different HTW/SCW plants with the conventional acetic acid system. The four hypothetical plants differed in reactor temperature and pressure and in the presence or absence of air separation. The HTW process (300 °C) appeared to be superior to the SCW process, and competitive with the existing technology, which uses acetic acid as the reaction medium. These results were encouraging, as they demonstrated the potential feasibility of HTW in productive industry.

Additional work has been done on the oxidation of other alkyl-substituted benzenes, naphthalenes, and pyridines to the corresponding acids [93, 100, 101] A range of aliphatic aromatics of varying sterics and electronics underwent oxidation to the acid derivative, showing that the synthesis was general, although usual nuances were revealed. For example, lower yields are observed for *m*-xylene oxidation (66%) than for *p*-xylene (90%) because there is no resonance stabilization of the radical intermediate at the *meta* position as there is for oxidation at the *ortho* and *para* positions [101].

Partenheimer and co-workers used extended X-ray absorption fine structure (EXAFS) and X-ray absorption near-edge structure (XANES) spectroscopic methods to study the coordination geometry of $MnBr_2$ solutions and to explain the importance of bromide concentration for this reaction [102]. Under dilute ambient conditions in water, Mn(II) was octahedral and fully hydrated with six molecules of water. Upon introducing a solution of 1.0 M or less $MnBr_2$ to the supercritical condition, the octahedral $[Mn^{II}(H_2O)_6]^{2+}$ transformed into the tetrahedral $[Mn^{II}(H_2O)_2(Br^-)_2]$. An excess of $Br^-$ ions led to more $Br^-$ ions occupying the first solvation shell and contact ion pairs with Mn, with the coordination number of water decreasing proportionally. Further, acetate deactivated the Mn catalyst through the formation of insoluble MnO [103].

### 11.3.9
### C−C Bond Cleavage

Bisphenol A decomposed in HTW to form equimolar amounts of phenol and p-isopropenylphenol [104]. The latter product, which is the desired one in this transformation, either hydrated to acetone and phenol or hydrogenated to p-isopropylphenol. Bisphenol A decomposition in HTW was first order in bisphenol A and occurred by specific-acid, specific-base, and general-water catalysis. A three-parameter model fitted to the experimental data based on a base-catalyzed mechanism portrayed the data well. Using their mechanistic understanding of this transformation, Hunter and Savage explained the reactivity observed for other biaryl groups linked by methylene bridges [43]. Savage and co-workers also modeled the decomposition of bisphenol E in HTW, which primarily produced phenol and 4-vinylphenol, the latter of which further reacted to form 4-ethylphenol [105]. Vinylphenol oligomers were also suspected.

### 11.3.10
### H–D Exchange

The deuterating ability of high-temperature deuterium oxide was recognized early on [106, 107] and efforts continue to exploit its potential. Studies have focused on the deuteration of 2-methylnaphthalene, eugenol, resorcinol, and phenol [108–110]. Although it was well known that the *ortho* and *para* positions of phenol were deuterated upon heating [111], it was found that the *meta* position could also be deuterated in supercritical deuterium oxide [110].

The kinetics of hexane deuteration were studied in supercritical water at 380 and 400 °C with acid catalysis by DCl. In contrast with the known pathway for deuteration in magic acid, no evidence was seen for hydride abstraction in supercritical $D_2O$; carbocation rearrangement products were not found, nor was hydrogen gas evolution detected. Hence it was concluded that hydride abstraction to form carbocations either does not occur at all, or is too slow to be measured by the applied experimental procedure [112]. Superacids, such as magic acid, are typically used for the deuteration of alkanes [111].

### 11.3.11
### Amidation

1-Hexanol and acetamide formed N-hexylacetamide in 75% yield after 10 min at 400 °C without catalyst by amination of 1-hexanol followed by amidation [113]. This constitutes a new method of producing amides from primary alcohols. Upon further study of the reaction mechanism, it was determined that yield could be increased further with the addition of ammonium acetate [114].

## Acknowledgments

We gratefully acknowledge support from the National Science Foundation (CTS-0625641) and the ACS-PRF (45642-AC4).

## References

1 Peterson, A.A. *et al.* (2008) Thermochemical biofuel production in hydrothermal media: a review of sub- and supercritical water technologies. *Energy and Environmental Science*, **1**, 32–65.
2 Savage, P.E. *et al.* (1995) Reactions at supercritical conditions: applications and fundamentals. *AIChE Journal*, **41**, 1723–1778.
3 Savage, P. (1999) Organic chemical reactions in supercritical water. *Chemical Reviews*, **99**, 603–622.
4 Akiya, N. and Savage, P. (2002) Roles of water for chemical reactions in high-temperature water. *Chemical Reviews*, **102**, 2725–2750.
5 Bröll, D. *et al.* (1999) Chemistry in supercritical water. *Angewandte Chemie International Edition*, **38**, 2998–3014.
6 Haar, L., Gallagher, J.S. and Kell, G.S. (1984) *NBS/NRS Steam Tables*, McGraw-Hill, New York.
7 Chandler, K. *et al.* (1998) Phase equilibria for binary aqueous systems from a near-critical water reaction apparatus. *Industrial and Engineering Chemistry Research*, **37**, 3515–3518.
8 Chen, H. and Wagner, J. (1994) An apparatus and procedure for measuring mutual solubilities of hydrocarbons + water:benzene + water from 303 to 373 K. *Journal of Chemical and Engineering Data*, **39**, 470–474.
9 Anderson, F. and Prausnitz, J. (1986) Mutual solubilities and vapor pressures for binary and ternary aqueous systems containing benzene, toluene, *m*-xylene, thiophene and pyridine in the region 100–200 °C. *Fluid Phase Equilibria*, **32**, 63–76.
10 Matubayasi, N., Wakai, C. and Nakahara, M. (1997) Structural study of supercritical water. I. Nuclear magnetic resonance spectroscopy. *Journal of Chemical Physics*, **107**, 9133.
11 Mizan, T., Savage, P. and Ziff, R. (1996) Temperature dependence of hydrogen bonding in supercritical water. *Journal of Physical Chemistry*, **100**, 403–408.
12 Dunn, J.B. and Savage, P.E. (2003) Economic and environmental assessment of high-temperature water as a medium for terephthalic acid synthesis. *Green Chemistry*, **5**, 649–655.
13 Connors, K.A. (1990) *Chemical Kinetics: the Study of Reaction Rates in Solution*, Wiley-VCH Verlag GmbH, Weinheim.
14 Patrick, H. *et al.* (2001) Near-critical water: a benign medium for catalytic reactions. *Industrial and Engineering Chemistry Research*, **40**, 6063–6067.
15 Klein, M.T., Mentha, Y.G. and Torry, L.A. (1992) Decoupling substituent and solvent effects during hydrolysis of substituted anisoles in supercritical water. *Industrial and Engineering Chemistry Research*, **31**, 182–187.
16 Huppert, G.L. *et al.* (1989) Hydrolysis in supercritical water: identification and implications of a polar transition state.

*Industrial and Engineering Chemistry Research*, **28**, 161–165.
17. Anikeev, V., Yermakova, A. and Goto, M. (2004) Decomposition and oxidation of aliphatic nitro compounds in supercritical water. *Industrial and Engineering Chemistry Research*, **43**, 8141–8147.
18. Iyer, S.D. and Klein, M.T. (1997) Effect of pressure on the rate of butyronitrile hydrolysis in high-temperature water. *Journal of Supercritical Fluids*, **10**, 191–200.
19. Townsend, S.H. et al. (1988) Solvent effects during reactions in supercritical water. *Industrial and Engineering Chemistry Research*, **27**, 143–149.
20. Oka, H. et al. (2002) Evidence for a hydroxide ion catalyzed pathway in ester hydrolysis in supercritical water. *Angewandte Chemie International Edition*, **41**, 623–625.
21. Mizan, T.I., Savage, P.E. and Ziff, R.M. (1994) Molecular dynamics of supercritical water using a flexible SPC model. *Journal of Physical Chemistry*, **98**, 13067–13076.
22. Mizan, T.I., Savage, P.E. and Ziff, R.M. (1997) Fugacity coefficients for free radicals in dense fluids: $HO_2$ in supercritical water. *AIChE Journal*, **43**, 1287–1299.
23. Matubayasi, N. and Nakahara, M. (2005) Hydrothermal reactions of formaldehyde and formic acid: free-energy analysis of equilibrium. *Journal of Chemical Physics*, **122**, 074509.
24. Balbuena, P.B., Johnston, K.P. and Rossky, P.J. (1994) Molecular simulation of a chemical reaction in supercritical water. *Journal of the American Chemical Society*, **116**, 2689–2690.
25. Akiya, N. and Savage, P. (2000) Effect of water density on hydrogen peroxide dissociation in supercritical water. 2. Reaction kinetics. *Journal of Physical Chemistry A*, **104**, 4441–4448.
26. Boero, M. et al. (2004) Hydrogen bond driven chemical reactions: Beckmann rearrangement of cyclohexanone oxime into ε-caprolactam in supercritical water. *Journal of the American Chemical Society*, **126**, 6280–6286.
27. Taylor, J. et al. (2002) Multiscale reaction pathway analysis of methyl *tert*-butyl ether hydrolysis under hydrothermal conditions. *Industrial and Engineering Chemistry Research*, **41**, 1–8.
28. Harano, Y., Sato, H. and Hirata, F. (2000) Solvent effects on a Diels–Alder reaction in supercritical water: RISM-SCF Study. *Journal of the American Chemical Society*, **122**, 2289–2293.
29. Morooka, S. et al. (2005) Hydrothermal carbon–carbon bond formation and disproportionations of $C_1$ aldehydes: formaldehyde and formic acid. *Journal of Physical Chemistry A*, **109**, 6610–6619.
30. Yoshida, K. et al. (2004) NMR spectroscopic evidence for an intermediate of formic acid in the water-gas-shift reaction. *Journal of Physical Chemistry A*, **108**, 7479–7482.
31. Brill, T. (2000) Geothermal vents and chemical processing: the infrared spectroscopy of hydrothermal reactions. *Journal of Physical Chemistry A*, **104**, 4343–4351.
32. Ikushima, Y. et al. (2000) Acceleration of synthetic organic reactions using supercritical water: noncatalytic Beckmann and pinacol rearrangements. *Journal of the American Chemical Society*, **122**, 1908–1918.
33. Fang, Z. et al. (1999) Phase behavior and reaction of polyethylene terephthalate–water systems at pressures up to 173 MPa and temperatures up to 490 °C. *Journal of Supercritical Fluids*, **15**, 229–243.
34. Fang, Z. and Kozinski, J.A. (2001) Phase changes of benzo[a]pyrene in supercritical water combustion. *Combustion and Flame*, **124**, 255–267.
35. Sugiyama, M. et al. (2002) Shadowgraph observation of supercritical water oxidation progress of a carbon particle. *Ind. Eng. Chem. Res*, **41**, 3044–3048.

36 Sue, K. et al. (2004) Apparatus for direct pH measurement of supercritical aqueous solutions. *Journal of Supercritical Fluids*, **28**, 287–296.

37 Liu, C., Snyder, S. and Bard, A. (1997) Electrochemistry in near-critical and supercritical fluids. 9. Improved apparatus for water systems (23–385 °C). The oxidation of hydroquinone and iodide. *Journal of Physical Chemistry B*, **101**, 1180–1185.

38 Mesmer, R.E. et al. (1988) Thermodynamics of aqueous association and ionization reactions at high temperatures and pressures. *Journal of Solution Chemistry*, **17**, 699–718.

39 Macdonald, D.D. et al. (1992) Measurement of pH in subcritical and supercritical aqueous systems. *Journal of Solution Chemistry*, **21**, 849–881.

40 Comisar, C. et al. (2008) Effect of pH on ether, ester, and carbonate hydrolysis in high-temperature water. *Industrial and Engineering Chemistry Research*, **47**, 577–584.

41 Hunter, S. and Savage, P. (2003) Acid-catalyzed reactions in carbon dioxide-enriched high-temperature liquid water. *Industrial and Engineering Chemistry Research*, **42**, 290–294.

42 Hunter, S.E. and Savage, P.E. (2008) Quantifying rate enhancements for acid catalysis in $CO_2$-enriched high-temperature water. *AIChE Journal*, **54**, 516–528.

43 Hunter, S. and Savage, P. (2004) Kinetics and mechanism of *p*-isopropenylphenol synthesis via hydrothermal cleavage of bisphenol A. *Journal of Organic Chemistry*, **69**, 4724–4731.

44 Hunter, S., Ehrenberger, C. and Savage, P. (2006) Kinetics and mechanism of tetrahydrofuran synthesis via 1,4-butanediol dehydration in high-temperature water. *Journal of Organic Chemistry*, **71**, 6229–6239.

45 Comisar, C.M. and Savage, P.E. (2004) Kinetics of crossed aldol condensations in high-temperature water. *Green Chemistry*, **6**, 227–231.

46 Comisar, C.M. and Savage, P.E. (2005) The benzil–benzilic acid rearrangement in high-temperature water. *Green Chemistry*, **7**, 800–806.

47 Reardon, P. et al. (1995) Palladium-catalyzed coupling reactions in superheated water. *Organometallics*, **14**, 3810–3816.

48 Crittendon, R.C. and Parsons, E.J. (1994) Transformations of cyclohexane derivatives in supercritical water. *Organometallics*, **13**, 2587–2591.

49 Yu, J. and Savage, P. (1998) Decomposition of formic acid under hydrothermal conditions. *Industrial and Engineering Chemistry Research*, **37**, 2–10.

50 Garcia-Verdugo, E. et al. (2006) *In situ* generation of hydrogen for continuous hydrogenation reactions in high temperature water. *Green Chemistry*, **8**, 359–364.

51 Morrison, R. and Boyd, R. (1966) *Organic Chemistry*, Allyn and Bacon, Boston.

52 Chandler, K. et al. (1997) Alkylation reactions in near-critical water in the absence of acid catalysts. *Industrial and Engineering Chemistry Research*, **36**, 5175–5179.

53 Sato, T. et al. (2001) Non-catalytic and selective alkylation of phenol with propan-2-ol in supercritical water. *Chemical Communications*, 1566–1567.

54 Sato, T. et al. (2002) Regioselectivity of phenol alkylation in supercritical water. *Green Chemistry*, **4**, 449–451.

55 Sato, T. et al. (2002) *Ortho*-selective alkylation of phenol with 2-propanol without catalyst in supercritical water. *Industrial and Engineering Chemistry Research*, **41**, 3064–3070.

56 Sato, T. et al. (2002) Dealkylation and rearrangement kinetics of 2-isopropylphenol in supercritical water. *Industrial and Engineering Chemistry Research*, **41**, 3124–3130.

57 Sato, T. *et al.* (2004) Control of reversible reactions in supercritical water: I. Alkylations. *AIChE Journal*, **50**, 665–672.

58 Sato, T., Ishiyama, Y. and Itoh, N. (2006) Non-catalytic anti-Markovnikov phenol alkylation with supercritical water. *Chemistry Letters*, **35**, 716–717.

59 Hunter, S.E. and Savage, P.E. (2004) Recent advances in acid- and base-catalyzed organic synthesis in high-temperature liquid water. *Chemical Engineering Science*, **59**, 4903–4909.

60 Diminnie, J., Metts, S. and Parsons, E.J. (1995) *In situ* generation and Heck coupling of alkenes in superheated water. *Organometallics*, **14**, 4023–4025.

61 Gron, L.U. and Tinsley, A.S. (1999) Tailoring aqueous solvents for organic reactions: Heck coupling reactions in high temperature water. *Tetrahedron Letters*, **40**, 227–230.

62 Gron, L.U. *et al.* (2001) Heck reactions in hydrothermal, sub-critical water: water density as an important reaction variable. *Tetrahedron Letters*, **42**, 8555–8557.

63 Zhang, R. *et al.* (2003) Noncatalytic Heck coupling reaction using supercritical water. *Chemical Communications*, 1548–1549.

64 Zhang, R. *et al.* (2004) Heck coupling reaction of iodobenzene and styrene using supercritical water in the absence of a catalyst. *Chemistry – A European Journal*, **10**, 1501–1506.

65 Leikoski, T. *et al.* (2005) Unusual Nazarov cyclization in near-critical water. *Organic Process Research and Development*, **9**, 629–633.

66 Hirano, S., Hiyama, T. and Nozaki, H. (1974) Acid-catalyzed cyclization of cross-conjugated dienone moiety to cyclopentenones. *Tetrahedron Letters*, **15**, 1429–1430.

67 Hirano, S. *et al.* (1980) Abnormal Nazarov reaction. A new synthetic approach to 2,3-disubstituted 2-cyclopentenones. *Bulletin of the Chemical Society of Japan* **53**, 169–173.

68 Nolen, S.A. *et al.* (2003) The catalytic opportunities of near-critical water: a benign medium for conventionally acid and base catalyzed condensations for organic synthesis. *Green Chemistry*, **5**, 663–669.

69 Dudd, L.M. *et al.* (2003) Synthesis of benzimidazoles in high-temperature water. *Green Chemistry*, **5**, 187–192.

70 Fraga-Dubreuil, J. *et al.* (2007) Rapid and clean synthesis of phthalimide derivatives in high-temperature, high-pressure $H_2O$/EtOH mixtures. *Green Chemistry*, **9**, 1067–1072.

71 Lu, X., Li, Z. and Gao, F. (2006) Base-catalyzed reactions in $NH_3$-enriched near-critical water. *Industrial and Engineering Chemistry Research*, **45**, 4145–4149.

72 Richter, T. and Vogel, H. (2001) The dehydration of 1,4-butanediol to tetrahydrofuran in supercritical water. *Chemical Engineering and Technology*, **24**, 340–343.

73 Nagai, Y., Matubayasi, N. and Nakahara, M. (2004) Hot water induces an acid-catalyzed reaction in its undissociated form. *Bulletin of the Chemical Society of Japan*, **77**, 691–697.

74 Lesutis, H.P. *et al.* (1999) Acid/base-catalyzed ester hydrolysis in near-critical water. *Chemical Communications*, 2063–2064.

75 Krammer, P., Mittelstädt, S. and Vogel, H. (1999) Investigating the synthesis potential in supercritical water. *Chemical Engineering and Technology*, **22**, 126–130.

76 Venardou, E. *et al.* (2004) On-line monitoring of the hydrolysis of acetonitrile in near-critical water using Raman spectroscopy. *Vibrational Spectroscopy*, **35**, 103–109.

77 Kramer, A., Mittelstädt, S. and Vogel, H. (1999) Hydrolysis of nitriles in supercritical water. *Chemical Engineering and Technology*, **22**, 494–500.

78 Duan, P., Wang, X. and Dai, L. (2007) Noncatalytic hydrolysis of

iminodiacetonitrile in near-critical water - a green process for the manufacture of iminodiacetic acid. *Chemical Engineering and Technology*, **30**, 265–269.

79 Ogawa, T., Watanabe, J. and Oshima, Y. (2008) Catalyst-free synthesis of polyorganosiloxanes by high temperature and pressure water. *Journal*, **45**, 80–87.

80 Yan, C. *et al.* (2008) The continuous synthesis of ε-caprolactam from 6-aminocapronitrile in high-temperature water. *Green Chemistry*, **10**, 98–103.

81 Yamabe, S., Tsuchida, N. and Yamazaki, S. (2007) Theoretical study of the role of solvent $H_2O$ in neopentyl and pinacol rearrangements. *Journal of Computational Chemistry*, **28**, 1561–1571.

82 Takahashi, H. *et al.* (2007) Novel quantum mechanical/molecular mechanical method combined with the theory of energy representation: free energy calculation for the Beckmann rearrangement promoted by proton transfers in the supercritical water. *Journal of Chemical Physics*, **126**, 084508/1–084508/10.

83 Comisar, C. and Savage, P. (2007) Benzil rearrangement kinetics and pathways in high-temperature water. *Industrial and Engineering Chemistry Research*, **46**, 1690–1695.

84 Wang, P., Kojima, H., Kobiro, K., Nakahara, K., Arita, T. and Kajimoto, O. (2007) Reaction behavior of secondary alcohols in supercritical water. *Bulletin of the Chemical Society of Japan*, **80**, 1828–1832.

85 Akiya, N. and Savage, P. (2001) Kinetics and mechanism of cyclohexanol dehydration in high-temperature water. *Industrial and Engineering Chemistry Research*, **40**, 1822–1831.

86 Ikushima, Y. and Sato, M. (2004) A one-step production of fine chemicals using supercritical water: an environmental benign application to the synthesis of monoterpene alcohol. *Chemical Engineering Science*, **59**, 4895–4901.

87 Aida, T.M. *et al.* (2007) Dehydration of D-glucose in high temperature water at pressures up to 80 MPa. *Journal of Supercritical Fluids*, **40**, 381–388.

88 Aida, T.M. *et al.* (2007) Reactions of D-fructose in water at temperatures up to 400 °C and pressures up to 100 MPa. *Journal of Supercritical Fluids*, **42**, 110–119.

89 Faisal, M. *et al.* (2005) Hydrolysis and cyclodehydration of dipeptide under hydrothermal conditions. *Industrial and Engineering Chemistry Research*, **44**, 5472–5477.

90 Dai, Z., Hatano, B. and Tagaya, H. (2004) Catalytic dehydration of propylene glycol with salts in near-critical water. *Applied Catalysis A: General*, **258**, 189–193.

91 Sato, N. *et al.* (2004) Reaction kinetics of amino acid decomposition in high-temperature and high-pressure water. *Industrial and Engineering Chemistry Research*, **43**, 3217–3222.

92 Dunn, J.B. *et al.* (2003) Hydrothermal stability of aromatic carboxylic acids. *Journal of Supercritical Fluids*, **27**, 263–274.

93 Fraga-Dubreuil, J. *et al.* (2006) The catalytic oxidation of benzoic acid to phenol in high temperature water. *Journal of Supercritical Fluids*, **39**, 220–227.

94 Zhai, P. *et al.* (2005) Deactivation of zeolite catalysts for benzene oxidation to phenol. *Chemical Engineering Journal*, **111**, 1–4.

95 Holliday, R.L., Jong, B. Y.M. and Kolis, J.W. (1998) Organic synthesis in subcritical water: oxidation of alkyl aromatics. *Journal of Supercritical Fluids*, **12**, 255–260.

96 Dunn, J. and Savage, P. (2002) Terephthalic acid synthesis in high-temperature liquid water. *Industrial and Engineering Chemistry Research*, **41**, 4460–4465.

97 Hamley, P.A. *et al.* (2002) Selective partial oxidation in supercritical water: the continuous generation of terephthalic

acid from *para*-xylene in high yield. *Green Chemistry*, **4**, 235–238.
98 Dunn, J., Urquhart, D. and Savage, P. (2002) Terephthlic acid synthesis in supercritical water. *Advanced Synthesis and Catalysis*, **344**, 385–392.
99 Dunn, J. and Savage, P. (2005) High-temperature liquid water: a viable medium for terephthalic acid synthesis. *Environmental Science and Technology*, **39**, 5427–5435.
100 Garcia-Verdugo, E. *et al.* (2005) Simultaneous continuous partial oxidation of mixed xylenes in supercritical water. *Green Chemistry*, **7**, 294–300.
101 Garcia-Verdugo, E. *et al.* (2004) Is it possible to achieve highly selective oxidations in supercritical water? Aerobic oxidation of methylaromatic compounds. *Advanced Synthesis and Catalysis*, **346**, 307–316.
102 Chen, Y., Fulton, J. and Partenheimer, W. (2005) A XANES and EXAFS study of hydration and ion pairing in ambient aqueous $MnBr_2$ solutions. *Journal of Solution Chemistry*, **34**, 993–1007.
103 Chen, Y., Fulton, J. and Partenheimer, W. (2005) The structure of the homogeneous oxidation catalyst, $Mn(II)(Br^{-1})_x$, in supercritical water: an X-ray absorption fine-structure study. *Journal of the American Chemical Society*, **127**, 14085–14093.
104 Hunter, S.E., Felczak, C.A. and Savage, P.E. (2004) Synthesis of *p*-isopropenylphenol in high-temperature water. *Green Chemistry*, **6**, 222–226.
105 Savage, P. *et al.* (2006) Bisphenol E decomposition in high-temperature water. *Industrial and Engineering Chemistry Research*, **45**, 7775–7780.
106 Kuhlmann, B., Arnett, E.M. and Siskin, M. (1994) Classical organic reactions in pure superheated water. *Journal of Organic Chemistry*, **59**, 3098–3101.
107 Kuhlmann, B., Arnett, E.M. and Siskin, M. (1994) H–D exchange in pinacolone by deuterium oxide at high temperature and pressure. *Journal of Organic Chemistry*, **59**, 5377–5380.
108 Kalpala, J. *et al.* (2003) Deuteration of 2-methylnaphthalene and eugenol in supercritical and pressurised hot deuterium oxide. *Green Chemistry*, **5**, 670–676.
109 Bai, S., Palmer, B. and Yonker, C. (2000) Kinetics of deuterium exchange on resorcinol in $D_2O$ at high pressure and high temperature. *Journal of Physical Chemistry A*, **104**, 53–58.
110 Kubo, M. *et al.* (2004) Noncatalytic kinetic study on site-selective H/D exchange reaction of phenol in sub-and supercritical water. *Journal of Chemical Physics*, **121**, 960.
111 March, J. (1985) *Advanced Organic Chemistry: Reactions, Mechanisms, and Structure*, John, Wiley and Sons, Inc., New York.
112 Yang, Y. and Evilia, R.F. (1999) Deuteration of hexane by DCl in supercritical deuterium oxide, *Journal of Supercritical Fluids*, **15**, 165–172.
113 Sasaki, M., Nishiyama, J., Uchida, M., Goto, K., Tajima, K., Adschirib, T. and Arai, K. (2003) Conversion of the hydroxyl group in 1-hexyl alcohol to an amide, group in supercritical water without catalyst, *Green Chemistry*, **5**, 95–97.
114 Tajima, K. *et al.* (2005) Amination of n-hexanol in supercritical water, *Environmental Science and Technology*, **39**, 9721–9724.
115 Osada, M. and Savage, P.E. (2009) Terephthalic acid synthesis at higher concentrations in high-temperature liquid water. 1. Effect of Oxygen Feed Method, *AIChE Journal*, **55**, 710–716.
116 Osada, M. and Savage, P.E. (2009) Terephthalic acid synthesis at higher concentrations in high-temperature liquid water. 2. Eliminating undesired byproducts, *AIChE Journal*, **55**, (in press).

# 12
# Water as a Green Solvent for Pharmaceutical Applications
*Peter Dunn*

## 12.1
### Introduction – Is Water a Green Solvent?

In principle, water is the ideal green chemistry solvent: it is benign, non-toxic, non-flammable, has a very low odor, has a high specific heat capacity to absorb energy from reactions, is available at a low cost, and is sustainable. A wide variety of chemical reactions can be performed in water and the groups of Breslow [1], Li [2], Kobayashi [3], Sharpless [4], and others have greatly expanded the number of reactions that can be performed in water. However, recently Blackmond *et al.* [5] challenged the often simply held view that all reactions performed in water must be green. They pointed out that sometimes large amounts of organic solvents can be required to extract the product from the water and, although pure water itself is non-toxic, the aqueous waste stream that is left after the reaction may be contaminated with toxic organic materials. They argue that on a manufacturing scale, "water is only truly a green solvent if it can be discharged directly to a biological effluent treatment plant." Nevertheless, many of the well-designed aqueous-based pharmaceutical processes described in this chapter, such as the pregabalin process (Section 12.2.1), should be considered green and, in the author's view, increasing the amount of pharmaceutical chemistry that is performed in water is a key part of the strategy of minimizing the reliance on carbon-based solvents. Of course, processes have to be designed in an efficient manner to avoid the problems outlined by Blackmond *et al.* This chapter covers water as a solvent in the preparation of small-molecule pharmaceuticals. The use of water to prepare large molecules (e.g. therapeutic proteins) is covered elsewhere in the series.

## 12.2
### Water-based Enzymatic Processes

Although enzymatic reactions can be carried out in organic solvents containing a small amount of water [5], the majority of enzymatic reactions are performed either in water or in water-rich solvent mixtures. In recent years, there has been an increase in the number of pharmaceutical processes that make use of enzymatic catalysis [6]. Two factors are driving this increase. The first is the growth in the numbers of enzymes available in particular through bioinformatics and cloning. The second is that reactions can now be screened using this increased enzyme pool using automated technology and using minimal amounts of substrate [7]. This automated screening has meant that enzymatic solutions to chemical problems can now be easily obtained within the accelerated timelines of pharmaceutical chemical development. Water-based biotransformations can be either whole cell-based, such as in the synthesis of LY300164, using a yeast-based catalyst, or as is now more common to use an isolated enzyme such as in the synthesis of pregabalin.

## 12.2.1
## The Pregabalin (Lyrica) Process

Pregabalin is the active ingredient in Lyrica, which was launched in the United States in September 2005 for the treatment of neuropathic pain. In 2008, sales of Lyrica reached $2.6 billion. Lyrica was initially launched using a final-stage classical resolution synthesis [8], which is shown in Scheme 12.1. In this route, the cyanodiester starting material **1** prepared using malonate chemistry is hydrolyzed and decarboxylated to give the nitrile **2**. Reduction of **2** with hydrogen gas using Raney nickel catalyst gives racemic pregabalin, which is resolved with (*S*)-mandelic acid to give the desired single-enantiomer pregabalin. There are two obvious issues with this first-generation synthesis; the first is the use of a late-stage classical resolution with over half of the product discarded at the end of the synthesis, and the second is that the wrong enantiomer could not be recycled.

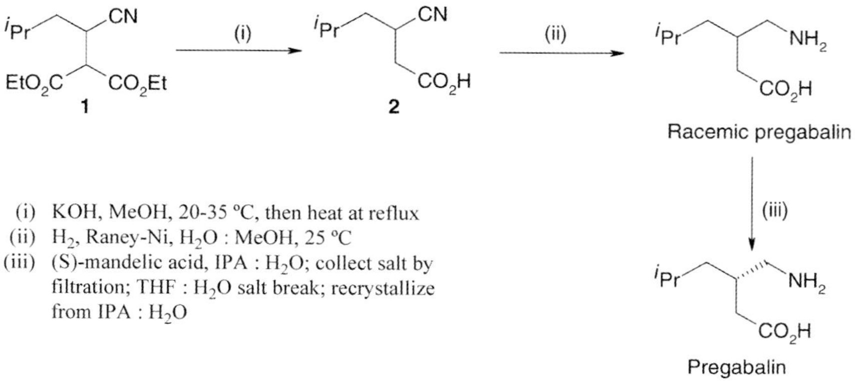

iPr
CN
EtO2C
CO2Et
1
(i)
2
CO2H
(ii)
NH2
Racemic pregabalin
(iii)
Pregabalin
(i) KOH, MeOH, 20-35 °C, then heat at reflux
(ii) H2, Raney-Ni, H2O : MeOH, 25 °C
(iii) (S)-mandelic acid, IPA : H2O; collect salt by filtration; THF : H2O salt break; recrystallize from IPA : H2O


**Scheme 12.1** The classical resolution process to make pregabalin.

In 2006, the route of manufacture was switched to an enzymatic process, shown in Scheme 12.2 [9]. In this route, the same starting material **1** is hydrolyzed using a lipase to give the (*S*)-monoester **3**. The *R*-enantiomer is left untouched by the lipase and stays as the diester, which can be removed by extraction and recycled by

## 12.2 Water-based Enzymatic Processes

a base-catalyzed epimerization (Scheme 12.2). The enzyme is more than 200-fold selective for the S-enantiomer compared with the R-enantiomer. The (S)-monoester stays in the aqueous layer, but when heated to 85 °C undergoes a decarboxylation reaction to give the ester 4. Having lost its carboxyl group, ester 4 has no water solubility and can easily be separated from the aqueous layer. Finally, the ester 4 is hydrolyzed and the potassium salt of the resulting acid is subjected to a hydrogenation reaction in water over a Raney nickel catalyst. The nitrile group is reduced to an amine, furnishing enantiomerically pure pregabalin. Hence all four chemical steps are performed in water and pregabalin is an extremely rare example of a synthetically produced drug in which every chemical reaction to produce that drug is carried out in water. The overall yield of the enzymatic process is 40–45% (allowing for one recycle), which compares very favorably with the classical resolution process where a 21–25% yield of pregabalin was obtained from the same starting material 1.

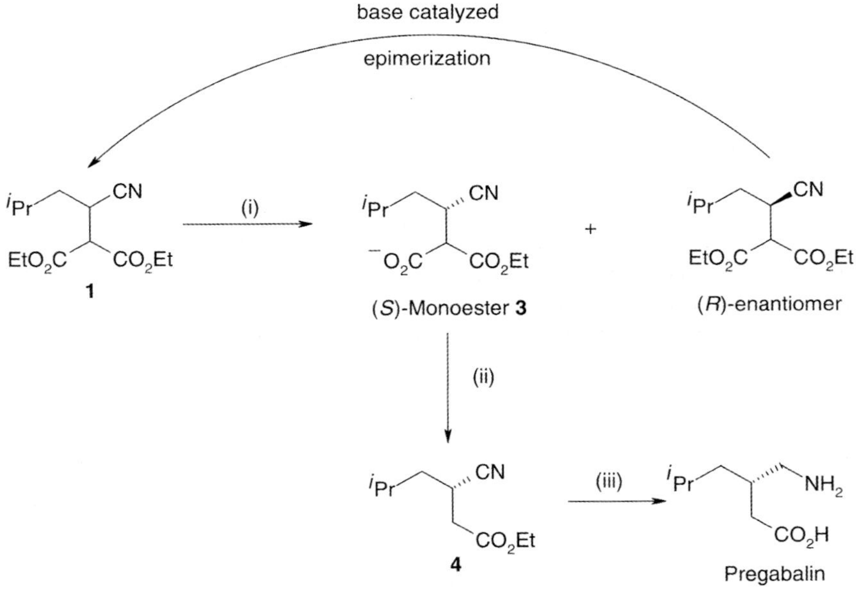

(i) Lipolase, Ca(OAc)$_2$, water, 20-25 °C, pH 7; (ii) water 85 °C, 3 h; (iii) KOH, water Raney-Ni, H$_2$

**Scheme 12.2** The enzymatic process to make pregabalin.

The extensive use of water as a reaction solvent and minimizing the solvents used in work-up result in a greatly reduced need for input materials, as summarized in Table 12.1 [9].

One of the criticisms of enzymatic processes is that at high concentrations, the substrate or product starts to inhibit the enzyme. In the pregabalin process, these issues were overcome and very high substrate concentrations are possible (765 g l$^{-1}$).

**Table 12.1** Comparison of classical resolution and enzymatic processes.

| Input | Classical resolution process (kg) | Enzymatic process (kg) |
| --- | --- | --- |
| 1 | 6212 | 4798$^a$ |
| Enzyme | 0 | 574 |
| (S)-Mandelic acid | 1135 | 0 |
| Raney nickel | 531 | 79.5 |
| Solvents | 50042 | 6230$^b$ |
| Total | 57920 | 11681.5 |

$^a$Not including the environmental savings due to recycle of the R-enantiomer.
$^b$After solvent recovery.

Pfizer has been able to scale up the process successfully to a 10 tonne batch size in its commercial facilities.

Pregabalin is a huge volume product by tonnage and by 2010 is expected to account for 1–2% of the current patent-protected prescription pharmaceutical market by tonnage. The achievement of performing the whole synthesis in water will make a useful reduction in the use of carbon-based solvents in the pharmaceutical industry.

## 12.2.2
### Enzymatic Routes to Statins

Rosuvastatin (Crestor) and atorvastatin (Lipitor) are important pharmaceutical agents which reduce low-density lipoprotein cholesterol. Sales in 2006 were $2 billion and $13 billion, respectively. It has been shown that both can be made by enzymatic processes in aqueous systems using enzymes from the deoxyribose-5-phosphate aldolase (DERA) class. Aldol reactions catalyzed by DERA were first reported by Wong and co-workers [10, 11]. Subsequently, it was shown that DERA can be used to prepare the lactol **5**. Oxidation of **5** gives the lactone **6**, which can be converted into either rosuvastatin or atorvastatin (Scheme 12.3) [12]. Hydrolysis of **6** to give the rosuvastatin intermediate **7** is particularly attractive and, indeed, the DERA-based chemistry from chloroacetaldehyde and acetaldehyde forms the basis of the commercial process to make rosuvastatin. The development of the aqueous-based enzymatic step into an industrial process is now described in more detail.

#### 12.2.2.1 The Enzymatic Process to Make Rosuvastatin (Crestor) Intermediate 5
Extensive optimization of the DERA-catalyzed chemistry to prepare lactol **5** was required to meet the normal targets of cost and throughput. The key to this improvement was the optimization of the enzyme itself. Detailed work by scientists at DSM [13] optimized the enzyme using a combination of directed evolution and site-specific mutagenesis approaches. Directed evolution was used to improve the resistance of DERA to chloroacetaldehyde. However, the resulting enzymes still lost activity when exposed to high concentrations of aldehydes, and this was attributed to imine formation between the aldehyde substrates and the amino groups of accessible lysine residues present in the enzyme. As part of the final enzyme optimization, site-specific

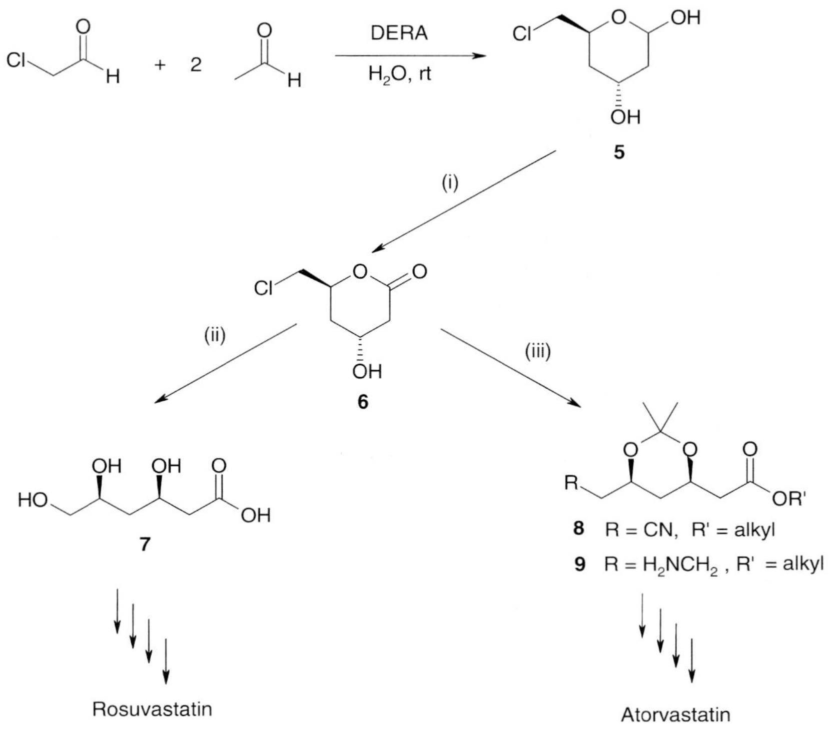

(i) NaOCl, HOAc, H₂O, rt, 3 h; (ii) NaOH, H₂O, 40 °C, 16 h; (iii) NaCN, DMF–H₂O, 40 °C, 16 h; dimethoxypropane, DMF, cat. H₂SO₄, then esterification

**Scheme 12.3** DERA-catalyzed enzymatic routes to statins.

mutagenesis was used to optimize the DERA and in particular the position of the lysine residues. The final DERA was about 10-fold more productive compared with the wild-type enzyme when used at high aldehyde concentrations [13, 14].

The process described by DSM [13] involves reacting 1 equiv. of chloroacetaldehyde with 2.3 equiv. of acetaldehyde using a catalytic quantity of DERA and 0.1 M sodium hydrogencarbonate buffer as the reaction medium. The reaction takes place at pH 7.2 and at room temperature over several hours.

#### 12.2.2.2 Enzymatic Routes to Atorvastatin (Lipitor)

There are also several published water-based enzymatic reactions which can be used as part of an atorvastatin synthesis. The lactone **6** can be reacted with sodium cyanide and, after protection of the diol and esterification, gave the nitrile **8** (Scheme 12.3). Reduction of the nitrile to the amine gives the methyl ester of the key atorvastatin intermediate **9**. Wong and co-workers also showed that a modified DERA enzyme was tolerant to 2-azidoethanal to give the equivalent lactol **10**, which could also be

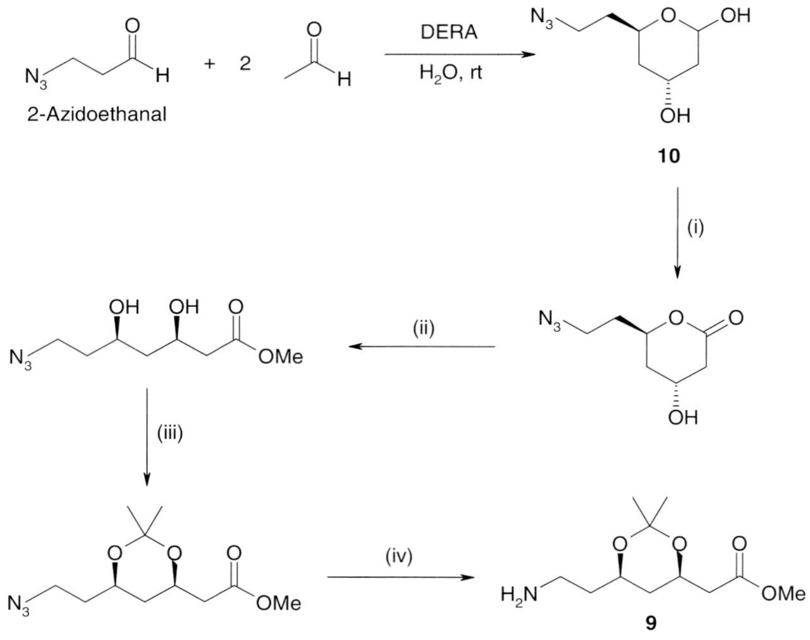

**Scheme 12.4** 2-Azidoethanal route to atorvastatin intermediate **9**.

(i) Br$_2$, BaCO$_3$; (ii) NaOMe, MeOH, 83%; (iii) camphorsulfonic acid, 2,2-dimethoxypropane, 76%; (iv) Ph$_3$P, 3 d, 88%

converted to **9** (Scheme 12.4) [15]. Obviously one downside of this route is the inherent hazards of high-energy functional groups such as azides.

Workers at Pfizer showed that other substrates could also undergo the DERA-catalyzed aldol reaction (Scheme 12.5) [16]. Thus, using Cbz-protected aminopropanal **11** (or other N-protected aminopropanals), the key intermediate **12** could be made

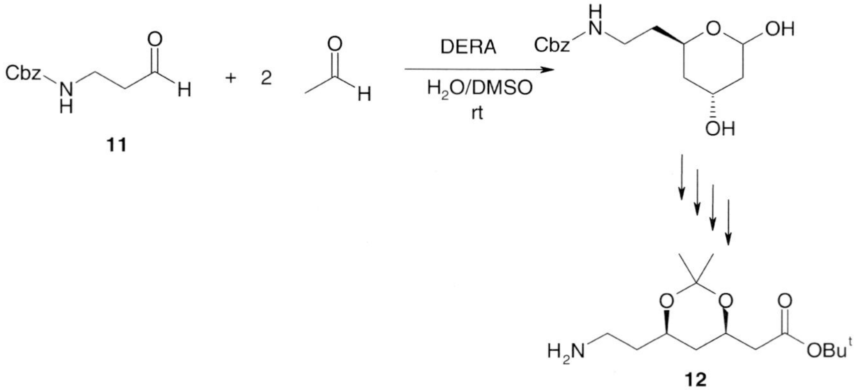

**Scheme 12.5** N-Protected routes to atorvastatin intermediate **12**.

more efficiently and more safely using the DERA technology as a key step. The enzymatic reaction is performed in water with a small amount of DMSO as cosolvent.

Another route to the intermediate **12** proceeds through the hydroxynitrile **13**. The hydroxynitrile can itself be made via enzymatic processes in water (Scheme 12.6). In the first step, a keto-reductase is used to set the stereochemistry. In the second step, a halohydrin dehalogenase is used to promote the cyanide displacement at pH 8. The halohydrin dehalogenase was optimized via a directed evolution approach [17].

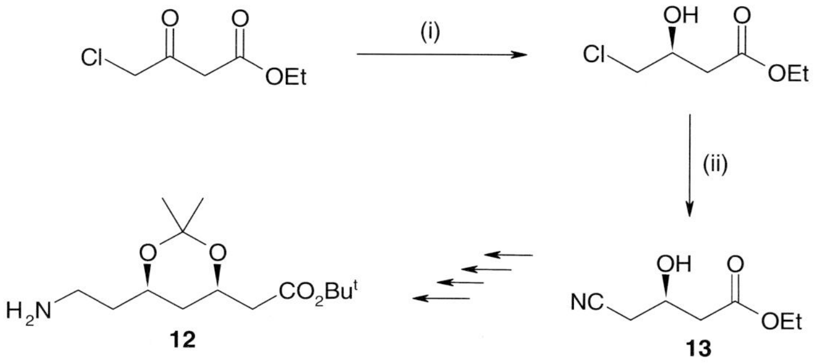

(i) Ketoreductase, NADPH, glucose, H$_2$O;
(ii) halohydrin dehalogenase, NaCN, H$_2$O, pH 8

**Scheme 12.6** An enzymatic process to prepare atorvastatin intermediate **13**.

## 12.2.3
### The Enzymatic Process to Make LY300164

LY300164 was an advanced clinical development candidate for the treatment of epilepsy and neurodegenerative disorders [18]. In the initial discovery synthesis (Scheme 12.7), the ketone **14** was reduced to give racemic alcohol **15**. Subsequently, an acid-catalyzed Prins-type cyclization gave racemic **16**. Jones oxidation of **16** gave the diketone **17**, which after further steps was converted to the benzodiazepine **18**. The desired chirality of LY300164 was introduced with a borane–dimethyl sulfide reduction in the presence of a chiral modifier. This gave **19**, which possessed the correct stereochemistry required for LY300164.

A modified version of the medicinal chemistry synthesis produced early kilogram quantities of LY300164 required for early clinical development, but the synthesis could not be considered as a manufacturing route for the following reasons:

- The initial stereoselectivity from the modified borane reduction was only 73% *ee*. This could be enriched to 96% *ee* by a single recrystallization, but in moderate yield (56%).
- Chromium trioxide is a carcinogen and the chromium by-products pose a major environmental concern. For each kilogram of LY300164 that was generated, 3 kg of chromium waste were produced.

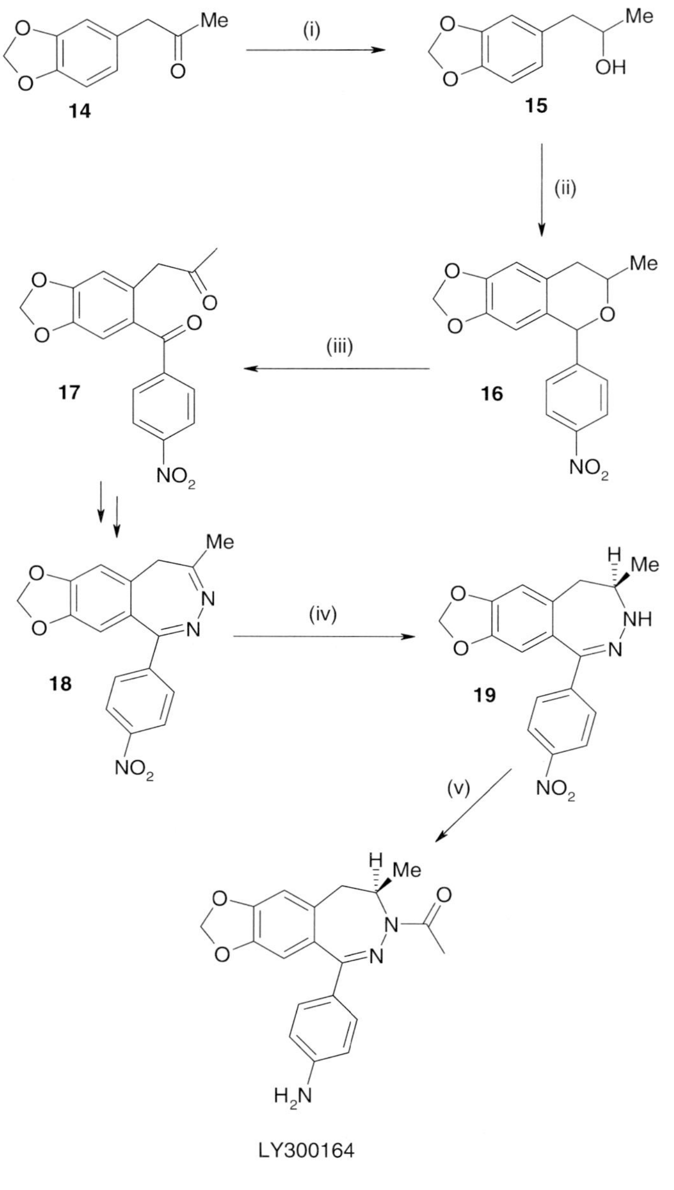

(i) NaBH₄; (ii) *p*-NO₂PhCHO, HCl; (iii) CrO₃, H₂SO₄;
(iv) BH₃·SMe₂, chiral modifier; (v) Ac₂O then H₂, Pd/C

**Scheme 12.7** Modified medicinal chemistry route to LY300164.

The solution to the problem was to use a bioreduction in water to set the stereochemistry of **20** in the final Lilly process (Scheme 12.8); this compound was not isolated but carried through into the Prins-type reaction to give **21** with the key chiral center in place. The chromium-based oxidation was replaced with an air

(i) *Zygosaccharomyces rouxii*, resin; (ii) *p*-NO$_2$PHCHO; (iii) air, NaOH, DMSO
**Scheme 12.8** The Lilly process to make LY300164.

oxidation to give **22**, which was converted through several further steps to give LY300164. The water-based bioreduction of the ketone is worthy of further comment.

The initial yeast used for pilot plant scale-up was *Candida famata*, identified by a screen of 50 microorganisms. This was initially scaled to a 1000 l reactor by Lilly, but there were two issues to be resolved. The first was that the maximum substrate concentration was 2 g l$^{-1}$ and the second was that there were work-up problems. After the reaction was complete, the broth was centrifuged to remove the cell solids, then the centrate was extracted with organic solvents, and it was during this extraction step that significant emulsions occurred, leading to yield and quality issues. Continued screening of yeast and fungi libraries led to the identification of *Zygosaccharomyces rouxii*, a yeast that had been used for centuries to prepare soy sauce. This yeast could tolerate higher substrate concentrations of 5–7 g l$^{-1}$ and, after extensive development work at Lilly, the process was successfully scaled up to give an impressive 95% yield [19, 20].

## 12.2.4
### The Enzymatic Process to Prepare 6-Aminopenicillanic Acid

6-Aminopenicillanic acid (6-APA) is a key intermediate in the synthesis of antibiotics. It is prepared from penicillin G, which is itself derived by fermentation. Converting penicillin G to 6-APA sets a significant chemical challenge; it requires the hydrolysis of a stable amide bond while leaving the reactive β-lactam bond

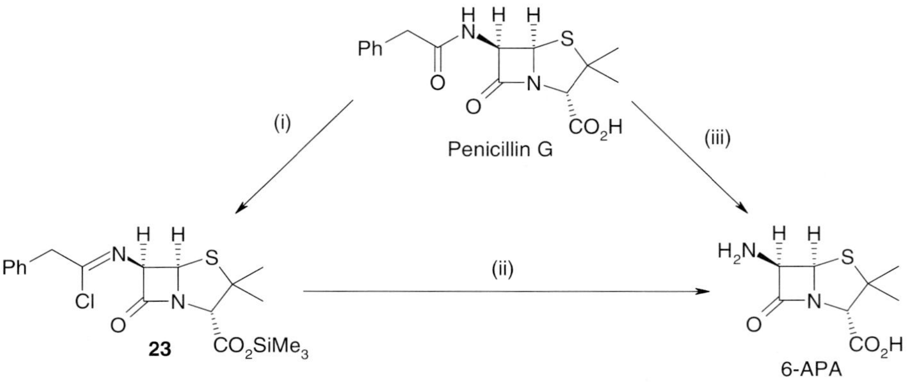

(i) Me$_3$SiCl then PCl$_5$, CH$_2$Cl$_2$, PhN(CH$_3$)$_2$; (ii) n-BuOH, -40 °C then H$_2$O, 0 °C; (iii) Pen-acylase, H$_2$O, 37 °C

**Scheme 12.9** Chemical and enzymatic processes to make 6-APA.

untouched. An ingenious chemical solution is first to protect the acid as the silyl ester with TMSCl, then react the amide group with phosphorus pentachloride to give the iminochloride **23**. Subsequent hydrolysis of both the iminochloride and the silyl ester groups gives 6-APA (Scheme 12.9). Ingenious though this chemical synthesis is, there are a number of drawbacks from a green chemistry perspective, notably the use of highly reactive and hazardous phosphorus pentachloride, the use of an undesirable chlorinated solvent (CH$_2$Cl$_2$) and the need to employ a reaction temperature of −40 °C, leading to high energy use [21]. Nevertheless the chemical process was successfully employed to manufacture 6-APA until the mid-1980s. This process was replaced by an enzymatic process which is carried out in water at 37 °C, leading to major environmental savings [22]. Several thousand tonnes of 6-APA are produced per year via the enzymatic process.

### 12.2.5
### Enzymatic Routes to Oseltamivir Phosphate (Tamiflu)

Oseltamivir phosphate is a neuramidase inhibitor that is the active ingredient in the drug Tamiflu, a key medicine in the prevention plan against an influenza pandemic. The current ten step commercial synthesis is summarized in Scheme 12.10 [23]. This process involves a water-based enzyme-mediated synthesis of shikimic acid using a genetically engineered *E. coli* strain [24]. Of course, one of the downsides of the current synthesis is that potentially explosive sodium azide is used to introduce the amino group.

Several azide-free syntheses have been reported and are summarized in a review [6]. One of the most promising approaches is to prepare aminoshikimic acid via an enzymatic process and then convert this intermediate through a number of steps to make oseltamivir phosphate (Scheme 12.11) [25]. In this process, glucose is converted to kanosamine in an aqueous fermentation reaction using the yeast *Bacillus pumilus*. A second aqueous fermentation reaction with *E. coli* converts kanosamine to

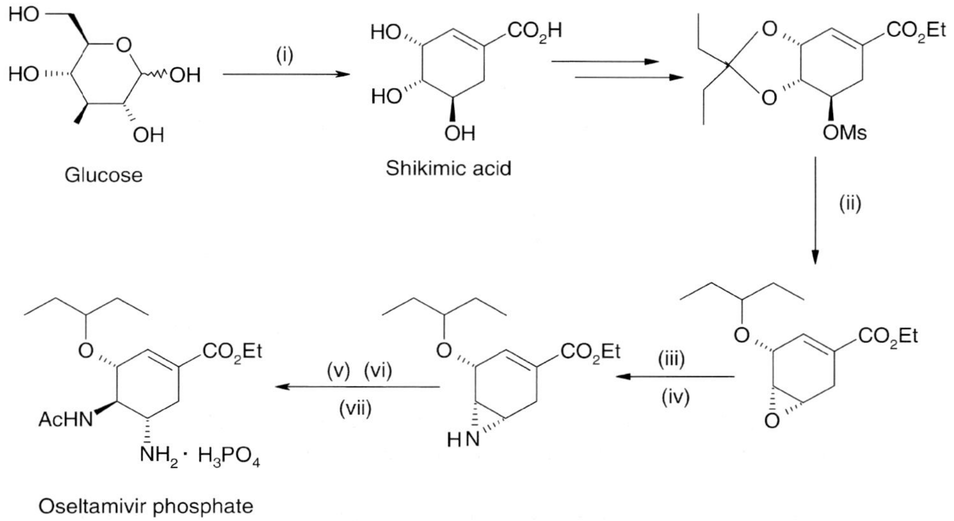

## 12.2 Water-based Enzymatic Processes

(i) *E. coli*, H$_2$O; (ii) Et$_3$SiH, TiCl$_4$ then aq. NaHCO$_3$; (iii) NaN$_3$, NH$_4$Cl;
(iv) Me$_3$P, CH$_3$CN; (v) Ac$_2$O, pyridine; (vi) RaNi, H$_2$; (vii) H$_3$PO$_4$

**Scheme 12.10** The current manufacturing route to oseltamivir phosphate.

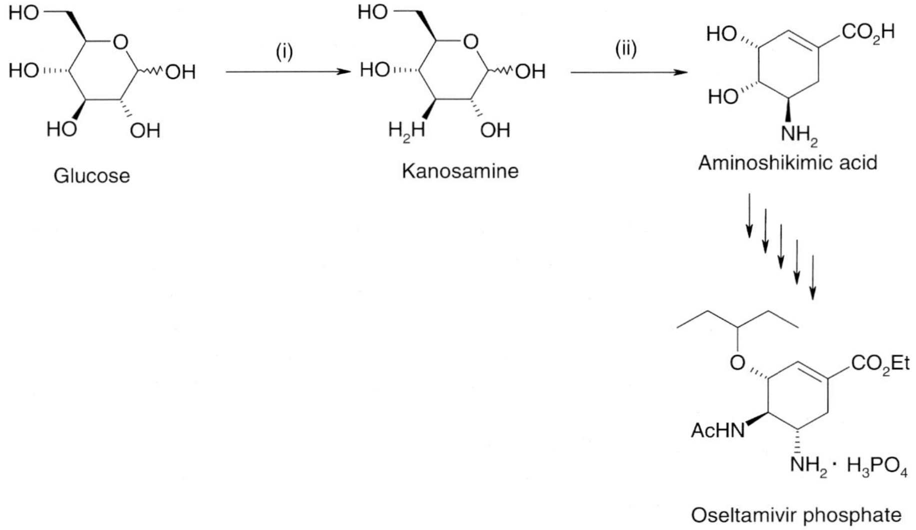

(i) *Bacillus pumilus*, H$_2$O, various additives, 30 °C; (ii) *E. coli*, H$_2$O, various additives, 37 °C

**Scheme 12.11** An azide-free synthesis of oseltamivir phosphate.

aminoshikimic acid. Although this method shows some promise, a significant improvement in process yields would be required to give a practical manufacturing process. Roche have been stockpiling oseltamivir phosphate and its intermediates in order to help protect human health in case there is an influenza pandemic. This is a major chemical program and it is reported that 15 fine chemical suppliers are involved [26]. With such a huge involvement, there is no doubt that there are some further chemistry improvements over what is currently reported in the literature.

## 12.3
### Processes in Which the Product is Isolated by pH Adjustment to the Isoelectric Point

Blackmond et al. [5] pointed out that some of the potential downsides of performing reactions in water include having to extract the product from the water with large amounts of organic solvent and the potential to leave an aqueous waste stream contaminated with toxic organic solvents and materials. One way to avoid these issues for substrates that have basic and acidic functional groups is to isolate the product from water by adjusting the pH to the isoelectric point of the product and collect the precipitated product by filtration, leaving a water-based filtrate. Two processes that illustrate this point are outlined below.

### 12.3.1
#### Process to Prepare the Sildenafil Citrate (Viagra) Intermediate 24

Selective chlorosulfonation of 2-ethoxybenzoic acid gives the sulfonyl chloride **25**, which is quenched and isolated as a water-wet filter cake. This material is resuspended in water and reacted with *N*-methylpiperazine to give the desired intermediate **24** (Scheme 12.12). At this point, extraction with copious amounts of an organic solvent would have given the sort of process much disliked by Blackmond et al. [5]. Instead, the pH is adjusted to the isoelectric point with sodium hydroxide solution (in this case pH 5.5). At this point, the product exists as the highly crystalline zwitterion **26** [27]. The product is collected by filtration and very minimal organic material is left in the aqueous layer.

### 12.3.2
#### The Sampatrilat Process

Sampatrilat is a joint inhibitor of the angiotensin-converting enzyme (ACE) and the (neutral endopeptidase) NEP enzyme, which was discovered by Pfizer and was out licensed in the late 1990s. Its synthesis has been reported (Scheme 12.13) [28] and the final synthetic step is carried out in water. The process gives another example where product isolation via the isoelectric point is used to facilitate an efficient environmentally benign work-up to an aqueous process.

The Cbz-protected precursor **27** was available as an ethyl acetate solution from the previous step. This solution is extracted with aqueous sodium hydroxide,

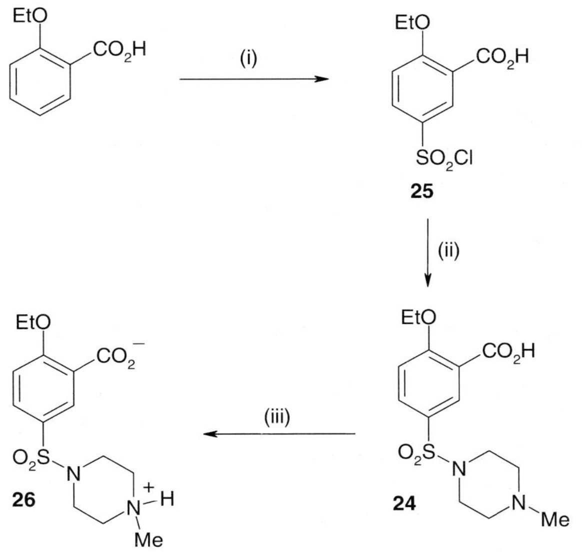

(i) ClSO₃H, 25 °C, water quench; (ii) N-methylpiperazine, water, 25 °C;
(iii) NaOH to pH 5.5, filter, dry

**Scheme 12.12** Process to prepare the sildenafil citrate intermediate **24**.

the disodium salt **28** partitions into the aqueous layer, and the ethyl acetate layer is available for recovery and reuse. The aqueous solution of **28** is hydrogenated over palladium on carbon. The use of water as the solvent rather than a flammable solvent gives additional safety advantages when using a pyrophoric catalyst. On completion of the reaction, a small toluene layer (from the Cbz group) is separated off and the catalyst removed by filtration. The palladium metal is recovered from this catalyst. The filtrate, which contains the disodium salt of sampatrilat, is acidified with hydrochloric acid to the isoelectric point (pH 4) and collected by filtration, presumably as the zwitterion (not drawn). Sampatrilat is a high-melting molecule (m.p. 256 °C) with a high crystal lattice energy. These properties mean that sampatrilat has low water solubility and therefore very little sampatrilat is present in the final filtrate that goes for treatment and disposal. The hydrogenation process is run at high concentration, less than 3 l of water per kilogram of substrate.

## 12.4
## Carbon–Carbon Bond-forming Cross-coupling Reactions in Water

Palladium catalyzed cross coupling reactions such as the Suzuki reaction [29] have become a key reaction for the assembly of pharmaceutical molecules and a recent survey by AstraZeneca, GSK and Pfizer indicted that the Suzuki reaction accounts for

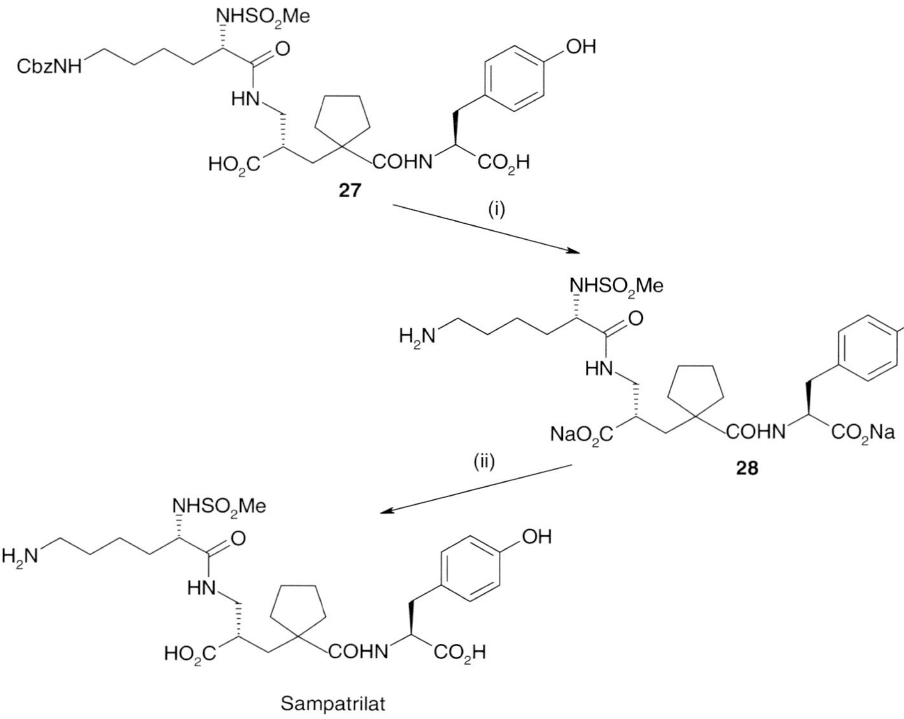

**Scheme 12.13** Process to make sampatrilat.

(i) NaOH (2 equiv.), H$_2$, H$_2$O, Pd/C, 25 °C;  (ii) HCl to pH 4

around 11% of all carbon–carbon bond forming reactions in the pharmaceutical industry [30]. Fortunately this reaction can sometimes be performed in water as illustrated by these two pharmaceutical examples.

### 12.4.1
### Process to Make Compound 29 an Intermediate for a Drug Candidate to Treat Depression

Ennis et al. [31] of Smith Kline and Beecham (now GSK) reported a three-step process to make **29** (Scheme 12.14). All three chemical steps are run in water-based solvents. In the first step, the aniline **30** is acylated using a Schotten–Baumann procedure using tetrabutylammonium chloride (TBAC) as catalyst; the resulting amide is not isolated but is cyclized to the pyrrolidinone **31** by warming the reaction mixture. In the third step, **31** is reacted with the boronic acid **32** in a Suzuki process using a palladium on carbon catalyst in a 1 : 1 mixture of water and methanol as solvent. The use of an aqueous solvent mixture in combination with Pd/C for the Suzuki coupling has a number of advantages from a green chemistry perspective:

- It minimizes the risks involved in handling the pyrophoric catalyst.

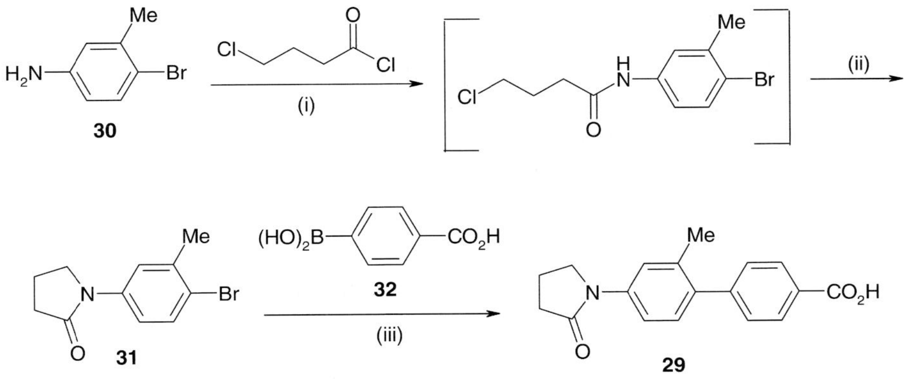

(i) NaOH, TBAC cat., THF–H₂O, 20-25 °C; (ii) 40-45 °C, 2 h;
(iii) Na₂CO₃, Pd/C (1.2 mol%) MeOH–H₂O (1:1), 78 °C, 5 h

**Scheme 12.14** Process to make **29**.

- By using a heterogeneous palladium source, the palladium metal can be easily recovered.
- As the solvent system is 50% water, this minimizes the use of carbon-based solvents.

The use of a heterogeneous palladium catalyst also meant that the amount of leaching of palladium on to **29** was very low (<6 ppm), and this is important in pharmaceutical processing as palladium and other heavy metals are controlled to very low levels in the final product.

## 12.4.2
### An Aqueous Suzuki Reaction to Prepare Diflusinal 33

Diflunisal is an FDA-approved non-steroidal anti-inflammatory agent [32]. Shaughnessy and co-workers reported the preparation of **33** and other similar analogs using a Suzuki reaction performed at room temperature in water and in excellent yield (Scheme 12.15) [33]. The catalyst ligand used to promote this reaction is *t*-Bu-Amphos **34**, which derives its water solubility from its tetraalkylammonium group.

In the same paper, the authors look into the reuse of the *t*-Bu-Amphos–Na₂PdCl₂ catalyst system using a simpler reaction system, the coupling of 4-bromotoluene with phenylboronic acid (Scheme 12.16). The reaction is performed in a 1:1 toluene–water mixture; at the end of the reaction, the product-containing toluene layer is separated off and further 4-bromotoluene and phenylboronic acid are added to the aqueous phase. It is very encouraging that the catalyst system can be reused, but there is a fall-off in catalyst performance, as shown in Table 12.2.

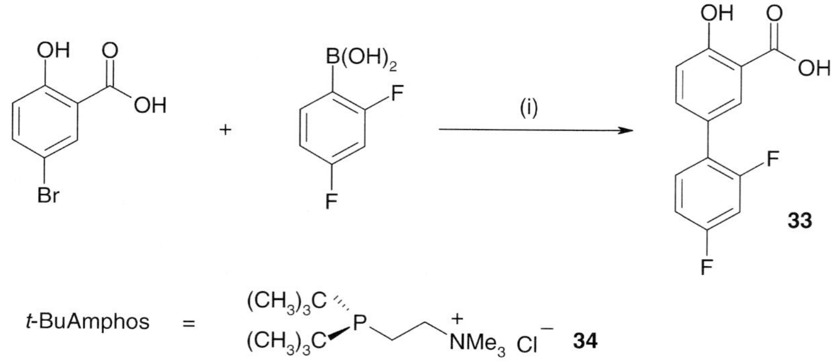

Na$_2$PdCl$_4$ (2 mol%), t-Bu-Amphos (2 mol%), Na$_2$CO$_3$ (2 equiv.), H$_2$O, rt, 6-8 h, 95% yield

**Scheme 12.15** The preparation of diflusinal **33**.

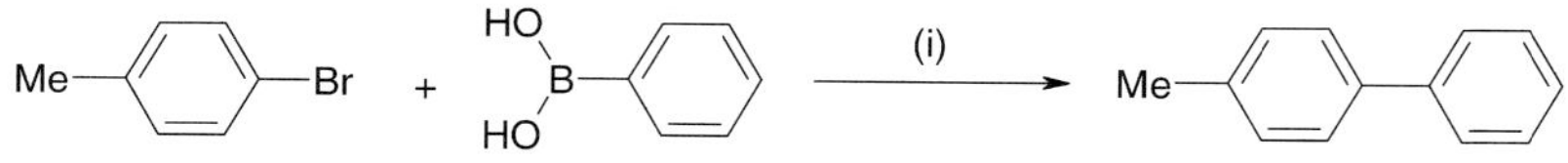

(i) Na$_2$PdCl$_4$ (2 mol%), t-Bu-Amphos (2 mol%), Na$_2$CO$_3$ (2equiv.), H$_2$O–toluene (1:1), 25 °C, 8 h

**Scheme 12.16** The Suzuki reaction between 4-bromotoluene and phenylboronic acid.

**Table 12.2** Impact of recycling the t-Bu-Amphos–Na$_2$PdCl$_4$ catalyst.

| Catalyst cycle | Pd (mol%) | Yield (%) |
| --- | --- | --- |
| Fresh catalyst | 2 | 92 |
| First recycle | 0[a] | 85 |
| Second recycle | 0[a] | 84 |
| Third recycle | 0[a] | 62 |
| Fourth recycle | 0[a] | 22 |

[a]The aqueous catalyst solution from the previous cycle was used in the reaction.

## 12.5
### Pharmaceutical Processes Using Mixed Aqueous Solvents

In many ways, these types of processes are less green as they can complicate solvent recovery and reuse. However, using aqueous mixtures helps to minimize the usage of carbon-based solvents. The two examples given below have further green advantages: the lumiracoxib process uses solvent mixtures from which the carbon-based solvent can be recovered, and the sampatrilat process uses a nicely designed procedure in which the solvent mixture can be reused for the same reaction.

## 12.5.1
### The Lumiracoxib Process

Lumiracoxib (the active ingredient in Prexige) is an orally active COX-2 inhibitor launched by Novartis in Brazil in 2005. Its preparation is summarized in Scheme 12.17. The benzylic alcohol **35** is converted to the benzylic bromide **36** with 48% aqueous HBr. The bromide **36** is displaced by sodium cyanide in a mixture of refluxing ethanol–water (4 : 1) to give the cyanide **37** in 97% yield from the alcohol **35**. The cyanide **37** is then hydrolyzed by sodium hydroxide in refluxing ethanol–water (7 : 4) in 81% yield (Scheme 12.15) [34]. The resulting acid **38** is converted to lumiracoxib. Novartis do not give any details of solvent recovery but the ethanol used in the conversion of **36** to **38** is likely to be easily recovered as the ethanol from alcohol-rich solvent mixtures such as these is generally easily recovered via its azeotrope.

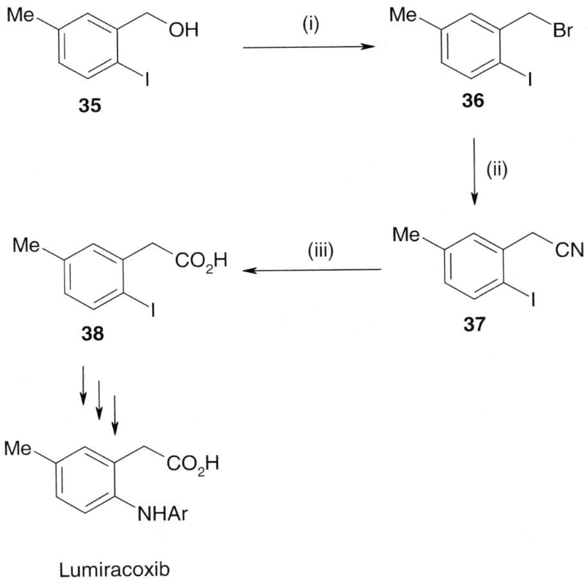

(i) 48% HBr, reflux, 4 h; (ii) NaCN, EtOH–H$_2$O, reflux, 3 h;
(iii) NaOH, EtOH–H$_2$O, reflux, 14 h

**Scheme 12.17** The synthesis of lumiracoxib.

## 12.5.2
### An Environmentally Friendly Baylis–Hillman Process

In principle, the Baylis–Hillman reaction [35] is an environmentally friendly reaction as each atom in the reagents is incorporated into the product and the reaction has 100% atom economy [36]. Workers at Pfizer wanted to design a process

(i) 3-Quinuclidinol (cat.), CH$_2$O (1.6 equiv.), CH$_3$CN–H$_2$O
**Scheme 12.18** The Baylis–Hillman reaction.

to make the hydroxymethacrylate **39** as an intermediate to the synthesis of a key chiral acrylate **40** (Scheme 12.18). This required the reaction of *tert*-butyl acrylate with formaldehyde, which can be catalyzed by tertiary amines such as DABCO or 3-quinuclidinol. Initial experiments were performed in organic solvents and these gave only a 30% yield of the desired compound **39**. The undesired ether **41** was formed in a similar yield along with a complex mixture of ethers and acetals. It was found that the presence of water was needed to give a high yield of hydroxymethacrylate **39**. Reactions in pure water were much less successful and only homogeneous mixtures of aqueous solvent mixtures such as *tert*-butanol–water or acetonitrile–water were successful. The acetonitrile–water reaction gave a slightly higher yield and therefore was selected for scale-up into the pilot plant.

In order to minimize the waste profile of the reaction and to allow for recycle of the 3-quinuclidinol catalyst, an efficient, elegant process was designed which is shown diagrammatically in Figure 12.1 [27]. The reaction takes place in the homogeneous aqueous acetonitrile. At the end of the reaction, toluene is added and the product is extracted into the toluene layer and used in the next process step (also carried out in toluene). The aqueous layer, which contains the 3-quinuclidinol catalyst, can then be used again in another Baylis–Hillman reaction. Some acetonitrile becomes lost into the toluene layer, so this has to be replenished before fresh *tert*-butyl acrylate and formaldehyde are added as it is critical to have enough acetonitrile to maintain a homogeneous reaction in order to achieve a favorable reaction profile.

## 12.6
## Conclusion

The use of water as a solvent in the synthesis of pharmaceutical agents will remain relatively low compared with the use of organic solvents; however, there is also likely to be a sharp increase in absolute terms. The reasons for this increase are (i) an

## 12.6 Conclusion

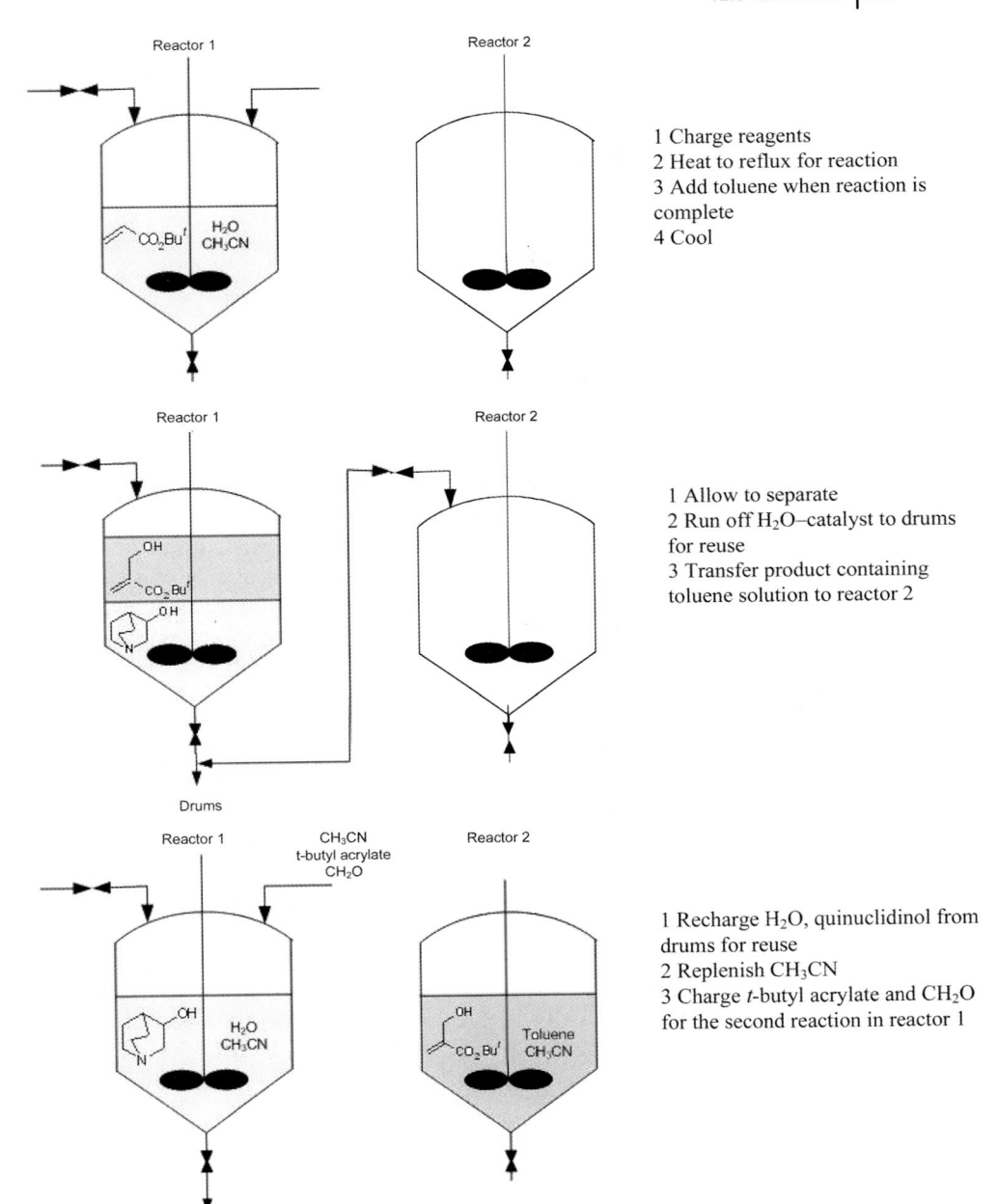

**Reactor 1** (contains CO₂Buᵗ, H₂O, CH₃CN)
**Reactor 2**

1. Charge reagents
2. Heat to reflux for reaction
3. Add toluene when reaction is complete
4. Cool

**Reactor 1** (contains OH-CO₂Buᵗ, quinuclidinol-OH)
**Reactor 2**

1. Allow to separate
2. Run off $H_2O$–catalyst to drums for reuse
3. Transfer product containing toluene solution to reactor 2

**Drums**

**Reactor 1** (quinuclidinol-OH, $H_2O$, $CH_3CN$) — charged with $CH_3CN$, t-butyl acrylate, $CH_2O$
**Reactor 2** (OH-CO₂Buᵗ, Toluene, $CH_3CN$)

1. Recharge $H_2O$, quinuclidinol from drums for reuse
2. Replenish $CH_3CN$
3. Charge t-butyl acrylate and $CH_2O$ for the second reaction in reactor 1

**Figure 12.1** The process to make hydroxymethacrylate **39**.

expected increase in biotransformation reactions, the majority of which will be carried out in water, and (ii) the work of the academic community in expanding the number of reactions that can be performed in water. To make full use of this uniquely green and useful solvent, processes will need to be carefully designed so that the overall process, both reaction and work-up, is green on a holistic basis.

## References

1 Breslow, R. (1991) *Acc. Chem. Res.*, **24**, 159–164; Breslow, R. (2004) *Acc. Chem. Res.*, **37**, 471–478.
2 Li, C.J. (1993) *Chem. Rev.*, **93**, 2023–2035; Li, C.J. (2005) *Chem. Rev.*, **105**, 3095–3165.
3 Kobayashi, S. (ed.) (2002) *Adv. Synth. Catal.* 344 (Special Issue on Water) 219–451; Ogawa, C. and Kobayashi, S. (2007) Acid catalysis in water *Organic Reactions in Water* pp 60–91. Kobayashi, S. (2007) *Pure Appl. Chem.*, **79**, 235–245; Boudou, M., Ogawa, C. and Kobayashi, S. (2006) *Adv. Synth. Catal.*, **348**, 2585–2589. Manabe, K. and Kobayashi, S. (2002) *Chem. Eur. J.*, **8**, 4094–4101; Ogawa, C. and Kobayashi, S. (2006) *Adv. Synth. Catal.*, **348**, 2585–2589; Manabe, K. and Kobayashi, S. (2002) *Chem. Eur. J.*, **8**, 4094–4101; Ogawa, C. and Kobayashi, S. (2008) Toward truly efficient organic reactions in water. In: *Process Chemistry in the Pharmaceutical Industry* (eds K. Gadamasetti, and T., Braish), pp 249–265.
4 Narayan, S., Forkin, V.V. and Sharpless, K.B. (2007) *Organic Reactions in Water*, pp 350–365; Narayan, S., Muldoon, J., Glynn, M.G., Fokin, V.V., Kolb, H.C. and Sharpless, K.B. (2005) *Angew. Chem. Int. Ed.*, **44**, 3275–3279.
5 Blackmond, D.G., Armstrong, A., Coombe, V. and Wells, A. (2007) *Angew. Chem. Int. Ed.*, **46**, 3798–3800.
6 Tao, J., Zhao, L. and Ran, N. (2007) *Org. Process Res. Dev.*, **11**, 259–267; Ran, N., Zhao, L., Chen, Z. and Tao, J. (2008) *Green Chem.*, **10**, 361–372.
7 Yazbeck, D.R., Tao, J., Martinez, C.A., Kleine, B.J. and Hu, S. (2003) *Adv. Synth. Catal.*, **4**, 524–532.
8 Hoekstra, M.S., Sobieray, D.M., Schwindt, M.A., Mulhern, T.A., Grote, T.M., Huckabee, B.K., Hendrickson, V.S., Franklin, L.J., Granger, E.J. and Karrick, G.L. (1997) *Org. Process Res. Dev.*, **1**, 26–38.
9 Martinez, C.A., Hu, S., Dumond, Y., Tao, J., Kelleher, P. and Tully, L. (2008) *Org. Process Res. Dev.*, **12**, 392–398.
10 Gijsen, H.J.M. and Wong, C.-H. (1994) *J. Am. Chem. Soc.*, **116**, 8422–8423.
11 Wong, C.-H., Garcia-Junceda, E., Chen, L., Blanco, O. and Gijsen, H.J.M. (1995) *J. Am. Chem. Soc.*, **117**, 3333–3339.
12 Greenberg, W.A., Varvak, A., Hanson, S.R., Wong, K., Huang, H., Chen, P. and Burk, M.J. (2004) *Proc. Nat. Acad. Sci. USA*, **101**, 5788–5793.
13 Jennewein, S., Schurmann, M., Wolberg, M., Hilker, I., Luiten, R., Wubbolts, M. and Mink, D. (2006) *Biotechnol. J.*, **1**, 537–548.
14 Dean, S.M., Greenberg, W.A. and Wong, C.H. (2007) *Adv. Synth. Catal.*, **349**, 1308–1320.
15 Lui, J., Hsu, C.-C. and Wong, C.-H. (2004) *Tetrahedron Lett.*, **45**, 2439–2441.
16 Hu, S., Tao, J. and Xie, Z. (2006) Patent WO 2006/134482.
17 Fox, R.J., Davis, S.C., Mundorff, E.C., Newman, L.M., Gavrilovic, V., Ma, S.K., Chung, L.M., Ching, C., Tam, S., Muley, S., Grate, J., Gruber, J., Whitman, J.C., Sheldon, R.A. and Huisman, G.W. (2007) *Nat. Biotechnol.*, **3**, 338–344.
18 Tarnawa, I., Berzsenyi, P., Andrasi, F., Botka, P., Hamori, T., Ling, I. and Korosi, J. (1993) *Bioorg. Med. Chem. Lett.*, **3**, 99–104.

19 Anderson, B.A., Hansen, M.H., Harkness, A.R., Henry, C.L., Vicenzi, J.T. and Zmijewski, M.J. (1995) *J. Am. Chem. Soc.*, **117**, 12358–12359.

20 For a detailed and very interesting account of the LY300164 program, see Anderson, B.A., Hansen, M.H., Vicenzi, J.T. and Zmijewski, M.J. (1999) Chemistry, biocatalysis and engineering: an interdisciplinary approach to the manufacture of the benzodiazepine drug candidate LY300164. In: *Process Chemistry in the Pharmaceutical Industry* (ed. K.G. Gadamasetti) Marcel Dekker, New York, pp. 263–282.

21 (1995) *Ullmann's Encyclopedia of Industrial Chemistry* 5th edn, VCH Verlag GmbH, Weinheim, Vol. B8, pp. 302–304.

22 Sheldon, R.A. (2007) *Green Chem.*, **9**, 1273–1283.

23 Federspiel, M., Fischer, R., Hennig, M., Mair, H.-J., Oberhauser, T., Rimmler, G., Albiez, T., Bruhin, J., Estermann, H., Gandert, C., Gockel, V., Gotzo, S., Hoffmann, U., Huber, G., Janatsch, G., Lauper, S., Rockel-Stabler, O., Trussardie, R. and Zwahlen, A.G. (1999) *Org. Process Res. Dev.*, **3**, 266–274.

24 Chandran, S.S., Yi, J., Draths, K.M., von Daeniken, R., Weber, W. and Frost, J.W. (2003) *Biotechnol. Prog.*, **19**, 808–814.

25 Guo, J. and Frost, J. (2004) *Org. Lett.*, **6**, 1585–1588.

26 Mullin, R. (2007) *Chem. Eng. News*, **85**, 39 110–111.

27 Dunn, P.J., Galvin, S. and Hettenbach, K. (2004) *Green Chem.*, **6**, 43–48; Dale, D.J., Dunn, P.J., Golightly, C., Hughes, M.L., Levett, P.C., Pearce, A.K., Searle, P.M., Ward, G. and Wood, A.S. (2000) *Org. Process Res. Dev.*, **4**, 17–22.

28 Dunn, P.J., Hughes, M.L., Searle, P.M. and Wood, A.S. (2003) *Org. Process Res. Dev.*, **7**, 244–253.

29 Miyaura, N. and Suzuki, A. (1995) *Chem Rev.*, **95**, 2457–2483; Suzuki, A. (1999) *J. Organomet. Chem.*, **576**, 147–168; Kotha, S., Lahiri, K. and Kashinath, D. (2002) *Tetrahedron*, **58**, 9633–9695.

30 Carey, J.S., Laffan, D., Thomson, C. and Williams, M.T. (2006) *Org. Biomol. Chem.*, **4**, 2337–2347.

31 Ennis, D.S., McManus, J., Wood-Kaczmar, W., Richardson, J., Smith, G.E. and Carstairs, A. (1999) *Org. Process Res. Dev.*, **3**, 248–252.

32 Hannah, J., Ruyle, W.V., Jones, H., Matzuk, K.W., Kelly, K.W., Witzel, B.E., Holtz, W.J., Houser, R.A., Shen, T.Y., Sarett, L.H., Lotti, V.J., Risley, E.A., Van Arman, C.G. and Winter, C.A. (1978) *J. Med. Chem.*, **21**, 1093–1100.

33 DeVasher, R.B., Moore, L.R. and Shaughnessy, K.H. (2004) *J. Org. Chem.*, **69**, 7919–7927.

34 Fujimoto, R.A., McQuire, L.W., Mugrage, B.B., van Duzer, J.H. and Xu, D.US Patent 6 291 523 (2001).

35 Baylis, A.D. and Hillman, M.A.D. (1972) German Patent 2 155 133; Drewes, S.E. and Roos, G.H.P. (1988) *Tetrahedron*, **44**, 4653–4670; Morita, K., Suzuki, Y. and Hirose, H. (1968) *Bull. Chem. Soc. Jpn.*, **41**, 2815.

36 Trost, B. (1991) *Science*, **254**, 1471–1477.

# 13
# Water as a Green Solvent for Bulk Chemicals
*Ferenc Joó and Ágnes Kathó*

## 13.1
## Introduction

The bulk chemicals industry relies mostly on heterogeneous catalysis and there is a good reason for that. When manufacturing several hundred thousand tonnes of chemicals, the separation of catalysts from the product mixture is of paramount importance and even a small loss of the catalyst may lead to unacceptable capital loss, product contamination, and health or environmental hazards. In addition, the catalyst must be physically and chemically robust and must retain its high activity and selectivity for long reaction times to allow the process to run without frequent interruptions. Heterogeneous catalysis has sophisticated chemical and technological developments to achieve these goals. However, in many homogeneous catalytic processes (in this chapter, homogeneous catalysis refers to reactions catalyzed by a dissolved catalyst), the recovery and reuse of the catalyst are tedious and costly or have not been solved at all. This can be tolerated in small-scale laboratory syntheses or in rare cases of the production of fine chemicals, but makes homogeneous catalysis less attractive for the bulk chemicals industry. This is especially so when the soluble catalyst is made of expensive platinum group metals and/or of costly ligands. Still, the molecular nature (when each metal atom may be involved in the chemical reaction as opposed to participation of the surface atoms only, as is the case in heterogeneous catalysis) and the excellent selectivity afforded by homogeneous catalysts make their use highly desirable even on a large scale. Moreover, there are processes for which no heterogeneous catalysts are known that are competitive for industrial dimensions.

Liquid–liquid biphasic catalysis is a highly successful approach to homogeneous catalysis in an economically and environmentally friendly manner. Several such processes use water as one of the liquid phases and the field is extensively referenced [1–9]. It should be noted that catalysis in biphasic systems may apply pairs of immiscible solvents other than water (fluorous solvents, ionic liquids, supercritical fluids, organic–organic mixtures) which serve as the basis for

*Handbook of Green Chemistry, Volume 5: Reactions in Water.* Edited by Chao-Jun Li
Copyright © 2010 WILEY-VCH Verlag GmbH & Co. KGaA, Weinheim
ISBN: 978-3-527-31591-7

important commercial processes [4–8]; however, these fall outside the scope of this chapter.

If we define bulk chemicals as those produced in quantities $\geq 5000$ t per year, there are only two industrial processes which apply aqueous organometallic catalysis for such products, namely hydroformylation with an Rh–sulfonated triphenylphosphine catalyst (Ruhrchemie–Rhône-Poulenc process), and telomerization of butadiene and water with a water-soluble Pd(II)–phosphine catalyst (Kuraray process); both will be described in detail later in this chapter. These two processes are good representative examples of the possibilities that water as solvent offers in the manufacture of bulk chemicals. The success of these processes has already stimulated the use of water as a solvent also in the fine chemicals industry [1–9] and it can be envisaged that this trend will continue.

This volume of the *Handbook of Green Chemistry* gives ample evidence of the favorable use of water as a solvent in green chemistry and there is no need to recapitulate all the advantageous properties of water. Nevertheless, with reference to the production of large quantities of chemicals, it should be noted that the nonflammability and high heat capacity of water are outstandingly important properties for safety and in designing technological process details. In this chapter, we describe those transition metal complex-catalyzed processes for bulk chemicals in which the use of water as solvent is a key element of technology and catalyst recovery.

## 13.2
## Hydroformylation – an Overview

### 13.2.1
### General Aspects of Hydroformylation

Hydroformylation (oxo synthesis) is the addition of hydrogen and carbon monoxide to an alkene resulting in the formation of linear (normal) or branched (iso-) aldehydes (Scheme 13.1).

**Scheme 13.1**

In most cases, hydrogen and carbon monoxide are applied in a 1 : 1 mixture called synthesis gas (syngas). Linear aldehydes are used in large quantities for the manufacture of plasticizer alcohols such as 2-ethylhexanol derived from *n*-butanal and therefore these are the desired products of hydroformylation, although there is also a small but constant market for branched aldehydes [10–12].

This seemingly simple reaction may be accompanied by several side reactions (Scheme 13.2) such as (a) isomerization of the alkene, (b) hydrogenation of the alkene (s), (c) hydrogenation of the product aldehydes, (d) condensation of aldehydes, and more. Obviously, isomerization and hydrogenation of the alkene feed lead to the loss of the starting material whereas hydrogenation and condensation of the aldehydes (formation of the so-called heavy ends) remove some of the product and lead to separation and purification problems. Consequently, these side reactions must be kept at minimum.

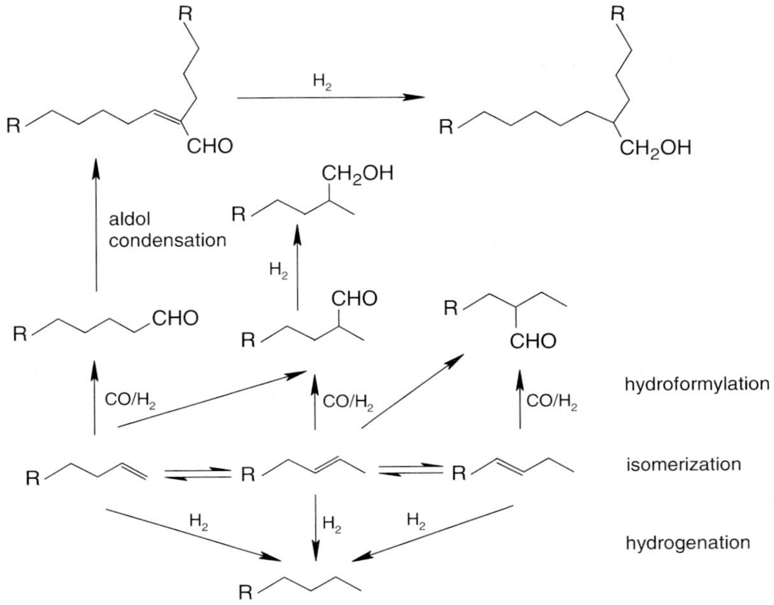

**Scheme 13.2**

Hydroformylation is a highly exothermic reaction [11], with an average reaction heat of 118 kJ mol$^{-1}$ (28 kcal mol$^{-1}$) for each double bond converted. In the case of propene hydroformylation, this converts to 2800 kJ (667 kcal) per kilogram of propene. Since the temperature of the reaction mixture should be strictly regulated to achieve the desired space–time yield and selectivity of the process, effective removal of this heat of reaction is of paramount importance and often requires special engineering considerations.

Today, hydroformylation belongs among the largest scale homogeneous catalytic industrial processes. Estimated total production is around $11 \times 10^6$ t annually, of which butanals, obtained by hydroformylation of propene, are by far the major products ($6 \times 10^6$ t in 2003) [10]. As will be seen below, in achieving this substantial output, industrial hydroformylation is characterized by a variety of feedstocks, numerous variations of catalysts and several different practical processes. Nevertheless, after its first 70 years, there are still constant developments in catalyst and process design to achieve the efficient use of new feedstocks (such as longer chain, branched or internal olefins) in order to serve emerging market needs.

## 13.2.2
## Industrial Hydroformylation Processes in Non-aqueous Systems

Following the original discovery (1938) of the oxo synthesis by Otto Roelen at Ruhrchemie, processes of hydroformylation developed in several well-recognizable steps characterized by the catalysts and process conditions used [1, 12, 13]. In order to acknowledge the "greening" of these processes in time, it is instructive to examine these characteristics and their impact on the environment.

### 13.2.2.1 Cobalt-based Hydroformylation Catalysts and Processes

The first industrially applied catalysts were based on cobalt. The various cobalt derivatives (or the finely divided metal itself) used as catalyst precursors are converted to [CoH(CO)$_4$] either under process conditions or in a separate process (pretreatment, catalyst formation). Since the catalytic activity of [CoH(CO)$_4$] is relatively low, the reaction conditions (temperature, pressure] are fairly harsh. In a process developed by BASF [high-pressure (HP) Co oxo process with unmodified cobalt catalyst], the reaction temperature is 120–180 °C and the pressure of syngas is 270–300 bar. The main reason for the use of such high pressures is to prevent the decomposition of [CoH(CO)$_4$], which is thermally unstable at the high process temperatures applied. The linear to-branched ratio of the resulting aldehyde mixture is typically 80 : 20. In this BASF process, aqueous extraction of the product mixture is performed under slightly acidic conditions in the presence of air and Co(II) salts (oxidative decobalting), and the aqueous extract is concentrated and recycled [14]. Since the formation of the actual catalytic species, [CoH(CO)$_4$], is relatively slow under hydroformylation conditions, instead of feeding this concentrated solution directly back to the reactor, it is advantageous to preform the catalyst with syngas separately at higher temperature. In a similar process developed by Exxon, the recovery of the catalyst includes aqueous extraction of the product mixtures under basic conditions (NaOH, Na$_2$CO$_3$); the extract is then acidified (H$_2$SO$_4$) and the resulting organosoluble [CoH(CO)$_4$] is extracted into the feed mixture and returned to the reactor (Kuhlmann process) [15]. In addition to the inevitable cobalt losses, this recovery process leads to the continuous generation of salt as a result of the neutralization of the basic extracts by an acid. Furthermore, all recovery steps should be done under syngas pressure.

The use of phosphine-modified cobalt catalysts [Shell, low-pressure (LP) Co oxo process] results in the formation of alcohols as the main products and the selectivity in favor of the linear alcohols increases to 88 : 12. In this process, phosphabicyclononane derivatives (L = 3) are used as ligands and the catalytically active species is [CoH(CO)$_3$(L)]. In contrast to tributylphosphine, often mentioned in the literature as a catalyst modifier, the former ligands have higher boiling points than PBu$_3$ and provide sufficient stability to the catalyst to pass through recovery by distillation. [CoH(CO)$_3$(L)] has a diminished catalytic activity relative to [CoH(CO)$_4$], so somewhat higher temperatures (150–190 °C) and higher catalyst concentrations are required to achieve sufficient space–time velocities. On the other hand, the process is run under relatively low syngas pressure (40–80 bar). The product alcohols (and aldehydes) are separated from the catalyst by distillation and then fractionated into individual components (if required) in a separate distillation step. The catalyst-containing fraction from the first distillation step is recycled.

It is worth recalling here that cobalt-based catalysts have high isomerization activity (especially in contrast to rhodium) and can be advantageously used for the hydroformylation to linear aldehydes of feeds with high internal olefin contents. As mentioned before, phosphine-modified cobalt catalysts ([CoH(CO)$_3$(L)]) have high activity for hydrogenating aldehydes, and therefore the main products of the Shell process are alcohols. This is a useful feature of the process in the case of hydroformylation of higher olefins, since the resulting longer-chain aldehydes should be hydrogenated anyway to obtain the target plasticizer alcohols. Conversely, hydrogenation should be avoided in propene hydroformylation, where the desired product is the $n$-butyraldehyde, which is later converted to the target 2-ethylhexanol by condensation and subsequent hydrogenation.

Cobalt-based technologies of hydroformylation present several problems which directly or indirectly affect the environment. High process temperatures and pressures have an energy demand and, although the chemical reaction itself is exothermic, considerable input of energy is required. With the unmodified Co catalyst, recovery of cobalt is effected by extraction; however, in this step, the catalyst is chemically destroyed or undergoes an acid dissociation/protonation cycle. This brings the need for pretreatment of the cobalt-containing extract with syngas before re-entering the catalytic reactor. With phosphine-modified Co catalysts, distillation is required to separate the products and the catalyst, which may lead to catalyst/ligand losses via thermal degradation.

Interestingly, the idea of recovering the catalyst without distillation or other destructive methods had been suggested many years ago (1973) in connection with phosphine-modified cobalt catalysts [16, 17]. Tris(aminoalkyl)phosphine complexes were examined as catalysts which were extracted from the product mixture without decomposition by an aqueous acid wash, and could be re-extracted to the organic (reaction) phase after neutralization. Formation of salt as byproduct could be avoided by the use of an aqueous solution of carbon dioxide under pressure for extraction of highly basic aminophosphine ligands such as P(CH$_2$CH$_2$CH$_2$NEt$_2$)$_3$ and their cobalt complexes. In this elegant method, re-extraction of the catalyst into a fresh organic phase of the substrate needs only decompression (CO$_2$ can be recycled). Although the feasibility of the method was demonstrated, perhaps the economic advantages of a better catalyst recovery were insufficient in the light of the relatively low price of cobalt and no real application of this early "green" concept is known in hydroformylation or in other reactions.

The problems of catalyst activity and selectivity, catalyst recycle, energy requirements, and environmental impact meant great challenges for the development of new technologies of hydroformylation. Major improvements were made possible by the use of rhodium-based catalysts.

### 13.2.2.2 Rhodium-based Hydroformylation Catalysts and Processes

Rhodium is expensive (in June 2008, the average monthly price was US$9800 per troy ounce, or US$315 000 kg$^{-1}$) [18], so the performance of a rhodium-based catalyst must justify its use [19]; furthermore, in bulk chemicals production, losses of rhodium above the ppb level cannot be tolerated. Consequently, extremely efficient recovery techniques in rhodium-catalyzed hydroformylation processes are key to economical operation. In addition, catalysts based on rhodium are more sensitive to catalyst poisons and oxygen than those containing cobalt, therefore the olefin and syngas must be carefully purified before entering the reactor in order to increase the lifetime of the expensive catalyst.

Similarly to cobalt, unmodified rhodium catalysts obtained from catalyst precursors such as rhodium acetate or 2-ethylhexanoate are often used for hydroformylation. In fact, the activity of such catalysts exceeds that of the unmodified cobalt catalyst and the hydroformylation reaction is completely chemoselective for the formation of aldehydes. Internal olefins are isomerized only slowly and therefore the product contains only a low proportion of linear aldehydes. Among the resulting aldehydes, especially at high pressure and temperature, the branched products may dominate (high *iso*-to-normal ratio). The starting olefins are hydrogenated to only a negligible extent. Recovery of rhodium is a more difficult problem than for cobalt in the BASF or Exxon processes since rhodium carbonyls are less stable than their cobalt analogs. Rhodium in the product stream can be separated by distillation; the spent catalyst residues contain metal aggregates, clusters, and so on. Alternatively, rhodium in the reactor effluent can be adsorbed on a basic ion exchanger which is finally incinerated. The catalyst residues or the ash are sent to external precious metal refiners. Processes with unmodified rhodium catalysts are only used for specific purposes such as the hydroformylation of mixtures of long-chain highly branched and internal olefins. They may also be part of a hydroformylation complex, such as that at the Ruhrchemie plant (now part of Oxea) in Oberhausen (Germany), where the HP oxo process with unmodified Rh catalyst is run concordantly with an HP process with a phosphine-modified rhodium catalyst and the Ruhrchemie–Rhône Poulenc aqueous biphasic hydroformylation (see later) [3]. Such an arrangement gives the plant flexibility in production and secures highly economical use (recovery) of rhodium.

Low-pressure oxo (LPO) processes use ligand-modified rhodium catalysts. Following the discovery of the extraordinary hydroformylation properties of [RhH(CO)(PPh$_3$)$_3$] by Wilkinson, several companies initiated the development of rhodium-based industrial processes. Of these, the two most widespread are known as UCC processes [20], which differ substantially in the separation of products and catalyst. In both cases the catalyst is prepared *in situ* from a suitable rhodium precursor (e.g. rhodium acetate) and a large excess of PPh$_3$ under syngas pressure.

In the UCC process with *gas recycle*, the catalyst remains in the stirred tank reactor. For this purpose, high-boiling aldehyde condensation products (which are formed in the process anyway) are used as solvent. The product aldehydes are removed from the reactor with an intense flow of recycled gas, and consequently the method can be applied only in the case of low-boiling aldehydes. This type of catalyst–product separation is an energy-demanding process but works well for the hydroformylation of propene. Typical process parameters are mild: temperatures around 90–95 °C and a syngas pressure of 15–18 bar.

In the UCC LPO process with *liquid recycle*, the catalyst again is dissolved in high-boiling aldehyde condensation products. In this case, however, part of the catalyst solution is continuously removed from the reactor together with the aldehyde products and, after pressure release, is degassed in a flash evaporator. The liquid part is subjected to distillation in order to separate the catalyst and solvent from the product aldehydes. The process operates typically at a temperature of 85–90 °C and 18 bar total pressure. It is suitable for hydroformylation of higher olefins provided that the product aldehydes have sufficiently different boiling points relative to the solvent.

### 13.2.2.3 Ligands Used for Catalyst Modification

The literature on hydroformylation contains reports on a countless number of ligands applied as modifiers in cobalt- and rhodium-based catalysts. However, only a few of them are applied industrially and all belong to the family of tertiary phosphines or phosphites (**1–9**).

**1** PPh$_3$

**2** BISBI

**3** 9-alkyl-9-phosphabicyclo[3.3.1]nonane (P—C$_n$H$_{2n+1}$)

**4** XANTPHOS

**5**

**6**

**7** *m*tppms

**8** *m*tppts

**9** n = 2, 3, 4

Triphenylphosphine (**1**), tributylphosphine, and phosphabicyclononanes (**3**) have already been mentioned. For the production of butanals (the largest volume hydroformylation products), selectivity requirements are limited to the linear/branched ratio. In this case, cheap PPh$_3$ gives catalysts with sufficient activity and selectivity, which makes it the ligand of choice. Phosphabicyclononanes are used in cobalt-catalyzed hydroformylation for two main reasons. They give stable [CoH(CO)$_3$(L)] catalysts, so that catalyst–product separation can be done by distillation. Furthermore, such catalysts have high isomerization activity so internal linear olefins can be converted to linear products with a regioselectivity of 75–85%. Rhodium complexes of monophosphite ligands such as **6** are useful for the hydroformylation of internal and branched olefins, whereas those with appropriate bisphosphite ligands have high isomerization activity and convert internal olefins to linear aldehydes with high selectivity. For example, in the hydroformylation of *trans*-4-octene with a catalyst obtained from [Rh(acac)(CO)$_2$] and a symmetrical bisphosphite, linear nonanal was obtained with 96% selectivity [12]. None of the ligands mentioned are water soluble (phosphites are even highly sensitive to water). For purposes of aqueous–organic biphasic catalysis, triphenylphosphine can be made water soluble by sulfonation. Depending on the conditions, sulfonation yields mono-, di-, and trisulfonated products [*m*tppms (**7**), *m*tppds, and *m*tppts (**8**), respectively], all sulfonated in the *meta*-position. Of these, *m*tppts and to a lesser extent *m*tppms are used in industrial hydroformylation processes.

### 13.2.3
#### Central Questions in Hydroformylation Processes

The short account on hydroformylation technologies in the preceding sections illustrates the central problems of hydroformylation processes. The ideal technology should be able to convert the given feedstock to the desired end product with high activity and selectivity, use a robust catalyst with minimum degradation, apply an efficient recycling of the catalyst (and unconverted educts), and operate in an environmentally friendly manner in terms of both low energy demand and minimum waste production. These requirements are sometimes contradictory and the final economics of a process depend on a delicate balance of the several factors, including also engineering aspects (size of reactors, technical simplicity of the process, etc.) and prices of the catalyst and modifying ligands (if any). All these are finally translated to capital expenditures and operational costs.

An obvious comparison can be made between cobalt- and rhodium-catalyzed hydroformylations. Cobalt is relatively cheap (US$100 kg$^{-1}$, June 2008 [21], which is approximately 3000 times cheaper than rhodium). Conversely, there is a factor of about 1000 in favor of rhodium when catalytic activities in hydroformylation are compared. Consequently, higher catalyst concentrations have to be used in cobalt-catalyzed processes than in those with rhodium catalysts. In general, cobalt catalysts require higher temperatures to obtain sufficient activities and higher syngas pressures to stabilize the less stable cobalt carbonyl derivatives. This requires higher energy input, so despite the exothermic nature of the hydroformylation reaction, such processes are net energy consumers. In the hydroformylation of short-chain linear olefins, rhodium catalysts give superior linear selectivities to cobalt-based catalysts. On the other hand, the high isomerization activity of the unmodified cobalt catalyst, [CoH(CO)$_4$], allows the hydroformylation of feedstocks such as dibutene (a mixture of linear octenes, 3-methylheptenes, and 3,4-dimethylhexenes, obtained by the oligomerization of 1-butene and 2-butene in the so-called raffinate II) to linear products with fairly high selectivity. In the case of dibutene (as with all longer chain olefins), the product mixture contains compounds (mainly alcohols and heavy ends) with high boiling points, so catalyst separation by distillation would require high temperatures and vacuum. Under such conditions, both cobalt- and rhodium-based catalysts decompose substantially but cobalt can be recycled by the well-established method of oxidative decobalting (Section 13.2.2.1).

Catalyst losses may be due to several factors. One is the formation of inactive clusters or metal (plating out) under process conditions. This may happen if the CO pressure is insufficient (unmodified catalysts). With phosphine- and phosphite-modified catalysts, oxidation of the ligands may take place due to the peroxide content of the olefin feed or to oxygen impurities in the syngas. Aldol condensation of the product aldehydes yields water, which leads to an autocatalytic acid-catalyzed hydrolysis of phosphite ligands. Thermal degradation of the complexes can be caused by the harsh conditions of catalyst recovery by distillation.

The general conclusion arises that the milder the conditions of hydroformylation processes and catalyst recovery are, the better. If for any reason harsh conditions cannot be avoided, the reaction mixture should stay under those conditions for the minimum time needed for acceptable space–time yields.

## 13.3
## Water as Solvent for Hydroformylation

### 13.3.1
### Aqueous–Organic Biphasic Catalysis

Aqueous–organic biphasic catalysis is one of the variants of liquid–liquid biphasic catalysis. In these systems, the educts of the catalyzed reaction reside in one of the liquid phases while the catalyst stays in the other phase. The purpose of such an arrangement is to facilitate the separation of the catalysts from products (and unreacted starting material). Traditional organic phase transfer-catalyzed reactions also take place in two-phase systems; however, in those cases the transition metal complex catalyst and the reactants are in the same phase while the other phase serves only as a reservoir of certain reactants and a sink of byproducts; therefore, such systems do not belong to the field of aqueous–organic biphasic catalysis.

Biphasic systems are made up from (at least) two mutually insoluble liquid phases. Since more often than not the products of petrochemistry are insoluble in aqueous solvents, water seems a perfect choice for constructing biphasic catalytic systems since it shows limited solubility in many of the commonly used organic solvents, educts, and products (and vice versa) [22, 23]. Other advantageous features which support its use include availability, relatively low price, non-flammability, high heat capacity, lack of toxicity, and in general a low level of workplace risk together with environmental compatibility. There are, however, several more aspects to consider.

First, because the substrates and products of most reactions of synthetic utility are insoluble in water and partition into the organic phase, the separation and recycle of the *catalyst* requires it to be soluble in the aqueous phase and insoluble in the organic phase. This requirement brings with it the need for chemical modification of the known organometallic catalysts in order to make them water soluble, or the design of new catalysts possessing inherent aqueous solubility. Second, water is very different from the organic phases in being highly polar, capable of forming hydrogen bonds, and of behaving both as a Brønsted acid and base, and not least in that it always contains $OH^-$ and $H^+$, with their actual concentrations governed by the solution pH. It is well known that many organometallic compounds are unstable in the presence of water; nevertheless, a rich organometallic chemistry/catalysis exists in fully or partially aqueous systems [1–9].

Solubility is the central parameter in biphasic catalysis. Although in the preceding discussion mutual *insolubility* of the solvents was mentioned, in practice this always means *limited mutual solubility*. For example, the solubility of ethyl acetate in water at 20 °C is 6.1 wt%, whereas that of water in ethyl acetate is 3.3 wt%. Very similar figures were determined for $Et_2O$ in $H_2O$ (6.9 wt%) and $H_2O$ in $Et_2O$ (3.3 wt%). Other mutual miscibility data with water as one of the phases can be found elsewhere [22, 23]. In addition, one or both phases may contain more than one component (e.g. water–ethylene glycol mixtures as the aqueous phase and 1-octene and nonanal as the organic phase). Perfect containment of the catalyst in the aqueous phase needs strongly lipophilic ligands with virtually no partitioning between the phases. In such a case, however, the catalyzed reaction cannot take place in the organic phase. Either the substrate has to show some limited solubility in the catalyst-containing aqueous phase, or the reaction should take place in the interphase region. Very low solubility of higher olefins severely hinders their catalytic reactions in aqueous–organic biphasic systems. Conversely, for clean catalyst recycling, the product should move to the organic phase, i.e. it should be immiscible with water.

### 13.3.2
**Aqueous–Organic Biphasic Hydroformylation**

Not long after the first reports on aqueous–organic two-phase reactions catalyzed by water-soluble phosphine complexes of rhodium and ruthenium [24, 25], in 1975 Kuntz reported that rhodium complexes of di- and trisulfonated triphenylphosphine (*m*tppds and *m*tppts) were active in biphasic hydroformylation of propene and 1-hexene [26, 27]. He observed virtually complete retention of rhodium in the aqueous phase, which could be easily separated from the organic phase of aldehydes by decantation. Based on the patents of Rhône-Poulenc, very intense research and development efforts by Cornils and co-workers at Ruhrchemie resulted in a new paradigm in industrial hydroformylation: an aqueous–organic biphasic process with a water-soluble rhodium–phosphine catalyst [8, 11, 28–33]. This is now known as the Ruhrchemie–Rhône-Poulenc (or RCH/RP) process, the catalyst for which is prepared *in situ* under syngas from rhodium acetate and trisulfonated triphenylphosphine (*m*tppts). What results during the preforming of the catalyst is a strict analog

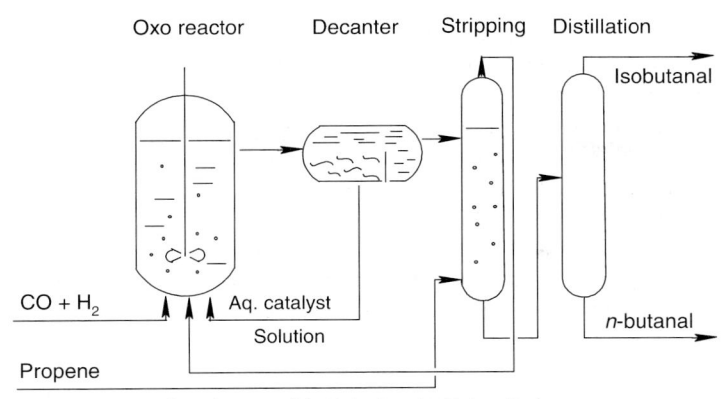

**Figure 13.1** Basic flow diagram of the Ruhrchemie–Rhône-Poulenc aqueous–organic biphasic propene hydroformylation process.

of [RhH(CO)(PPh$_3$)$_3$], that is, [RhH(CO)(mtppts)$_3$], although under process conditions [RhH(CO)(mtppts)$_2$] is the main catalytic species (Section 13.3.2.1). This new process started operation in 1984 at the Oberhausen site of Ruhrchemie with a capacity of 100 000 t per year, and after expansion at that site and together with a plant in South Korea, the total amount of butanals produced by this process now amounts to 800 000 t per year [3].

### 13.3.2.1 The Ruhrchemie–Rhône-Poulenc Process

In its original version, the Ruhrchemie–Rhône-Poulenc process (Figure 13.1) was developed for the hydroformylation of propene. The reaction takes place in a continuously stirred tank reactor. Propene and syngas are introduced into a liquid mixture of the aqueous catalyst solution and butanals. By the time the organic phase leaves the reactor, conversion of propene is practically complete. Part of the reaction mixture is continuously transferred to a separator where the organic and aqueous phases are separated, and the aqueous catalyst solution is taken back to the reactor. The organic phase is stripped with fresh syngas and finally the product is fractionated into *n*- and *iso*-butanals. The basic characteristics of the RCH/RP process are shown in Table 13.1.

**Table 13.1** Characteristic data for the Ruhrchemie–Rhône-Poulenc process [3, 11, 31–33].

| Conditions and products | Typical value |
| --- | --- |
| Temperature | 120 °C |
| Pressure of synthesis gas | 50 bar |
| [mtppts]: [Rh] | ≥60 |
| Conversion | 95% |
| Selectivity towards C$_4$ aldehydes | 99% |
| *n*-Butanal yield | 96% |
| Isobutanal yield | 4% |
| Heat recovery | 99% |
| E-factor | 0.04 |

Although the process is extremely simple, many important points should be considered for continuous and economic operation.

Kinetic and speciation studies revealed that the mechanism of the aqueous biphasic propene hydroformylation with Rh–*mtppts* catalyst is very similar to that of the reaction proceeding in homogeneous organic solution with Rh–PPh$_3$ catalyst. Depending on the conditions, the actual catalytic species can be either [RhH(CO)(*mtppts*)$_2$] promoting the formation of *n*-butanal or [RhH(CO)$_2$(*mtppts*)] leading to the branched product (Scheme 13.3). As usual in rhodium-catalyzed hydroformylations, the RCH/RP process also uses a large excess of the phosphine ligand in order to achieve high *n/iso* selectivity. The ligand of choice is trisulfonated triphenylphosphine, since of the sulfonated triphenylphosphines only *mtppts* has a sufficiently high solubility in water. (There are fairly large differences in solubility data in the literature [2]; the room temperature solubility of the sodium salt of *mtppts* has been reported as 1100–1400 g l$^{-1}$ solution and that of the sodium salt of *mtppms* as 12–200 g l$^{-1}$ solution.) An interesting phenomenon was discovered concerning the [RhH(CO)(*mtppts*)$_2$] vs [RhH(CO)$_2$(*mtppts*)] equilibrium. High-pressure $^{13}$C and $^{31}$P NMR measurements showed no formation of any new species in a solution of [RhH(CO)(*mtppts*)$_3$] + 3 *mtppts* up to 200 bar CO–H$_2$ (1 : 1) [34]. This is in sharp contrast to the case with [RhH(CO)(PPh$_3$)$_3$], which quantitatively gives [RhH(CO)$_2$(PPh$_3$)$_2$] under 30 bar CO–H$_2$ (1 : 1), in the presence of 3 equiv. of PPh$_3$. These observations indicate a less probable dissociation of *mtppts* from [RhH(CO)(*mtppts*)$_3$] than that of PPh$_3$ from [RhH(CO)(PPh$_3$)$_3$]. The activation energy of phosphine exchange, calculated from the linewidth of variable-temperature $^{31}$P NMR spectra, was indeed higher for *mtppts* than for PPh$_3$, namely 125 ± 4 vs 79 ± 4 kJ mol$^{-1}$. The fact that in the aqueous system formation of [RhH(CO)$_2$(*mtppts*)$_2$] the immediate precursor of [RhH(CO)$_2$(*mtppts*)] was not detected explains the high *n/iso* selectivity of the RCH/RP process.

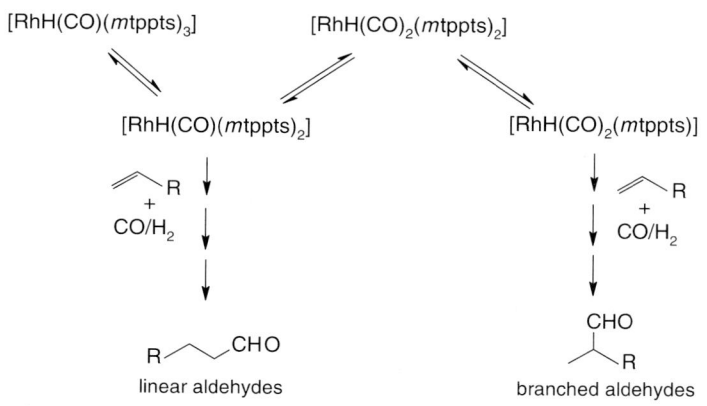

**Scheme 13.3**

The enormous solubility of *mtppts* in water and its practical insolubility in the organic phase ensure the essentially complete retention of rhodium in the aqueous phase. As a consequence, the rhodium content of crude aldehyde is in the ppb range. Another aspect is that aqueous solutions of the triply charged *mtppts* do not froth. Consequently, the whole reactor volume can be used efficiently and the phase separation in the decanter is fast and clean. Conversely, *mtppms* is a surfactant and its use in aqueous biphasic catalysis leads to frothing and less efficient phase separation.

During the hydroformylation process, gradual decomposition of the sulfonated triphenylphosphine ligand takes place [11, 30]. Apart from the possible oxidation by feed impurities, the main decomposition route involves splitting of Ar groups (Ar = 3-$C_6H_4SO_3Na$) from the phosphine, which eventually leads to the formation of benzenesulfonic acid (ArH), 3-formylbenzenesulfonic acid (ArCHO), and $Ar_2$-POH. Propyldi(3-sulfonatophenyl)phosphine is also formed by replacement of an Ar group by $-C_3H_7$. The rhodium complexes of this mixed aromatic–aliphatic phosphine are much less active catalytically than [RhH(CO)(*mtppts*)$_3$]. Formation of phosphido-bridged rhodium clusters also contributes to a decrease in the overall activity of the catalyst solution. Incremental additions of fresh *mtppts* to the aqueous phase prevent the decline of the catalytic activity and increase the catalyst lifetime.

Several of the degradation steps are catalyzed by hydroxo complexes of rhodium, such as [Rh(OH)(CO)(*mtppms*)$_2$] and [Rh(OH)(CO)$_2$(*mtppms*)]. Formation of these complexes is obviously aided by higher pH, hence careful control of pH is essential. Similarly, the formation of the active catalyst from its precursors is also pH dependent. In a model study, it was shown that below pH 4 [Rh(CO)(OAc)(*mtppms*)$_2$] and [RhCl(CO)(*mtppms*)$_2$] do not react with $H_2$ to give [RhH(CO)(*mtppms*)$_2$], but the extent of formation of the latter species increases steeply with increase in pH [35]. For optimum results of hydroformylation, the pH of the aqueous phase in the RCH/RP process is kept between 5 and 6. Keeping the pH of the solution in the slightly acidic regime also effectively eliminates unwanted side reactions such as condensation of aldehydes.

Careful studies established that under actual process conditions, the reaction takes place at the interphase region and not in the bulk aqueous phase [36]. Considering the mutual solubilities of the components in the reactor, this means that very efficient stirring is required to increase this interphase surface area; therefore, specially designed, highly sophisticated stirrers are used. CO and $H_2$ have lower solubilities in water than in organic solvents (the room temperature solubility of $H_2$ in water differs by a factor of 2–5 from those in customary organic solvents); however, this does not seem to be a problem in achieving fast propene hydroformylation with the water-soluble rhodium catalyst.

In order to ensure that most of the catalyst remains in the reactor and participates in the oxo reaction, only a small part of the aqueous phase is taken to the separator together with the butanals. Liquid–liquid phase separation is a highly developed operation of chemical technology, so the catalyst solution is recovered efficiently and under mild conditions. The hot aqueous solution is fed back to the reactor through a heat exchanger generating steam. This makes possible the efficient use of the reaction heat (the plant is a net steam exporter) while the cooled solution contributes to the temperature regulation of the reactor. Note also that to some extent water is soluble in isobutanal (but virtually insoluble in *n*-butanal), and for that reason some water is taken out from the reactor with the product mixture. This water loss is continuously made up for with the recycling catalyst solution.

Part of the product $n$-butanal circulates between an internal cooler in the reactor and the distillation column. Again, the reactor temperature is regulated by this flow while the heat generated in the hydroformylation reaction is used for the separation of $n$- and isobutanal by distillation.

Stripping of the crude oxo products with fresh syngas and the final distillation of butanals are carried out after their separation from the aqueous phase, that is, in the absence of the rhodium catalyst. Therefore, no further reactions occur which would change the composition of the reaction mixture.

In summary, the data in Table 13.1 show that aqueous biphasic operation proved highly successful. In addition to the outstanding performance in terms of mild reaction conditions, selectivity, and energy management, the process is characterized by extremely low rhodium losses, which total a few kilograms over a 20 year period, and the production of approximately $5 \times 10^6$ t of $n$-butanal [3]. It is estimated that if license fees are not included, then the total operational costs of the RCH/RP process are about 10% lower than those of conventional low-pressure hydroformylation with a phosphine-modified rhodium catalyst.

#### 13.3.2.2 Green Features of the Ruhrchemie–Rhône-Poulenc Process

In principle, hydroformylation is a reaction of 100% atom economy, since all atoms of the olefin, $H_2$, and CO are found in the product aldehydes. Formation of byproducts and waste can heavily damage this picture. Therefore, the various processes are better compared in terms of the so-called environmental factor (E-factor), which is defined as the mass ratio of total amount of waste to the target product [37]. Typically, in the bulk chemicals industry processes are characterized by an E-factor of 1–5. In the evaluation of hydroformylation processes, the waste from ligand manufacture should also be included in the total amount of waste produced by the process. Calculations show that for a conventional oxo process with a Co-based catalyst, the E-factor is >0.6, whereas in the RCH/RP process it is only 0.04 (calculated for $n$-butanal in both cases). This reflects the very high $n/iso$ selectivity and also the enormous reduction in producing waste (heavy ends, salts, products of catalyst degradation, spent solvents, etc.) compared with earlier technologies.

The E-factor only accounts for the waste produced by a process. There are also other important aspects. Heat recovery in the RCH/RP process is 99%; it neither requires input of external energy (e.g. for the separation of $n$- and isobutanal by distillation) nor is the heat allowed to dissipate. The very mild process conditions and the simple technological design of the plants also lead to large energy savings.

The advantages using water as a solvent are clearly demonstrated by the process. The catalyst is recovered with virtually no loss due to the fast and clean separation of the aqueous and organic phases in the decanter. The volume ratio of the aqueous catalyst phase to the organic phase is 6, so the majority of the reaction mixture is a non-flammable liquid. This greatly reduces the risk of fires.

In summary, based on the principle of aqueous–organic biphasic catalysis, the Ruhrchemie–Rhône-Poulenc propene hydroformylation process set a new industry standard, not only in process design but also in protection of the environment already

Table 13.2 Room temperature solubility of n-alkenes in water [39].

| Alkene | Solubility (ppm) |
| --- | --- |
| Ethene | 131 |
| Propene | 200 |
| 1-Butene | 222 |
| 1-Pentene | 148 |
| 1-Hexene | 50 |
| 1-Octene | 2.7 |
| 1-Decene | 0.6 |

at a time when E-factors and the 12 principles of green chemistry [38] were not even conceived.

### 13.3.2.3 Hydroformylation of Longer Chain Alkenes in Aqueous–Organic Biphasic Systems

The solubility of alkenes in water decreases rapidly with increasing chain length, as shown in Table 13.2. Since these values were determined at room temperature, the data for ethene, propene, and 1-butene are solubilities of gases, whereas the rest refer to solubilities of liquid alkenes.

The Ruhrchemie–Rhône-Poulenc process is suitable for the hydroformylation of propene to pentene. In addition to propene, in practice it is used for the hydroformylation of 1-butene in the so-called raffinate II, a mixture of 1-butene, 2-butene, and butane in a roughly 40 : 40 : 20 ratio [31, 40]. Since the Rh–*m*tppts catalyst is highly selective for the hydroformylation of terminal olefins, the product is 95% *n*-pentanal and 5% isopentanal while 2-butene remains unreacted. After separation of the gases from the aldehydes, 2-butene is hydroformylated and hydrogenated in a high-pressure Co-catalyzed process to a mixture of isomeric amyl alcohols. The reaction of 1-butene is slower than that of propene, therefore higher catalyst concentrations are used. Longer chain alkenes react even slower and, coupled with their very low solubility in water, this prevents their hydroformylation by the RCH/RP process. Since the feedstocks used in the various hydroformylation processes comprise a large and increasing proportion of alkenes with carbon numbers >5, this is a serious limitation of the RCH/RP process. Several ingenious solutions have been suggested (use of cosolvents, surfactants, additives such as cyclodextrins, etc.); however, none of these proved viable under real-life conditions. All manipulations that increase the solubility of the alkene in the aqueous catalyst solution inevitably lead to increased partitioning of the catalyst to the organic phase; moreover, they are usually costly.

Interestingly, although the solubility of ethene is high enough for effective hydroformylation with the [RhH(CO)(*m*tppts)$_3$] catalyst dissolved in water, propanal is not produced by this method. The reason is in that propanal is fairly miscible with water. Consequently, the water content of the product has to be removed by distillation; moreover, the wet propanal dissolves and removes some of the catalyst from the reactor, necessitating a tedious catalyst recovery. This draws attention to the importance of the *solubility of water* in the organic phase (and not only vice versa).

### 13.3.2.4 Developments in Reactor Design for Aqueous–Organic Biphasic Hydroformylations

As mentioned earlier, the heart of the Ruhrchemie–Rhône-Poulenc process is a specially designed tank reactor with multiple stirrers and internal cooling. In such a reactor, even under the best available conditions, low solubilities of higher olefins ($>C_5$) prevent their hydroformylation with industrially acceptable rates. In their recent study, Wiese and co-workers applied a tubular reactor with static mixers of the Sulzer SMV type [12, 41]. The hydrodynamic characteristics of such mixers are well understood. Hydroformylation of alkenes with aqueous Rh–*m*tppts catalyst was carried out with a high catalyst mass flow (100–400 kg h$^{-1}$) compared with the mass flow of olefins ($\geq 3$ kg h$^{-1}$) and syngas ($\geq 1$ kg h$^{-1}$), other conditions being equal to the industrial process. At a low mass flow (100 kg h$^{-1}$) of catalyst solution (800 ppm Rh, 30 wt% *m*tppts in water) and with propene as starting material, a space–time yield of 0.17 t m$^{-3}$ h$^{-1}$ was obtained, which is in the range usually achieved with stirred reactors. Increasing the catalyst mass flow to 400 kg h$^{-1}$ increased the space–time yield to 0.92 t m$^{-3}$ h$^{-1}$, which means an enhancement by a factor of 5. Taking the dimensions of the tubular reactor applied, such high mass flows correspond to very high linear velocities (up to 0.5 m s$^{-1}$) and very short residence times (2 s m$^{-1}$). This means that the catalyst and the products spend only seconds in the reactor, so the consequences of heat stress and the chance of side reactions are minimized. As a result, selectivity higher than 99% in *n*-butanal could be achieved. Detailed studies have shown that compared with the case of the industrially applied stirred reactor, in the tubular reactor with static mixers the reaction takes place in the bulk of the aqueous phase rather than at the boundary layer. In other words, the process has been shifted from the diffusion-controlled regime near to a kinetic regime, that is, to a quasi-homogeneous reaction, and this resulted in the 5-fold increase in the overall rate. Addition of ethylene glycol to the aqueous phase (up to 36 wt%) led to a substantial further increase in the reaction rate, with a space–time yield for propene hydroformylation of 3 t m$^{-3}$ h$^{-1}$. This is an increase in the rate by a factor of 20 compared with the conventional two-phase process in a stirred tank reactor. Under such conditions, 1-octene was also hydroformylated with high space-time yield (0.1 t m$^{-3}$ h$^{-1}$).

So far, the concept of aqueous–organic hydroformylation of alkenes in tubular reactors with static mixers has been applied in a miniplant, i.e. the step before a pilot plant [13]. Concerning the possibility of hydroformylating longer chain olefins (up to $C_8$–$C_{10}$), the results are very encouraging. Extensive and detailed further research and development work is needed to clarify the effects of the reactor design on the important process characteristics in long-term continuous operation on a large scale (catalyst degradation, accumulation of the products of side reactions in the aqueous phase, etc.). The use of ethylene glycol as cosolvent also has to be justified, especially concerning a possible increase in rhodium leaching, the extraction of ethylene glycol by the product organic phase, and not least the additional cost incurred by its use. Nevertheless, the findings show that knowledge of the kinetics and mechanism of a reaction (e.g. mass transport control vs kinetic control) makes changes in process parameters and reactor designs possible, with substantial benefits in productivity.

### 13.3.3
### Catalyst Recovery by Water-induced Phase Separation

Water can be advantageously used for the recovery of the catalyst also in cases when it is not the solvent of the catalyst as in the Ruhrchemie–Rhône-Poulenc process. One such example is the Kuhlmann process (Section 13.2.2.1). Strictly, these processes do not belong to the field of aqueous–organic biphasic catalysis since the hydroformylation itself takes place in a homogeneous organic phase. However, in some cases there are several technological similarities to the RCH/RP process, so a short discussion of these methods seems justified.

One such procedure is the separation of the catalyst and products (plus unreacted starting material and inert components of the feed) by extraction. In the case of the few water-soluble substrates it is the product aldehydes which can be extracted into water, leaving behind the water-insoluble catalyst, such as $[RhH(CO)(PPh_3)_3]$, and excess phosphine in the organic phase. This concept dates back to the origin of aqueous biphasic catalysis [25]. Industrially, the hydroformylation of allyl alcohol (Scheme 13.4) is carried out in this way since 4-hydroxbutyraldehyde (and the undesired 3-hydroxy-2-methylpropionaldehyde byproduct) are readily soluble in water. Further hydrogenation of 4-hydroxbutyraldehyde yields 1,4-butanediol (1,4-BDO). 1,4-BDO is used mainly for the production of γ-butyrolactone, N-methylpyrrolidone, and tetrahydrofuran, and plays an important role in the manufacture of elastomers.

**Scheme 13.4**

The process for the hydroformylation of allyl alcohol was developed by Kuraray and Daicel Chemical Industries [42] and commercialized by ARCO (now LyondellBasell) (Figure 13.2). It uses a conventional Rh–PPh$_3$ catalyst together with a small amount of

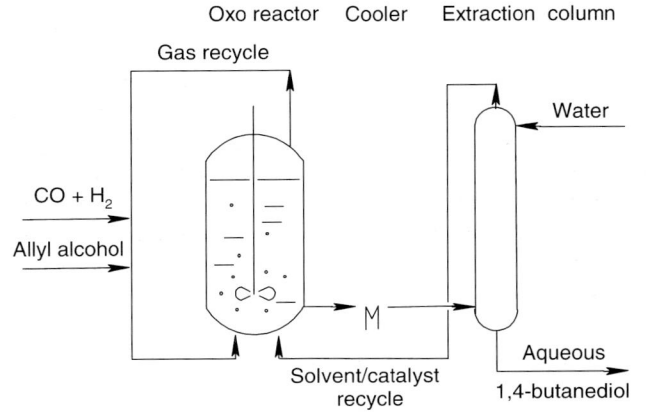

**Figure 13.2** Basic flow diagram of the ARCO allyl alcohol hydroformylation process with product extraction by water.

bis(diphenylphosphino)butane (with an Rh: PPh$_3$: dppb ratio of 1: 150: 0.2), which prevents the deactivation of the catalyst. Allylic alcohol is much more reactive than unfunctionalized alkenes, therefore the process uses mild conditions (60–65 °C and 2.0–2.5 bar syngas). The reaction takes place in a homogeneous organic phase in an aromatic solvent such as toluene, and the products are extracted with water. The rhodium-containing apolar phase is recycled with negligible rhodium loss (10 ppb). Conversion of allylic alcohol is around 98% and the $n/iso$ ratio of the aldehydes is 7. Iron and ruthenium complexes as cocatalysts [43] and recently *trans*-1,2-bis[bis(3,5-di-*n*-alkylphenyl)phosphinomethyl]cyclobutane ligands [44] were suggested to increase further the selectivity in 4-hydroxbutyraldehyde. Although most 1,4-DBO is manufactured by other processes (in reactions of acetylene and formaldehyde, or starting from propylene oxide), hydroformylation of allyl alcohol is practiced on the hundred thousand tonnes scale.

Another important means of catalyst recovery by phase manipulation is used in the process developed by Union Carbide for the hydroformylation of higher olefins [45, 46]. In this process, the phosphine-modified Rh catalyst is prepared from [Rh(CO)$_2$(acac)] (acac = 2,4-pentanedionate) and ω-diphenylphosphinoalkylsulfonates, such as Ph$_2$P(CH$_2$)$_n$SO$_3$Na ($n$ = 2, 3, 4), as ligands (**9**). The better known *m*tppms also has suitable solubility properties; however, under hydroformylation conditions it undergoes substantial phenyl scrambling so that PPh$_3$, *m*tppds, and *m*tppts are also found in the catalyst solution after some time of operation. The rhodium catalyst is dissolved in *N*-methylpyrrolidone (NMP) and the reaction takes place in a one-phase system. Following the reaction, a sufficient (but small) amount of water is added to induce phase separation, upon which the catalyst moves exclusively to the aqueous phase and the aldehydes are found in the organic phase. The phases are separated and the aldehydes are washed free from NMP impurities with water. The catalyst-containing polar phase is subjected to distillation in order to separate it into a water fraction and an NMP solution of the catalyst, which is then recycled to the reactor (Figure 13.3). The distilled water obtained in this step is recycled for washing the aldehydes and to induce phase separation. The process works efficiently for longer chain olefins such as dodecene. Since water has a high heat of evaporation, the process has the drawback that distillation of water from the NMP-containing phase consumes large amounts of energy. It seems, however, that the fast and efficient hydroformylation of longer chain olefins offsets this disadvantage.

Still another variation of catalyst recovery was demonstrated in the hydroformylation of 1-tetradecene [47]. The reaction, catalyzed by an Rh–ω-diphenylphosphinoalkylsulfonate catalyst, was run in a homogeneous methanolic solution and gave slightly better results than the Rh–PPh$_3$ catalyst under identical conditions. After the reaction, most of the methanol was distilled off, and the rhodium–phosphine catalyst together with the free phosphine ligand present was extracted into water from the remaining organic phase of aldehydes and unreacted alkenes. The aqueous phase was evaporated to dryness and the catalyst was taken up in methanol and reused. No loss of activity and selectivity was observed in three recycles.

### 13.4 Water as Solvent in the Production of 2,7-Octadien–1-ol (Kuraray Process)

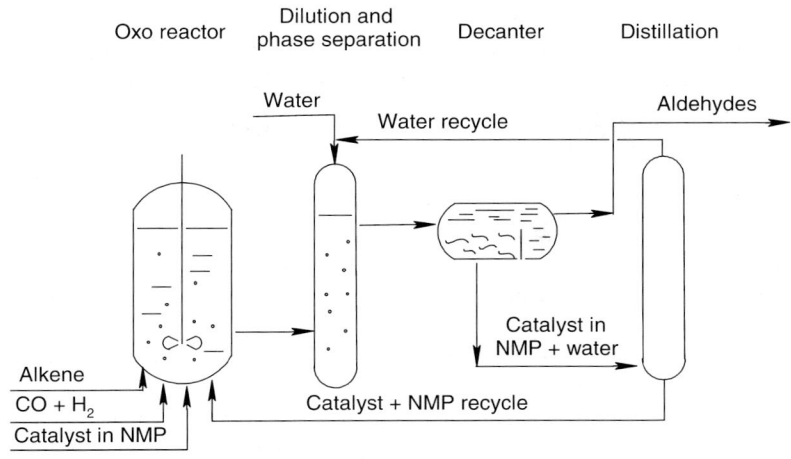

**Figure 13.3** Basic flow diagram of the UCC process for hydroformylation of higher olefins with water-induced phase separation.

The success of the last two methods for catalyst recycling is due to the complete separation of the product-containing apolar phase and the catalyst-containing aqueous phase. It should be noted that there is no need for alterations in the chemical composition of the solutions other than dilution. Evaporation of water from the aqueous extracts (and that of the methanol in the second case) requires considerable energy, and this adds to the process costs, but catalyst degradation during this stage does not seem a problem. Since water (and methanol) are also recycled, there is no inherent generation of waste in the chemistry of these processes (other than formation of the byproducts of hydroformylation).

## 13.4
### Water as Solvent in the Production of 2,7-Octadien–1-ol (Kuraray Process)

Dienes react with nucleophiles (water, alcohols, amines, carboxylic acids, active methylene compounds, etc.) to give oligomerization products incorporating the nucleophile. When the reaction involves specifically 2 mol of diene and 1 mol of the nucleophile, the reaction is also called hydrodimerization (Scheme 13.5).

NuH = HOH, ROH, AcOH, $R_2$NH etc.
**Scheme 13.5**

Telomerization of butadiene with water leads to the formation of 2,7-octadien-1-ol (Scheme 13.6). In fact, this reaction was among the processes disclosed in the first patents on the use of *mtppts* in biphasic solvent mixtures [48]. The product 2,7-octadien-1-ol is a very useful intermediate. It is hydrogenated to 1-octanol, which is a raw material for plasticizers for poly(vinyl chloride). Alternatively, it can be isomerized to 1-octenal [49], further reactions of which allow the production of various important compounds such as 1,9-nonadiol (via hydroformylation followed by hydrogenation), caprylic acid (via hydrogenation and subsequent oxidation of the resulting *n*-octanal), bromo- and cyanocaprylic esters, and 8-aminocaprylic acid.

The catalyst for such telomerizations usually consists of palladium (0) and an excess of *mtppts*, *mtppms*, or other water-soluble phosphines [50]. It has been observed that a large excess of the phosphine ligands diminished the reaction rate to impracticably low values. Conversely, low [P]:[Pd] ratios were insufficient to protect and keep the palladium catalyst in solution.

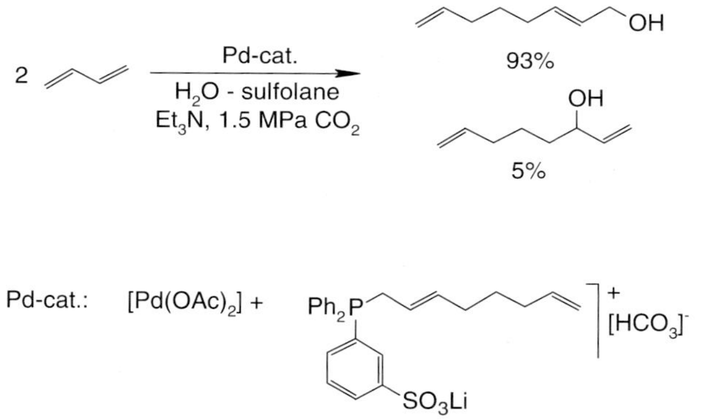

$$2 \text{ (butadiene)} \xrightarrow[\text{Et}_3\text{N, 1.5 MPa CO}_2]{\text{Pd-cat.} \atop \text{H}_2\text{O - sulfolane}} \text{2,7-octadien-1-ol (93\%)} + \text{1,7-octadien-3-ol (5\%)}$$

Pd-cat.: $[Pd(OAc)_2]$ + [Ph$_2$P-octadienyl-C$_6$H$_4$-SO$_3$Li]$^+$ [HCO$_3$]$^-$

**Scheme 13.6**

The telomerization of butadiene with water was developed into an industrial process by Kuraray Industries [51]. Key to the success was the discovery that phosphonium salts such as that shown on Scheme 13.6 could be applied in high excess relative to palladium without any adverse effect on the rate of telomerization and at the same time eliminated catalyst degradation [52]. Such phosphonium salts can be synthesized in a variety organic solvents from the appropriate tertiary phosphine such as *mtppms* (as the Na$^+$ or Li$^+$ salt) and allylic alcohols in the presence of an amine with aqueous basicity $pK \geq 7$ under $CO_2$ pressure [53]. As will be seen, these parameters are close to those of the telomerization process, so it is likely that in the presence of the product 2,7-octadienol (an allylic alcohol) any free phosphine which might form in the reverse reaction is converted to the corresponding phosphonium salt with an octadienyl substituent (Scheme 13.6).

For the telomerization process (Figure 13.4), the catalyst is prepared *in situ* from the ligand and $[Pd(OAc)_2]$ [50, 51]. It is assumed that under the reaction conditions the phosphonium salt reacts to yield the corresponding tertiary phosphine to some extent, which then becomes protected by strong coordination to palladium. In any

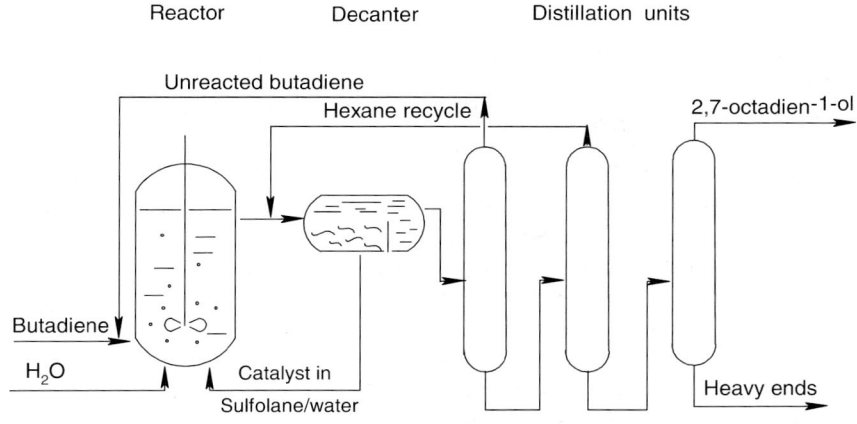

**Figure 13.4** Basic flow diagram of the Kuraray process for telomerization of butadiene and water.

case, with a large excess of the phosphonium salt ligand (P: Pd = 40), a stable catalyst is obtained, which, in the presence of triethylamine and $CO_2$ (1.5 MPa), actively catalyzes the telomerization of butadiene. The best solvent for the process is a mixture of water and sulfolane (tetrahydrothiophene-1,1-dioxide) (45 : 55 wt%). Selectivity to 2,7-octadien-1-ol is typically 90–93%; the main byproduct is 2,7-octadien-3-ol (4–5%). The final reaction mixture is extracted with hexane. Very clean phase separation occurs and the catalyst-containing water–sulfolane phase is recycled to the reactor. The palladium-free organic phase can be separated to unreacted butadiene, hexane, and the products by distillation, with no danger of side reactions which could be catalyzed by residual palladium at the distillation temperature. The production level of 2,7-octadien-3-ol by this process is in the region of 5000 t per year (which is regarded as the borderline between bulk and fine chemicals).

Although it has not been commercialized, telomerization of butadiene with water can also be carried out in a genuine biphasic process, using Pd–*m*tppts or Pd–*m*tppms catalysts [54]. Amines containing one long alkyl chain ($C_6$–$C_{18}$) facilitate the telomerization without the need for added solvent (such as sulfolane in the Kuraray process). Presumably the quaternary ammonium salts formed under $CO_2$ in the aqueous solution, namely $[HN(R)Me_2]^+[HCO_3]^-$, act as phase-transfer agents and solubilize the diene in the aqueous phase. Depending on the amine used, reactions at 85 °C under 1 MPa $CO_2$ afforded 50–60% octadienol yield at 60–80% butadiene conversion, with 87–91% selectivity to 2,7-octadien-1-ol and dimers and dioctadienyl ethers as major byproducts.

## 13.5
## Conclusion

The developments in aqueous–organic biphasic catalysis have shown that water can be advantageously used for the manufacture of certain bulk chemicals. Most outstanding is the Ruhrchemie–Rhône-Poulenc process for propene hydroformylation, which can also be used with minor modifications for the hydroformylation of 1-butene and even for 1-pentene. This process is the prototype aqueous–organic two-phase catalytic technology. In the Kuraray process of telomerization of butadiene and water to yield 2,7-octadienol, the reaction is carried out in a partly aqueous solvent under homogeneous conditions and the product is extracted with hexane. Hydroformylation of long-chain alkenes is applied by Union Carbide in slightly aqueous N-methylpyrrolidone and the catalyst is recovered by a water-induced phase separation. Finally, hydroformylation of allyl alcohol is done in a homogeneous organic solution and the water-soluble products are obtained by extraction into an aqueous phase (Kuraray/ARCO process). The four processes involve the use of water on different levels, from being the sole solvent to its use as an extractant. Nevertheless, in all four cases water is essential for catalyst–product separation. It should be pointed out that all four processes are run with excellent recycle of the costly rhodium and palladium catalysts (with losses in the lower ppb range) and also the ligands used in excess to the metal.

It seems that for the moment the basic needs of the petrochemical industry are all well served by the available processes. Nevertheless, even in the field of bulk chemicals, feedstocks and product demand change over time. One such obvious change is the need to convert long-chain, internal, and branched olefins to valuable products by hydroformylation. This may require new types of catalyst; however, the results of reactor engineering described in Section 13.3.24 may well contribute significantly to technological advancement in the use of these feedstocks. Another field is the catalytic transformation of functionalized olefins and, indeed, aqueous–organic biphasic hydroformylation of pentenoic acid (for the manufacture of nylon derivatives) [55] or acrylate esters (into pharmaceutical raw materials) [56] may be developed in the future into viable processes.

## Acknowledgment

The support of the Hungarian National Research and Technology Office – National Research Fund (NKTH-OTKA K 68482) is gratefully acknowledged.

## References

1 Cornils, B. and Herrmann, W.A. (eds) (2004) *Aqueous-phase Organometallic Catalysis*, 2nd edn, Wiley-VCH Verlag GmbH, Weinheim.
2 Joó, F. (2001) *Aqueous Organometallic Catalysis*, Kluwer, Dordrecht.
3 Wiebus, E. and Cornils, B. (2006) in: *Catalyst Separation, Recovery and Recycling* (eds D.J. Cole-Hamilton and R.P. Tooze), Springer, Dordrecht, pp. 105–143.
4 Adams, D.J., Dyson, P.J. and Tavener, S.J. (2004) *Chemistry in Alternative Reaction Media*, John Wiley & Sons, Ltd, Chichester.
5 Cornils, B., Herrmann, W.A., Horváth, I.T., Leitner, W., Mecking, S., Olivier-Bourbigou, H. and Vogt, D. (eds) (2005)

*Multiphase Homogeneous Catalysis*, Wiley-VCH Verlag GmbH, Weinheim.
6 Joó, F., Papp, É. and Kathó, Á. (1998) *Top. Catal.*, **5**, 113–124.
7 Joó, F. (2003) Biphasic catalysis – homogeneous, in: *Encyclopedia of Catalysis* (ed. I.T. Horváth), John Wiley & Sons, Ltd, Chichester, Vol. 1, pp. 737–805.
8 Cornils, B. (2003) Biphasic catalysis – industrial, in: *Encyclopedia of Catalysis* (ed. I.T. Horváth), John Wiley & Sons, Ltd, Chichester, Vol. 1, pp. 805–846.
9 Joó, F. and Kathó, Á. (2007) in: *Handbook of Homogeneous Hydrogenation* (eds J.G. de Vries and C.J. Elsevier), Wiley-VCH Verlag GmbH, Weinheim, pp. 1327–1359.
10 Röper, M. (2006) *Chem. Unserer Zeit*, **40**, 126–135.
11 Frohning, C.D. and Kohlpaintner, C.W. (1996) in: *Applied Homogeneous Catalysis* (eds B. Cornils and W.A. Herrmann), VCH Verlag GmbH, Weinheim, pp. 29–104.
12 Wiese, K.-D. and Obst, D. (2006) *Top. Organomet. Chem.*, **18**, 1–33.
13 Behr, A. (2008) *Angewandte Homogene Katalyse*, Wiley-VCH Verlag GmbH, Weinheim, pp. 302–320.
14 Grenacher, A.V. and Stepp, H. (2004) US Patent 6 723 884, to BASF AG.
15 Kamer, P.J.C., Reek, J.N.H. and van Leeuwen, P.W.M.N. (2004) in: *Aqueous-Phase Organometallic Catalysis* (eds B. Cornils and W.A. Herrmann), 2nd edn, Wiley-VCH Verlag GmbH, Weinheim, pp. 686–698.
16 Gregorio, G. and Andreetta, A. (1973) Patent Ger. Offen. 231 302, to Montecatini Edison SpA.
17 Andreetta, A., Barberis, G. and Gregorio, G. (1978) *Chim. Ind. (Milann)*, **60**, 887–891.
18 *Platinum Today* (July 15, 2008) available at http://www.platinum.matthey.com.
19 van Leeuwen, P.W.N.M. and Claver, C. (eds) (2000) *Rhodium Catalyzed Hydroformylation*, Kluwer, Dordrecht.
20 Tudor, R. and Ashley, M. (2007) *Platinum Metals Rev.*, **51**, 116–126 and 164–171.

21 *Recycle INme* (July 15, 2008) available at http://www.recycleinme.com.
22 Reichardt, C. (2003) *Solvents and Solvent Effects in Organic Chemistry*, 3rd edn., Wiley-VCH, Weinheim.
23 Stoye, D. (1993) in (eds B. Elvers, S. Hawkins, W. Russey and G. Schulz), *Ullmann's Encyclopedia of Industrial Chemistry*, 5th edn, VCH Verlag GmbH, Weinheim. p. 437.
24 Joó, F. and Beck, M.T. (1975) *React. Kinet. Catal. Lett.*, **2**, 257–263.
25 Dror, Y. and Manassen, J. (1977) *J. Mol. Catal.*, **2**, 219–222.
26 Kuntz, E. (1987) *CHEMTECH*, **17**, 570–575.
27 Kuntz, E. (1976) Patent Ger. Offen. 2 627 354, to Rhône-Poulenc.
28 Cornils, B., Hibbel, J., Lieder, B., Much, J. and Wiebus, E. (1984) German Patent DE 3 234 701, to Ruhrchemie AG.
29 Kohlpaintner, C.W., Fischer, R.W. and Cornils, B. (2001) *Appl. Catal. A*, **221**, 219–225.
30 Beller, M., Cornils, B., Frohning, C.D. and Kohlpaintner, C.W. (1995) *J. Mol. Catal. A*, **104**, 17–85.
31 Herwig, J. and Fischer, R. (2000) in *Rhodium Catalyzed Hydroformylation* (eds P.W.N.M. van Leeuwen and C. Claver), Kluwer, Dordrecht. pp. 189–202.
32 Cornils, B. (1999) *Top. Curr. Chem.*, **206**, 133–152.
33 Cornils, B. (1997) *J. Mol. Catal. A*, **116**, 27–33.
34 Horváth, I.T., Kastrup, R.V., Oswald, A.A. and Mozeleski, E.J. (1989) *Catal. Lett.*, **2**, 85–90.
35 Kovács, J., Joó, F. and Frohning, C.D. (2005) *Can. J. Chem.*, **83**, 1033–1036.
36 Wachsen, O., Himmler, K. and Cornils, B. (1998) *Catal. Today*, **42**, 373–379.
37 Sheldon, R.A. (1994) *CHEMTECH*, **24**, 38–47.
38 Anastas, P.T. and Warner, J.C. (1998) *Green Chemistry: Theory and Practice*, Oxford University Press, Oxford.
39 McAuliffe, C. (1966) *J. Phys. Chem.*, **70**, 1267–1275.

**40** Bahrmann, H., Frohning, C.D., Heymanns, P., Kalbfell, H., Lappe, P. and Wiebus, E. (1997) *J. Mol. Catal. A*, **116**, 35–37.

**41** Wiese, K.-D., Möller, O., Protzmann, G. and Trocha, M. (2003) *Catal. Today*, 79–80, 97–103.

**42** Matsumoto, M., Miura, S., Kikuchi, K., Tamura, M., Kojima, H., Koga, K. and Yamashita, S. (1986) US Patent 4 567 305, to Kuraray Co. and Daicel Chemical Industries.

**43** Pitchai, R., Gaffney, A.M., Nandi, M.K. and Han, Y.-Z. (1994) US Patent 5 276 210, to ARCO Chemical Technology.

**44** White, D.F. (2007) US Patent 7 294 602, to Lyondell Chemical Technology.

**45** Abatjoglou, A.G., Bryant, D.R. and Peterson, R.R. (1993) US Patent 5 180 854, to Union Carbide Corporation.

**46** Abatjoglou, A.G., Peterson, R.R. and Bryant, D.R. (1996) in *Catalysis of Organic Reactions* (ed. R.E. Malz) Chemical Industries Series, Vol. 68, Marcel Dekker, New York, pp. 133–139.

**47** Kanagasabapathy, S., Xia, Z., Papadogianakis, G. and Fell, B. (1995) *J. Prakt. Chem.*, **337**, 446.

**48** Kuntz, E. (1978) Patent Ger. Offen. 2 733 516, to Rhône-Poulenc.

**49** Tsuda, T., Tokitoh, Y., Watanabe, K. and Hori, T. (1999) US Patent 5 994 590, to Kuraray Co.

**50** Yoshimura, N. (2004) in *Aqueous-Phase Organometallic Catalysis*, 2nd edn (eds B. Cornils and W.A. Herrmann), Wiley-VCH Verlag GmbH, Weinheim, pp. 540–549.

**51** Matsumoto, M. and Tamura, M. (1980) US Patent 4 215 077, to Kuraray Co.

**52** Tokitoh, Y., Higashi, T., Hino, K., Murasawa, M. and Yoshimura, N. (1992) US Patent 5 118 885, to Kuraray Co.

**53** Maeda, T., Tokitoh, Y. and Yoshimura, N. (1990) US Patent 4 927 960, to Kuraray Co.

**54** Monflier, E., Bourdauducq, P., Couturier, J.-L., Kervennal, J. and Mortreux, A. (1995) *J. Mol. Catal. A*, **97**, 29–33.

**55** Gelling, O.J. and Tóth, I. (1995) PCT Int. Patent Appl. WO 95/18738, to DSM.

**56** Frémy, G., Monflier, E., Grzybek, R., Ziolkowski, J.J., Trzeciak, A.M., Castanet, Y. and Mortreux, A. (1995) *J. Organomet. Chem.*, **585**, 11–16.

# Index

## a

AAC2 mechanism   349
Abraham–Kamlett–Taft (AKT) model   253
absorbents, super-   78
absorption, selective   278
acceleration, hydrophobic   17, 18
acceptor number   246
2-acetamidoacrylic acid   118
acetate, ammonium   356
acetophenones   276
– benchmark substrate   129, 130
achiral hydrogenation   109, 124–126
achiral water-soluble ligands   107
acids   315, 378
– 2-acetamidoacrylic   118
– acid–base equilibrium   139, 140
– acid-facilitated equilibration   121
– acrylic   301
– amino, see amino acids
– 6-aminopenicillanic   371, 372
– benzoic   347, 353
– Brønsted, see Brønsted acids
– camphorsulfonic   169
– carboxylic   354, 355
– chlorobenzoic   286
– coumarin-3-carboxylic   66
– diiodobenzoic   183
– dodecylbenzenesulfonic   45–48, 51
– formic   116
– glyoxylic   246
– green catalysis   31–54
– keto, see keto acids
– magic   355
– phenylcinnamic   283
– phosphonic   302
– Shikimic   373
– styrenesulfonic   279, 281, 282
– sulfonic   53
– super-   355
– ulosonic   322
acridizinium bromide   242
acrylates
– alkyl   239
– 2-(bromomethyl)-   301
– fluorous-tagged   154
– hydroxymethacrylate   380
acrylic acid   301
activated complex, cyclic   229
activated olefins   305, 306
activation
– dual   211–213
– energy of   236, 237
– Gibbs energy   245
activity
– biological   48
– isomerization   389, 392
acyclonucleosides   275
acylation, cyclodextrins   4, 5
additions
– aqueous cyclo-   322
– bis-aza-Michael   283, 284
– conjugate   219
– cyclo-, see cycloadditions
– 1,3-dipolar cycloaddition   210, 230, 249–251
– direct   158
– endo   18
– heteronucleophiles   211–213
– nucleophilic   161
– photocyclo-   248–267
– α,β-unsaturated derivatives   171, 172
adducts, Diels-Alder   320
aerobic oxidation   76
– direct   221
– palladium-catalyzed   86–89
aggregates, surfactant   261–266
*Agrobacterium tumefaciens*   319

AIDS, HIV-1 integrase inhibitor  49
alcohols
– allyl  401, 402
– ferrocenyl  224
– hemiterpene  352
– oxidation  85–94
– primary  86
– propargyl  183
– secondary  87
aldehydes
– aromatic  286
– branched/linear  396
– 'on water' oxidation  222
– oxidations in water  94–96
– sugar  305, 306
aldol reactions
– condensation  16, 17
– cross-aldol condensations  64
– direct  213–215
– Mukaiyama  60, 308, 309
– nitro-  61
aldolase, deoxyribose-5-phosphate  366–369
aldoses  299, 300
alkaline ion-exchange silicates  58
alkenes
– conjugated  170–172
– coupling reactions  163–172
– electron-deficient  63
– isomerization  165, 166
– longer chain  399, 400
– microwave-assisted Heck coupling  275
– room temperature solubility  399
– unconjugated  163–170
alkyl acrylates  239
alkyl groups, direct addition  158
alkyl isocyanates  316
alkylated polystyrene-supported sulfonic acids  53
alkylations  25, 157, 158
– double N-  277, 278
– Friedel–Crafts  344
– phenol  344
alkynes
– coupling reactions  163–172
– 1,3-dipolar cycloaddition  251
– oxidative dimerization  181–186
– terminal  181, 182, 185
– unconjugated  163–170
alkynylation  161, 162
allenylation  156, 157
allyl alcohol  401, 402
allyl scavenger  187
allyl vinyl ether  231

allylation  152–156
– indium-mediated  298
– zinc-mediated  153
allylic substitutions  186
amidation  355, 356
amine ligands  82, 83, 128
amines
– alkylation  277, 278
– anchored  58
– poly-  13
– silica-grafted  68
amino acids  10
– racemization  12
2-aminobenzothiazole  212
6-aminopenicillanic acid  371, 372
aminopropanals  369
ammonia, aqueous  310
ammonium acetate  356
ammonium substituents, quaternary  175
amphiphiles  264, 265, 296
– catalysts  113
anchored amines  58
anion-modified hydrotalcites  70
anions, chorismate  252
anti-inflammatory agent  377
anti-malaria drugs  282
antibodies  258
antihydrophobic material  22
antiviral/antitumor compound  157
applications, pharmaceutical  363–383
approximate transition state (TS) theory  226
aprotic solvents  141, 236
aqueous ammonia  310
aqueous binary mixtures  252, 253
aqueous cycloaddition  322
aqueous–ionic liquid biphasic medium  60
aqueous Krohnke reaction  283
aqueous media
– alcohol oxidations  85–94
– Lewis acids  31, 32
– microwave-assisted organic transformations  274–284
– olefins  78–85
– pericyclic reactions  234–267
– Suzuki/Stille coupling  59
aqueous–organic biphasic
– catalysis  393, 394
– hydroformylation  394–400
aqueous pericyclic reactions, microwave-assisted  266, 267
aqueous phase, reusable basic  64
aqueous solvents, mixed  378

aqueous sugar solutions  319–324
aqueous suspensions  207
aqueous Suzuki reaction  377, 378
arenes
– diazonium tetrafluoroborates  177
– hydrogenation  114
aromatic aldehydes  286
aromatic chlorination  3, 4
aromatic iodides  183
aromatic rings, hydrogenation  114, 115
aromatic substitution, electrophilic  306–308
aromatics, electron-rich  306
artificial enzymes  13, 15, 16
artificial metalloenzymes  134
artificial odorant  43
aryl chlorides, cross-coupling  179
aryl halides, microwave-assisted Heck coupling  275
aryl iodides  173–180, 217
arylation  158–161
$N$-arylation, intramolecular  284
$O$-arylation, intermolecular  216
aryldiazonium salts  181
arylimines  162
aryliodinium salts  181
arylmercuric chlorides  152
arylsiloxanes, cross-coupling  190
5-aryltriazole acyclonucleosides  275
*Aspergillus melleus*  119
asymmetric catalysis  151
asymmetric hydrogenation  118–122
asymmetric Mannich-type reaction, catalytic  61
asymmetric reactions, chiral  36–44
asymmetric synthesis, aza sugars  312
asymmetric transfer hydrogenation  126–135
– biomimetic catalysts  133–135
atom economy  398
atorvastatin  367–369
axial position  304
aza-Diels–Alder reactions  323
aza-Michael addition, bis-  283, 284
aza sugars  311, 312
azide, phenyl  249
2-azidoethanal  368
azodicarboxylate, dimethyl  208, 209
– dimethyl  254

## b

β-$C$-glucosidic ketone  295
β-hydroxy sulfides  282
Baeyer–Villiger oxidation  94–96
Barbier-type reactions  297–305

barbiturates, $C$-glycosyl  292
bases
– acid–base equilibrium  139, 140
– base-catalyzed epimerization  365
– green  57–73
– Lewis-  60
– non-ionic  62
basic aqueous phase, reusable  64
basic ionic liquid  65
Baylis–Hillman reactions  62, 305, 306
– environmentally friendly process  379, 380
beads, glass  188
Beckmann rearrangement  338, 350
benchmark substrate, acetophenone  129, 130
benzaldehyde  215
benzene  333, 345
benzil  350
benzofurans  182
benzoic acid  347, 353
benzoin  23, 351
benzonitrile oxide  250
1,4-benzoquinone  218
benzoxazines  280
bidentate dienophiles  256, 264
bidentate phosphane ligand  168
bimetallic nanoclusters  90
bimolecular Diels–Alder reaction  237, 243
binary mixtures, aqueous  252, 253
binding of species  2, 3
bioactive heterocycles  280
biocatalysis  255–259
bioconjugation reactions  184
bioconversions  318, 319
biological activity  48
biological effluent treatment plant  363
biomimetic catalysts  133–135
bioreduction  370, 371
biotinylated metal catalyst  134
biphasic catalysis  385, 393, 394
biphasic hydroformylation  394–400
biphasic reactions  126–128
biphasic systems  60, 394
bipyridine ligand, 4,4′-dimethyl-2,2″-  259
bis-aldol reaction, tandem  281
bis-aza-Michael addition  283, 284
bisimidazolecyclodextrin  17
bisphenol A decomposition  342
bleach, household  91–94
bleaching booster  310
bond formation

– carbon–carbon, *see* carbon–carbon-forming reactions
– carbon–nitrogen 309–311
– hydrogen 213
bound ester 5, 6
branched aldehydes 396
bromides 173–180
– acridizinium 242
– cetyltrimethylammonium 37, 130, 133, 263, 264
bromination 223, 224
– oxidative 224
bromoenopyranosides 302–304
2-(bromomethyl)acrylates 301
4-bromotoluene 378
– Suzuki reaction 378
Brønsted acids 75
– catalysis 44–53
Brønsted bases 75
BSA-catalyzed Diels–Alder reaction 258
bulk chemicals, green solvents 385–408
butadiene, telomerization 404
butadienyl ethers 317
1,4-butanediol 348

## c

$C_1$ interconversion pathways 337
Cadiot–Chodkiewicz coupling reaction 166
cages, molecular 255
camphorsulfonic acid 169
camphorsultam derivative 157
carbamation 313–317
carbenes 16, 165
carbocyclic acids 354, 355
carbohydrates
– amphiphiles 296
– chiral dienes 320, 321
– functionalization 291–330
carbon–carbon bond cleavage 355
carbon–carbon-forming reactions 34, 47
– cross-coupling 375–378
– functionalization of carbohydrates 292–309
– HTW 344
– microwave-assisted 274–277
– pericyclic 229
– stereoselective 53
carbon dioxide
– $CO_2$-enriched HTW 340, 341
– hydrogenation 116, 117, 141
– supercritical 1
carbon nanotubes, single-walled 225
carbon–nitrogen-forming reactions 309–311
carbonates 66, 314

carbonyl compounds
– achiral transfer hydrogenation 124–126
– asymmetric hydrogenation 121, 122
– coupling reactions 151–163
– hydrogenation 113, 114
– unsaturated 213–215
carbonyl reductions 18, 19
carboxylates, 1,3-dipolar cycloaddition 252
carboxylic acids 66, 354, 355
carboxypeptidase 6
catalysis
– aqueous–organic biphasic 393, 394
– asymmetric 151
– asymmetric Mannich-type reaction 61
– base-catalyzed epimerization 365
– Brønsted acid 44–53
– Cu(II) 257
– green acid catalysis in water 31–54
– heterogeneous 385
– homogeneous 191
– industrial homogeneous 387
– intra-complex transfer 4
– liquid–liquid biphasic 385
– 'on water' 227
– organometallic 133
– oxidations 76–78
– pericyclic reactions 255–259
– phase-transfer 80
– (pre)micellar 231
– turnover 14
– yield 32–53
catalysts
– amphiphilic 113
– biomimetic 133–135
– biotinylated metal 134
– BSA 258
– chiral organic 214
– cobalt-based 388, 389
– cyclodextrins 21, 259–261
– heterogeneous 122
– hydrophilic 174, 175
– indium 304
– iron 84
– lanthanides 323
– LASCs 34
– Lewis acids 153–155
– Lewis bases 60
– loading 42
– losses 393
– manganese 80
– metal complex 106
– modification by ligands 391, 392
– montmorillonite 306

- nano-sized magnesium oxide   281
- nickel   180
- palladium   86–89, 173–180, 317
- PEG-immobilized   136
- phase-transfer   133
- phosphine-modified rhodium   398
- polymer-supported Brønsted acid   51–53
- PSSA   279, 281, 282
- pyrophoric   376
- quasi-homogeneous   90
- Raney nickel   364
- recovery   401, 402
- recyclable   93
- rhodium   115, 116
- rhodium-based   389–391
- ruthenium   83, 117, 180
- scandium   294
- separation   106, 135–137, 191
- 'sole solvent'   32–36
- surfactant aggregates   261–266
- surfactant-type Brønsted acid   51–53
- thermoresponsive   91
- transition metals   124, 215, 216
- Yamamoto–Yanagisawa   155
cation-exchanged zeolites   70
cationic amphiphiles   265
cells, electrochemical   339
cetyltrimethylammonium bromide (CTAB)   37, 130, 133, 263, 264
chain-elongated sugars   299
chain elongation   299–301
chalcones   63
chelating transition state   298
chemical kinetics, solution-phase   336
chemical potentials   237, 238
chemoselective deprotection   187
chiral amine ligands   83
'chiral-at-metal' species   135
chiral cationic amphiphile   265
chiral dienes, carbohydrate-based   320, 321
chiral Lewis acid-catalyzed asymmetric reactions   36–44
chiral organic catalysts   214
chiral sulfur ylides   63
chiral water-soluble ligands   108
chlorides
- aryl   179
- arylmercuric   152
- fatty acid   315
chlorination, aromatic   3, 4
$o$-chlorobenzoic acid   286
chloroformates   314, 315

chorismate anion   252
*cis*-hydroxylation   81
Claisen rearrangement   207, 230, 231, 251, 252
- 1,3-diols   323
- monomolecular   261
Claisen–Schmidt condensation   346
classical resolution process   366
cleavage, carbon–carbon bond   355
cobalt-based hydroformylation catalysts   388, 389
cocatalysts   96
coenzymes   9–15
cohesive energy density   233
colloidal dispersions   188
colloids, protective-colloid agents   164
complexes
- coordination   82
- cyclic activated   229
- half-sandwich   110
- half-sandwich metal   123
- hydrophobic   19
- hydroxo   397
- intra-complex transfer   4
- metal complex catalyst   106
- model dihydrogen   139
- monoaqua   138
- organometallic   129, 130
- tethered   132
- tetranuclear   110
computational modeling   337–339
condensations
- aldol   16, 17
- Claisen–Schmidt   346
- cross-aldol   64
- HTW   346–348
- Knoevenagel   292–297
- nano-sized magnesium oxide-catalyzed   281
conductivity, thermal   332
conformational behavior, sucrose   313
conjugate addition, Friedel–Crafts-type   219
conjugated alkenes, coupling reactions   170–172
control
- diastereoselectivity   303, 304
- end-of-pipeline   232
- geometric   9
cooling, internal   400
coordination complexes   82
coordination to metals, water   138, 139
copolymerization, diiodobenzoic acid   183

copper, Cu(II) catalysis  257
copper-mediated coupling  161
cosolvents  316
coumarin  249
coumarin-3-carboxylic acids  66
coupling
– alkenes and alkynes  163–172
– aqueous media  59
– carbonyl compounds  151–163
– copper-mediated  161
– cross-  179–181, 190
– cross-dehydrogenative  219
– dehydrative  218
– dehydrogenative  216–221
– direct  218
– Glaser, Eglinton and Cadiot–Chodkiewicz  166
– Hartwig–Buchwald  189
– Heck  345, 346
– in water  151–206
– 'on water'  215–221
– organic halides  172–191
– oxidative  219
– pinacol  162, 163, 286
– Sonogashira  182–184
– Stille  59, 180, 181
– Suzuki  59, 60, 66, 176
– Suzuki–Miyaura  215
Crestor  366, 367
critical values  331, 334
cross-aldol condensations  64
cross-coupling
– aryl chlorides  179
– arylsiloxanes  190
– carbon–carbon  274
– carbon–carbon-forming reactions  375–378
– Hiyama  276
– Sonogashira  276
– transition metal-catalyzed  215, 216
cross-dehydrogenative coupling  219
CTAB, see cetyltrimethylammonium bromide
Cu(II) catalysis, copper  257
cyclic activated complex  229
cyclic ketones  95
cyclic peptides  287
cyclic transition state  303, 304
cyclization  167–170
– Nazarov  346–348
– Prins-type  369
– RNA  7
cycloadditions  207, 320, 321
– aqueous  322

– 1,3-dipolar  210, 230, 249–251
– photo-  248–267
cyclodextrins  6
– acylation  4, 5
– bisimidazole-  17
– catalysts  21
– imidazole rings  8
– modified  164
– pericyclic reactions  259–261
– pyridoxal unit  11
– thiazolium salt  16
cyclohexanol, dehydration  340
cyclohexanone  215, 288
cyclohexanone oxime  338
cyclopentadiene  236, 237, 242
cyclopropanation, styrene  165
cycloreversion  247
cyclotrimerization  167–170

d
DBSA, see dodecylbenzenesulfonic acid
decanting  334
decarboxylation  283
– oxidative  353
decarboxylative transaminations  14, 15
decomposition, bisphenol A  342
dehalogenase, halohydrin  369
dehydration
– 1,4-butanediol  348
– cyclohexanol  340
– D-glucose  352
– HTW  351–353
dehydrative coupling  218
dehydrogenative coupling  216–221
density, water  332
2-deoxyaldoses  299
deoxybenzoin derivatives  216
deoxyribose-5-phosphate aldolase (DERA)  366–369
depression  376, 377
deprotection, chemoselective  187
deprotonation  139
derivatives
– carbonyl  151–163
– furan  293
– organic halides  172–191
– phenol  219
– polyfunctional unsaturated  167–170
– $\alpha,\beta$-unsaturated  171, 172
detergents, bleaching booster  310
deuterium, H–D exchange  140, 141, 355
DFT calculations  139–141, 226
diamine-enabled protocol  130
diarylmethanols  160

diastereoisomers  300
diastereoselectivity  156, 162
– functionalization of carbohydrates  303, 304
– hydroxyl functionalization  169
– 'on water' reactions  214
diazoacetate, ethyl  69
diazocarbonyl  251
diazonium salts  173–180
diazonium tetrafluoroborates, arene  177
dibenzyl ether hydrolysis  341
dielectric constant, water  332
– water  336
Diels-Alder adducts  320
Diels–Alder reactions  23, 209, 210
– aza-  323
– bimolecular  237, 243
– BSA-catalyzed  258
– enantioselective  258
– forward  244–248
– hydrophobic acceleration  17, 18
– intramolecular  242, 243
– inverse electron-demand  234–242
– retro-  243, 244
– retro-hetero-  244–248
dienes  403
– carbohydrate-based chiral  320, 321
– hydrophobicity  240
– polarity  263
– telomerization  170, 171
dienophiles
– bidentate  256, 264
– dienophile affect  239
– glyoxylic acid  246
– hydrophobicity  240
diflusinal  377, 378
dihydrogen complex, model  139
dihydropyrimidinones  279
dihydroxylation  78–85
diiodobenzoic acid, copolymerization  183
diketones  297
dimedone  294
dimerization
– hemiterpene alcohols  352
– oxidative  181–186
– photo-  248
dimethyl azodicarboxylate  208, 209, 254
4,4′-dimethyl-2,2″-bipyridine ligand  259
dimethyl sulfoxide (DMSO)  24
1,3-diols  323
dioxanes  280
dipeptides  3

diphosphine ligands  122
1,3-dipolar cycloaddition  210, 230, 249–251
dipolarophiles  250
direct addition, alkyl groups  158
direct aerobic oxidation  221
direct aldol reaction  213–215
direct coupling, indole compounds  218
$C$-disaccharides  305
discrimination, facial  322
diseases
– AIDS  49
– depression  376, 377
– malaria  282
dispersions, colloidal  188
DMSO, see dimethyl sulfoxide
dodecyl sulfate, sodium  120, 130, 165, 263, 264
dodecylbenzenesulfonic acid (DBSA)  45–48, 51
double $N$-alkylation  277, 278
double-tailed amphiphile  264
droplet size  255
drugs
– anti-depression  377, 378
– anti-inflammatory agents  377
– antiviral/antitumor  157
– Prexige  379
– Tamiflu  372–374
– Viagra  374, 375
– see also pharmaceutical compounds
dual activation  211–213

**e**

electrochemical cells  339
electrochemical methods  155
electrodes, graphite  155
electron-deficient alkenes, epoxidation  63
electron-demand Diels–Alder reactions  234–242
electron-donating substituents  88
electron-poor dipolarophiles  250
electron-rich aromatics  306
electron-withdrawing groups  171
electrophiles, functionalized  177
electrophilic aromatic substitution  306–308
elimination  353
'emulsion-like' system  121
emulsions  164
– micro-  266
enantiomers  10, 11
enantioselectivity
– Diels–Alder reactions  258
– direct aldol reactions  213–215

- hydrogenation   119, 120, 127
- sulfoxidation   97
end-of-pipeline control   232
*endo* addition   18
*endo/exo* isomers   235
ene reactions   207
energy
- non-conventional sources   273–290
energy curve   22
energy density, cohesive   233
energy of activation, Gibbs   236, 237
enforced hydrophobic interaction   238, 241, 250
engineering, HTW process engineering   334, 335
enol ether, silyl   32
- silyl   309
enolate, silicon   41
enolization   16, 17
enone   346
environmentally friendly Baylis–Hillman process   379, 380
enzymatic hydrolysis   119
enzymatic processes   363–374
enzymes
- aldolase   366–369
- artificial   13, 15, 16
- artificial metallo-   134
- asymmetric transfer hydrogenation   133
- carboxypeptidase   6
- DERA   366–369
- glucose dehydrogenase   318
- halohydrin dehalogenase   369
- HIV-1 integrase   49
- metallo-   5, 6
- neuramidase   372
- ribonuclease   7–9
- tyrosine kinase   178
epimerization   365
epoxidations
- electron-deficient alkenes   63
- iron-catalyzed   84
- manganese-catalyzed   80
- olefins   78–85
- ruthenium-catalyzed   83
- solvent- and halide-free   79
epoxide hydrolysis   317
epoxide ring-opening reaction   39, 40
epoxides
- fatty   317
- *meso-*   39
equatorial position   303
equilibration, acid-facilitated   121

equilibrium, acid–base   139, 140
esterification   313–317
esters
- bound   5, 6
- hydrazono   156
- sucrose   314
ethene   399
etherification   313–317
ethers
- allyl vinyl   231
- butadienyl   317
- dibenzyl   341
- silyl enol   32, 309
ethyl diazoacetate   69
extractor column   402

## f

facial discrimination   322
fatty acid chlorides   315
fatty diketones, symmetrical   297
fatty epoxides   317
ferrocenyl alcohols   224
flash heating   273
fluorous-tagged acrylates   154
formaldehyde (HCHO)   41–44
formic acid   116
formyl *C*-glucoside   308
forward Diels–Alder reactions   244–248
free radicals
- intermediates   223
- stable   87, 91–94
Friedel–Crafts-type reactions   34, 51
- alkylation   344
- conjugate addition   219
D-fructose   296
functionalization
- carbohydrates   291–330
- diastereoselective   169
- electrophiles   177
- hydroxyl groups   169, 312–317
- olefins   109
- SWNTs   225
furan derivatives   260
- trihydroxyalkyl-substituted   293

## g

γ-adduct   155
gasless approach to hydrogenation   116
geometric control   9
geometry
- rigid   5
- transition states   21–25
Gibbs energy of activation   236, 237, 245

Gibbs energy of transfer 241
Glaser, Eglinton and Cadiot–Chodkiewicz
    coupling 166
glass beads 188
D-glucosamine 294
D-glucose, dehydration in HTW 352
glucose dehydrogenase 318
C-glucosidation 307
C-glucoside, formyl 308
glycamines 309–311
– surfactants 311
glyceraldehyde, derivatives 321
glycine 12
glyco-organic substrates 319–324
C-glycolipids 297
C-glycosyl barbiturates 292
glycosylamines 309–311
N-glycosylation 311
– yields 312
glyoxylic acid, dienophiles 246
(+)-goniofurfurone 157
grafting 68
graphite electrodes 155
green acid catalysis 31–54
green bases 57–73
Green Chemistry
– Handbook of 232, 386
– holistic concept 232
– non-conventional energy sources
    273–290
– 'on water' conditions 207–228
– 12 Principles of 152, 232, 399
– water as solvent 1–29
– water under extreme conditions
    331–361
green oxidation 75–103
green reduction 105–149
– role of water 137
green solvents 232–234
– bulk chemicals 385–408
'greenest' solvent 142
Grignard-type reactions 151–162
groups
– alkyl 158
– electron-withdrawing 171
– oxime 5
– polar 107
– pyridoxal unit 11
guanidinylated phosphanes 184

## h

H–D exchange 140, 141
– HTW 355

half-sandwich complexes 110
– metal 123
halide-free epoxidation 79
halides
– aryl 275
– organic 172–191
– pseudo- 178
– see also bromides, chlorides, iodides
halobenzene 345
halohydrin dehalogenase 369
Handbook of Green Chemistry 232, 386
Hartwig–Buchwald coupling 189
Hattori's classification 58
HCHO 41–44
heat exchange 335
heating, flash 273
Heck coupling
– HTW 345, 346
– microwave-assisted 274
hemiterpene alcohols 352
heptoses 300
hetero-Diels–Alder reactions, retro-
    244–248
heteroaryl ketones 130, 131
heteroatom-based nucleophiles 211
heterocoupling 172, 173
heterocycles 277–284
– bioactive 280
– heterocyclic hydrazones 282
– heterocycloreversion 247
– nitrogen-containing 277–280
– oxygen-containing 280, 281
– sonochemical reactions 285
heterogeneous catalysis 122
– bulk chemicals 385
heteronucleophiles 213–215
hexadentate imidazolium salts 159
high-temperature water (HTW) 331–355
– benzene solubility 333
– $CO_2$-enriched 340, 341
– organic synthesis 335
– properties 332–334
higher olefins 403
highly hindered diarylmethanols 160
highly polar substrates 122
Hildebrand solubility parameter 253
hindered diarylmethanols 160
HIV-1 integrase inhibitor 49
Hiyama–Heck reactions, one-pot 190
Hiyama reaction 189–191, 276
homocoupling 172, 173
homogeneous catalysis 191
– industrial processes 387

household bleach, hypochlorite   91–94
HTW, see high-temperature water
hydration   233, 351–353
hydrazines, alkylation   277, 278
hydrazones, heterocyclic   282
hydrazono esters   156
hydride transfer   117
hydroarylation   166, 167
hydrocarbons, hydrophobic effect   2
hydroformylation   163, 164, 386–393
– aqueous–organic biphasic   394–400
– cobalt-based catalysts   388, 389
– industrial processes   388–392
– longer chain alkenes   399, 400
– propene   395, 396
– rhodium-based catalysts   389–391
hydrogen
– bond formation   213
– bonding interactions   141
– H–D exchange   140, 141, 355
hydrogen-bonding solvents   244
hydrogen peroxide ($H_2O_2$)   77–86, 90–97
– photochemical reactions   287
hydrogenations   105
– achiral   109
– arenes   114
– aromatic rings   114, 115
– asymmetric   118–122
– carbon dioxide   116, 117
– carbonyl compounds   113, 114
– enantioselectivity   119, 120, 127
– gasless approach   116
– HTW synthesis   343
– hydroformylation   387
– in water   108–122
– olefins   109–113
– selective   110
– transfer, see transfer hydrogenation
hydrolysis
– dibenzyl ether   341
– enzymatic   119
– epoxide   317
– HTW   348–350
– nitrile   336
hydrophilic catalysts   174, 175
hydrophilic starches   77, 78
hydrophilic substrates   40, 41
hydrophobic acceleration, Diels–Alder reactions   17, 18
hydrophobic association   243
hydrophobic complexes   19
hydrophobic dienes and dienophiles   240

hydrophobic effect   2, 3, 5, 6
– carbonyl reductions   18, 19
– geometries of transition states   21–25
– hydrogenation   141
– 'on water' conditions   207, 212
hydrophobic interaction, enforced   238, 241, 250
hydrophobic ketones   17
hydrotalcites   67
– anion-modified   70
'hydrothermal' fluids   267
hydroxo complexes   397
hydroxycarbonylation   164
hydroxyl-bridged palladium(II) dimer   88
hydroxyl functionalization   312–317
– diastereoselective   169
hydroxylation   20
– cis-   81
hydroxymethacrylate   380, 381
hydroxymethylation   40–44
hypervalent iodine compounds   91
hypochlorite (household bleach)   91–94

*i*

imidazole rings   8
imidazolium salts, hexadentate   159
imines
– aryl-   162
– asymmetric transfer hydrogenation   132, 133
– hydrogenation   115, 116
– reduction   132
indium   304
indium-mediated allylation   298
indoles   182
– direct coupling   218
industrial processes
– bulk chemicals   385–408
– homogeneous catalytic   387
– hydroformylation   388–392
– paint industry   2
inhibitors
– HIV-1 integrase   49
– neuramidase   372
– tyrosine kinase   178
insolubility   106
interface, oil–water   226
intermolecular *O*-arylation   216
internal cooling   400
intracomplex transfer   4
intramolecular Diels–Alder reactions   242, 243
intramolecular *N*-arylation   284

inverse electron-demand Diels–Alder reactions 234–242
iodides
– aromatic 183
– aryl 173–180, 217
– vinyl 173–180
iodine compounds, hypervalent 91
ion-exchange silicates, alkaline 58
ion product, water 332
ion-supported TEMPO 93
ionic liquids 34–36
– basic 65
– hydrogenation 111
– room temperature 239
ionization potential, metals 298
iron-catalyzed epoxidations 84
*iso*-to-normal ratio 390
isocyanates, alkyl 316
isoelectric point 374, 375
isomerization 165, 387
isomerization activity 389, 392
isomers, *endo/exo* 235

## j
Juliá–Colonna method 63

## k
kanosamine 373
KDN 300–302, 321
keto acids 10
– transamination 15
ketol intermediates 346
ketones
– asymmetric transfer hydrogenation 126–132
– cyclic 95
– β-*C*-glucosidic 295
– heteroaryl 130, 131
– hydrophobic 17
– oxidations in water 94–96
– tandem bis-aldol reaction 281
– α,β-unsaturated 347
3-ketosucrose 319
kinetics, solution-phase 336
Kirkwood plot 337
Knoevenagel reactions 61, 64, 65, 292–297
Krohnke reaction, aqueous 283
Kuhlmann process 388
Kuraray process 402

## l
lanthanide-catalyzed aza-Diels–Alder reactions 323

lanthanide triflates 258
leaching, palladium 176
Lewis acid–surfactant combined catalyst (LASCs) 33, 34
Lewis acids 75, 153–155
– chiral catalysts 36–44
– in water 31–36
– pericyclic reactions 256, 257
Lewis bases 75
– catalysts 60
life-cycle assessment 336
ligands
– achiral 107
– catalyst modification 391, 392
– chelating 126
– chiral 108
– chiral amine 83
– 4,4′-dimethyl-2,2″-bipyridine 259
– diphosphine 122
– phosphane 168
– phosphonium salt 404, 405
– (S)-proline amide 128
– pyridylamine 82
– soft 75
– tetraamido macrocyclic 76–78
– trisulfonated triphenylphosphine 396
– water-soluble 76, 77, 107
light, UV 287
Lilly process 371
linear aldehydes 396
linear multi-parameter model, AKT 253
Lipitor 367–369
liquid–liquid biphasic catalysis 385
liquid–liquid processes 59–65
liquids
– biphasic media 60
– ionic, *see* ionic liquids
– non-polar 210
– solid–liquid processes 65–70
loading, catalyst 42
long-chain alkenes, hydroformylation 399, 400
long-chain olefins 163
low-pressure oxo (LPO) processes 390
lumiracoxib process 379
LY300164 369–371
Lyrica process 364–366

## m
macrocyclic ligand, tetraamido 76–78
magic acid 355
magnesium oxide, nano-sized 281
maleimides 267
manganese-catalyzed epoxidation 80

Mannich-type reactions  36, 37, 47, 48
– asymmetric  61
medium effects  251
*meso*-epoxides  39
metal catalysts, biotinylated  134
metal complex catalyst  106
metal oxide bases  58
metal salts  32
metallo-ene reactions  168
metalloenzymes  5, 6, 134
metallophthalocyanines  76–78
metalloporphyrins  76–78
metals
– Barbier-type reactions  298
– half-sandwich complexes  123
– ionization potential  298
– transition, *see* transition metals
metathesis  165
methacrolein  321
methyl carboxylate  252
methylation, hydroxy-  40–44
2-methylindole  218
micellar catalysis  231
micellar effect  265
micellar systems  91
– nano-  112
Michael addition, bis-aza-  283, 284
Michael reactions  36–38
– non-ionic bases  62
– silica-grafted amines  68
microemulsions  164, 266
microorganisms, asymmetric transfer hydrogenation  133
micropores  85
microwave, flash heating  273
microwave-assisted reactions  274–284
– pericyclic  266, 267
microwave conditions  176
mimics
– biomimetic catalysts  133–135
– enzymes  7–17
– metalloenzymes  5, 6
– ribonuclease  7–9
– thiamine pyrophosphate  15, 16
mixed aqueous solvents, pharmaceutical processes  378
mixtures, aqueous binary  252, 253
Mizoroki–Heck reaction  166, 167
model
– AKT  253
– pseudo-phase  261, 262
model dihydrogen complex  139
model reaction  47, 48
modeling
– DFT calculations  139–141, 226
– molecular and computational  337–339
modified cyclodextrins  164
mole fraction of water  238
molecular area  34
molecular cages  255
molecular modeling  337–339
molecules, self-assembly  179
monoaqua complexes  138
monometallic nanoclusters  90
monomolecular Claisen rearrangement  261
monomolecular sigmatropic reaction  230
Montanari protocol  92
montmorillonite  306
Mukaiyama aldol reaction  60, 308, 309
multi-parameter model, AKT  253
multiple stirrers  400

**n**
*N*-methylpyrrolidone (NMP)  402, 403
Nafion resin  84
nano-sized magnesium oxide  281
nanoclusters  90
nanomicelles  112
nanotubes, single-walled carbon  225
natrium, *see* sodium
Nazarov cyclization  346–348
neocuproin  89, 90
neuramidase inhibitor  372
nickel-catalyzed reactions  180
nitrile hydrolysis  336
nitriles, hydrogenation  115, 116
nitro compounds, hydrogenation  115, 116
nitroaldol reactions  61
nitrogen, carbon–nitrogen-forming reactions  309–311
nitrogen-containing heterocycles  277–280
'no catalyst added' procedure  177
non-aqueous systems, industrial hydroformylation processes  388–392
non-conventional energy sources, green chemistry  273–290
non-ionic bases  62
non-polar liquids  210
norbornene  249
nucleophiles  403
– heteroatom-based  211
nucleophilic addition, Grignard-type  161
nucleophilic oxime groups  5
nucleophilic ring opening  207
nucleophilic substitution, 'on water'  224, 225

## o

2,7-octadien-1-ol   402
octoses   301
odorant, artificial   43
oil–water interface   226
olefins   78–85
– activated   305, 306
– asymmetric hydrogenation   118–121
– functionalized   109
– higher   403
– hydrogenation   109–113
– long-chain   163
– substituted   86
– terminal   86
oligomerization   393, 403
'on water' catalysis   227
'on water' conditions   207–228
– rate enhancement   226
– solvents   209
'on water' pericyclic reactions   254, 255
one-electron oxidation   94
one-pot Hiyama–Heck reactions   190
Oppolzer camphorsultam derivative   157
organic biphasic catalysis, aqueous–   393, 394
organic catalysts, chiral   214
organic halides, coupling reactions   172–191
organic solutes, hydration   233
organic solvents   209, 210
– polar   115
organic synthesis in HTW   335
organic transformations, microwave-assisted   274–284
organoaqueous conditions   151
organometallic catalysis   133
organometallic complexes   129, 130
organometallic systems, recyclable   151
oseltamivir phosphate   372–374
oxidations
– aerobic   76, 86–89
– alcohol   85–94
– aldehydes   94–96, 222
– Baeyer–Villiger   94–96
– catalyzed   76–78
– direct aerobic   221
– green   75–103
– hydrophobic effect   19–21
– hydroxyl functionalization   318, 319
– ketones   94–96
– 'on water'   221–223
– one-electron   94–96
– partial   354, 355
– 1-phenyl-1-trimethylsiloxypropene   222

– sulf-   96, 97
oxidative bromination   224
oxidative coupling   219
oxidative decarboxylation, benzoic acid   353
oxidative dimerization, alkynes   181–186
oxides
– benzonitrile   250
– carbon dioxide, see carbon dioxide
– metal   58
– nano-sized magnesium   281
oxidoreductases   76, 97
oximes
– cyclohexanone   338
– nucleophilic groups   5
oxo processes, low-pressure   390
oxo synthesis, see hydroformylation
oxygen-containing heterocycles   280, 281
ozonolysis   300

## p

paint industry   2
palladium-catalyzed aerobic oxidation   86–89
palladium-catalyzed reactions   173–180
palladium-catalyzed telomerization   317
palladium leaching   176
palladium(II) dimer, hydroxyl-bridged   88
PAMAM   14
partial oxidation   354, 355
Passerini reaction   223
PEG-immobilized catalyst   136
penicillin G   371, 372
pentane-2,4-dione   294–296
pericyclic reactions   208–210, 229–271
– aqueous media   234–267
– microwave-assisted   266, 267
– 'on water'   254, 255
– stereochemistry   261
petrochemistry   393, 406
pH adjustment   374, 375
pH effects   339–343
pharmaceutical compounds   173
– acyclonucleosides   275
– anti-inflammatory   377
– anti-malaria drugs   282
– antiviral/antitumor   157
– water as green solvent   363–383
phase separation, water-induced   401, 402
phase-transfer catalysis   133
– reaction-controlled   80
1,10-phenanthroline ligands   126
phenol alkylation   344
phenol derivatives   219

1-phenyl-1-trimethylsiloxypropene, oxidation 222
phenylazide 249
phenyl–phenyl overlap 24
phenylalanine 10
phenylboronic acid 378
α-phenylcinnamic acid derivatives 283
2-phenylthiazole 217
phenyltrimethyltin 160
phloroacetophenone 307
phosphabicyclononane 391, 392
phosphanes
– bidentate ligands 168
– guanidinylated 184
phosphate
– oseltamivir 372–374
– pyridoxamine/pyridoxal 9–15
phosphenylenes 174
phosphine-modified rhodium catalyst 398
phosphines 107
– water-soluble 114
phosphonic acid 302
phosphonium salt 404, 405
photochemical reactions, non-conventional energy sources 287, 288
photocycloadditions 248–267
photodimerization 248
pinacol coupling 162, 163
– sonochemical 286
pinacol rearrangement 350
PIPO (polymer-immobilized piperidinyloxyl) 92, 93
polar groups 107
polar organic solvents 115
polar substrates 122
polar transition state 278
polarity, dienes 263
pollution prevention 232
polyamines 13
polyfunctional unsaturated derivatives 167–170
polymer-immobilized piperidinyloxyl 92, 93
polymer-supported Brønsted acid catalysts 51–53
polymer-supported carbonates 66
polymer-supported scandium triflate 34
polymerization reactions 165
polymers
– styrene-based 51
– unsaturated 112
polyorganosiloxanes 349

polystyrene-supported sulfonic acids, alkylated 53
poly(styrenesulfonic acid) (PSSA) 279, 281, 282
polyvinylsaccharides 293
porphyrin ring 20
potential, chemical 237, 238
pregabalin process 364–366
(pre)micellar catalysis 231
Prexige 379
primary alcohols 86
Prins-type cyclization 369
process engineering, HTW 334, 335
processes
– Baylis–Hillman 379, 380
– classical resolution 366
– enzymatic 363–374
– industrial 387–392
– Kuhlmann 388
– Kuraray 402
– Lilly 371
– liquid–liquid 59–65
– low-pressure oxo 390
– lumiracoxib 379
– pregabalin 364–366
– Rhône-Poulenc 395–399
– sampatrilat 374, 375
– solid–liquid 65–70
– UCC 403
– see also reactions
prochiral sulfides 97
(S)-proline amide ligand 128
propargyl alcohols 183
propargylation 156, 157
propene hydroformylation 395, 396
protective-colloid agents 164
protic solvents 235, 236
protocols
– diamine-enabled 130
– Montanari 92
proton inventory 8
pseudo-halides 178
pseudo-phase model 261, 262
PSSA, see poly(styrenesulfonic acid)
pure water 44–53
pyridoxal phosphate 9–15
pyridoxal unit 11
pyridoxamine phosphate 9–15
pyridylamine ligands 82
pyrophoric catalyst 376
pyrophosphate, thiamine 15, 16

## q

quadricyclane   208, 209, 254
quasi-homogeneous catalysts   90
quaternary ammonium substituents   175

## r

racemic pregabalin   364
racemization, amino acids   12
Raney nickel catalyst   364
rate constants, second-order   262
rate enhancment, 'on water' conditions   226
reactants, water-insoluble   207
reaction-controlled phase-transfer catalysis   80
reactions
– acylation   4, 5
– addition, see additions
– aldol, see aldol reactions
– alkylation   25, 157, 158
– alkynylation   161, 162
– allenylation   156, 157
– allylation   152–156
– amidation   355, 356
– aqueous Suzuki   377, 378
– aromatic chlorination   3, 4
– arylation   158–161
– asymmetric   36–44
– Barbier-type   297–305
– Baylis–Hillman   62, 305, 306, 379, 380
– benzoin   23
– bioconjugation   184
– bioreduction   370, 371
– biphasic   126–128
– bisphenol A decomposition   342
– bromination   223, 224
– carbon–carbon-forming, see carbon–carbon-forming reactions
– carbon–nitrogen-forming   309–311
– chiral   36–44
– Claisen rearrangement, see Claisen rearrangement
– condensations, see condensations
– coupling, see coupling
– cyclization   7, 167–170, 346–348
– cycloaddition, see cycloadditions
– cyclopropanation   165
– cycloreversion   247
– cyclotrimerization   167–170
– decarboxylation   283
– decarboxylative transaminations   14, 15
– dehydration   340, 348, 351–353
– deprotonation   139
– Diels–Alder, see Diels–Alder reactions
– dihydroxylation   78–85
– dimerization   181–186, 352
– double $N$-alkylation   277, 278
– elimination   353
– ene   207
– enolizations   16, 17
– enzymatic hydrolysis   119
– epimerization   365
– epoxidation, see epoxidations
– epoxide ring-opening   39, 40
– Friedel–Crafts-type   34, 51, 219, 344
– geometries of transition states   21–25
– $C$-glucosidation   307
– grafting   68
– Grignard-type   151–162
– heterocoupling   172, 173
– heterocycloreversion   247
– Hiyama   189–191
– Hiyama–Heck   190
– homocoupling   172, 173
– hydration   351–353
– hydroarylation   166, 167
– hydroformylation   163, 164, 386–393
– hydrogenation, see hydrogenations
– hydrolysis   119, 317, 336, 341, 348–350
– hydroxycarbonylation   164
– hydroxylation   20
– hydroxymethylation   40–44
– in aqueous sugar solutions   319–324
– isomerization   165, 166, 387
– Knoevenagel   61, 64, 65, 292–297
– Krohnke   283
– Kuhlmann process   388
– Kuraray process   402
– Lewis acid-catalyzed   36–44
– liquid–liquid processes   59–65
– Mannich-type, see Mannich-type reactions
– metallo-ene   168
– metathesis   165
– Michael, see Michael reactions
– Mizoroki–Heck   166, 167
– monomolecular sigmatropic   230
– $N$-glycosylation   311
– Nazarov cyclization   346–348
– nickel-catalyzed   180
– nitroaldol   61
– oligomerization   393, 403
– oxidation, see oxidations
– oxidative decarboxylation   353
– oxidative dimerization   181–186
– ozonolysis   300
– palladium-catalyzed   173–180
– Passerini   223

- pericyclic, *see* pericyclic reactions
- photochemical   287, 288
- photodimerization   248
- polymerization   165
- Prins-type cyclization   369
- propargylation   156, 157
- reduction, *see* reductions
- ruthenium-catalyzed   180
- saponification   57
- Schotten–Baumann procedure   376
- solid–liquid processes   65–70
- sonochemical   285
- Sonogashira   181–186
- stereoselective C–C-bond-forming   53
- substitution, *see* substitutions
- Suzuki   378
- Suzuki–Miyaura (S–M)   173–180
- tandem bis-aldol   281
- telomerization   170, 171, 317, 404
- thiolysis   282
- Tsuji–Trost   186–188
- Ullmann–Goldberg   286
- vinylation   158–161
- *see also* processes
reactivity, carbenes   165
reactors
- design   400–402
- supramolecular   179
- tank   400
- tubular   400
reagents
- stoichiometric   75
- trapping   247
rearrangements
- Beckmann   338, 350
- Claisen   207, 230, 231, 251, 252, 261, 323
- HTW   350, 351
- pinacol   350
recovery, catalysts   401, 402
recyclable catalysts   93, 106, 135–137, 191
recyclable organometallic systems   151
recycle, aqueous   335
reducing sugars   292
reductions
- bio-   370, 371
- carbonyl   18, 19
- green   105–149
- imines   132
regioselectivity   265
relative standard chemical potentials   237, 238
resin, Nafion   84
resolution process, classical   366
retro-Diels–Alder Reactions   243, 244

retro-hetero-Diels–Alder reactions   244–248
reusable basic aqueous phase   64
rhodium, phosphine-modified catalysts   398
rhodium catalysts   115, 116, 389–391
ribonuclease, mimics   7–9
D-ribose   294
rigid geometry   5
rings
- aromatic   114, 115
- epoxide   39, 40
- imidazole   8
- nucleophilic opening   207
- porphyrin   20
RNA, cyclization   7
rod–coil molecules, self-assembly   179
room temperature ionic liquids   239
room temperature solubility, $n$-alkenes   399
rosuvastatin   366, 367
Ruhrchemie–Rhône-Poulenc process   395–399
ruthenium catalysts   83, 117, 180

**s**

saccharides
- C-di-   305
- polyvinyl-   293
- *see* sugars
salts
- aryldiazonium   181
- aryliodinium   181
- diazonium   173–180
- imidazolium   159
- metal   32
- phosphonium   404, 405
- thiazolium   16
sampatrilat process   374, 375
sandwich complexes, half-   110
saponification   57
scandium, silica gel-supported   34–36
scandium triflate   294, 308
- polymer-supported   34
scandium tris(dodecyl sulfate)   33
Schotten–Baumann procedure   376
SCW, *see* supercritical water
SDS, *see* sodium dodecyl sulfate
second-order rate constants   262
secondary alcohols   87
selectivity
- absorption   278
- hydrogenation   110
- in water   18–21
- *syn*   298
self-assembly, rod–coil molecules   179
separation

- catalysts   106, 135–137, 191
- water-induced phase   401, 402
Shikimic acid   373
sigmatropic reaction, monomolecular   230
sildenafil citrate   374, 375
silica gel-supported scandium   34–36
silica-grafted amines   68
silicates, alkaline ion-exchange   58
silicon enolate   41
silyl enol ether   32, 309
single-walled carbon nanotubes (SWNTs)   225
sodium carboxylate   252
sodium dodecyl sulfate (SDS)   120, 130, 165, 263, 264
soft heteroatom-based nucleophiles   211
soft ligands   75
'sole solvent'   32–36
solid–liquid processes   65–70
solubility
- benzene in HTW   333
- biphasic systems   394
- Hildebrand parameter   253
- $n$-alkenes   399
solutes, organic   233
solutions
- aqueous sugar   319–324
- chemical kinetics   336
solvent-free epoxidation   79
solvents
- acceptor number   246
- aprotic   141, 236
- green   232–234, 385–408
- 'greenest'   142
- hydrogen-bonding   244
- mixed aqueous   378
- 'on water' conditions   209
- organic   209, 210
- polar organic   115
- protic   235, 236
- 'sole solvent'   32–36
- volatile organic compounds (VOCs)   1, 2
- water   1–29
solvophobicity   253
sonochemical reactions   285
Sonogashira coupling   182–184
Sonogashira cross-coupling   276
Sonogashira reaction   181–186
'stabilized carbenes'   16
stable free radical   87, 91–94
standard chemical potentials   237, 238
starches   77, 78
statins   366–369
Staudinger-type conditions   367
stereoselectivity

- C–C-bond-forming reactions   53
- 'on water' reactions   214
Stille coupling   180, 181
- aqueous media   59
stirring   207, 208
- multiple   400
stoichiometric reagents   75
stripping   395
styrene   70
- cyclopropanation   165
styrene-based polymers   51
substituent effects   246
substituents, electron-donating   88
substituted α-phenylcinnamic acid derivatives   283
substituted olefins   86
substituted tetrazines   245
substitutions
- allylic   186
- electrophilic aromatic   306–308
- nucleophilic   224, 225
substrates
- benchmark   129, 130
- highly polar   122
- hydrophilic   40, 41
sucrose   313–317
- conformational behavior   313
sugar aldehydes   305, 306
sugar solutions, aqueous   319–324
sugars
- aza   311, 312
- chain-elongated   299
- indium-mediated allylation   298
- reducing   292
- unprotected   295, 296
- see also saccharides
sulfates, dodecyl   120, 130, 165, 263, 264
sulfides
- β-hydroxy   282
- prochiral   97
sulfonic acids   53
sulfoxidation   96, 97
- enantioselective   97
sulfur ylide, chiral   63
superabsorbents   78
superacids   355
supercritical carbon dioxide   1
supercritical water (SCW)   267, 331–355
supported bases   58
supramolecular reactor   179
surfactants
- aggregates   261–266
- Brønsted acid catalysts   45–51
- CTAB   37, 130, 133, 263, 264

– glycamine-based   311
– SDS   120, 130, 165, 263, 264
suspensions, aqueous   207
Suzuki coupling   66, 176
– aqueous media   59, 60
– microwave-assisted   274
Suzuki–Miyaura coupling   215
Suzuki–Miyaura reaction   173–180
Suzuki reaction   377, 378
SWNTs, see single-walled carbon nanotubes
symmetrical fatty diketones   297
syn selectivity   298
syngas   398, 402

*t*
Tamiflu   372–374
TAML, see tetraamido macrocyclic ligand
tandem bis-aldol reaction   281
tank reactor   400
taxol   162
telomerization
– butadiene   404
– dienes   170, 171
– palladium-catalyzed   317
TEMPO (2,2,6,6-tetramethyl-
   piperidinoxyl)   87, 91–94
terminal alkynes   181, 182, 185
terminal olefins   86
tethered complex   132
tetraamido macrocyclic ligand (TAML)
   76–78
tetrafluoroborates, arene diazonium   177
tetrahedral structure   333
tetrahydrofuran   342, 348
tetranuclear complex   110
tetrazines   245
TfOH   38
theoretical studies
– HTW   335–339
– 'on water' conditions   226, 227
– transition states   226
thermal conductivity, water   332
thermoresponsive catalysts   91
thiamine pyrophosphate   15, 16
thiazolium salt   16
thiolysis   282
*trans*-stilbene   248
transaminations   14, 15
transfer
– Gibbs energy   241
– intra-complex   4
transfer hydrogenation   105, 123–137
– achiral   124–126
– asymmetric   126–135

– transition metal-catalyzed   124
transition metals
– cross-coupling reactions   215, 216
– transfer hydrogenation   124
transition states
– approximate theory   226
– chelating   298
– cyclic   303, 304
– geometries   21–25
– participation of water   141, 142
– polar   278
trapping reagent   247
treatment plant, biological effluent   363
tributylphosphine   391
triflates   173–180
– Lanthanide   258
– scandium   34, 294, 308
trihydroxyalkyl-substituted furan
   derivatives   293
triphenylphosphine   391
– trisulfonated   396
trisulfonated triphenylphosphine   396
Tsuji–Trost reaction   186–188
tubular reactors   400
turnover catalysis   14
12 Principles of Green Chemistry   152, 232,
   399
tyrosine kinase inhibitors   178

*u*
UCC process   403
Ullmann–Goldberg reaction   286
ulosonic acids   322
ultrasound, non-conventional energy
   sources   285, 286
ultraviolet (UV) light   287
unconjugated alkenes and alkynes, coupling
   reactions   163–170
unprotected sugar   295, 296
unsaturated carbonyl compounds   213–215
α,β-unsaturated derivatives, 1,4-addition
   171, 172
unsaturated derivatives, polyfunctional
   167–170
α,β-unsaturated ketones   347
unsaturated polymers   112
unsymmetrical diketones   297
uridyluridine   7

*v*
variable hydrophobicity   240
vesicle-forming amphiphile   264
Viagra   374, 375
vicinal dihydroxylation   78

vinyl ether, allyl 231
vinyl iodides 173–180
vinylation 158–161
viscosity, water 332
volatile organic compounds (VOCs) 1, 2

## w
water
- -based enzymatic processes 363–374
- bulk chemicals 385–408
- chiral Lewis acid-catalyzed asymmetric reactions 36–44
- coordination to metals 138, 139
- coupling reactions 151–206
- extreme conditions 331–361
- functionalization of carbohydrates 291–330
- green acid catalysis 31–54
- green bases 57–73
- green oxidation 75–103
- green reduction 105–149
- H–D exchange 141, 142, 355
- high-temperature, *see* high-temperature water
- hydrophobic effect, *see* hydrophobic effect
- Lewis acids 31–36
- mole fraction 238
- oil–water interface 226
- 'on water' conditions 207–228
- participation in transition states 141, 142
- pharmaceutical applications 363–383
- pure 44–53
- selectivities in 18–21
- solvent for green chemistry 1–29, 232–234
- superabsorbents 78
- supercritical 267, 331–355
- water–cosolvent media 316
- water-facilitated catalyst separation and recycle 135–137
- water-in-oil microemulsions 266
- water-induced phase separation 401, 402
- water-insoluble reactants 207
- water-soluble ligands 76, 77, 106–108
- water-soluble phosphenylenes 174
- water-soluble phosphines 114

## x
XANTPHOS 391

## y
Yamamoto–Yanagisawa catalyst 155
yield
- catalysis 32–53
- chain elongation 300
- *C*-glucosidation 307
- *N*-glycosylation 312
ylide, sulfur 63

## z
zeolites 96
- cation-exchanged 70
zinc-mediated allylation 153